# ELEMENTARY PROCESSES IN DENSE PLASMAS

The Proceedings of the
Oji International Seminar

Tomakomai, Japan, June 27–July 1, 1994

*Setsuo Ichimaru and Shuji Ogata, Editors*

ADDISON-WESLEY PUBLISHING COMPANY
*The Advanced Book Program*

Reading, Massachusetts • Menlo Park, California • New York
Don Mills, Ontario • Wokingham, England • Amsterdam • Bonn
Sydney • Singapore • Tokyo • Madrid • San Juan
Paris • Seoul • Milan • Mexico City • Taipei

Many of the designations used by manufacturers and sellers to distinguish their products are claimed as trademarks. Where those designations appear in this book and Addison-Wesley was aware of a trademark claim, the designations have been printed in initial capital letters.


Copyright © 1995 by S. Ichimaru and S. Ogata


All rights reserved. No part of this publication may be reproduced, stored in a retrieval system, or transmitted, in any form or by any means, electronic, mechanical, photocopying, recording, or otherwise, without the prior written permission of the publisher. Printed in the United States of America. Published simultaneously in Canada.

Cover design by Lynne Reed
Typeset by the authors using the TeX programming language.

ISBN 0-201-41019-2

1 2 3 4 5 6 7 8 9 10-MA-98979695
First printing, May 1995

# Contents

## I. NEUTRON STARS, WHITE DWARFS, AND SUPERNOVAE

1. The quark-gluon plasma
   G. Baym ............................................................................. 3

2. Hot neutron stars at birth: A realistic treatment
   T. Takatsuka ....................................................................... 17

3. Global oscillations of neutron stars as potential probes of microscopic physics at ultrahigh densities
   H.M. Van Horn, U. Lee, R.I. Epstein, and T.J.B. Collins ............ 25

4. Explosion mechanism of collapse-driven supernovae
   K. Sato, T. Shimizu, and S. Yamada ...................................... 35

5. Role of magnetized plasmas in production of cosmic x-rays
   K. Makishima ...................................................................... 47

6. Intense magnetic field phenomena
   J. Weisheit ......................................................................... 61

## II. NEUTRINO ASTROPHYSICS

1. Neutrinos in cosmology and astrophysics: An introductory overview of theoretical aspects
   M. Fukugita ........................................................................ 73

2. Neutrino emission from protoneutron stars
   H. Suzuki and K. Sumiyoshi .................................................. 85

3. Solar neutrinos
   Y. Suzuki ........................................................................... 93

4. Neutrino emission processes in stars
   N. Itoh .............................................................................. 101

## III. NUCLEAR PROCESSES

1. Nuclear reactions in dense matter
   S. Ichimaru and H. Kitamura — 113
2. Experimental approach to nucleosynthesis in the universe
   S. Kubono — 125
3. Cosmological phase transition and nucleosynthesis
   T. Kajino — 135

## IV. ATOMIC AND OPTICAL PROCESSES IN STELLAR MATTER

1. Equation of state, metallization, and energy transport in dense stellar matter
   S. Ogata, H. Kitamura, and S. Ichimaru — 145
2. Thermodynamics, kinetics, and phase transitions of dense plasmas
   W. Ebeling and A. Förster — 165
3. New astrophysical opacities and their effect on stellar models
   F.J. Rogers and C.A. Iglesias — 181
4. Atmospheric structure of very low mass stars: M dwarfs, subdwarfs, and brown dwarfs
   T. Tsuji and K. Ohnaka — 193

## V. HELIOSEISMOLOGY

1. Helioseismology and solar models
   H. Shibahashi — 203
2. Equation-of-state issues in helioseismology
   W. Däppen — 215

## VI. PLANETARY SCIENCES

1. Current uncertainties in the interior physics of brown dwarfs and giant planets
   W.B. Hubbard, A. Burrows, J.I. Lunine, and D. Saumon — 227
2. Constitution and evolution of the earth
   E. Ito — 239

## VII. HIGH-PRESSURE METAL PHYSICS

1. The dense-hydrogen plasma: Translational, orientational, and electronic structure
   N.W. Ashcroft   251

2. Progress on hydrogen at ultrahigh pressures
   R.J. Hemley and H.-k. Mao   271

3. The potential of high-power beams for studying megabar matter, including low-entropy hydrogen compression
   J. Meyer-ter-Vehn and A. Oparin   283

## VIII. CONDENSED-MATTER PLASMAS

1. Interatomic interactions in dense mercury vapors
   F. Hensel and M. Yao   295

2. Liquids near the critical point
   H. Endo and M. Yao   307

3. Volume dependence of the structure of liquid metals
   K. Tsuji   317

4. Electron-hole plasmas in elemental semiconductors
   J.P. Wolfe   325

## IX. COMPLEX FLUIDS AND SOLIDS

1. Complex fluids: Anomalous relaxation, percolation, and wetting
   F. Yonezawa, S. Fujiwara, S. Gomi, and K. Omata   339

2. Density functional theory of freezing of soft-core systems
   M. Hasegawa   351

3. Computer simulation of materials on parallel architectures: Glasses, solid $C_{60}$, and graphitic tubules
   P. Vashishta, R.K. Kalia, W. Jin, J. Yu, and A. Nakano   359

## X. LASER AND SHOCK COMPRESSED PLASMAS

1. High density plasma physics in laser produced plasmas
   K. Mima, H. Takabe, Y. Kato, S. Miyamoto, and S. Kato   375

2. Shock compressed nonideal plasmas
   V.E. Fortov   389
3. Spectroscopic analysis of hot dense laser produced plasmas: Review and update
   C.F. Hooper, Jr., R.C. Mancini, D.A. Haynes, Jr., and D.T. Garber   403
4. Electron transport phenomena and dense plasmas produced by ultra-short-pulse laser interaction
   R.M. More   415
5. Internal structure of a compressed atom and molecule in dense plasmas
   Y. Furutani and A. Fukuyama   425

## XI. MAGNETIC RECONNECTION

1. Recent experiments on magnetic reconnection in laboratory plasmas: A review
   M. Yamada   435
2. Dynamics of the solar corona observed with Yohkoh
   S. Tsuneta   447

## XII. POSITRON PLASMAS

1. Physics with trapped positron gases and plasmas
   C.M. Surko and R.G. Greaves   463
2. Production of slow positrons with a linac as a source of positron plasma formation
   A. Mohri, H. Tanaka, T. Michishita, Y. Yuyama, Y. Kawase, and T. Takami   477

PROGRAM OF THE OJI INTERNATIONAL SEMINAR   489
LIST OF PARTICIPANTS   493
INDEX   499

# Preface

Astrophysical dense plasmas are found in the interiors, surfaces, and outer envelopes of such astronomical objects as neutron stars, white dwarfs, the sun, brown dwarfs, and giant planets. Condensed plasmas in laboratories include metals and alloys (solid, amorphous, liquid, and compressed), semiconductors (electrons, holes, and their droplets), and various realizations of plasmas (shock-compressed, diamond-anvil cell, metal vaporization, pinch-discharges, and so on).

The elementary processes of interest in astrophysical dense plasmas include radiative transfer and opacities, electronic transport, electromagnetic (magnetic reconnection) and nuclear (fusion reactions, $\beta$ captures, and neutrino) processes. The rates of these processes depend sensitively on a change in plasma states, such as the freezing transition, chemical separation, supercritical fluids, ionization or insulator-to-metal transition, magnetic transitions, and transitions between the normal and superconducting states.

Given the importance of understanding elementary processes in dense plasmas and the range of topics required for the understandings, it seemed appropriate to organize an international seminar devoted to this subject. The aim of the seminar was to invite contributions, with an equal emphasis, from experimental and observational activities in these various disciplines involved as well as from their theoretical counterparts, in order to elucidate the physical relationship between elementary processes and states of matter, and thereby to promote understanding of those astrophysical and laboratory objects containing condensed plasmas. This interdisciplinary seminar would provide links between the fields of atomic, molecular, and nuclear physics dealing with single- or few-particle problems, those of the condensed-matter and statistical physics concerning many-particle or mesoscopic systems, and those of astrophysics and planetary sciences where these fundamental processes have important applications to such specific issues as stellar structure and evolution, generation and transport of energy, and stellar magnetism.

# PREFACE

The Fujihara Foundation of Science, endowed jointly by New Oji Paper Co., Nippon Paper Industries Co., Honshu Paper Co., and 16 other related companies, which awards one to two international symposia a year in the fields of basic science, through the office of Japan Society for the Promotion of Science (JSPS), agreed to support such a seminar, and the Oji International Seminar on *ELEMENTARY PROCESSES IN DENSE PLASMAS* was held in Tomakomai, Hokkaido, Japan, for June 27 – July 1, 1994.

This is the Conference Volume, consisting of the articles contributed by the leading scientists who participated in the Seminar. The authors were invited to contribute state-of-the-art reviews of the research activities in the fields of their specialties, leading to expositions of the frontal developments in individual disciplines. In this way, a whole spectrum of the subjects distributed in various interdisciplinary fields may be covered in a collective manner by amalgamating these individual contributions.

As the chairperson of the local organizing committee, it is my privilege and pleasant duty to express sincere thanks to all the participants, members of the organizing committees, officials of the Fujihara Foundation of Science and JSPS, and many other colleagues and friends for their support, cooperation, and assistance which have made the Seminar a successful one. The co-editors of this volume would like to thank Mr. Jack Repcheck and Ms. Heather Mimnaugh, Advanced Book Program, Addison-Wesley Publishing Company for their pertinent collaboration for its publication.

*Setsuo Ichimaru*

*Tokyo*
*January, 1995*

# I. Neutron Stars, White Dwarfs, and Supernovae

# THE QUARK-GLUON PLASMA

Gordon Baym

Department of Physics, University of Illinois at Urbana-Champaign
Urbana, IL 61801, U. S. A.

Ordinary nuclear matter, when heated or compressed sufficiently, is expected to turn into a new state—a *quark-gluon plasma*—in which the fundamental degrees of freedom are the quarks that compose neutrons and protons, and, at finite temperature, antiquarks and gluons as well. We describe here the physical situations, such as the early universe and ultrarelativistic heavy-ion collisions, in which such a plasma is expected to appear; review our current understanding of the phase transition from ordinary matter to the quark-gluon plasma; and discuss microscopic properties of the plasma, including its elementary modes and transport properties.

## I. INTRODUCTION

The quark-gluon plasma is a new phase of matter—in the sense that it has not yet been detected in the laboratory—whose elementary constituents are the quarks, antiquarks, and gluons that make up the strongly interacting particles. It is also the oldest phase of matter, the form of the matter that filled up the early universe, until the first few microseconds after the big bang. Appropriate for this seminar on dense plasmas, quark-gluon plasmas are the densest plasmas in the universe. They may possibly be present deep in the cores of neutron stars, and even be made in stellar collapse. To probe the densest states of nuclear matter, the nuclear physics community has embarked on a large-scale program of studying collisions of ultrarelativistic heavy-ions, at Brookhaven National Laboratory and CERN. A major step in the program is the construction of a large colliding beam accelerator at Brookhaven, the Relativistic Heavy Ion Collider (RHIC), which will provide the capability of colliding nuclei as heavy as Au on Au at 100 GeV per nucleon in the center of mass (equivalent to 20 TeV per nucleon in a fixed target experiment), and should, by the turn of the century, enable one to produce and study quark-gluon plasmas in the laboratory.[*]

The basis for expecting a quark-gluon plasma at high densities or temperatures is that quite generally as matter is heated or compressed its degrees of freedom change from composite to more fundamental. For example, by heating or compressing a gas of atoms, one eventually forms a plasma in which the nuclei become stripped of the electrons, which go into continuum states forming an electron gas. Similarly, when nuclei are squeezed (as happens in the formation of neutron stars in supernovae where the matter is compressed by gravitational collapse) they merge into a continuous fluid

---

[*] Useful general references on quark-gluon plasmas and relativistic nucleus-nucleus collisions are the proceedings of the ongoing conferences on quark matter [Stenlund et al. 1994, and earlier].

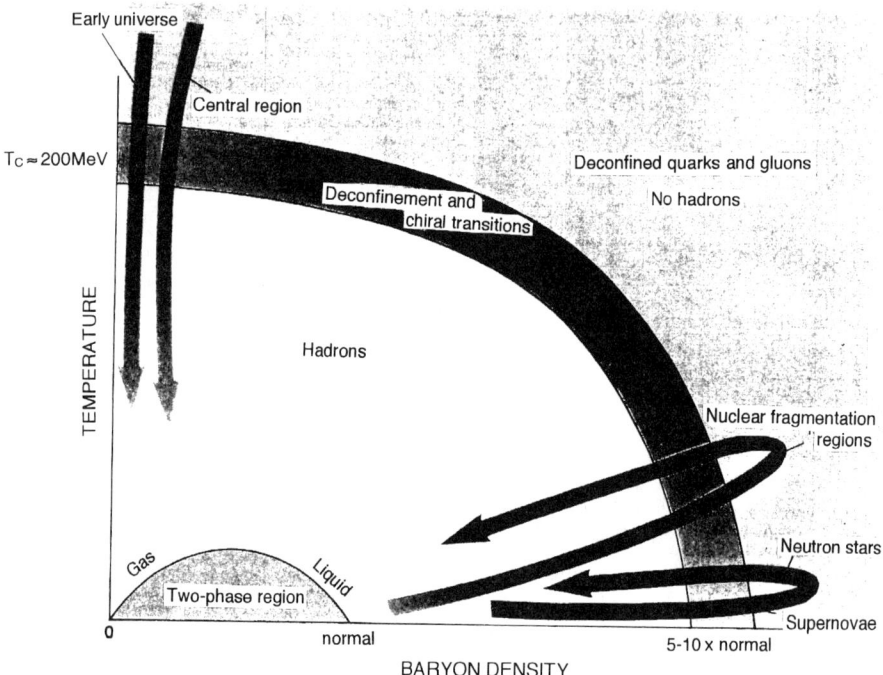

Figure 1. Phase diagram of equilibrated nuclear matter in the baryon density, temperature plane.

of neutrons and protons—nuclear matter liquid. Nucleons, and the other strongly interacting particles, or hadrons, are made of quarks, which are confined to the individual hadrons. One can thus go one step further and predict that a gas of nucleons, when squeezed or heated, turns into a gas of uniform quark matter, composed of quarks, and at a finite temperature, antiquarks and gluons—the mediators of the strong interaction—as well, which are no longer confined in individual hadrons but are free to roam over the entire volume of the deconfined region. The physics is basically the same as that leading to formation of ordinary plasmas.

The regions in the phase diagram of matter in the temperature-baryon density plane where quark-gluon plasmas are expected to occur are shown in fig. 1. In the low temperature–low baryon density region the basic degrees of freedom are hadronic, those of nucleons, mesons and internally excited states of the nucleon, while in the high temperature—high baryon density regions (temperatures above $\sim$ 200 MeV equivalent to a few times $10^{12}$K, and densities of order 5-10 times the density of matter inside a

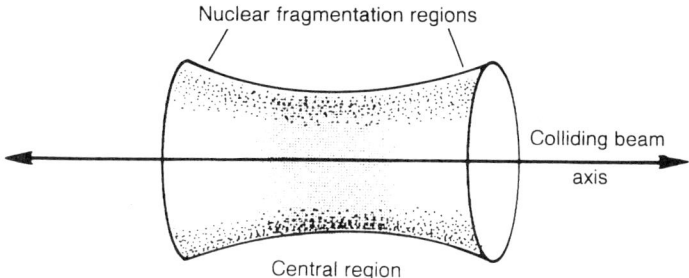

Figure 2. An ultrarelativistic heavy-ion collision.

large nucleus, $\rho_0 \sim 0.16$ particles/fm$^3$, equivalent to $2 \times 10^{14}$ g/cm$^3$), the basic degrees of freedom should become those of quarks and gluons. The transition between these regions may, as we shall discuss, be first or second order, or not a sharp transition at all, but washed out. Under any circumstances, the physics between the two regimes changes strongly.

Let us consider in more detail where quark-gluon plasmas are expected to occur in nature. In the first microseconds of the early universe the temperature falls as

$$T \simeq \frac{0.5 \text{ MeV}}{[t(\sec)]^{1/2}}, \qquad (1)$$

so that prior to $\sim$ 5-10 microseconds after the big bang, when temperatures are above hundreds of MeV, matter is in the form of a quark-gluon plasma, and follows the downward trajectory practically along the vertical axis of the phase diagram. The matter of the early universe has a relatively small net baryon density, of order one part in $10^9$ (as inferred from the present photon/baryon ratio). As the universe expands it cools and matter hadronizes, undergoing the reverse of the deconfinement transition and emerging primarily in the form of pions, with a slight baryon excess. If the transition is first order, one expects formation of bubbles of hadrons—the ordinary neutrons, protons, and pions—in the middle of the plasma.

A second possible astrophysical situation in which quark-gluon plasmas might play a role is in neutron stars. Measured masses of neutron stars are $\sim$ 1.4 solar masses, with radii calculated to be $\sim$ 10 km. Typical temperatures are very low, less than one MeV. The central density in a neutron star, indicated on the phase diagram, can reach five to ten times the density inside an ordinary atomic nucleus, and may be in the form of a quark plasma. Not only may neutron stars have quark cores [Kutschera and Kotlorz, 1993, and references therein], but one cannot positively rule out the possibility that there exists a distinct family of quark stars with higher central densities than neutron stars [see, e.g., Fechner and Joss, 1978; Alcock and Olinto, 1988].

A snapshot of the matter in a central collision of ultrarelativistic heavy-ions, above $\sim 10$ GeV per nucleon c.m. (the energy in CERN fixed target experiments at 200 GeV), is shown in fig. 2. At low energies, $\sim$ a few GeV per nucleon in the center-of-mass, as studied in the AGS experiments at Brookhaven, one can picture the two Lorentz-contracted colliding nuclei as nearly stopping each other, with reasonable probability of forming, crudely, a fireball. Such heavy ion collisions may reach energy densities of order a few GeV/fm$^3$ and baryon densities several times $\rho_0$; the matter may indeed cross into the deconfined region, as shown in fig. 1, in the curve labelled "nuclear fragmentation regions," and then expand out. The high density matter produced in such collisions will be relatively baryon rich. As the beam energy is increased, the nuclei begin to pass through each other, becoming highly excited internally, and at the same time, leaving the vacuum between them in a highly excited state. The fragments of the original nuclei, which recede from each other at the speed of light, contain essentially all the initial baryons. The central region, to a first approximation, has little baryon excess, and at sufficiently high energy density, should be a quark-gluon plasma, similar to that in the early universe. The matter in this region is essentially manufactured in the collision. After its production and thermalization, the matter cools and expands, on a time scale of order $10^{-22}$ sec, and if having initially reached a quark-gluon plasma state, it undergoes a hadronization transition.

## II. ELEMENTARY PROPERTIES OF QUARK-GLUON PLASMAS

The basic degrees of freedom of a quark-gluon plasma are deconfined quarks, antiquarks, and gluons. The up ($u$) and down ($d$) quarks, the fundamental ingredients of nucleons, are very light particles ($m_u \sim m_d \sim 10$ MeV), and are the dominant ingredient in plasmas expected in heavy-ion collisions. The strange ($s$) quark, with a mass of order 150 MeV, will be less prevalent. The heavier charm, top and bottom quarks should play a significant thermodynamic role only in the very hot plasmas of the early universe. One important feature of the quark-gluon plasma is that it has many more internal degrees of freedom (helicity states) than ordinary hadronic matter. Ordinary nuclear matter has four helicity states (two for spin times two for isospin), while pi mesons, if present, contribute three helicity states, one per charge state. On the other hand, a quark of a given flavor ($u$, $d$, etc.) has six helicity states, two for spin, times three for the internal color degree of freedom. Adding in antiparticles ($\bar{u}$, $\bar{d}$, etc.) and summing over flavors present, we see that the quarks in a plasma contribute from 24 to 36 helicity states. depending on whether strange quarks are also present in addition to the light up and down quarks. The gluon is a massless vector particle, like the photon; in addition to two polarizations, its internal color degree of freedom can take on eight values. Thus gluons contribute a total of 16 helicity states.

From a thermodynamic point of view a quark-gluon plasma at a given energy density has a high entropy compared with hadronic matter. Hot free quark matter is similar to ordinary black-body radiation with energy density $E \sim T^4$, where $T$ is the temperature; in a system with equal number of $u$, $\bar{u}$, $d$, and $\bar{d}$ quarks, as well as gluons,

$$T \simeq 160 \text{ MeV} E^{1/4}, \qquad (2)$$

and the total density of excitations per fm$^3$ is

$$n_{exc} \simeq 2.25 E^{3/4}, \qquad (3)$$

with $E$ measured in GeV/fm$^3$. Since the transition to a quark-gluon plasma is believed to occur at $T$ of order 150–200 MeV, we see from (2) that the scale of energy densities that must be deposited in collisions to excite a plasma is of order several GeV/fm$^3$. Because of the slow dependence of $T$ on $E$, it will not be easy to heat a plasma in a nuclear collision much beyond hundreds of MeV.

Let us recall a few features of qcd which are relevant for understanding the elementary physics of the quark-gluon plasma. In quantum electrodynamics, photon exchange produces the basic force between charges, which between static point charges is simply the Coulomb interaction, $\sim e^2/r$. The force between charges of opposite sign is opposite to that between like charges; thus, in qed one can form electrically neutral systems, such as positronium or hydrogen atoms, which do not give rise to long-range Coulomb fields. Qcd has a similar structure, in that the forces between quarks arise from exchange of gluons, and the color degree of freedom of the quarks functions as a three-valued charge, rather than simply $\pm$, as in qed. Again one can form color singlet or neutral systems which do not give rise to long-range color Coulomb fields. Such a charge scheme requires eight gluons—rather than a single photon—having a color charge and hence coupling directly to themselves producing the rich non-linear structure of qcd. Because qcd allows color neutrality, quark matter in equilibrium in its state of lowest energy, or free energy at finite temperature, will on average have no long-range color Coulomb fields, as in an ordinary electrically neutral plasma.

Qcd is also asymptotically free. In qed an electron gathers around it a polarization cloud of electron-positron pairs in the vacuum which decreases the net charge seen at large distances; the effective charge of an electron at large distances is given by $e^2/4\pi\hbar c \equiv \alpha_{\rm em} = 1/137$. [Note the units of $e$.] At short distances, inside the polarization cloud, the effective charge on the electron grows, diverging at zero distance—one of the troublesome divergences of qed. In qcd a rather different behavior occurs. Because the gluons themselves carry color, they also screen the bare charges, but their net effect is opposite that of quark-antiquark pairs; the result is that close to a quark, the effective charge does not become infinite, but rather goes to zero—the property of asymptotic freedom. At short distances, corresponding to large momentum scales, interactions become arbitrarily weak, with the effective coupling

$$\alpha(p) = \frac{g^2}{4\pi\hbar c} = \frac{6\pi}{(33 - 2N_f)\ln(cp/\Lambda)}, \qquad (4)$$

where $N_f$ is the number of relevant quark flavors, $\Lambda$ the qcd scale parameter is of order 100–200 MeV, and $p$ is the momentum scale. At large distances however quite the opposite happens; as colored particles are separated, the forces between them become larger and larger, giving rise to confinement.

At very short distance scales, small compared with $\hbar c/T$, the interparticle spacing, the plasma can be treated as a non-interacting gas of relativistic quarks, antiquarks, and gluons. The reason is that, as in an ordinary plasma, any small region of the matter at high densities will on average be color neutral and not produce long-range (color) Coulomb fields, while the residual short distance forces in the region become weak as the interparticle separation becomes small, due to asymptotic freedom. However, at larger distances this simple behavior breaks down. In a weakly interacting plasma, one can distinguish several length scales: The next larger scale beyond the interparticle spacing $\hbar c/T$ is $(\hbar c)^{3/2}/gT$, a regime where one can effectively do perturbation theory, albeit with difficulties. At still larger scales, of order $(\hbar c)^2/g^2T$, the non-Abelian features of qcd—are self-couplings of the gluon field—begin to enter, and the system become non-perturbative (a regime not encountered in electromagnetic plasmas). Generally one does not understand how to calculate properties of the plasma in this regime.

The structure at even larger distance scales reflects the existence of confinement, in the sense that at long wavelengths the basic excitations of the plasma are color neutral, rather than simple colored quark and gluon excitations (see DeTar [1988] and Weiss [1989], and references therein). An example of such structure is the picture of Polonyi [1989], in which screening of color occurs in a plasma through the action of monopole-like objects constructed from the gluon field, and the basic excitations become quarks accompanied by monopoles. However, the detailed large scale structure has yet to be determined.

Such a division as above into different regimes is true only if the coupling $g$ is small, which is unlikely to obtain in practice. Certainly, as one goes earlier in time in the early universe, the temperature becomes higher and the fine structure constant smaller. However, even back at the grand unification scale of the universe, at a temperature of about $10^{15}$ GeV, where all the strong, electrodynamic and weak interactions have the same strength, $g/\sqrt{\hbar c}$ is of order 1/2. In fact, real quark-gluon plasmas are likely to be essentially non-perturbative.

## III. THE TRANSITION TO A QUARK-GLUON PLASMA

We can set the scale for the transition from hadronic matter to a quark-gluon plasma by asking, for example, at what density the nucleons in nuclear matter begin to fill space and become squeezed together. If $r_n \sim 0.8$ fm is the radius of a nucleon, then nucleons begin to touch at a baryon density $\rho_b \equiv (3/4\pi r_n^3) \sim 3\rho_0$. Similarly the point at which the hot vacuum undergoes a deconfining transition can be estimated by calculating where thermally created pi mesons begin to fill space. The density of thermal pions is $\rho_\pi \sim 3T^3\zeta(3)/\pi^2$ (neglecting $m_\pi \simeq 140$ MeV, and all interactions); thus pions fill space for $T \sim \hbar c/r_\pi \sim 250$ MeV, where $r_\pi \sim 0.6$ fm is the radius of the pion.

One can roughly estimate the location of the deconfinement transition by by asking, at finite baryon density, for example, whether nuclear matter or a quark-gluon plasma has a lower free energy per baryon, as a function of baryon density [Baym and Chin, 1976]. Such an approach necessarily implies a first-order transition, with a discontinuity in the baryon density; the onset of deconfinement at $T = 0$ is typically at $\rho \sim 5-10\rho_0$. However, phenomenological theories of nuclear matter and of quark-gluon plasmas are the question of whether neutron stars can have quark matter cores remains open, as is the issue of whether a distinct family of quark stars with higher central densities than neutron stars can exist.

## IV. DEBYE SCREENING AND PLASMA OSCILLATIONS

The elementary physics of a weakly interacting quark-gluon plasma, such as Debye screening and plasma oscillations, is to a first approximation, physics similar to that in an electromagnetic plasma. A non-interacting plasma has fermion excitations, the quarks and antiquarks, with an energy-momentum relation, $\varepsilon(p) = \left(p^2c^2 + m_q^2c^4\right)^{1/2}$, and transverse gluon excitations, similar to free photons, with a linear energy-momentum relation, $\omega(k) = ck$. When interactions are included, the color charge of quarks and gluons is screened over a wavelength $1/q_D$, where $q_D$ is the Debye screening wavevector. Neglecting for the moment screening by gluons, one finds,

$$q_D^2 = \frac{g^2}{2} \sum_{\vec{p},i} \left(-\frac{\partial n_{p_i}}{\partial \varepsilon_{p_i}}\right). \tag{5}$$

where $n_{p_i}$ is the density of quarks of momentum $p$ and quark flavor and spin $i$; this expression differs from that for a qed plasma by the replacement of $e^2$ by $(g/2)^2 \overline{tr \lambda_\alpha^2}$ = $g^2/2$, where the $\lambda_\alpha$ are the eight color SU(3) generators, and the bar denotes an average over the $\alpha$. For a plasma of massless quarks at temperature $T$, $q_D^2$ is of order $g^2$ times a characteristic excitation density, $\sim T^3$, divided by a typical excitation energy, $\sim T$, so that the characteristic screening length is of order $(\hbar c)^{3/2}/gT$, or $(\hbar c)^{1/2}/g$ times a typical quark-quark separation. Taking screening by gluons into account, one finds for equal numbers of quarks and antiquarks that

$$q_D^2 = (N_f + 2N_c)\frac{g^2 T^2}{6}, \tag{6}$$

where $N_f$ is the number of quark flavors, and $N_c$ (= 3 in the real world) is the number of colors. As in qed, $q_D^2$ for a quark-gluon plasma is proportional to a compressibility, now a color compressibility.

Quark-gluon plasmas oscillations also develop plasma oscillations analogous to those in an electromagnetic plasma. The transverse gluonic excitations have a spectrum

$$\omega(k) = \sqrt{c^2 k^2 + \omega_{pl}^2}. \tag{7}$$

For massless quarks one finds

$$\omega_{pl}^2 = \frac{q_D^2}{3}, \tag{8}$$

compared with the result for ordinary plasmons in a non-relativistic plasma, $\omega_{pl}^2 = ne^2/m$, where $n$ is the electron density and $m$ the electron mass. The plasma oscillations here represent counteroscillations of different color degrees of freedom, rather than of electrical charge.

generally based on inequivalent physical descriptions of the two phases and thus cannot be expected to describe the transition accurately.

The only reliable approach at present to determining the properties of strongly interacting quark-gluon plasmas and the deconfinement transition is through Monte Carlo calculations of lattice gauge theory (for recent reviews see Karsch [1994] and Gottlieb et al. [1994], as well as the earlier review by Satz [1985]). The calculations are typically done on a lattice with equal number of sites $N$ ($\sim$ 16) in each spatial direction and, to describe finite temperature, a different number $N_t$ ($\sim$ 8) in the time direction, related to the temperature by $T = \hbar c N_t / a$, where $a$ is the lattice constant. The calculations require very large computing capabilities to achieve good statistics on large lattices. Lattice gauge theory is approaching the point where it will be able to give quantitatively good information on the properties of quark matter over large ranges of temperature and also baryon density. Calculations have been successful so far only for the case of zero baryon density at finite $T$. The results depend strongly on the masses assumed for the quarks. For infinitely heavy quarks, only gluons play a dynamic role; lattice calculations in this case predict a sharp first-order phase transition associated with deconfinement [Satz, 1985]. The test for deconfinement is to separate a test quark-antiquark pair, $Q\bar{Q}$, far apart: the energy in the confined phase required to do so grows linearly with separation, while it remains finite in the deconfined phase.

However, with finite mass quarks it always takes only finite energy to separate a test quark-antiquark pair, $Q\bar{Q}$, since at sufficient separation, it becomes favorable to create a $q\bar{q}$ pair in the system, which screens out the interaction between the test pair; the created quark $q$ binds to the test antiquark $\bar{Q}$, and the created anti-quark $\bar{q}$ to the test quark $Q$, creating effectively a pair of mesons which can be separated to infinity with finite energy. Once light quarks degrees of freedom are included there no longer exists a good measure of whether the system is in a confined or deconfined state, and there need not be a sharp phase transition between the confined and deconfined phases. The transition between the two phases can be smooth, as in ionization of a gas being heated, where the system goes gradually from a gas of molecules to electrons and nuclei; the two states are qualitatively different and there is a reasonably rapid onset of ionization, but it is not sharp.

For massless quarks, on the other hand, the transition is again first-order, associated now with the spontaneous breaking of the SU(3)⊗SU(3) chiral symmetry of strong interactions. In the hadronic world chiral symmetry is spontaneously broken, analogous to the breaking of rotational symmetry in a ferromagnet, while in the deconfined phase it is restored. Since chiral symmetry is not exact for finite mass quarks, the situation with realistic light $u$ and $d$ quarks and strange quarks must be determined by detailed calculations [Karsch, 1994]; these presently indicate that rather than a sharp transition there is a rapid crossover from the hadronic to the deconfined phase at a temperature $\sim$ 150–200 MeV.

Lattice gauge calculations at non-zero baryon density are beset by technical problems; to date we do not have a reliable estimate of the transition density at $T = 0$ from nuclear to quark matter, or even compelling evidence that there is a sharp phase transition. In the absence of a good theory of the equation of state at very high densities,

As in a qed plasma, static transverse (magnetic) interactions are not screened at the perturbative level. As we discuss below, this lack of screening leads to problems in calculating finite transport cross-sections, as well as determining the quasiparticle spectrum of quark-gluon plasmas. There is good evidence, however, from a detailed analysis of the structure of perturbation theory in qcd, that transverse excitations in qcd systems at finite temperature should develop a magnetic "mass," $m_T \sim g^2 T/\hbar c^3$, [Gross, Pisarski and Yaffe, 1981], so that analogous to Debye screening of longitudinal (Coulombic) interactions, the long wavelength transverse interaction, for momentum transfer $q$ would behave as

$$V_{tr} \sim \frac{g^2}{q^2 + m_T^2 c^2}. \qquad (9)$$

Long range color magnetic fields, and the interaction between color currents, would then be screened on a distance scale $\sim (\hbar c)^2/g^2 T$, similar to the screening of magnetic fields in a superconductor (the Meissner effect). The magnetic mass $\sim g^2 T$ is higher order in the coupling than $q_D \sim gT$.

## V. TRANSPORT PROPERTIES

In attempting to calculate elementary transport processes in quark-gluon plasmas, or even in relativistic electromagnetic plasmas, one immediately encounters the problem that the cross-sections for elementary scattering processes between quarks or gluons at the perturbative level, or between electromagnetic charges, diverge strongly at small angles, $d\sigma/d\Omega \sim (1 - \cos\theta)^{-2}$ as in Rutherford scattering, rendering naive calculations unphysical. In non-relativistic plasmas, where magnetic interactions can be ignored, one controls the divergences by including the fact that Coulomb-like interactions are Debye screened at distances of order the Debye screening length, $1/q_D$. Screening of longitudinal interactions leads to finite cross-sections; e.g., the momentum-transfer cross-section, $\sigma_{tr} \equiv \int d\Omega(1 - \cos\theta)d\sigma/d\Omega$, becomes for longitudinal interactions $\sim \alpha^2 \ln T/q_D \sim \alpha^2 |\ln\alpha|$, where $\alpha$ is the qcd or qed fine-structure constant.

However, magnetic-like interactions, from exchange of transverse photons or gluons, are not similarly screened; for example, the magnetic interaction of two current-carrying wires surrounded by plasma is simply the long-ranged bare interaction, multiplied by the magnetic permeability, $\mu$, of the plasma. Scattering via magnetic interactions leads, in lowest order, to divergent transport scattering rates.

Several possible physical effects could be responsible for making the transport coefficients finite. First, in heavy-ion collisions for example, plasmas formed would be limited in spatial extent, giving a small momentum cutoff of order the inverse of the system size. Such a cutoff would not solve the problem for for large-scale astrophysical plasmas, including the early universe. A second effect would be the magnetic mass in qcd. However, this effect cannot solve the problem for electromagnetic plasmas, nor does it provide the correct leading order effect in weakly coupled quark-gluon plasmas, as we shall see.

The essential physics that one must include in any relativistic plasma is the dynamical screening of magnetic interactions, known in solid state physics as the "anomalous skin effect" [Reuter and Sondheimer, 1948] in which a magnetic field of finite frequency $\omega$ imposed from the outside on a very pure (and thus "anomalous") metal falls off at distances $\sim q_D^{-2/3}\omega^{-1/3}$ in from the surface. The finite energy transfers inherent in scattering processes at finite temperature thus imply a falloff of the magnetic interactions with distance, adequate to make the transport coefficients finite.

To see how this works, recall that the bare interaction between two transverse currents is proportional to the transverse photon or gluon propagator, $D_0 = (\omega^2 - q^2)^{-1}$, where $\omega$ and $\vec{q}$ are the energy transfer and momentum transfers in the process (we take $\hbar=c=1$ from now on). Including the self-energy in a plasma to lowest order, the simple "particle-hole bubble," we find [Baym et al., 1990] that the propagator assumes the form, at small $q$ and $\omega$,

$$D = \frac{1}{\omega^2 - q^2 - \zeta q^2 + i\pi q_D^2 \omega/4q}, \qquad (10)$$

where $\zeta$ is a constant $\sim \alpha$. In the static limit, $D \to -\mu/|\vec{q}^2|$, where $\mu = (1+\zeta)^{-1}$ is the magnetic permeability of the plasma. The imaginary part of the denominator, arising from the imaginary part of the bubble, gives rise to Landau damping of plasma oscillations and at the same time screening of magnetic fields in plasmas at finite frequency. (Its role in qcd plasmas was first noted by Weldon [1982a].) To see the falloff note that at small given $\omega$ the inverse propagator has a zero at complex wavevector $(i\pi\omega/4q_D)^{1/3}q_D$. The basic physics of the screening is that the particles of the plasma "ride" the finite frequency wave, absorbing energy resonantly over a wavelength, $\sim 1/q$. With (10), the interaction matrix element squared at small $q$ and $\omega$ is proportional to

$$|D|^2 \simeq \frac{q^2}{q^6 + (\pi q_D^2 \omega/4)^2}. \qquad (11)$$

For small frequencies, $\omega \lesssim q$, the imaginary part gives an effective cutoff for small $q$ of order $gT$, or $eT$ in the electromagnetic case, or in space at the Debye screening distance $\lambda_D$ in these systems, although the physics of the screening is qualitatively different from the static screening of longitudinal interactions. This screening scale, $gT$ in a weakly interacting qcd plasma, dominates over that, $g^2T$, provided by a magnetic mass.

With the physics of the screening of transverse interactions included, one can begin to calculate various transport properties of plasmas [Baym et al., 1990, 1994]. Solving the Boltzmann equation exactly to leading logarithmic order in the coupling $\alpha$, we find that the characteristic lifetimes entering the transport coefficients are of the form

$$\frac{1}{\tau} \sim \alpha^2 T \int_0^q q^2 dq \int_{-q}^{q} \frac{d\omega}{|q^2 - i\pi q_D^2 \omega/4q|^2}. \qquad (12)$$

These relaxation times are $\gtrsim 1$ fm/c in a quark-gluon plasma at temperature $\sim 200$ MeV. However, plasmas formed at the energy densities achievable in ultrarelativistic heavy-ion collisions are expected to be strongly interacting, and thus are likely to experience considerably faster relaxation than the result (12) which is calculated in the weak-coupling limit. In fact, relaxation should be rapid enough to permit the plasmas formed in ultrarelativistic collisions to achieve a measure of local thermal equilibrium during their fleeting existence.

As an example of computed transport coefficients, the first viscosity, $\eta$, of a quark-gluon plasma at zero baryon chemical potential, to leading order in $\alpha$, is an additive sum of the gluon and quark viscosities, $\eta = \eta_g + \eta_q$, where each $\eta_i$ ($i = q, g$) is of the form $w_i \tau_{\eta i}/5$, with $w_i$ the enthalpy density of the quarks or gluons, and

$$\frac{1}{\tau_{\eta g}} = 4.11 T \left(1 + \frac{N_f}{6}\right) \alpha^2 |\ln \alpha|. \qquad (13)$$

The "1" is the result of scattering of gluons on gluons, while the term $\propto N_f/6$ is from scattering of gluons on quarks; these two processes are additive contributions to the gluon scattering rate. The viscous relaxation rate for quarks is somewhat smaller than that for gluons: $\tau_{\eta q}^{-1} = 0.39 \tau_{\eta g}^{-1}$.

It is also interesting to compute the electrical conductivity, $\sigma_{el}$, of the early universe, after the electroweak phase transition at $T \sim 200$ GeV. Conduction is primarily by electrons (as well as muons), limited by electromagnetic scattering on other charged particles; strongly interacting particles have much shorter mean-free paths. Generally $\sigma_{el} = ne^2 \tau_{el}/4\pi m$, where $n$ is the density of charges, here $\sim T^3$, $m$ is their effective mass, $\sim T$ here, and $\tau_{el}$ is the scattering time. To leading order one finds $\tau_{el}^{-1} \sim T \alpha_{em}^2 |\ln \alpha_{em}|$, where $\alpha_{em}$ is the electromagnetic fine-structure constant, so that $\sigma_{em} \sim T/(\alpha_{em}|\ln \alpha_{em}|)$. The electrical conductivity of the plasma in the early universe is sufficiently large that large-scale magnetic flux present in this period should not diffuse significantly over expansion timescales.

The calculations of transport properties carried out so far have been in the weak coupling limit for systems near thermal equilibrium. It is important to extend these calculations in the future to strongly coupled plasmas, as well as plasmas far from equilibrium, as are expected in the early stages of ultrarelativistic heavy-ion collisions.

## VI. QUASIPARTICLE STRUCTURE

Let us turn now to the question of the effect of long wavelength unscreened magnetic interactions on the quasiparticle structure, a very serious problem in both relativistic electromagnetic and quark-gluon plasmas, for which there is no solution yet. To see the nature of the problem consider the the degenerate non-relativistic electron gas. As a result of the Coulomb interaction, the electron quasiparticle spectrum is given to first order in $e^2$ by [Dirac, 1929]

$$\varepsilon_p = \frac{p^2}{2m} - \sum_{\vec{q}} \frac{4\pi e^2}{q^2} n_{\vec{p}-\vec{q}}, \qquad (14)$$

where $n_p$ is the electron distribution function. In a zero-temperature electron gas near the Fermi surface, $\varepsilon_p$ is thus $\sim e^2(p_F - p) \ln(p - p_F)$, where $p_F$ is the Fermi momentum, indicating that the effective mass, $m^* = p/(\partial \varepsilon_p/\partial p)$, varies as $-1/\ln(p - p_F)$. As a consequence of this singular structure the low-temperature specific heat of the electron gas would vary as $T/\ln T$; however, when Debye screening is included the singularity is removed, and the specific heat behaves as $T$, as expected for a normal Fermi liquid.

Transverse interactions, which are normally neglected in non-relativistic systems since they are of order $(v_F/c)^2$ as large as the Coulomb interaction, have the opposite sign from the Coulomb interaction, and would, for a filled Fermi sea, lead to a quasiparticle spectrum near the Fermi surface

$$\varepsilon_p = \varepsilon_{p_F} + v_F(p - p_F) + ae^2(v_F/c)^2(p - p_F)\ln(p - p_F), \tag{15}$$

where $a$ is a positive constant of order unity. Such a spectrum has infinite *negative* slope at $p = p_F$, and would in fact make the Fermi surface unstable. With this spectrum it becomes favorable to populate the low energy states just above the Fermi momentum, $p_F$, rather than the higher energy states just below, and so the system would have three nesting Fermi surfaces in momentum space. However, if one repeats the argument for the new Fermi surfaces, one again finds that any piece of sharp Fermi surface breaks into three further surfaces, and so on. The Fermi surface, as a consequence of the lack of static transverse screening at long wavelengths, breaks down and the system becomes a "marginal Fermi liquid," whose structure is a subject of great current interest, particularly in two-dimensional systems [e.g., Castro Neto and Fradkin, 1993; Stamp, 1993; Khveshchenko and Stamp, 1993]. Similar effects are expected in the quasiparticle spectrum of degenerate relativistic electromagnetic and quark-gluon plasmas as well.

Long-range magnetic interactions also dramatically affect the excitation spectrum of high-temperature relativistic plasmas. Consider the simple question of the spectrum of a single electron in a black-body cavity at temperatures large compared to the electron mass (or the analogous question for a quark in a hot gluon gas). The energy of a zero momentum electron is given by the solution of

$$\omega - \Sigma(\vec{p} = 0, \omega) = 0, \tag{16}$$

where the leading term of the self-energy $\Sigma(\vec{p} = 0, \omega)$ due to the electron Compton scattering with ambient photons is given to lowest order in $e^2$ by

$$\Sigma(\vec{p} = 0, \omega) \simeq \frac{e^2}{\omega}\int \frac{d^3k}{(2\pi)^3}\frac{n_k}{2k} = \frac{e^2 T^2}{12\omega}; \tag{17}$$

here $n_k$ is the photon distribution function. Equation (16) has the solutions $\omega = \pm eT/\sqrt{12}$; in other words, the spectrum develops a gap of order $eT$. The negative sign solution corresponds to the positron.

One can simply extend this calculation to include a thermal bath of electrons. The full fermion spectrum, as a function of momentum, has the remarkable structure shown in fig. 3. The solution, discussed originally by Klimov [1981] and Weldon [1982b], not only has a gap ($eT/\sqrt{8}$, and in qcd, $= gT/\sqrt{6}$), but is split at finite momentum, as a consequence of the chiral symmetry of massless electrons. The way this structure

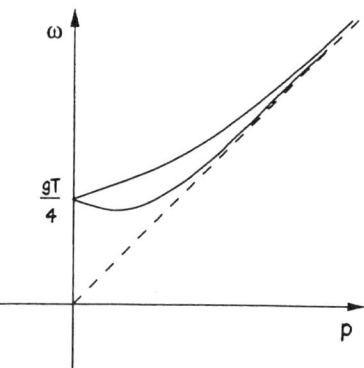

Figure 3. Spectrum of electron excitations in an ultrarelativistic plasma. The dashed line is $\omega = p$. The value of the gap shown $gT/4$ is for electrons coupled to scalar photons [Baym et al., 1992]; in qed the gap is $eT/\sqrt{8}$, and in qcd, $= gT/\sqrt{6}$.

emerges as the temperature increases is described in Baym et al. [1992], where calculational details as well as further references can be found. Similar structure in degenerate relativistic plasmas is discussed by Blaizot and Ollitrault [1993].

## VII. CONCLUSION

In this talk we have primarily focused on quark-gluon plasmas at the level at which they resemble electromagnetic plasmas. Both relativistic electromagnetic and quark-gluon plasmas present difficult problems related to the effects of long-ranged unscreened transverse or magnetic interactions on both the transport properties and the quasiparticle structure. Furthermore, quark-gluon plasmas provide a whole range of new challenges related to the non-Abelian structure of qcd, including non-perturbative effects, as well as the more complicated structure of the perturbation expansion. Substantial progress has been made in recent years in understanding how to sum higher order processes systematically (see particularly Braaten and Pisarski [1992], Pisarski [1993], and references therein). In addition, quark-gluon plasmas have unique modes arising from the non-Abelian structure [Blaizot and Iancu, 1994]. Detailed predictions of the behavior of quark-gluon plasmas in heavy ion collisions and other physical situations, must await a clear understanding of these problems.

This research was supported in part by the U. S. National Science Foundation under Grants DMR88-18713 and PHY89-21025.

## REFERENCES

Alcock, C. and A. Olinto, 1988, Ann. Rev. Nucl. Part. Sci. **38**, 161.
Baym, G., J.-P. Blaizot and B. Svetitsky, 1992, Phys. Rev. **D46**, 4043.
Baym, G. and S. A. Chin, 1976, Phys. Lett. **62B**, 241.
Baym, G., H. Heiselberg, H. Monien, C. J. Pethick, and J. Popp, 1994, University of Illinois preprint.
Baym, G., H. Monien, C. J. Pethick and D. G. Ravenhall, 1990, Phys. Rev. Letters **64**, 1867.
Blaizot, J.-P., and E. Iancu, 1994, Phys. Rev. Letters **72**, 3317.
Blaizot, J.-P., and J.-Y. Ollitrault, 1993, Phys. Rev. **D48**, 1390.
Braaten, E. and R. Pisarski, 1992, Phys. Rev. **D46**, 1829.
Castro Neto, A. and E. Fradkin, 1993, Phys. Rev. **B49**, 10877.
DeTar, C. 1988, Phys. Rev. **D37** (1988).
Dirac, P. A. M., 1929, Proc. Roy. Soc. Lond. **A123**, 714.
Fechner, W. B. and P. C. Joss, 1978, Nature **274**, 347.
Gottlieb, S. et al. 1993, Phys. Rev. **D47**, 3619.
Gross, D. J., R. D. Pisarski and L. G. Yaffe, 1981, Rev. Mod. Phys. **53**, 43.
Karsch, F., 1994, Nucl. Phys. B (Proc. Suppl.) **34**, 63.
Khveshchenko, D. V., and P. C. E. Stamp, 1993, Phys. Rev. Letters **71**, 2118.
Klimov, V., 1981, Yad. Fiz. **33**, 1734 [Sov. J. Nucl. Phys. **33**, 934 (1981)].
Kutschera, M. and A. Kotlorz, 1993, Ap. J. **419**, 752.
Pisarski, R., 1993, Phys. Rev. **D47**, 5589.
Polonyi, J., 1989, Nucl. Phys. **A498**, 313c.
Reuter, G. E. H. and E. H. Sondheimer, 1948, Proc. Roy. Soc. **A195**, 336.
Satz, H., 1985, Ann. Rev. Nucl. Part. Sci. **35** 245.
Stamp, P. C. E., 1993, J. Phys. I France **3**, 625.
Stenlund, E., H.-Å. Gustaffson, A. Oskarsson and I. Otterlund, eds., 1994, *Quark Matter '93*, (North-Holland, Amsterdam) [Nucl. Phys. **A566** (1994)].
Weiss, N., 1989, Nucl. Phys. **A498**, 313c.
Weldon, H. A., 1982a, Phys. Rev. **D26**, 1394.
Weldon, H. A., 1982b, Phys. Rev. **D26**, 2789.

# I.2

# HOT NEUTRON STARS AT BIRTH: A REALISTIC TREATMENT

Tatsuyuki Takatsuka

College of Humanities and Social Sciences, Iwate University
Morioka 020, Japan

Dense supernova matter is investigated by a realistic approach in which a set of finite-temperature Hartree-Fock equations with effective nucleon interaction are solved self-consistently. On the basis of these results, issues on the hot neutron stars at birth, such as the fat density profile and spin up, are discussed; the energy release in the subsequent cooling stage is also considered. Constraints on the maximum mass and maximum rotation rate of neutron stars are commented through consideration of their hot stage.

## I. INTRODUCTION

Hot neutron stars at birth are composed of "supernova matter" instead of neutron star matter responsible for usual cold neutron stars. Supernova matter is a new form of nucleon matter distinguished in the participation of degenerate neutrinos as well as electrons. It is characterized by almost constant entropy per baryon $S(=1.0\text{--}1.5)$ throughout the density $\rho$ [Bethe et al., 1979] and also by a high and almost constant lepton fraction $Y_l(=0.3\text{--}0.4)$ [Bethe et al., 1979; Epstein and Pethick, 1981; Lattimer et al., 1985], in different from neutron star matter with $S \simeq 0$ and $Y_l \lesssim 0.05$. These characteristics are caused by the effect that the neutrino-trapping in supernova core takes place for $\rho \gtrsim 10^{12}$ g/cc and the collapse proceeds adiabatically [Sato, 1975; Freedman, 1974; Mazurek, 1975, 1976].

The purposes of this paper are as follows. Firstly, we make a realistic calculation of supernova matter and investigate the properties of supernova matter such as the composition and the equation of state (EOS). Secondly, on the basis of the EOS, we obtain the hot neutron star models. In our previous works [Takatsuka, 1991, 1992; Muto et al., 1993], we studied hot neutron stars at birth. But the studies were made by adopting simplified EOS. Here we intend to show results based on a more realistic approach. We discuss how hot neutron stars are different from cold ones. We ask how much energy they can release in a subsequent cooling stage due to neutrino diffusion and to what extent their rotations are accelerated. Finally, we give some comments to the new constraint on the maximum mass and the maximum rotation rate of neutron stars, by taking into account their hot and lepton-rich birth stage.

## II. DENSE SUPERNOVA MATTER

A. Calculation

We concentrate on dense supernova matter with the density $\rho$ exceeding the standard nuclear density $\rho_0(=0.17 \text{ nucleons/fm}^3 \simeq 2.8 \times 10^{14} \text{ g/cc})$. In order to get results at a realistic level, we pay attention to the following points [Takatsuka et al.,

1993, 1994]: (i) A set of finite-temperature ($T > 0$) Hartree-Fock (HF) equations for thermally equilibrated matter composed of $n, p, e^-, \nu_e, e^+$, and $\bar{\nu}_e$ is solved selfconsistently. (ii) The conditions such as the $\beta$-equilibrium under the $\nu$-degeneracy, the conservation of total baryon and lepton numbers and the charge neutrality are taken into account. (iii) An effective two-nucleon interaction $\tilde{V}_{RSC}$ is introduced in order to deal with the short-range correlation. The $\tilde{V}_{RSC}$ is based on the G-matrix calculation in asymmetric nuclear matter at $T = 0$ with the Reid-soft-core potential and depends on the asymmetry parameter $x(=(\rho_n - \rho_p)/\rho)$ as well as $\rho$ and the two-nucleon state. (iv) In addition, to assure the empirical saturation property of symmetric nuclear matter, a phenomenological two-nucleon interaction $\tilde{V}_{TNI}$ which simulates the effect of three-body force in medium is introduced. The $\tilde{V}_{TNI}$ is based on the idea of Lagaris and Pandharipande [1981].

In actual calculations, the "simplified treatment" [Nishizaki et al., 1994] is used. There the HF equations at $T > 0$ are solved for the case with $\tilde{V}_{RSC}$ only and the effect of $\tilde{V}_{TNI}$ is treated in terms of the direct-energy contribution calculated at $T = 0$. This method gives us an advantage to see transparently how the results depend on the phenomenological part ($\tilde{V}_{TNI}$). We consider three cases for $\tilde{V}_{TNI}$ which leads to the nuclear incompressibility $\kappa = 200$, 250 and 300 MeV. As well known, the $\kappa$ provides an important measure for the stiffness of the EOS of neutron star matter. For details of the present approach, references cited above are referred to.

B. Composition and equation of state of dense supernova matter

Results for the fraction $Y_i (= \rho_i/\rho)$ of each component are illustrated in Fig. 1(a). We remark the following point:

(i) The fraction of anti-particles ($e^+, \bar{\nu}_e$) is extremely small and not figured; (5–6) orders of magnitude smaller than that of particles ($e^-, \nu_e$). (ii) The proton (hence electron) population is very large; $Y_p = Y_e \simeq (31-32)\%((24-25)\%)$ for $Y_l = 0.4(0.3)$, in contrast with the small $Y_l(= Y_e \lesssim 5\%)$ of neutron star matter. (iii) The population of degenerate neutrinos is not small but is rather comparable with $Y_e$; $Y_\nu \simeq (8-9)\% \sim Y_e/3 (Y_\nu \simeq (5-6)\% \sim Y_e/4)$ for $Y_l = 0.4(0.3)$. (iv) The population of each component is almost independent of $\rho$.

To summarize, we propose an empirical formula for the $Y_p$ in dense supernova matter (note that $Y_e = Y_p$ and $Y_\nu = Y_l - Y_e$) as

$$Y_p = \frac{2}{3}Y_l + \Delta Y_p \simeq \frac{2}{3}Y_l + 0.05, \qquad (1)$$

which reproduces our results within 3% accuracy [Takatsuka et al., 1994]. The reason why so large and so constant $Y_p$ is realized is essentially in the participation of degenerate neutrinos. For transparency, we consider the $T = 0$ case. For supernova matter, the increase of $Y_p (= Y_e)$ means at the same time the decrease of $Y_\nu (= Y_l - Y_e)$ and hence the energy per baryon of lepton part $E_l = E_e + E_\nu$ has itself a minimum at large $Y_p$; exactly at $Y_p = \frac{2}{3}Y_l (\simeq 0.27)$ irrespectively of $\rho$ since the $Y_p$-dependence is given by $Y_p^{2/3} + 2^{1/3}(0.4 - Y_p)^{4/3}$. The effect of nucleon part $E_N (\propto (1-2Y_p)^2)$ pushes the minimum to higher $Y_p$ depending on $\rho$, but the shift $\Delta Y_p (\simeq 0.05)$ is small and almost constant because the effect of $E_N$ is not large enough to change the situation that the

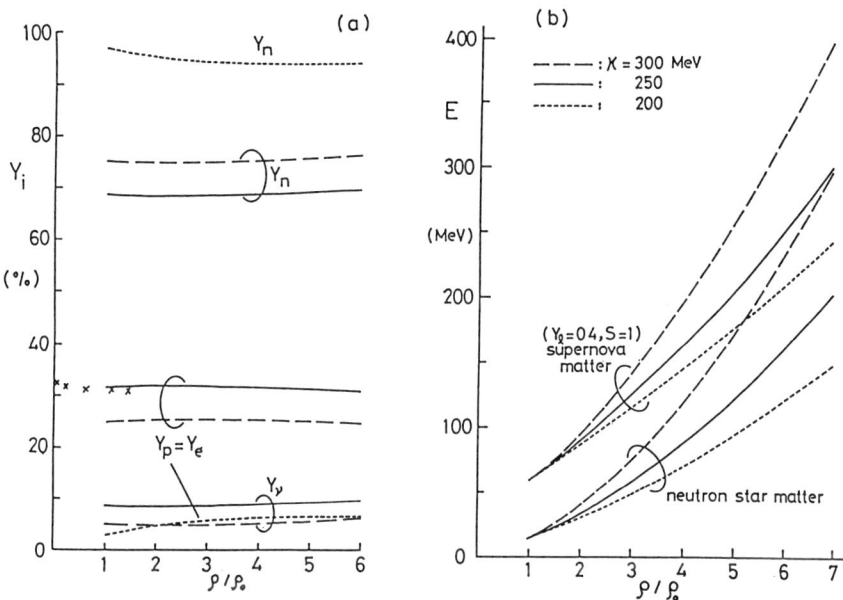

Figure 1: Properties of dense supernova matter; (a) Percentage fraction $Y_i$ of each component in supernova matter with {lepton fraction $Y_l =0.4$, entropy per baryon $S =1.0$}(solid lines) and {$Y_l =0.3$, $S =1.0$}(dashed lines) as a function of the density $\rho$ ($\rho_0$ being the standard nuclear density). Dotted lines are for neutron star matter; (b) Equation of state (EOS; Internal energy per baryon $E$ versus $\rho$) for nuclear EOS specified by the nuclear incompressibility $\kappa$.

lepton part governs the determination of the minimum point. On the other hand, for neutron star matter, the $E_\nu$ part is absent ($E_l = E_e$) and so the lepton energy increases monotonously with $Y_p$, leading to a minimum point of a small $Y_p(\simeq 0.04)$. Even for the case with $T > 0$, the situation is found to be very similar although the discussion is made in terms of the free energy per baryon.

The EOS of dense supernova matter is remarkably stiffer than that of neutron star matter, irrespectively of $\kappa$, as shown in Fig. 1(b). The mechanism of the stiffening is understood as follows [Takatsuka et al., 1994]. As far as the nucleon part $E_N$ is concerned, the effect of symmetry energy lowers the $E_N$ of supernova matter as compared with that of neutron star matter, since $Y_p \simeq 0.3$ for the former whereas $Y_p \lesssim 0.05$ in the latter. However, this energy gain in the former is well overwhelmed by the kinetic energy increase due to the abundant leptons ($Y_l = 0.4$ in the former whereas $Y_l \lesssim 0.05$ in the latter). This is the reason why a higher $E$ results for supernova matter. In the figure the case with larger $S(=1.5$, a dash-dotted line), namely, higher $T$ case, is also shown for comparison. The energy increase by $S = 1.0 \to 1.5$ is considerable but is smaller than $E_l$. This indicates that the temperature-effect is relatively small as compared with the lepton-effect. The case with smaller $Y_l(=0.3$, a dashed line) is also shown there for comparison with $Y_l=0.4$ case. The $E$ is decreased considerably by $Y_l = 0.4 \rightarrow 0.3$ but a remarkably higher $E$ for supernova matter remains unaltered.

## III. DISCUSSION OF HOT NEUTRON STARS

Our EOS of hot neutron stars is constructed as in the following; the EOS of dense supernova matter in II for $\rho \geq \rho_0$, the isentropic EOS constructed from the results of Lattimer *et al.* [1985] for $\rho \simeq (0.001\text{--}1)\rho_0$ and the EOS given by Baym, Bethe and Pethick [1971] for $\rho \leq 0.001\rho_0$. These three EOS's are smoothly matched in the regions $\rho \simeq (0.6\text{--}1)\rho_0$ and $\rho \simeq (0.001\text{--}0.002)\rho_0$.

A. Hot and fat features

Corresponding to the stiffened EOS, newborn neutron stars are "fat" as well as hot. Fig. 2(a) shows the density profiles $\rho(r)$ with $r$ being the distance from the center of a star. The stiffer the nuclear EOS (larger $\kappa$), the fat features are all the more

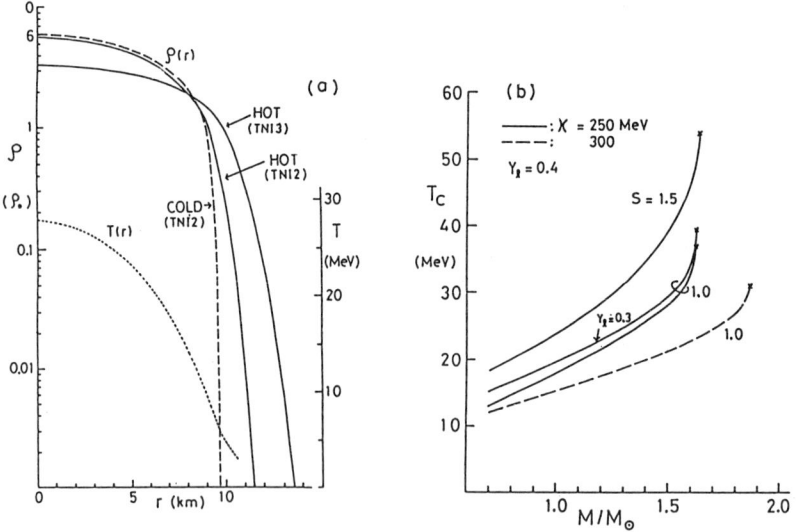

Figure 2: Characteristics of hot neutron stars; (a) Comparison of density profile $\rho(r)$ between hot and cold neutron stars. Hot neutron stars with $\{Y_l = 0.4, S = 1.0\}$ are considered for two cases; $\kappa = 250$ and 300 MeV. Temperature-profile $T(r)$ for a hot neutron star with $\{Y_l = 0.4, S = 1.0, \kappa = 250 \text{MeV}\}$ is shown by a dotted line; (b) Central temperature $T_c$ of hot neutron stars as a function of the gravitational mass $M$.

distinguished. The temperature profile $T(r)$ is also illustrated in Fig. 2(a) for a hot neutron star with $\{Y_l = 0.4, S = 1, \kappa = 250\text{MeV}\}$, which shows that the central temperature $T_c$ amounts to about 30 MeV. Fig. 2(b) shows how $T_c$ depends on hot neutron star models. It is known that $T_c$ gets higher as $\kappa$ decreases (softer EOS), $M$ increases (massive star) and $S$ increases. In some cases with $M = (1.5 - 1.6)M_\odot$, the central temperature is as high as $T_c = (40 - 50)$ MeV. It is noted that higher entropy $(S = 1.0 \to 1.5)$ raises the $T_c$ remarkably.

B. Releasable energy and spin-up

Newborn hot neutron stars contract in the subsequent cooling stage due to the $\nu$-diffusion and release the binding energy $E_B$ by an amount $\Delta E_c = E_B(\text{hot}) - E_B(\text{cold})$ as shown in Fig. 3. For example, in the case of a cold neutron star with $M = 1.4 M_\odot$ based on $\kappa = 250$ MeV EOS, $\Delta E_c \simeq 1.3 \times 10^{53}$ erg can be released, which is comparable to the released energy in the formation stage $\Delta E_f = -E_B(\text{hot}) \simeq 1.9 \times 10^{53}$ erg. The

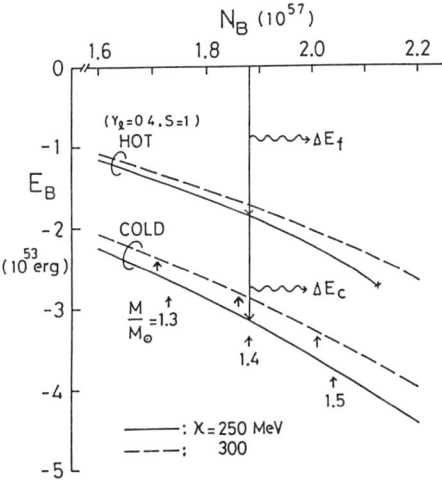

Figure 3: Binding energy $E_B$ of hot and cold neutron stars as a function of the total baryon number $N_B$. The $\Delta E_f$ and $\Delta E_c$ denote the releasable energy in the formation and the cooling stages, respectively.

contraction of hot neutron stars also means the acceleration of their rotation. If we assume that the total angular momentum of the star is conserved in the cooling stage, the ratio of angular frequency $\Omega$ (cold)/$\Omega$(hot) is extracted by the ratio of their moment of inertia $I(\text{hot})/I(\text{cold})$ from neutron star models. The spin-up is noticeable; the extent of spin-up is 20, 15 and 10% according to $M = 1.3, 1.4$ and $1.5 M_\odot$, respectively, in the case of $\{Y_l = 0.4, S = 1, \kappa = 250 \text{ MeV}\}$ as an example. This spin-up is of interest in relation to the maximum rotation rate of cold neutron stars to be discussed in the next section.

## IV. NEW CONSTRAINT ON MAXIMUM MASS AND MAXIMUM ROTATION RATE

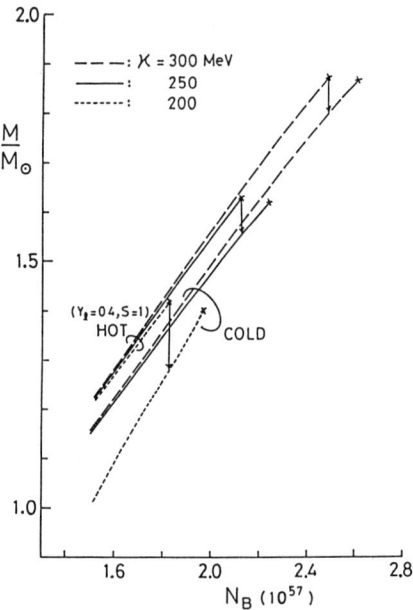

Figure 4: Gravitational mass $M$ versus total baryon number $N_B$ for hot and cold neutron stars. Crosses denote the critical mass for a given EOS and arrows indicate the decrease of a critical mass for cold neutron stars by taking account of the hot stage.

A. Constraint on maximum mass

In Fig. 4, the neutron-star mass $M$ is plotted as a function of total baryon number $N_B$.

Corresponding to the stiffened EOS, the maximum mass (indicated by the cross) is larger for hot case than cold case ($M_{max}(\text{hot}) > M_{max}(\text{cold})$). On the other hand, the total baryon number $N_{B,max}$ of the maximum-mass star is smaller for hot case than for cold case ($N_{B,max}(\text{hot}) < N_{B,max}(\text{cold})$). This means that cold neutron stars with $N_B > N_{B,max}(\text{hot})$ cannot exist. Consequently, the maximum mass of cold neutron stars is given not by $M_{max}(\text{cold})$ but by $M'_{max}(= M(\text{cold})$ at $N_B = N_{B,max}(\text{hot}))$. Fig. 4 shows that the true maximum mass is decreased from $M_{max}(\text{cold})$ by an amount $\Delta M_{max} = M_{max}(\text{cold}) - M'_{max} \simeq 0.065, 0.07$ and $0.12 M_\odot$ according to $\kappa=300, 250$, and 200 MeV, respectively, when the star evolved from the hot star with $\{Y_l = 0.4, S = 1\}$. The $\Delta M_{max}$ gets larger for smaller $\kappa$ (softer EOS). It also depends on to what extent the EOS of hot stars is stiffened as compared with that of cold stars. For the case with $\kappa=250$ MeV, the $\Delta M_{max} \simeq 0.03 M_\odot$ for $\{Y_l = 0.3, S = 1\}$ (weak stiffening case) and $0.075 M_\odot$ for $\{Y_l = 0.4, S = 1.5\}$ (strong stiffening case). Roughly speaking, the maximum mass (or critical mass) of neutron stars should be taken smaller by about $0.05 M_\odot$ than that currently taken from cold neutron star model. We want to stress that the consideration of hot stage gives more stringent constraint on the maximum mass of neutron stars.

B. Constraint on maximum rotation rate

Keplerian angular frequency $\Omega_K$ of a rotating neutron star places an upper limit for the rotation rate. The maximum rotation rate $\Omega_{max}$ of cold neutron stars is currently discussed by this $\Omega_K$. According to the relativistic Roche model proposed by Shapiro, Teukolsky and Wasserman [1983], the $\Omega_K$ is given by

$$\Omega_K \simeq 6.3 \times 10^3 (M_B/M_\odot)^{1/2}(R/10\text{km})^{-3/2} \quad \text{rad/s}, \qquad (2)$$

with $M_B$ being the baryon mass of a neutron star ($M_B = N_B \times$nucleon mass). Since a hot star has a larger radius $R$ compared to a cold one with the same $N_B$ (hence same $M_B$), Eq. (2) means that $\Omega_K(\text{hot}) < \Omega_K(\text{cold})$. Therefore the $\Omega_{max}$ of a given neutron star (given $N_B$ and $\kappa$) should be discussed in terms of the $\Omega_K(\text{hot})$ and by taking account of the spin-up in the cooling stage. That is, if the $\Omega'_K$ defined by

$$\Omega'_K = \Omega_K(\text{hot}) \times \text{spin-up ratio}, \qquad (3)$$

is smaller (larger) than $\Omega_K(\text{cold})$, the $\Omega_{max}$ should be taken as $\Omega_{max} = \Omega'_K(\Omega_K(\text{cold}))$. Fig. 5 shows the results. It is observed that $\Omega'_K < \Omega_K$ for all the cases considered. We remark that the maximum rotation rate is more severely constrained by $\Omega'_K$ than usual $\Omega_K(\text{cold})$.

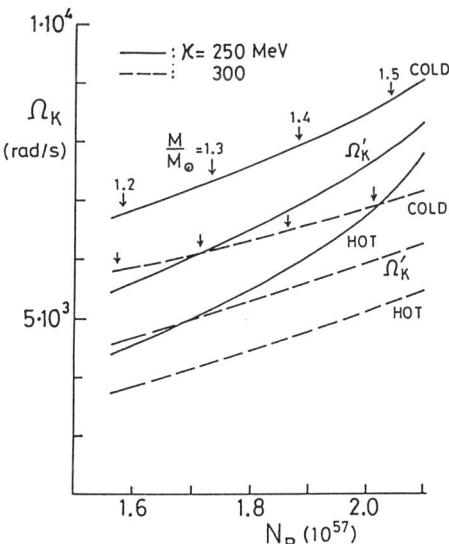

Figure 5: Keplerian angular frequency $\Omega_K$ versus $N_B$ for hot and cold neutron stars. $\Omega'_K(< \Omega_K(\text{cold}))$ defined by $\Omega'_K = \Omega_K(\text{hot}) \times$ spin-up ratio constrains more severely the maximum rotation rate of cold neutron stars than $\Omega_K(\text{cold})$.

## V. CONCLUDING REMARKS

Hot neutron stars at birth are "fat" as well as hot. For medium-mass neutron stars ($M = (1.4\text{--}1.5)M_\odot$) based on the EOS with $\kappa = (250\text{--}300)$MeV, the radius is larger by $(20\text{--}30)\%$ than that of cold stars. The central temperature $T_c$ is as high as $T_c \sim (20\text{--}40)$ MeV. By the contraction in the cooling stage, they release the binding energy $\Delta E_c \sim (1.1\text{--}1.4) \times 10^{53}$ erg, which is rather comparable to the energy $\Delta E_f \sim (1.7 \sim 2.4) \times 10^{53}$ erg released in the formation stage. From the observation of $\nu$-burst from SN1987A, the total energy released $\Delta E = \Delta E_c + \Delta E_f$ has been estimated as $(0.9\text{--}3.5) \times 10^{53}$ erg [Sato and Suzuki, 1987] and $(1.6\text{--}3.1) \times 10^{53}$ erg [Kahana et al., 1987]. Although the $\Delta E$ is not well determined, a constraint $\Delta E \lesssim (3.1\text{--}3.5) \times 10^{53}$ erg from observation and the results in Fig. 5 suggest that the mass of a neutron star in SN1987A would be smaller than $(1.4\text{--}1.5)M_\odot$. If the information of $\Delta E_c$ or $\Delta E_f$ could be extracted from the time profile of $\nu$-events, it would be more useful for the determination of the mass.

Cold neutron stars evolved from hot neutron stars. We want to stress that some new constraints on usual cold neutron stars are obtained if the hot stage is taken into account. The critical mass should be taken smaller by $\sim 0.05 M_\odot$ than the maximum mass from cold-neutron star models and the maximum rotation rate would be smaller by $(10\text{--}15)\%$ than the Keplerian angular frequency of cold neutron stars. The observation of $M(\text{PSR}1913+16) = 1.44 M_\odot$ suggests that the EOS of cold neutron stars should be so stiff as to sustain at least the mass of about $1.5 M_\odot$.

## ACKNOWLEDGEMENT

The author wishes to thank Dr. S. Nishizaki for cooperative discussion. This work is indebted to the Grant-in-Aid for Scientific Research from the Ministry of Education, Science and Culture (04640279,05243103).

## REFERENCES

Baym, G., Bethe, H.A., and Pethick, D.J., 1971, Nucl. Phys., **A175**, 225.
Bethe, H.A., Brown, G.E., Applegate, J., and Lattimer, J.M., 1979, Nucl. Phys., **A324**, 487.
Epstein, R.I., and Pethick, C.J., 1981, Astrophys. J., **243**, 1003.
Freedman, D.Z., 1974, Phys. Rev., **D9**, 1389.
Kahana, S.H., Cooperstein, J., and Baron, E., 1987, Phys. Lett., **B196**, 259.
Lagaris, I.E., and Pandharipande, V.R., 1981, Nucl. Phys., **A359**, 349.
Lattimer, J.M., Pethick, C.J., Ravenhall, D.G., and Lamb, D.Q., 1985, Nucl. Phys., **A432**, 646.
Mazurek, T.J., 1975, Astrophys. and Space Sci., **35**, 117; 1976, Astrophys. J. Lett., **207**, L87.
Muto, T., Takatsuka, T., Tamagaki, R., and Tatsumi, T., 1993, Prog. Theor. Phys. Suppl., No. 112, 221.
Nishizaki, S., Takatsuka, T., and Hiura, J., 1994, Prog. Theor. Phys., **92**, No.1.
Sato, K., 1975, Prog. Theor. Phys., **53**, 595; **54**, 1325.
Sato, K., and Suzuki, H., 1987, Phys. Lett., **B196**, 267.
Shapiro, S.L., Teukolsky, S.A., and Wasserman, I., 1983, Astrophys. J., **272**, 702.
Takatsuka, T., Nishizaki, S., and Hiura, J., 1993, Prog. Theor. Phys., **89**, 55l; 1994, submitted to Prog. Theor. Phys.
Takatsuka, T., 1991, in *Unstable Nuclei in Astrophysics*, (World Scientific), p.291; 1992, in *Structure and Evolution of Neutron Stars*, (Addison Wesley), p.257.

# I.3

# GLOBAL OSCILLATIONS OF NEUTRON STARS AS POTENTIAL PROBES OF MICROSCOPIC PHYSICS AT ULTRAHIGH DENSITIES

H.M. Van Horn,[1,2] U. Lee,[3] R.I. Epstein,[4] and T.J.B. Collins[1]

[1] Department of Physics and Astronomy and C. E. Kenneth Mees Observatory
University of Rochester, Rochester, NY 14627-0171, U. S. A.
[2] Division of Astronomical Sciences, National Science Foundation
Arlington, VA 22230, U. S. A.
[3] Laboratoire de Physique, Ecole Normale Superieure, Lyon, France
[4] Los Alamos National Laboratory, MS D436, Los Alamos, NM 87545, U. S. A.

The global oscillations of neutron stars have been studied theoretically since the early 1960s, at first primarily in an effort to understand the effects of strong, relativistic gravity on the oscillation spectrum. With the discovery of pulsars in 1968, and the realization that they are indeed neutron stars, theories of the microscopic physical properties of neutron stars underwent rapid improvement. We currently understand a neutron star to be a very dense body containing a core of degenerate, superfluid neutrons with a small admixture of superfluid, superconducting protons and degenerate, normal electrons. Overlying the core is a solid crust consisting of a strongly coupled plasma of highly neutron-enriched nuclei embedded in a sea of degenerate electrons, and at the very surface may be a thin, superficial "ocean" of molten crustal material – a strongly coupled, *fluid* plasma. The theory of the global oscillations of these complicated bodies has undergone a renaissance during the past decade, and the spectrum of non-radial global oscillation modes is now understood to be far more complex than originally realized. In this paper, I review the current understanding of these oscillations, focusing on their potential use as probes of the properties of matter under conditions of ultra-high density, and I highlight a number of cases in which neutron-star oscillations may be observable.

## I. INTRODUCTION

The existence of neutron stars was predicted theoretically by Landau (1932). Subsequent investigations focused primarily on using these dense bodies as testbeds for studying gravitation in the strong-field limit. Neutron stars remained objects of purely theoretical interest until the discovery of pulsars by Hewish *et al.* (1968) and the recognition (Gold 1968) that they are rapidly rotating neutron stars with strong surface magnetic fields.

In a series of papers in the late 1960s and early 1970s, Thorne and his coworkers laid the foundation for the investigation of neutron-star oscillations (Thorne and Campolattaro 1967; Price and Thorne 1969; Thorne 1969a, b; Campolattaro and Thorne 1970; Ipser and Thorne 1973). From a perturbation analysis of Einstein's equations for general relativistic gravity, Thorne *et al.* derived the equations describing the infinitesimal oscillations of completely fluid, zero-temperature, non-rotating, spherical models for neutron stars. They solved these equations to obtain the periods and pulsation eigenfunctions for the acoustic oscillations of neutron stars. In the language of stellar pulsation theory (Cowling 1941; Unno *et al.* 1979; Cox 1980), these are the $p$-mode oscillations.

Real neutron stars, however, are much more complex than these early models suggest. They are rapidly rotating. They have solid crusts. They usually possess strong magnetic fields which alter the properties of matter at the stellar surface. They contain superfluid neutrons and protons in their cores. Each of these complications affects the oscillation spectrum of a neutron star. In turn, if neutron-star oscillations can ever be observed, they offer the potential for making direct seismological measurements of many of the physical properties of the dense plasmas in these objects.

In the present paper, I briefly review the current state of investigations of neutron-star oscillations. In §II I first summarize our current understanding of the microphysics of neutron-star matter, emphasizing those aspects that are expected to affect the properties of neutron-star oscillations. The current state of neutron-star oscillation theory is discussed in §III, and in §IV I consider some possibilities for exciting and perhaps observing neutron-star oscillations. In §V I summarize the main conclusions of this paper.

## II. NEUTRON STAR MICROPHYSICS

A typical neutron star has a mass of about $1.4 M_\odot$, where $M_\odot = 1.989 \times 10^{33}$ g is the mass of our Sun. This mass is compressed into a volume of radius $R_* \approx 10$ km, about the size of a small city. The mean density of a neutron star is thus $\sim 2 \times 10^{15}$ g cm$^{-3}$ – greater than the density of matter in an atomic nucleus! Internal temperatures in neutron stars are believed to be $\sim 10^7$ K, ranging from $\sim 10^8$ K in X-ray burst sources down to surface temperatures perhaps as low as $\sim 10^5$ K in old pulsars.

At the surface, a neutron star at finite temperature has a layer of molten crustal material we have termed the "ocean." This is actually a fully ionized, dense, degenerate, fluid plasma very like that inside a white dwarf star. If the neutron star has a strong $\sim 10^{12}$ G magnetic field, as is the case in a pulsar, the properties of this plasma will be very different from those of a non-magnetized plasma. In the strong-field case, the incompletely ionized matter at the very surface consists of "magnetic matter" (Ruderman 1974) – matter in which the magnetic forces completely dominate the Coulomb force, forming "atoms" in which electrons fill cylindrical Landau levels. Recent calculations of the properties of magnetic matter (Fushiki, Gudmundsson, and Pethick 1989; Lai and Shapiro 1991) indicate that it has a finite mass density $\rho \sim 10^4$ g cm$^{-3}$ at zero pressure when the magnetic field strength is $B \sim 10^{12}$ G. Magnetic matter is expected to exist at the surfaces of pulsars, at least near the magnetic poles. The existence of magnetic matter is expected to affect the properties of neutron-star oscillation modes that are concentrated near the neutron star surface, but calculations of this effect have not yet been carried out, and I shall not discuss them further here.

If a neutron star is non-magnetic, as may be the case in the X-ray burst sources, for example, the base of the "ocean" of molten crustal material is formed by a frozen solid consisting of fully stripped ions of charge $Z$ embedded in a nearly uniform, neutralizing sea of increasingly degenerate electrons. The transition from the liquid to the solid phase occurs at the depth where the Coulomb parameter is $\Gamma \equiv (Ze)^2/akT \sim 180$. It has generally been assumed that the solid crust is crystalline, but it is conceivable that the neutron star may cool sufficiently rapidly that the crust will be "glassy" instead (Ichimaru et al. 1983). In either case, the existence of the fluid-solid boundary within the neutron star permits the existence of surface waves, analogous to Love waves in the Earth.

Deeper in the crust of a neutron star, the electron Fermi energy becomes increasingly large, forcing electrons back into the nuclei and creating increasingly neutron-enriched nuclei. Some of the transitions between different nuclear species are believed to occur abruptly, with associated discontinuities in the mass density (cf. Baym, Pethick, and Sutherland 1971). Surface waves can also exist at these interfaces (see §III.B below).

Still deeper, where the mass density is $\sim 4.3 \times 10^{11}$ g cm$^{-3}$, it becomes energetically favorable for free neutrons to "leak" out of the nuclei. This level marks the boundary between the outer crust and the inner crust of the neutron star. In the inner crust, matter consists of a solid lattice (or glass) of highly neutron-enriched ions immersed in a sea of free, highly relativistic electrons as well as a sea of free neutrons. It is believed (cf. Shapiro and Teukolsky 1983 and references therein) that the neutrons form Cooper pairs and exist in the superfluid state. Because the neutron superfluid is only weakly coupled to the ion solid, the neutron superfluid and the solid crust move essentially independently of each other. This permits the existence of still further global oscillation modes of the neutron star (cf. §III.C).

At a density equal to the density of nuclear matter, $\sim 2 \times 10^{14}$ g cm$^{-3}$, the remaining nuclei in the crust finally dissolve into a sea predominantly of neutrons, with a small admixture of protons and electrons, that forms the core of the neutron star and comprises the bulk of the stellar mass. Both the neutrons and protons are thought to exist as superfluids throughout much of the core. The boundary between the inner crust and the core of a neutron star has been assumed to be another fluid/solid boundary. If so, it provides yet another site for surface waves.

The properties of matter in the cores of neutron star are still very uncertain. If the nuclear equation of state is "stiff," meaning that the repulsion between the nucleons at very high densities is relatively strong, the central density of a typical neutron star may be comparatively low, only a few times the density of nuclear matter, and the radius of the neutron star may be relatively large, perhaps $\sim 15$ km. Conversely, if the equation of state of matter in the neutron star core is relatively "soft," the radius of the neutron star may be much smaller, perhaps less than 10 km, and the central density may be so high that exotic states of matter may exist there. Possibilities that have been suggested include mixtures of free hyperons, pion condensates, neutron crystals, and even free quarks (cf. Baym 1994, these proceedings). Nothing is known about the possible effects of such exotic forms of matter on the oscillatory properties of neutron stars, and I shall not discuss them further.

## III. NEUTRON-STAR OSCILLATION THEORY

The possibility that neutron stars might sustain observable oscillations was essentially ignored until Boriakoff (1976) published his suggestive paper on observations of the pulsar PSR 2016+28. In part, the reason for this neglect seems to have been due to our collective ignorance of the oscillation spectra of neutron stars. The only oscillation modes about which anything was known at the time were the neutron-star $p$-mode oscillations. Thorne and his collaborators (cf. references in §I) had found the periods of these modes to be $\lesssim$ 1 ms, with amplitudes damped by gravitational radiation within just a few cycles. Thus, there was an intellectual bias among astronomers against the observability of neutron-star oscillations. Boriakoff's observations, using the 300-m Arecibo radiotelescope, resolved the structure of individual pulses from PSR 2016+28 at a timescale of a fraction of a millisecond. He found substantial quasiperiodic fine structure within each individual drifting subpulse, and he showed that some residue of this quasiperiodicity persisted, at a timescale of about 0.9 ms, even in the autocorrelation function averaged over all the pulses in his sample. Boriakoff accordingly suggested that he had actually observed oscillations of the neutron star.

Boriakoff's observations stimulated Van Horn (1980) to reconsider the question of the spectrum of neutron-star oscillations. He found that neutron stars can sustain a wide diversity of non-radial oscillation modes in addition to the acoustic modes (cf. Table 1). These modes fall into two broad classes, termed "spheroidal" and "toroidal," depending upon which of two possibilities is satisfied for the angular separation of variables in the system of eigenvalue equations. The familiar acoustic ($p$-) modes fall in the spheroidal class. There are no toroidal modes in ordinary, non-rotating, fluid stars. Motions associated with toroidal modes are purely transverse, and a separate restoring force – such as, e.g., the shear strength of the solid crust – is necessary for such modes to exist. More detailed information about the oscillation modes of spherical, non-rotating, non-magnetic neutron stars is provided in McDermott, Van Horn, and Hansen (1988).

Table 1. Oscillation Modes of Neutron Stars

| Type | Period | Sensitive to: |
|---|---|---|
| Spheroidal Modes | | |
| $p$-modes | $\lesssim$ 0.1 – 1 ms | Mean density |
| $g$-modes | $\gtrsim$ 100 ms | Internal temperature |
| Interfacial modes | $\sim$ 10 ms | Density discontinuities |
| Alfvén modes | $\sim$ 150 s? | Magnetic field strength |
| Tkachenko modes | $\sim$ years | Superfluidity |
| Toroidal Modes | | |
| Shear modes | $\lesssim$ 20 ms | Shear strength, crust thickness |
| $r$-modes | $\lesssim$ 1 s | Rotation period |

In the following subsections I describe in a bit more detail the results of some relatively recent calculations demonstrating the effects of various improvements in the microphysics upon the non-radial oscillation spectra of neutron stars.

A. Effects of a glassy crust

As noted above, Ichimaru et al. (1983) have suggested that in some circumstances, the strongly coupled, one-component plasma (OCP) in a dense, degenerate star may form a glass rather than a crystalline solid when sufficiently cool. The crust of a neutron star provides one physical environment in which this may occur, and for this reason, Strohmayer et al. (1991) undertook an investigation of this possibility. They performed a new, *ab initio* calculation of the shear modulus of an OCP in the solid phase for both crystalline and glassy solids and found appreciable departures of the shear modulus $\mu$ from the estimate obtained from the old Fuchs value for a Coulomb solid. They attributed this to their use of a more accurate expression for $\mu$. They also found that $\mu$ drops considerably as the solid is warmed to the melting point, with a larger difference for the glassy solid than for the crystal.

Strohmayer et al. also computed new eigenfunctions and eigenvalues for the non-radial oscillation modes of neutron stars using the new shear modulus. Not surprisingly, the modes that are most affected are those which are most sensitive to $\mu$: the toroidal shear modes and their spheroidal counterparts (the $s$-modes – cf. McDermott et al. 1988) as well as the interfacial modes associated with the top and bottom of the crust. The effects can produce shifts as large as $\sim$ 20 to 30% in the periods of some of these modes. This difference, however, is essentially that produced by the use of the improved expression for the shear modulus. The effects of the difference between a crystalline and a glassy solid are almost completely undetectable.

B. Discontinuity $g$-modes and core $g$-modes

As noted above, the density discontinuities associated with composition transitions in the crust of a neutron star allow the existence of associated interfacial oscillation modes. These modes, for which the restoring force is gravity acting upon perturbations in the density distribution in the vicinity of the composition discontinuity, were first investigated by Finn (1987). They were subsequently studied by McDermott (1990) and more recently by Reisenegger and Goldreich (1992). The essential characteristics of these modes are that they exhibit a cusp in the radial displacement eigenfunction associated with the mode and that the eigenfunction declines sharply away from the interface. For a given value of the spherical harmonic index $l$ associated with the eigenfunction, there is one single "discontinuity $g$-mode" (to use the term introduced by Finn 1987) associated with each composition discontinuity. For example, the $g_1^d$ discontinuity $g$-mode, which is associated with the transition from $^{56}$Fe$\to^{62}$Ni in the outer crust of the neutron star, has a period randing from about 5.1 ms to 5.5 ms depending upon the particular model used. Discontinuity $g$-modes associated with composition transitions deeper in the crust have longer periods.

Reisenegger and Goldreich (1992) have also investigated a class of modes they termed "core $g$-modes," for which the restoring force is provided by the buoyantly stable, density-dependent compositional stratification of the core. These modes have large amplitudes only in the core of the neutron star, and according to Reisenegger and Goldreich, they have periods of a few tens of milliseconds, depending upon the particular neutron star model used. It is not clear what relationship, if any, these modes bear to those with periods of some tens of *seconds* that had previously been termed "core *g*-modes" by McDermott *et al.* (1988). Both calculations suffer from the neglect of superfluidity in the neutron-star core, which will change the results in as yet unknown ways. In addition, *any* modes associated primarily with the neutron-star core will almost certainly be unobservable: not only are their amplitudes small at the surface, but also the enormous energy needed to excite such a mode to a potentially observable amplitude is unlikely to be available from any plausible source, except perhaps a collision between two neutron stars.

C. Effects of superfluidity

Epstein (1988) pointed out that superfluidity affects neutron-star oscillations in a variety of ways, and he performed a short-wavelength analysis to elucidate the different mode classes. Van Horn and Epstein (1990) subsequently extended this analysis to the global *toroidal* oscillation modes of neutron stars. They found that superfluidity in the inner crust of a neutron star effectively reduces the density of *normal* matter, thus increasing the propagation speed $c_t \propto (\mu/\rho)^{1/2}$ of the transverse shear waves. This causes the shear waves to propagate mainly in the outer part of the outer crust, in effect causing the modes computed with superfluidity taken into account to resemble higher overtones of similar modes computed without superfluidity. The toroidal modes that include superfluid effects thus have shorter periods than those without.

In several recent papers, Mendell (1991) and his collaborators (Lindblom and Mendell 1994) have formulated the vector equations of motion for neutron stars, including the effects of superfluidity. Their equations of motion are very similar to those derived independently by Lee *et al.* (1993). Lindblom and Mendell (1994) have shown that calculations which include the effects of superfluidity permit a new type of oscillation mode in which the superfluid and normal matter oscillate against each other, as well as a mode in which the normal fluid and the superfluid move essentially in phase. Similar results have been obtained by Lee *et al.* (1993, 1994), and additional work is currently in progress.

## IV. CAN OSCILLATIONS BE EXCITED AND OBSERVED?

As the calculations described above show, a neutron star is a rich resonant system, capable of sustaining a wide range of different types of non-radial oscillation modes. Before it is possible to use such modes as probes of the microphysics of neutron stars, however, it is essential that neutron star oscillations actually be detected. There is no agreement that this has yet occurred; indeed the general view at this time is that neutron star oscillations have *not* yet been observed. One of the difficulties is that we have not yet identified mechanisms capable of exciting neutron star oscillations to potentially observable amplitudes. This should not be construed to mean that such excitation is impossible; rather, with few exceptions we have not yet been sufficiently clever to think of appropriate physical processes. In the present section, we consider separately two types of systems in which we know that neutron stars are being subjected to very energetic processes that may be capable of inducing oscillations: X-ray burst sources and pulsars.

A. X-ray bursters

An X-ray burst source is a binary star system in which a low-mass companion transfers matter through an accretion disk onto a neutron star. The high temperatures encountered as mass flows through the inner part of the accretion disk and onto the neutron star are believed to burn H more-or-less quiescently through ordinary thermonuclear processes, leading to a gradual accumulation of He at the surface of the neutron star. When the He layer becomes sufficiently deep, He-burning is ignited under highly degenerate conditions, leading to a violent thermonuclear outburst. Calculations of this process (*cf.* Joss 1977) have demonstrated that it can (i) provide the integrated total energy observed in an X-ray burst, (ii) produce the observed luminosities and keV temperatures for an X-ray burst source, and (iii) produce X-ray light curves that closely resemble the observed light curves. This has been accepted as the correct model for Type I X-ray bursts for more than a decade and a half.

Because an X-ray burst delivers a strong mechanical and thermal pulse to a neutron star ($E \sim 10^{38}$ to $10^{39}$ ergs in $\lesssim 10$ s), it seems highly probable that it can excite oscillations of the neutron star, particularly of those modes which are concentrated primarily near the surface: surface $g$-modes, $r$-modes, and perhaps some of the interfacial modes or crustal oscillations. With this in mind, McDermott and Taam (1987) computed several of the low-order $g$-modes for a neutron star model at different stages during an X-ray burst. They found that the strong temperature increase associated with the thermonuclear outburst caused the $g$-mode periods first to decrease sharply at the beginning of the X-ray burst and then gradually to return to normal during the decline of the burst. For example, the $l = 2$ $g_1$ mode period dropped from a normal level of about 20 ms prior to the X-ray burst to a minimum of about 10 ms at the peak of the burst. These periods are comparable to periodicities of $\sim 12$ to 15 ms reported for some outbursts of the sources 1H 1909+096 and XB 1728-34. Furthermore, McDermott and Taam computed the excitation of these oscillation modes by thermonuclear driving in the He-burning shell. They found that the modes were strongly damped *except* near the time of peak burning, when the lower-order $g$-modes were strongly excited.

Although this appears to be an excellent example of a mechanism for exciting $g$-mode oscillations in neutron stars (and it may indeed be so), two notes of caution are appropriate. First, the computed damping for models away from peak burning is so strong that even if the oscillation modes can be excited by during the thermonuclear outburst, they may be damped too rapidly to permit observational detection. Second, because the thermal structure of the neutron star evolves rapidly during the X-ray burst, the adiabatic pulsation periods of the $g$-modes exhibit very rapid changes in time. This makes it very difficult, perhaps impossible, to employ conventional power-spectrum analyses or auto-correlation function techniques to identify and follow $\sim 10$ ms oscillations through a $\sim 10$ second X-ray burst. Thus, even if oscillations are excited in such systems, it may not be possible to demonstrate this convincingly. Quasiperiodic oscillations *are* observed in a number of low-mass X-ray binary systems (*cf.* van der Klis 1989), but the favored current explanation for this phenomenon is the "beat frequency" model, which relies upon the difference in rotation rates between the inner edge of the accretion disk and the surface of the neutron star, rather than upon oscillations of the star.

One other process that may be capable of exciting oscillations in the surface layers of X-ray burst sources is mechanical excitation resulting from the intersection of the inner edge of the accretion disk with the surface of the neutron star. To date, there has been no serious investigation of the excitation of time-dependent oscillations of the neutron star by this process, so that we are unable to say whether or not it is effective.

B. Pulsars

Strohmayer, Cordes, and Van Horn (1992) have recently reinvestigated the quasiperiodic oscillations first observed in PSR 2016+28 by Boriakoff (1976). Their purpose was to conduct a power-spectrum analysis of the signal from this pulsar and to compare it with the results of time-series simulations in order to place constraints on the oscillations. They reaffirmed the existence of two main features in the power spectrum: (i) a broad feature associated with the quasiperiodic microstructure and (ii) a fairly sharp feature produced by the drifting subpulses in this pulsar. The broad microstructure feature peaks at a period of $\sim 0.9$ ms and is the feature Boriakoff originally suggested as indicative of neutron-star oscillations. However, Strohmayer et al. found this feature to be essentially uncorrelated from one pulse to the next, with $Q \sim 6$. These characteristics make it very unlikely that this feature is actually associated with a high-$Q$ oscillation of the neutron star, in agreement with conclusion reached by most other investigators.

Conversely, the feature in the power spectrum that is associated with the drifting subpulses *does* maintain coherence over several pulses, corresponding to $Q \sim 90$. This is sufficiently great that it may indicate that an oscillation of the neutron star provides the underlying "clock" mechanism for the drifting subpulses. Again, there are difficulties with this hypothesis. On the theoretical side, we still have *no* mechanism identified that is capable of tapping some of the energy associated with pulsar emission and using it to drive oscillations. In consequence, theoreticians cannot advise observers as to probable ranges of periodicities that could plausibly be associated with pulsar emission. The difficulty on the observational side is similar to that afflicting observations of oscillations in X-ray burst sources: any neutron-star oscillations in pulsars must be observed through the time-"window" provided by the main pulse of the pulsar. Thus, any $\sim 1$ to 10 ms oscillations that might be present will be broadened by the $\sim 100$ ms window.

## V. SUMMARY

I have tried to show in this paper that neutron-star oscillations can in principle probe many different physical properties of neutron stars, *if* they can ever be observed.

The key observational question is: *Can neutron star oscillations be detected?* Analyses of the drifting subpulses in pulsars indicates that neutron-star oscillations may provide the "clock" mechanism underlying this phenomenon, but it remains to be confirmed whether or not this is actually the case. Theory also suggests that at least some of the quasiperiodic oscillations observed in the X-ray burst sources may be associated with neutron-star oscillations, but this, too, remains to be confirmed. These are difficult observational problems.

The key theoretical question is: *How can neutron star oscillations be excited?* Calculations suggest that the thermonuclear outburst that produces an X-ray burst may also excite neutron-star oscillations, but calculations have so far been performed for only one model, and additional questions remain to be answered. It is also conceivable that mechanical excitation by the surrounding accretion disk may excite oscillations of the neutron star, but to date this hypothesis is purely speculative. Finally, we have no idea at present whether or not there is a mechanism capable of exciting oscillations in pulsars.

If these observational and theoretical challenges can be answered positively, neutron star seismology may in the future provide a tool to enable us to measure the physical properties of the dense plasmas in neutron stars.

## REFERENCES

Baym, G. 1994, these proceedings.
Baym, G., Pethick, C. J., and Sutherland, P. G. 1971, Astrophys. J., **170**, 299.
Boriakoff, V. 1976, Astrophys. J. Letters, **208**, L43.
Campolattaro, A., and Thorne, K. S. 1970, Astrophys. J., **159**, 847.
Cowling, T. G. 1941, M. N. R. A. S., **101**, 367.
Cox, J. P. 1980, *Theory of Stellar Pulsation*, (Princeton, Princeton Univ. Press).
Epstein, R. I. 1988, Astrophys. J., **333**, 880.
Finn, L. S. 1987, M. N. R. A. S., **227**, 265.
Fushiki, I., Gudmundsson, E. H., and Pethick, C. J. 1989, Astrophys. J., **342**, 958.
Gold, T. 1968, Nature, **218**, 731.
Hewish, A., Bell, S. J., Pilkington, J. D. H., Scott, P. F., and Collins, R. A. 1968, Nature, **217**, 709.
Ichimaru, S., Iyetomi, H., Mitake, S., and Itoh, N. 1983, Astrophys. J., **265**, L83.
Ipser, J. R., and Thorne, K. S. 1973, Astrophys. J., **181**, 181.
Joss, P. C. 1977, Nature, **270**, 310.
Lai, D., and Shapiro, S. L. 1991, Astrophys. J., **383**, 745.
Landau, L. D. 1932, Phys. Z. Sowjetunion, **1**, 285.
Lee, U., Collins, T. J. B., Epstein, R. I., and Van Horn, H. M. 1993, in *Proc. IAU Coll. 147: The Equation of State in Astrophysics*, ed. G. Chabrier, in press.
Lee, U., Collins, T. J. B., Epstein, R. I., and Van Horn, H. M. 1994, in preparation.
Lindblom, L., and Mendell, G. 1994, Astrophys. J., **421**, 689.
McDermott, P. N. 1990, M. N. R. A. S., **245**, 508.
McDermott, P. N., and Taam, R. E. 1987, Astrophys. J., **318**, 278.
McDermott, P. N., Van Horn, H. M., and Hansen, C. J. 1988, Astrophys. J., **325**, 725.
Mendell, G. 1991, Astrophys. J., **380**, 515.
Price, R., and Thorne, K. S. 1969, Astrophys. J., **155**, 163.
Reisenegger, A., and Goldreich, P. 1992, Astrophys. J., **395**, 240.

Ruderman, M. A. 1974, in *IAU Symp. 53: Physics of Dense Matter*, ed. C. J. Hansen (Dordrecht, Reidel), p. 117.
Shapiro, S. L., and Teukolsky, S. A. 1983, *Black Holes, White Dwarfs, and Neutron Stars* (New York, Wiley & Sons).
Strohmayer, T. E., Ogata, S., Iyetomi, H., Ichimaru, S., and Van Horn, H. M. 1991, Astrophys. J., **375**, 679.
Strohmayer, T. E., Cordes, J. M., and Van Horn, H. M. 1992, Astrophys. J., **389**, 685.
Thorne, K. S. 1969a, Astrophys. J., **158**, 1.
Thorne, K. S. 1969b, Astrophys. J., **158**, 997.
Thorne, K. S., and Campolattaro, A. 1967, Astrophys. J., **149**, 591.
Unno, W., Osaki, Y., Ando, H., and Shibahashi, H. 1979, *Nonradial Oscillations of Stars*, (Tokyo, Univ. Tokyo Press).
van der Klis, M. 1989, Ann. Revs. Astron. Ap., **27**, 517.
Van Horn, H. M. 1980, Astrophys. J., **236**, 899.
Van Horn, H. M., and Epstein, R. I. 1990, Bull. A. A. S., **22**, 748.

**I.4**

# EXPLOSION MECHANISM OF COLLAPSE-DRIVEN SUPERNOVAE

K. Sato, T. Shimizu, and S. Yamada

Department of Physics, University of Tokyo, Bunkyo, Tokyo 113, Japan

Recent theoretical development of collapse-driven supernova explosion is reviewed. In particular, we discuss in detail i) convection in the hot bubble region above the proto-neutron star as the source of the large amplitude velocity fluctuations which is necessary to explain large scale mixing in SN1987A, and ii) jet-like explosion induced by axisymmetric neutrino emission from a rotating oblate proto-neutron star, which might account for asymmetry of expanding envelope of SN1987A.

## I. INTRODUCTION

According to theories of stellar evolution, iron cores are formed in stars whose masses are greater than $8M_\odot$. The core begins to collapse when the central density becomes higher than $10^{9.5}$g/cm$^3$, or the temperature becomes higher than $10^{9.7}$K, due to the disintegration of iron nuclei or electron capture of nuclei. In the iron core, electrons are strongly degenerate. The Fermi energy of degenerate electrons is of the order of 5MeV. Because the temperature of the core ($\sim 0.7$MeV) is much lower than the Fermi energy, the thermal pressure of electrons and ions is negligible compared with the degenerate electron pressure, i.e., the iron core is sustained by the degenerate electron pressure against gravity. If the number of electrons in the core is decreased by electron captures of nuclei (hereafter denoted by e-capture), the iron core becomes gravitationally unstable and begins to collapse more rapidly. The e-capture of iron nuclei begins when the Fermi energy of electrons becomes higher than the threshold energy, 3.7MeV.

In the early stage of gravitational collapse, neutrinos emitted by e-captures can escape from the core without scattering, because neutrino interactions with matter are very weak. However, when the density of the core becomes higher than about $10^{10.5}$g/cm$^3$, the mean free path of neutrinos $\ell_{mfp}$ becomes shorter than the radius of the iron core $R$, i.e., $\ell_{mfp} < R$.

The most effective process of neutrino scattering in the core is the coherent scattering with nuclei due to neutral current interaction. In the core, the typical energy of neutrinos is about 10MeV and the wave length $\lambda_\nu$ is about 20fm. On the other hand, the typical size of nuclei in the core is $A \sim 100$ and the radius $R_N$ is about 6fm. Therefore, the condition for coherent scattering $\lambda_\nu > R_N$ is sufficiently satisfied. An important point of the coherent scattering is that the cross section of coherent scattering is proportional to the square of the mass number of nucleus $A$, $\sigma = \sigma_0 a^2 A^2 \varepsilon_\nu^2/4$, where $\sigma_0 = 1.6 \times 10^{-44}$cm$^2$ and $a^2 = \sin^2\theta_W$, $\theta_W$ being the Weinberg angle. This means that the neutrino mean free path depends sensitively on the mass number of nuclei which exist in the core. With increasing the density of the core, the mean free path becomes shorter and shorter. When the density $\rho$ becomes greater than about

$10^{12}$g/cm$^3$, the diffusion time of neutrinos, $\tau = R^2/(c\ell_{mfp})$, becomes longer than the characteristic collapsing time scale $\tau_{ff}$, which is defined by $\tau_{ff} = (G\rho)^{-1/2}$, $G$ being the gravitational constant. This means that neutrinos cannot escape from the core and are trapped in the core (Sato 1975). When the central density of the supernova core becomes about $10^{14.5}$g/cm$^3$ (the nuclear density), the diffusion time is of the order of $\sim 10$ sec and the hydrodynamic time scale of the core, $\tau_{ff}$, is about 0.1msec. It is, therefore, expected that neutrinos are strongly degenerate in the core as well as electrons, because neutrinos are also fermions. Then e-captures of nuclei (or neutronization of nuclei) are greatly suppressed because of the Fermi sea of neutrinos. In this situation, the lepton number per a baryon in the core $Y_\ell$ never decreases but $Y_\ell$ is conserved irrespective of the process of e-captures. Soon after the neutrino trapping, the beta equilibrium including neutrinos $e^- + (Z, N) \rightleftharpoons \nu_e + (Z - 1, N + 1)$ is established, i.e., $\mu_e = W + \mu_\nu$, where $W$ is the threshold energy of e-captures, $W = M(Z - 1, N + 1) - M(Z, N)$. Note that $W = \mu_n - \mu_p$, because the reactions of neutron and proton are much faster than the hydrodynamical processes and the chemical equilibrium between nuclei and nucleon gas is established. Once the neutrino trapping begins, neutrinos are strongly trapped more and more by the effect of trapped neutrinos themselves (Sato 1975), since the increase of the Fermi energy of degenerate neutrinos suppresses e-captures more strongly. As the neutronization of nuclei is suppressed, the nuclei neither drip neutrons nor melt.

Until now many simulations of gravitational collapse were carried out. However the explosion mechanism is still unclear (See for example, Woosley 1991). At present following three mechanisms are proposed.

A. Prompt explosion by shock generated by core bounce

If the strong shock wave is generated by core bounce and it propagates out into the mantle, the envelope of the star is blown off. Most of the simulations, however, showed that the shock wave stalls on its exits from the core and becomes accretion shock. This is due to the neutrino energy loss from the high temperature post shock matter and energy consumption for disintegration of heavy elements passing through the shock to free baryons. Then, is the possibility of explosion by prompt bounce ruled out? Baron et al. (1985) showed that if the equation of state at the region higher than nuclear density is extremely soft, i.e., the increase of pressure is very slow, violent explosion occurs by the strong shock wave generated by core bounce. This is because large amount of gravitational energy is released by the contraction due to the softness of the equation of state. A difficulty of this explosion scenario is that the maximum mass of neutron stars $R_c$ becomes very small ($M_c \ll 1.4 M_\odot$) due to the softness of the equation of state. Independently of Baron et al., Takahara and Sato (1985) investigated the effect of equation of state on the supernova explosion. They showed 1) the explosion energy increases as the equation of state becomes softer and softer, but until the inner core goes into a black hole without bouncing, however 2) extremely soft equations of state, which induce strong explosion, conflict with the condition that the critical mass of neutron stars $M_c \geq 1.4 M_\odot$. In order to make clear whether prompt explosion occurs or not, more systematic and realistic investigations are necessary.

B. Delayed explosion by neutrino deposition

In supernova cores, the diffusion time of neutrinos is $\sim 10$ sec, which is ten thousand times longer than the characteristic hydrodynamical time scale (Sato 1975, Burrows et al. 1981). Neutrinos, however, begin to diffuse out at the late times after the core bounce. Wilson (1985, 1986) showed in his collapse simulations that a half second after bounce, the accretion shock can be revived. Energy is deposited in the region behind the shock by neutrinos, thereby strengthening the shock wave and allowing it to propagate out into the envelope. The important problem of this model is that the explosion energy is small, $\sim 10^{50}$erg (Wilson 1986). Mayle (1985) showed that the explosion energy is enhanced by convection (Epstein 1979) and becomes almost $10^{51}$erg. The results, however, depend on the treatment of convection greatly. Recently simulations of convection were carried out by Herant et al. (1992), Janka (1992), Burrows (1992), Miller et al. (1993) and Yamada et al. (1993). In the next section, II., we review these works and show the recent result of our three dimensional simulation, Shimizu et al. (1993).

C. Jet-like explosion by the effect of rotation

Massive stars on the main sequence have large angular velocity. If we define the normalized angular momentum, $q \equiv J_z / \frac{2GM^2}{c}$, the value of $q$ is the of the order of 10, where $J_z$ is the total angular momentum and $M$ is the mass of the star. It is expected spinning cores are formed if the efficiency of the angular momentum transfer from the core to the envelope is not extremely high during the stellar evolution. LeBlanc and Wilson (1970) carried out numerical simulation of rotational collapse of stellar cores, and suggested that Jet-like explosion occurs in the direction of rotational axis. Müller and Hillebrandt (1981), Symbalisty (1984) and Mönchmeyer and Müller (1989) investigated it more in detail and observed that very hot regions are formed on the rotational axis by shock wave, but explosion or mass ejection from cores is very hard. Recently we carried out the same simulation by using more accurate method, and investigated the explosion more systematically by changing $q$ (Yamada et al. 1993). We found that the shock front became prolate and that the more rotational energy the core had, the more asymmetric the shock became. However, we found that the explosion energy is decreased as the angular momentum is increased or the degree of differential rotation is greater. It is, therefore, concluded that the rotation of a core is harmful for successful prompt explosion. In spite of this result, Jet-like explosion is necessary to explain asymmetry of expanding envelope of SN1987A which was observed by speckle and linear polarization (Cropper et al. 1988, Papaliolios et al. 1989). We recently carried out three dimensional simulation of convection and explosion induced by axisymmetric neutrino emission from a central rotating oblate proto-neutron star (Shimizu et al. 1993, 1994). In the section III., we show our results.

## II. THREE DIMENSIONAL SIMULATION OF CONVECTION IN THE SUPERNOVA CORE

Hydrodynamical instability in a collapse-driven supernova has been a current topic among supernova researchers since SN1987A appeared. The recent interest is focused on the convective instability in the central core (Herant et al. 1992, Janka 1992, Burrows

1992, Miller et al. 1993, Yamada et al. 1993, Shimizu et al. 1993). There are several reasons for it. First, we are required to specify the seed of Rayleigh-Taylor instability in the outer envelope. As well known, there are many observations of SN1987A which tell us that heavy elements were synthesized deep inside the star and mixed up to the outer envelope during the expansion. Most researchers in this field agree that the mixing of these heavy elements were caused by the Rayleigh-Taylor instability. Although the quantitative numerical studies made clear that at least 10% fluctuations were initially required at the contact surfaces to reproduce the many observations (Hachisu et al. 1990, Fryxell et al. 1991), no plausible seed of the initial fluctuation can be found in the envelope. If we then consider that the initial fluctuations were prepared in the shock wave itself in some way before coming out of the core, the required amplitude becomes as large as 30% at the Si shell (Den et al. 1990). We must in turn answer the question that such a large fluctuation is really generated during the core collapse and the core bounce. The second reason is that we wonder if this convective instability in the core plays an important role in the explosion mechanism itself. Since in the delayed mechanism the stalled shock is revived by neutrinos diffused out of the inner core, it may be possible that the shock becomes more energetic if the convection behind the stalled shock transports energy very efficiently. In fact, recent one-dimensional calculations incorporate this effect through mixing length theory and emphasize its significance in the delayed mechanism (Wilson and Mayle 1988). The use of the mixing length theory, however, comes from its numerical convenience and its validity should be confirmed by multi-dimensional numerical simulations. Thirdly, it is because the hot bubbles, which cause the convective instability, are plausible sites of r-process nucleosynthesis; this possibility was recently discussed by Meyer et al. (1992). In the former r-process scenarios such as dynamical r-process scenario (Schramm 1973, Sato 1974), there is a problem to extract the synthesized r-process nuclei out of the supernova star. Thus it is important to investigate how such hot bubbles are ejected from the vicinity of the neutron star in multi-dimensional calculations.

We have carried out multi-dimensional simulations of convection above the proto-neutron star in supernova explosions; this convection is driven by a negative entropy gradient due to heating of shocked matter by high energy neutrinos radiated from the proto-neutron star. In particular, we have carried out three-dimensional simulations of convection for the first time, because we would like to investigate how two-dimensional calculations are different from three-dimensional ones. All the simulations of convection have been, so far, carried out on the assumption of axial symmetry. However, it is easy to imagine that a mass element in the axisymmetric system is in the shape of a torus and that, obviously, such a convective motion is not realistic. Thus it is required to study the convection without assuming any symmetry. We also carried out two-dimensional simulations including rotation.

We begin our simulations from a few 100msec after bounce when the shock wave stalls at a radius of about 500km. There is a proto-neutron star whose surface is at a radius of 50km. It is assumed that it is emitting thermal neutrinos with temperature of 4.5MeV, and that the neutrino flow is spherical. A high entropy region called the hot bubble, where the radiation is dominant, forms around the radius of 300km owing to heating by the neutrinos. Convective motion is induced at the region between those high entropy bubbles and the matter above them.

We constructed an initial model which reproduces the stalled shock in order to make the initial model consistent with our equation of state and our treatment of neutrinos. The equation of state consists of contributions from radiation, gas particles, and degenerate electrons at zero temperature as follows:

$$p = \frac{1}{3}aT^4 + \frac{\rho}{\mu}k_B T + K\rho^{4/3} ,$$

where $k_B$ is the Boltzmann constant, $\mu$ the mean molecular weight, and $K = 1.24 \times 10^{15} \, Y_e^{4/3}[\text{g}^{1/3}\text{cm}^3\text{s}^{-2}]$. $Y_e$ is the electron fraction and assumed to be 0.45 in the following calculations. We included neutrino heating due to absorption of neutrinos by free nucleons and scattering by electrons, and also neutrino cooling due to electron capture and photo, pair, plasma neutrino emission. We adopted the same rates of neutrino heating and cooling as Herant et al. (1992) used in their calculation, except that we further neglected the dependence on the chemical potential of electrons. The initial model was constructed on the assumption that matter is hydrostatic under the gravitational force of the central neutron star (1.4$M_\odot$) and in radiative equilibrium with respect to neutrinos, where ram pressure terms were neglected. The both sides of the shock was connected by Rankine-Hugoniot relations.

Hydrodynamical calculations were carried out by second order accurate Roe method (Eulerian description). We used Cartesian coordinates with $70 \times 70 \times 70$ grids and spherical $(r, \theta, \phi)$ coordinates with $300 \times 50$ grids, respectively, in three-dimensional and two-dimensional calculations. Onto the shock wave, cool matter of the outer core, consisting chiefly of iron nuclei, is assumed to infall with a velocity of 80% of the free fall velocity. The gravitational field due to the central neutron star with mass of $1.4M_\odot$ was included and assumed to be spherical, and we didn't solve the motion of the neutron star actually. The effect of neutrino heating was treated as an energy source terms in the hydrodynamical equations.

The initial perturbations are added to the radial velocity and their profiles of two-dimensional perturbation and three-dimensional perturbation are $\alpha c_s \sin(n\theta)$ and $\alpha c_s \sin(n\theta)\sin(n\phi)$, respectively, where $c_s$ is the sound velocity, $\alpha$ is the amplitude of the initial perturbation. A small mode number ($n = 8$) was adopted because of the limit of resolution in three-dimensional calculations, while the mode number was selected as $n = 20$ in two-dimensional calculations. We also carried out two-dimensional simulations with rotation, in which the initial rotation was given not only as rigid rotation but also as differential rotation. In two-dimensional simulations, we performed a series of calculations to check the validity of our results mentioned below; we carried out a calculation with small amplitude of initial fluctuations ($\alpha = 1\%$), a calculation with greater mode number of $n = 40$, a calculation with larger mesh number of ($300 \times 200$), and a calculation continued until the shock reached the radius of 2000km. For reference, some calculations were also carried out without perturbation or without the neutrino heating, in two-dimensional calculations as well as in three-dimensional calculations. In each model, the computation was continued for a time of $\sim$50msec.

Figure 1 is for comparison between the density distribution (Figure 1(a)) and the entropy distribution (Figure 1(b)) at 45msec for the model with three-dimensional perturbation. In this figure it is observed that the initial velocity perturbation grew, and that low density, high entropy regions (i.e., hot bubbles) formed at the place where the outward velocity perturbation was added initially. Such a behavior was common to all models in which initial perturbation was added.

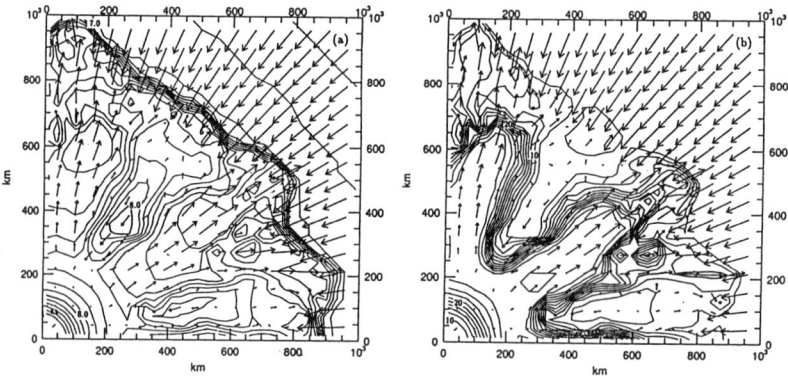

Figure 1. (a) Density distribution (the contour is spaced on a log scale) and (b) entropy distribution (linear), at 45msec for the model with 3-D initial perturbation. Sections parallel to the equatorial plane. Arrows represent the velocity field.

The following is the chief result we obtained. There was no difference in the explosion energy for each model whether initial perturbation was added or not, except for models without neutrino heating. That is, the energy transport by the convective motion does not contribute to the explosion energy. Even if no heating by neutrinos was incorporated, there was not much difference in the explosion energy within 20%, in other words, neutrino heating itself was not so efficient. There is also no difference with respect to the amount of escaping mass (where $e + \max(0, v_r)^2/2 > GM/r$). Those results are obtained in every model of two-dimensional calculations as well as three-dimensional ones. In all the perturbed models, however, the velocity perturbation grew into nonlinear one, that is, 100% velocity perturbation. Moreover, the same result was obtained even in the model with 1% initial perturbation. Since nonlinear fluctuations are large enough to explain all the observations of mixing, we can conclude that the convective instability of the hot bubbles becomes a seed of Rayleigh-Taylor instability in the envelope. We checked the above results by a similar calculation with finer mesh ($300 \times 200$) and a calculation with larger mode number($n = 40$). Furthermore, we continued one of two-dimensional calculations until the shock reached $\sim 2000$km (at $\sim 200$msec). The hot and less dense bubbles continued to move outwards together with the shock and did not merge with one another except for the inner region. We confirmed that the nonlinear fluctuations survived until the shock went out of the Si shell.

Although we carried out two-dimensional calculations with rotation, we found that rotation (the centrifugal force is at most 5% of the gravitational force) does not affect the way of growth of the instability seriously either no matter what the initial rotational law is, let alone the explosion energy. We found, however, the difference in detailed distributions of the velocity field between three-dimensional and two-dimensional calculations. Figure 2 shows distributions of velocity for the region of hot bubbles for two models, one of which 3-D perturbation was given initially and the other 2-D perturbation; each model was carried out by 3-D code in order to make the difference clear. It is seen that, at the velocity of $1.2 \times 10^9 \mathrm{cm\ s}^{-1}$, there is a sharp cutoff for the model with 2-D perturbation as shown in Figure 2(b), whereas such a cutoff is not seen but there is a high velocity tail for 3-D model as Figure 2(a). This implies that more high-velocity regions are produced in a realistic 3-D calculation than a 2-D calculation. The reason for it is considered that there is more room for the surrounding fluid to get out of the way in 3-D calculations than in 2-D calculations. Thus the 3-D convective motion can have higher velocity components than the 2-D one. It should be stressed that such high-velocity regions are plausible sites of the r-process nucleosynthesis because they have high entropy and escape easily from the gravitational potential of the neutron star. It is also noted that it is three-dimensional motion that is favorable for the ejection of r-process nuclei.

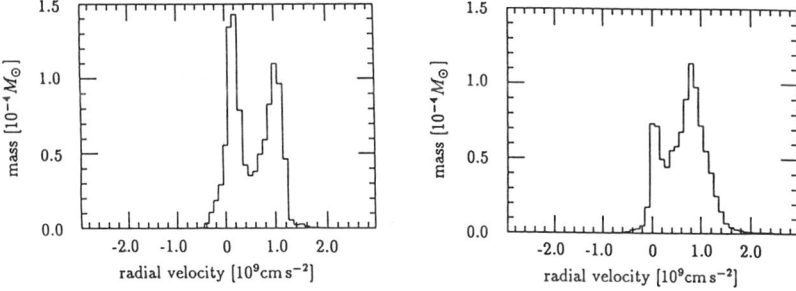

Figure 2. The distribution of the convection velocity in the hot bubble region, (a) for the model with 3-D perturbation, (b) for the model with 2-D perturbation. Both were calculated by 3-D code.

In summary, we, first, found that the growth of the convective motion was so large that the convection in the central core becomes a seed of the initial perturbation of matter mixing in the envelope. Only 1% initial perturbation of the convection is sufficient in the iron core. Secondly, we found in our simulations that convection induced by the negative entropy gradient did not contribute to increase the explosion energy. Thirdly, we also found that three-dimensional motion has an advantage for the ejection of r-process nuclei.

### III. JET-LIKE EXPLOSION BY AXISYMMETRIC NEUTRINO FLOW

We have carried out three-dimensional simulations of convection in the shocked matter of the supernova core, assuming the neutrino radiation from the proto-neutron star is radial but axisymmetric. Since there are many ambiguities in what to be assumed, we, as a first step, adopted a simple model as follows: the interaction of neutrinos with the matter were included in the simulations as heating and cooling terms, where $\nu_e$, $\bar{\nu}_e$, $e^-$, $e^+$ captures on free nucleons are dominant terms. The neutrino flux at each point was estimated on the assumption of black body radiation from the surface of Maclaurin spheroid, which models the rotating hot neutron star. Its dimensions were determined so that the total luminosity and the black body temperature ($L_\nu$ and $T_\nu$, respectively, for each individual neutrino species; $T_\nu$ is constant on the spheroid surface) would be fixed at the same values as in a spherically symmetric case. Then the local neutrino flux $l_\nu$ for each species at the point $(r, \theta)$ in spherical coordinates is given by a geometrical consideration as:

$$l_\nu = (7/16)\sigma T_\nu^4 a_x \sqrt{a_x^2 \cos^2\theta + a_z^2 \sin^2\theta}/r^2 ,$$

where $\sigma$, $a_x$, and $a_z$ are the Stefan-Boltzmann constant, the semi-major and semi-minor axes, respectively. Note that the neutrino fluxes along the rotational axis and the equatorial plane are proportional to the projected areas of the spheroid, $\pi a_x^2$ and $\pi a_x a_z$, respectively. One can easily recover the total luminosity as $L_\nu = (7/16)S(a_x, a_z)\sigma T^4$ by integrating equation (1) over the whole solid angle, where $S(a_x, a_z)$ is the area of the spheroidal surface.

We used the same numerical techniques and initial model as those developed to investigate convection in the core (Shimizu et al. 1993); the numerical code and the physical inputs were unchanged except for the increased grid number of $100^3$ and the aspherical neutrino flow. The initial model was produced consistently with the adopted physical inputs on the assumption that the shocked matter is in a steady state and spherically symmetric. These are sufficient ingredients in order to see what would happen as a result of such axisymmetric neutrino emission.

We have done four calculations including a convergence test for three models with different parameters. The calculation of the first model was carried out on the assumption that $T_\nu = 5.0$MeV and the axis ratio, $a_x/a_z = 1.5$. Note that this ratio corresponds to the ratio of rotational energy to gravitational energy, $T/|W| = 0.105$ for the uniformly rotating Maclaurin spheroid. The values of $a_x$ and $a_z$ are determined in order that the total luminosity should take the same value as a spherical case in which $a_x = a_z = 50$km. Figure 3 shows the density distribution and the velocity field on one of meridian planes for the first model at the time of 27msec; the spherical stalled shock front is initially placed at the radius of 200km. Since neutrinos heat up matter around the rotational axis more intensively than around the equatorial plane, powerful and global convection of matter between the shock front and the proto-neutron star is induced and also jet-like motion along the axis pushes the shock front into the prolate form. Such a jet-like explosion is desirable to explain the asymmetry observed in SN 1987A as well as SN 1993J. Note that Ishikawa et al. (1992) found that the observed asymmetry (the axis ratio is around 1.5) requires either a jet-like explosion (the axis ratio of explosion energy should be more than 2) or an extremely distorted envelope and, furthermore, that the latter is unlikely. Figure 4a is the same figure as Figure 3 except for the entropy distribution. It is clearly seen from this entropy contour map that the hot bubble (namely, high entropy region) is distorted due to the quadrupolar convective motion and, consequently, that turbulence-like instability of hot bubbles occurs behind the shock wave. Figure 4b is the entropy distribution on another section parallel to the equatorial plane at the height of 125km. It shows that the hot bubble is so unstable that the same instability also occurred under azimuth-angle dependent perturbation, numerical grids, not given by hand. Then such instability is considered to provide the initial fluctuation of Rayleigh-Taylor instability in the envelope, i.e., the seed of matter mixing in SN 1987A. Although the initial fluctuation as much as 30% was needed in the central core region, it is easily found in Figures 4a and b that much larger fluctuation is automatically provided by the instability of hot bubbles. Furthermore, note that both the large fluctuations (over 30 %) in the hot bubble and the jet-like explosion are just what is needed to explain the high velocity ($\sim$ 4000km s$^{-1}$) of mixed $^{56}$Ni according to Yamada and Sato (1991). These results were confirmed with a convergence test in which the simulated space was restricted in 80 % of the original space with the same grid number.

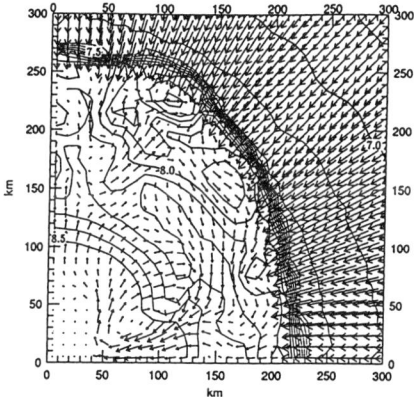

Figure 3. Density contour map for the model of $a_x/a_z =1.5$ at the time of 27 msec on a section that corresponds to one of meridian planes. The contour lines are spaced on a log scale. It shows that global convective motion is produced through the aspherical neutrino heating. The maximum convection velocity is about $2 \times 10^4$km s$^{-1}$.

In the second model, the same assumption was made as in the first model except that the axis ratio, $a_x/a_z = 2.0$; this corresponds to the case of the faster rotating neutron star. This model is found to yield more enhanced jet-like motion and rapider growth of entropy in the hot bubbles compared with the first model, which is favorable for the r-process nucleosynthesis in the bubbles (Meyer et al. 1992).

Although the above jets are not so collimated as cosmological jets, they are expected to be further enhanced and collimated by several other mechanisms, e.g., steep density gradients along the rotational axis due to rotation (in fact, this also yields a jet in the simulation of Yamada and Sato 1994), aspherically convective and diffusive neutrino transfer inside the neutron star, temperature distribution on the neutrino spheroid, and neutrino-antineutrino pair annihilation just above the pole of the spheroid (note that its cross section is larger for head-on collisions).

Only slight energy gains due to the global convective motion were achieved for the first and second models, unlike the hot-bubble convection mechanism of Herant et al. (1992); this is because, we consider, the interior of neutron star should be included in our simulations. Nevertheless, our successful results with respect to other subjects should lead us to expect that a neutrino jet may be produced under some physical process inside the neutron star. One possible, important mechanism may be convective and diffusive neutrino transfer inside the neutron star, which is also induced by the effect of rotation as seen above. Burrows and Fryxell (1993) recently demonstrated a remarkable effect of convective neutrino transfer on the enhancement of neutrino luminosity as well as explosion energy. The same effect is expected for the explosion with the neutrino jet It should be noted, however, that only radial neutrino transfer was included in their simulations, and thus the smoothing effect by the transverse neutrino diffusion was neglected. This effect does not seem to be negligible since it is shown from the random walk and Brünt-Väisälä analyses that the time scale of diffusion is shorter than that of convective instability for the small-scale perturbation. On the other hand, it is considered that the jet created by the large-scale convective motion is free from such a difficulty, and that neutrino convective transfer and diffusion would cooperate with each other, producing a luminous neutrino jet

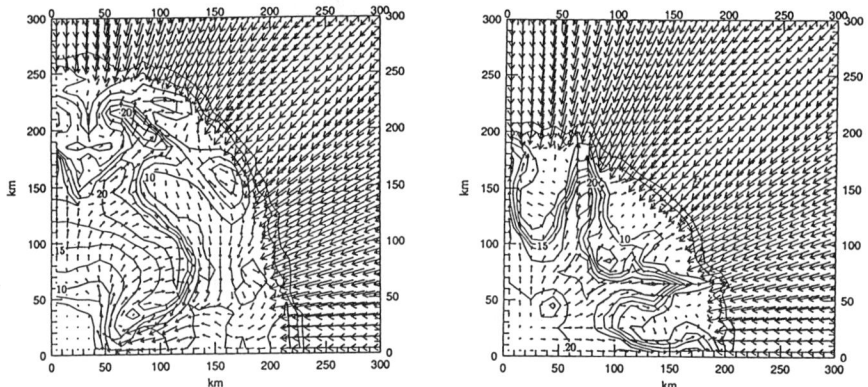

Figure 4. (a) The same as in Fig. 3, except for the entropy distribution (linearly spaced). The high entropy hot-bubbles are highly deformed due to the global convective motion. (b) The same as in Fig. 4a, except for a section parallel to the equatorial plane at the height of 125km. The hot bubble is also unstable under azimuth-angle dependent perturbation.

The hot bubble jet in a successful explosion is simulated in the third model in which only the black body temperature of neutrinos was increased by a factor of 4% compared with the second model. We confirmed that the global convective motion is enhanced rather than smoothed out, and that our above results are not altered even in the energetic explosion.

It is suggested from these results that three unsettled subjects in type II supernovae are solved and unified naturally on the only one assumption that the proto-neutron star is rotating fast and thus the aspherical neutrino flow is induced by its rotation. (1) The seed of matter mixing is supplied by the turbulent instability induced in the global hot-bubble convection. (2) The jet-like explosion together with the large initial fluctuations in the core yields the high mixing velocity of elements in the envelope. (3) The jet-like explosion is the most reasonable interpretation for the asymmetry observed in both SN 1987A and 1993J.

## IV. CONCLUSIONS

It was 1934, Baade and Zwicky pointed out that supernova explosion is resulted from the collapse of stellar core. As discussed in preceding sections, we have not known the definite mechanism of collapse-driven supernovae yet a half century after the Baade and Zwicky's foresight. We can, however, say that we are now realizing the mechanism from observations of SN1987A and recent results of numerical simulations of gravitational collapse. At present stage, a reasonable unified picture of the mechanism would be the delayed explosion enhanced by convection in slowly rotating cores ($q < 0.5$). This scenario is expected to explain, 1) large amplitude velocity fluctuations for the seed of Rayleigh-Taylor instability to mix $^{56}$Ni, 2) asymmetry of expanding envelope observed by speckle and linear polarization, and 3) the site of r-process nucleosynthesis (Sato et al. 1993).

At present, SUPER KAMIOKANDE is under construction. If gravitational collapse of stellar core occurs at the Galactic center (8.5kpc from the sun), almost 10,000 $\bar{\nu}_e$ events would be detected. We would be also able to detect about 40 $\nu_e$ events from the "neutronization Burst". Gravitational wave detector with high sensitivity, LIGO, is also under construction. This detector can detect the gravitational wave from Galactic supernova even if the the total energy of gravitational wave emitted from the supernova is small as $\sim 3 \times 10^{-7} M_\odot c^2$. Observations of supernova remnants by X-rays (Koyama et al. 1986) and radio (Sofue 1987) suggest that the rate of supernova explosion in our Galaxy was 1/10years to 1/30years. It seems very reasonable to expect large amount of data on supernova explosion are obtained. It is important and necessary to make reliable model and theoretical prediction, which can be compared with observations.

This work was partially supported by Grant-in-Aids for Scientific Research from the Ministry of Education, Science and Culture of Japan (05243104, 04234104 and 3013). Some parts of the calculations were also performed on a HITAC at KEK (National Laboratory for High Energy Physics, Japan). T. Shimizu acknowledges Cray Research de Mexico, S.A. de C.V., and Universidad Nacional Autonoma de Mexico for Cray time allocation.

## REFERENCES

Baron, E., Cooperstein, J. and Kahana, S. 1985, Phys. Rev. Letters, **55**, 126.
Burrows, T. Mazurek and A. and Lattimer, J. M. 1981, Astrophys. J., **251**, 325.
Burrows, A. and Fryxell, B. 1992, Science, **258**, 430.
Burrows, A., and Fryxell, B. A. 1993, Astrophys. J., **418**, L33
Cropper, M. et al. 1988, M. N. R. A. S., **231**, 695.
Den, M., Yoshida, T., and Yamada, Y. 1990, Prog. Theor. Phys., **83**, 723.

Epstein, R. I. 1979, M. N. R. A. S., **188**, 305.
Fryxell, B. A., Müller, E., and Arnett, D. 1991, Astrophys. J., **367**, 619.
Hachisu, I. et al. 1990, Astrophys. J., **358**, L57.
Herant, M., Benz, W., and Colgate, S. 1992, Astrophys. J., **395**, 642.
Ishikawa, S., Yamada, S., Kiguchi, M., and Sato, K. 1992, Astron. Astrophys., **258**, 415
Janka, H.-T. 1992, in *Frontiers of Neutrino Astrophysics*, ed. Y. Suzuki and K. Nakamura (Tokyo: Universal Academy Press), 203
Koyama, K. et al. 1986, Publ. Astron. Soc. Japan, **38**, 121.
LeBlanc, J. M. and Wilson, J. R. 1970, Astrophys. J., **161**, 541.
Mayle, R. W. 1985, Ph D Thesis.
Meyer, B. S. et al. 1992, Ap. J., **399**, 656.
Miller, D. S., Wilson, J. R., and Mayle, R. W. 1993, Astrophys. J., **415**, 278.
Mönchemeyer, R. M. and Müller, E. 1989, NATO ASI series.
Müller, E. and Hillebrandt, W. 1981, Astron. Astrophys., **103**, 358.
Papaliolios, C. et al. 1989, Nature, **338**, 565.
Sato, K. 1974, Prog. Theor. Phys., **51**, 726.
Sato, K. 1975, Prog. Theor. Phys., **53**, 595 and **54**, 1325.
Sato, K., Shimizu, T., and Yamada, S. 1993, in *Frontiers of Neutrino Astrophysics*, ed. Y. Suzuki and K. Nakamura (Tokyo: Universal Academy Press), 191
Schramm, D. N. 1973, Astrophys. J., **185**, 293.
Shimizu, T., Yamada, S., and Sato, K. 1993, Publ. Astron. Soc. Japan., **45**, L53.
Shimizu, T., Yamada, S., and Sato, K. 1994, to appear in Astrophys. J.
Sofue, Y. 1988, Talk in the Workshop of Particle Astrophysics, KEK.
Symbalisty, E. M. 1984, Astrophys. J., **285**, 729.
Takahara, M. and Sato, S. 1985, Phys. Lett., **159B**, 17.
Wilson, J. R. 1985, in *Numerical Astrophysics*, ed. J. Centrella, 422p.
Wilson, J. R. et al. 1986, Ann. N.Y. Aca. Sci., **470**, 267.
Wilson, J.R. and Mayle, R. 1988, Phys. Rep., **163**, 63.
Woosley, S. E. 1991, The Supernovae, The 10th Santa Cruz Summer Workshop, Springer-Verlag.
Yamada, S. and Sato, K. 1991, Astrophys. J., **382**, 594.
Yamada, S., Shimizu, T., and Sato, K. 1993, Prog. Theor. Phys., **89**, 1175.
Yamada, S. and Sato, K. 1994, to appear in Astrophys. J.

# I.5

# ROLE OF MAGNETIZED PLASMAS IN PRODUCTION OF COSMIC X-RAYS

Kazuo Makishima

Department of Physics, University of Tokyo, Bunkyo-ku, Tokyo 113, Japan

## I. INTRODUCTION

Production of cosmic X-rays is always closely related to astrophysical plasmas of various characteristics. Here we wish to focus upon two topics with astrophysical plasmas, one is very dense while the other is very tenuous. These topics are both based on original observational results obtained with Japanese X-ray astronomy satellites.

A neutron star in a close binary system is a preferred site of intense X-ray emission, where the matter from the companion star accretes onto the neutron star and gets heated by the gravity to form a dense ($10^{20-22}$ cm$^{-3}$) and hot ($T = 10^{7-8}$ K) plasma around the neutron star. When the neutron star is strongly magnetized, the plasma is funneled onto magnetic polar regions thus producing two hot spots. These objects are called X-ray pulsars (White *et al.* 1983; Nagase 1989). The plasma in the magnetic polar regions undergoes a strong interaction with the magnetic field via electron-cyclotron resonance, which produces a significant feature in the X-ray spectrum. This subject has been studied extensively by the *Ginga* (meaning *the galaxy* in Japanese) satellite, launched by the Institute of Space and Astronautical Science on 1987 February 5 and reentered the atmosphere on 1991 November 1. We review the *Ginga* results in II.

X-ray observations provide a unique way of detecting optically-thin hot plasmas in a variety of astrophysical environment, including solar and stellar coronae, supernova remnants, hot gaseous components associated with galaxies and galaxy clusters, and so on. In III we cast new light upon magnetohydrodynamic (MHD) aspects of these plasmas, mainly based on the data obtained with the new Japanese X-ray astronomy satellite *Asca* (meaning a flying bird in Japanese, and also an acronym to Advanced Satellite for Cosmology and Astrophysics; Tanaka *et al.* 1994) which succeeded *Ginga*. *Asca*, launched on 1993 February 20 under an extensive US-Japan collaboration, is extremely powerful for the plasma diagnostics through spatially-resolved X-ray spectroscopy.

## II. SURFACE MAGNETIC FIELDS OF NEUTRON STARS

A. Overview

In addition to the mass and radius, the magnetic field strength is one of the key parameters characterizing a neutron star (NS). While the mass and radius are useful to discriminate among different nuclear force models, the magnetic field is expected to reflect solid-state characteristics of the NS.

Various classes of celestial objects are known to involve NSs. In some cases the NS is magnetized strongly, up to $10^{11-13}$ G, while in other cases at most weakly ($< 10^9$ G) (see Chanmugam 1992 for a review). The former examples include ordinary radio pulsars and binary X-ray pulsars, both being young objects, and possibly some gamma-ray burst sources (Murakami et al. 1988). The latter group is exemplified by millisecond pulsars and low-mass X-ray binaries, both belonging to the old population. A popular scenario has been to assume that a NS is born with a strong field which is sustained by a persistent electric current flowing inside the NS, and the Ohmic loss leads to a gradual decay of the current, hence of the magnetic field, on a time scale of $\sim 5 \times 10^6$ years.

To examine the physics of the NS interior, and to understand the relation between the strong-field and weak-field NSs, it is essential to measure the NS surface magnetic field $B$ with a reliable technique. So far $B$ of radio pulsars has been estimated on the assumption that their spin-down occurs via emission of magnetic dipole radiation, but this method in fact gives only upper limits to $B\sin\alpha$, where $\alpha$ is the angle between magnetic and rotational axes, because the spin-down luminosity may go into other channels as well.

A more reliable estimate on $B$ can be obtained, for mass-accreting X-ray pulsars, by detecting cyclotron resonance scattering feature (CRSF), i.e. a spectral feature due to electron-cyclotron resonance scattering that is supposed to take place in the accretion column above the magnetic polar regions. This utilizes the relation of

$$E_c(\mathrm{keV}) = 11.6 B(\mathrm{G})$$

that holds for the cyclotron resonance energy $E_c$. Before *Ginga*, this technique has been applied only to two cases; Her X-1 in which a CRSF was seen either at $\sim 35$ keV in absorption, or at $\sim 60$ keV in emission (Trümper et al. 1978; Voges et al. 1982), and a transient X-ray pulsar 4U 0115+63 in which harmonic CRSFs were detected at $\sim 12$ keV (White, Swank & Holt 1983) and at $\sim 20$ keV (Wheaton et al. 1979).

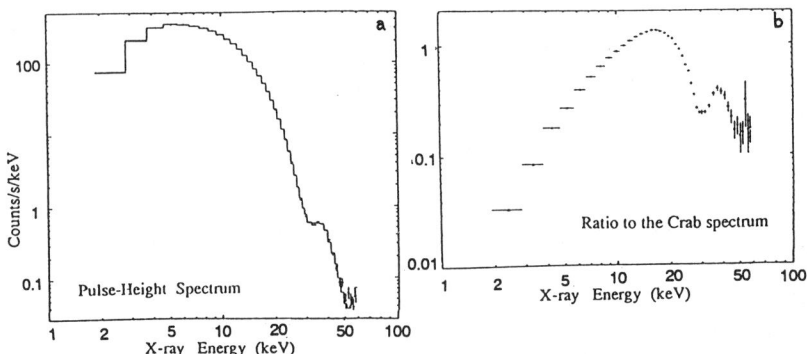

Fig. 1: The prominent CRSF observed at 28.5 keV from the transient X-ray pulsar X0331+53 with *Ginga* (Makishima et al. 1990). (a) The observed raw pulse-height spectrum, fitted with an empirical model for cyclotron resonance. (b) The spectrum normalized to that of the Crab nebula.

B. Observations

Using *Ginga*, we have detected CRSFs from at least 9 X-ray pulsars including the two previously known examples, and possibly another. Individual detections have already been reported (Nagase 1989; Clark *et al.* 1990; Mihara *et al.* 1990, 1991; Makishima *et al.* 1990; Nagase *et al.* 1992; Koyama *et al.* 1989a) and a review on this subject has been given in Makishima (1992) and Makishima & Mihara (1992). Therefore we here summarize only the essence of these results.

Figure 1 shows the best example of CRSF as observed with *Ginga*. Although the CRSFs are generally less prominent than this, they are statistically significant, and exhibit strong dependence on the pulse phase indicating that they are not instrumental. In all cases the CRSF has been observed to appear in absorption rather than in emission. This is reasonable because the accretion column is considered to have a very high optical depth for the cyclotron resonance scattering. The resonance width is typically a fraction of $E_c$, while the apparent optical depth at the resonance is in the range 1.5–3. For very luminous X-ray pulsars the resonance width tends to increase, possibly due to the thermal Doppler effect, thus making the CRSF harder to detect. In a few cases we have also obtained evidence for the 2nd harmonic resonances.

The X-ray pulsars showing CRSFs amount to roughly 50% of the pulsars that have been observed in detail with *Ginga*. We therefore conclude that the presence of CRSF is the rule rather than the exception among X-ray pulsars. We also presume that the cyclotron resonance plays an essential role in the formation of X-ray pulsar spectra, which are typically very flat (photon index = 0.5–1.0) up to 10–20 keV and then turn over very steeply. The steep turn over is most likely caused by the cyclotron resonance, since for the 9 pulsars we have confirmed that the spectral turn-over always occurs at an energy of $(0.5 - 0.7)E_c$. As a consequence, we can estimate the values of $E_c$ for other pulsars simply based on their spectral turn-over energies, which are much easier to detect than the CRSFs themselves.

Theoretical account of the X-ray pulsar spectrum, including the CRSF and its pulse-phase dependence, is an interesting challenge (Bulik *et al.* 1992). It is a complex problem, in which we have to deal with anisotropic radiative transfer in an optically-thick plasma under a strong magnetic interaction. We must take into account thermal Doppler effects and general relativistic effects as well. Such theoretical investigation may clarify whether the NS magnetic field has a dipole geometry or not, whether the dipole center is coincident with the NS center or not, whether the emission region has a significant vertical extent or not, and so on. However this subject is beyond the scope of the present article.

C. Results and discussion

Figure 2 shows occurrence distribution (for logarithmic interval) of all the observed CRSF energies. We have also included three detections of cyclotron absorption lines in the gamma-ray burst spectra (Murakami *et al.* 1988; Yoshida *et al.* 1992). Thus the distribution is clearly concentrated at $E_c = 10$–40 keV, or $B = (1-4) \times 10^{12}$ G. Since it is difficult to explain this concentration solely in terms of selection effects, we conclude that it is real.

We can estimate $E_c$ of other X-ray pulsars which do not exhibit detectable CRSFs, or for which the *Ginga* data are of relatively poor quality. These pulsars generally exhibit spectral turn-over at energies of 7–20 keV, which are very similar to those of the pulsars with positive CRSF detections. Since the spectral turn-over is closely related to the cyclotron resonance, we conclude that the surface magnetic fields of the pulsars without CRSFs are also concentrated in a similar narrow range.

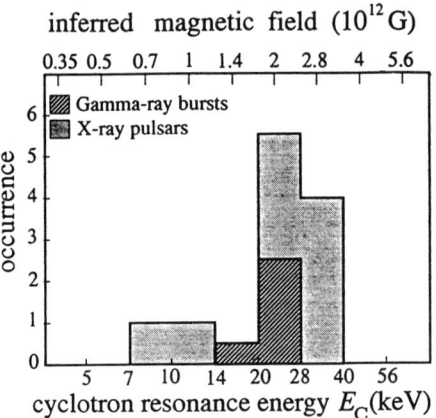

Fig. 2: Occurrence distribution of cyclotron resonance energies for the 9 X-ray pulsars and the 3 gamma-ray burst events. For one particular gamma-ray burst event the resonance energy changed in the course of burst, so that it is split into to energy bins (each 0.5).

Another important conclusion from our study is that there is no indication of magnetic field decay among X-ray pulsars, over a time scale up to $\sim 10^8$ years (Makishima & Mihara 1992; Makishima 1992). This conclusion is reached because the observed values of $E_c$ do not correlate with the system age, which is estimated from the mass of the companion star (the heavier, the younger). This makes a clear contradiction to the field-decay hypothesis that was generally believed before the *Ginga* era.

In spite of these results, there clearly exist strong-field NSs and weak-field ones. Then we are left with three possibilities to explain the overall picture. One is to assume that the strong-field and the weak-field NSs are genetically different; e.g. produced by the collapse of strong-field stars and weak-field stars, respectively. Another possibility is to assume that the field decay takes place on a very long time scale (Wakatsuki *et al.* 1992), e.g. $\sim 10^9$ years. The other alternative is to assume that the magnetic field makes an abrupt transition, between $\sim 10^{12}$ G and $10^{8-9}$ G.

For further investigation of the issue, we must clarify how the NS magnetic fields are maintained. One obvious possibility is the presence of a persistent electric current in the NS crust. However we are not sure if this idea can explain the narrow scatter in the value of $B$. Alternatively the field may be produced by nuclear spin alignment among a small fraction ($10^{-4}$–$10^{-3}$) of neutrons in the NS. If this idea works, the two types of NSs can be naturally explained by the magnetic vs non-magnetic phase transitions in the NS interior. This possibility is being investigated by Iida & Ichimaru (1994).

## III. LARGE-SCALE COSMIC PLASMAS WITH HIGH TEMPERATURE

A. Overview

Variety of large-scale thin hot plasmas are distributed in the interstellar space within galaxies, and in the intergalactic (intracluster) space within clusters of galaxies. Those associated with the galaxy clusters are particularly important, as they are thought to contain a large fraction of baryonic matter in the universe. Radiation from these plasmas appear predominantly in thermal Bremsstrahlung continuum as well as atomic K- and L-emission lines from abundant elements, making the X-ray frequency best suited for the study of these plasmas. Before introducing a few *Asca* results of interest, here we first review basic aspects of these large-scale cosmic plasmas from physical rather than astronomical viewpoint.

Since the plasma considered here is usually so thin (contrary to the subject of this conference!) and hot ($10^{6-8}$ K) that it can be regarded as an ideal magnetohydrodynamic (MHD) fluid. Taking therefore the Boltzmann equation as the basic equation, its zero-th order moment describes mass continuity, as well as sources and sinks for the plasmas. Specifically, supernovae (SNe) and starburst activities (successive occurrence of a large number of SNe in star forming regions) provide the most important plasma sources. On the other hand, accretion onto gravitating objects, such as neutron stars, stellar-mass black holes, active galactic nuclei (AGNs), and some giant galaxies, acts as the plasma sink. However we notice that the compact gravitating objects sometimes recycle the plasma by converting a small fraction of the accreting matter into collimated jets, which usually have very high specific free energies.

The 1st moment of the Boltzmann equation describes plasma dynamics, including plasma outflow, gravitational pull, bulk plasma acceleration, plasma confinement, pressure balance, and so on. Although the gravity plays a leading role in these dynamics, the plasma interaction with electromagnetic fields is another important aspect. In particular we emphasize the role of the plasma interaction with magnetic fields, because in cosmic plasmas that have high hydrodynamic and magnetohydrodynamic Reynolds numbers, large-scale magnetic fields can easily develop in an almost autonomous manner. Cosmic plasmas without significant magnetic field would be highly unlikely. Nevertheless, elementary processes of X-ray emission are usually little affected by the presence of magnetic fields of order $10^{-6}$ G or so. Consequently the contemporary non-solar X-ray astrophysics have almost completely ignored the magnetic aspects of the plasmas so far. This tendency may have introduced a certain bias into our understanding of the plasma universe. The *Asca* observations provide us with a unique opportunity to reduce such a bias.

The 2nd moment equation describes energy-related aspects of the plasma, e.g. heating, cooling, thermal conduction, particle/energy transport, shock phenomena, and so on. While the radiation provides a dominant mode of plasma cooling, various plasma heating mechanisms include gravitational compression, shock transition, Joule heating, magnetic constriction, magnetic reconnection, and so on. In particular, it is often the case that the kinetic/gravitational energy of a dense matter is once converted into electromagnetic field energy, which is then utilized in the plasma heating and acceleration. In the solar corona, for example, the kinetic energy associated with the differential rotation and convection of the sun generates the coronal magnetic field, which then produces the coronal plasma as well as the solar wind. Observations with the *Yohkoh* satellite (Tsuneta 1994) have clearly demonstrated the leading role of magnetic reconnection in the production, heating, and sustainment of the solar corona. Through such a hierarchical energy transfer, the *energy density* usually decreases (from kinetic, electromagnetic to thermal), whereas the *specific energy* often increases drastically (e.g. from the solar photosphere to the solar corona). We expect that such a scheme serves as an important mechanism to maintain the universe away from the overall thermal equilibrium.

Another important aspect associated with the large-scale cosmic plasma is the elemental abundance, which can be estimated by measuring equivalent widths of atomic emission lines in the X-ray spectra. Employing the standard view of big-bang nucleosynthesis, heavy elements in the galaxy and cluster plasmas must have been produced via the nuclear reactions in the stellar interior, and then somehow distributed over the interstellar and intergalactic volume. Therefore the heavy elements serves as a valuable tracer of both cosmic elemental evolution and structure formation. So far the iron K-lines have been the almost only mean to study the elemental abundance of cosmic plasmas (Makishima 1986), but *Asca* has such an improved spectral capability over an energy range of 0.5–10 keV, that K-lines from O, Ne, Mg, Si, and S, as well as Fe-K and Fe-L lines, can be studied with a high sensitivity.

B. X-ray emission along the milky way

1. The galactic ridge X-ray emission (GRXE)

As revealed by *HEAO-1*, *EXOSAT* (Warwick *et al.* 1985), *Tenma* (Koyama *et al.* 1986; Koyama 1989) and *Ginga* (Koyama *et al.* 1989b; Yamauchi *et al.* 1990) missions, an apparently diffuse X-ray emission is distributed along the Milky Way, at least within $\sim \pm 60°$ from the Galactic center with a typical latitudinal scale height of 2–3° or $(0.3 - 0.5)$ kpc. This phenomenon is called Galactic ridge X-ray emission (GRXE). Figure 3 shows the GRXE spectrum obtained with *Asca*. As the X-ray spectrum clearly exhibits K-lines from helium-like Fe (at 6.7 keV), S (at 2.46 keV), and Si (at 1.86 keV), the GRXE is thought to originate from thin hot plasmas of temperature $kT = 5$–10 keV and iron abundance of (0.5–1.0) times solar. The integrated 2–10 keV luminosity is $(1-2) \times 10^{38}$ ergs s$^{-1}$.

The GRXE could be synthesized by a collection of a large number of faint compact X-ray sources, such as stellar coronae, each exhibiting plasma emission. However there is no corresponding population of X-ray objects that can altogether account for the observed GRXE properties (Koyama *et al.* 1986). Furthermore the preliminary *Asca* results (H. Kaneda and S. Yamauchi, a private communication) directly shows that the GRXE has a quite smooth surface brightness, with little evidence of small-scale fluctuation which would be observed if the GRXE is indeed of discrete source origin. Therefore we presume that the GRXE originates from a truly diffuse interstellar plasma. Assuming that the plasma is distributed within a disk of radius 10 kpc and a scale height of 0.5 kpc with a filling factor of $f$, the electron density is estimated to be $0.7 \times 10^{-3} f^{-1/2}$ cm$^{-3}$.

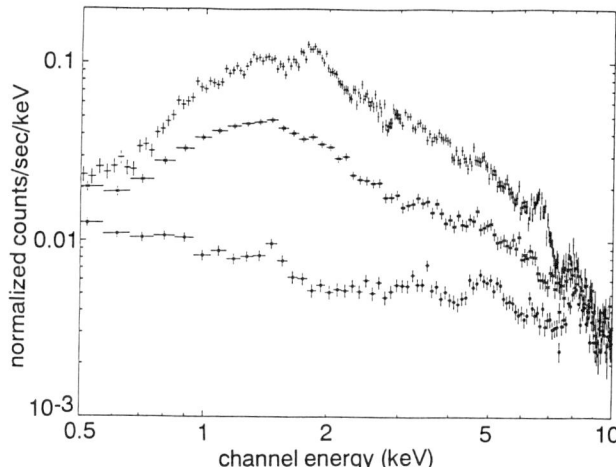

Fig. 3: The X-ray spectrum acquired on the Galactic plane at a latitude of $\sim 30°$ (top), at a high Galactic-latitude region (middle), and while the X-ray telescope is pointing to the night earth (bottom). They respectively consist of; GRXE+CXB+NXB, CXB+NXB, and NXB, where CXB is the cosmic X-ray background and NXB is non X-ray background due to charged particles and gamma-rays.

2. Is the plasma confined or not?

The most puzzling aspect of the GRXE is that its temperature of 5–10 keV far exceeds the gravitational escape temperature of the Galaxy, $\sim 0.5$ keV. Therefore the plasma cannot be gravitationally bound to the Galaxy. The estimated electron density further implies that the plasma pressure for the GRXE is of order $\sim 10^{-11}$ ergs cm$^{-3}$, exceeding by $\sim 2$ orders of magnitude those typical of interstellar media. Therefore we conclude that the plasma cannot be confined to the Galaxy by external pressure from various components of interstellar media, either.

If the plasma emitting the GRXE is *not* confined, the plasma should be escaping from the Milky Way with a sound velocity on a time scale of $5 \times 10^5$ yr. This plasma loss must be replenished by some plasma source, unless we are observing an afterglow of a very energetic and transient event that occurred in the past within $\sim 10^6$ years. (This may be the case for a strong enhancement in the X-ray emission around the Galactic center; Yamauchi *et al.* 1990.) Although the most likely plasma source is the SNe as we mentioned earlier, the required SN rate becomes as high as one SN every $10 \sim 20$ years which makes us feel very uncomfortable. Moreover, although supernova remnants (SNRs) form an important class of hot interstellar plasma, their temperature (typically 2 keV or less) is considerably lower than that of the GRXE emission. The surface brightenss of the GRXE above $\sim 3$ keV, as revealed by *Asca*, is also too smooth to be dominated by individual SNe. We therefore conclude that the Galactic SNe alone cannot fully explain the GRXE.

3. Possible magnetized corona in Our Galaxy

We may alternatively assume that the GRXE plasma is nearly confined within numerous magnetic loops in the Galaxy disk, and even heated by the dissipation of Galactic rotational energy through magnetic compression and magnetic reconnection. More quantitatively, the overall GRXE luminosity can be explained if the magnetic loops have a filling factor of $f \sim 0.1$, and inside each loop the plasma has a density $\sim 2 \times 10^{-3}$ cm$^{-3}$. The luminosity can in turn be sustained if only a minor fraction ($\sim 1/2000$) of the huge gravitational/rotational energy of the Galaxy has been released over the Hubble time through electromagnetic channels as we mentioned earlier.

In this picture, the *local* equipartition magnetic field within the loops becomes as high as 20–30 $\mu$Gauss. However its average over the filling factor gives a large-scale field intensity of several $\mu$G that is perfectly consistent with recent measurements (Ohno & Shibata 1993). In short, we are here proposing a Galactic analogue of the solar corona.

Although the above idea can naturally explain the overall properties of the GRXE, it is not trivial to prove (or disprove) this picture. Detailed imaging with *Asca* may not help very much, since a large number of magnetic loops embedded within the Galaxy disk will overlap in our line of sight along the Galactic plane, to make the emission apparently quite smooth rather than filamentary.

4. Future prospects

Spectroscopy with *Asca* may help in assessing the reality of the proposed Galactic corona. If the plasma is continuously supplied within the Galactic plane with a filling factor $f \sim 1$ and freely escaping on a time scale of $5 \times 10^5$ yr $= 1.5 \times 10^{13}$ s, the $nt$ (density-time) parameter of the plasma will be $\sim 10^{10}$ cm·s. Then the plasma is expected to be significantly deviated from ionization equilibrium, and we should be able to detect the effect as a deviation of the atomic K-line centroid energy from the ionization equilibrium value. The preliminary *Asca* results suggest an iron K$_\alpha$-line energy around 6.65 keV, implying $nt > 3 \times 10^{10}$ cm·s. Therefore the plasma may not be in a full ionization equilibrium, but is likely to stay within the Galaxy longer than the escaping time, supporting the magnetic confinement picture.

*Asca* results also suggest that the continuum brightness and temperature vary quite gradually on angular scales of $> 1°$, while the Fe-K$_\alpha$ line equivalent width (EW) and its centroid energy vary significantly on smaller angular scales. There is a hint of positive correlation between the EW and the line-center energy. We presume that the plasma confinement time, hence the $nt$ parameter, differ considerably from direction to direction along the Milky Way.

Another important finding with *Asca* is the detection of Mg-K$_\alpha$, Si-K$_\alpha$, and S-K lines in the GXRE spectra (Fig. 3). Since these elements would be completely ionized in the $kT > 5$ keV plasma if close to the ionization equilibrium, the plasma must either be in an extreme departure from the ionization equilibrium, or there co-exists cooler plasma ($kT < 1$ keV). However since the Fe-K line does not suggest an extreme ionization non-equilibrium, we presume that the latter is more likely. The cooler plasma may be related to the SNRs, and may be confined by the gravity of our Galaxy. The co-existence of hotter and cooler plasmas can readily be explained if they are separated by magnetic fields.

Finally, using *Asca* we attempt to search nearby edge-on spiral galaxies for similar galactic coronae which may stick out of the galaxy plane. We have already observed two such galaxies, NGC4565 and NGC4631, and the data are currently being analyzed.

C. Hot plasmas associated with galaxies and clusters

1. Gravitational confinement of intra-cluster medium

A cluster of galaxies (CLG) is a self-gravitating bound system, containing $10^{2-3}$ galaxies and a large-scale hot plasma called intra-cluster medium (ICM; see Sarazin 1988 for a review). The ICM exhibits a temperature of $kT = 2$–$10$ keV and an electron density profile (as a function of radius $r$) of the form

$$n(r) = n_0 \left[1 + (r/a)^2\right]^{-3\beta/2},$$

where $n_0 \sim 10^{-3}$ cm$^{-3}$ is the central density, $a = (0.1$–$0.5)$ Mpc is the core radius, and $\beta = (0.5$–$0.7)$ is so called beta parameter. We usually estimate the total ICM mass to be $M_{\rm ICM} = 10^{13-14} M_\odot$ within several core radii, although it diverges for larger values of $r$. This is comparable to, or often larger than, the integrated galaxy mass $M_*$ (which may be dominated by stars) in the cluster. Therefore we regard the ICM as being the most dominant component of baryonic matter in the universe.

The ICM is thought to be confined within the cluster volume by the gravity, which balances the ICM pressure of $p \sim 1 \times 10^{-11}$ ergs cm$^{-3}$. Assuming a spherical symmetry, the hydrostatic balance equation becomes

$$\frac{dp}{dr} = -\mu m_p n \frac{d\phi}{dr}$$

where $\mu \sim 0.6$ is the mean molecular weight of the plasma in unit of the proton mass $m_p$, and $\phi$ is the gravitationalpotential.

Through imaging spectroscopic X-ray observations of a CLG, we can estimate the temperature profile and the electron density profile of the ICM. Then we can derive gravitational potential of the CLG, hence the total gravitating mass $M_{\rm G}$ which is necessary to confine the ICM. Specifically, $M_{\rm G}$ within a radius $R$ is given from the above equation as

$$M_{\rm G}(R) = -\frac{kTR}{\mu m_p G}\left(\frac{d\log n}{d\log r} + \frac{d\log T}{d\log r}\right)$$

where $G$ is the constant of gravity. The value of $M_{\rm G}$ thus estimated is usually 5–15 times larger than the baryonic mass $M_{\rm ICM} + M_*$, estimated within the same radius. This provides a strong evidence for the dark matter. However our current knowledge from X-ray observations is far inadequate to establish a concrete picture of the dark matter. We expect that the new *Asca* observations will produce an innovation in this research field.

Origin of the ICM is another subject of importance. The main constituent of the ICM is thought to be the primordial gas that has been compressed and heated during the course of gravitational structure formation. However the X-ray spectrum of a CLG exhibits iron K-line emission (at about 6.7 keV), and the derived iron abundance is typically 0.2–0.4 times solar. Therefore the ICM also contains a certain amount of secondary gas, which has been processed in the nuclear furnace of the stellar interior in each member galaxy, and then ejected into interstellar and intergalactic spaces. Clearly, spatial gradient of the heavy elements and evolution of the heavy element abundance are of essential importance when we try to clarify the structural as well as chemical evolution of the universe.

2. Magnetic fields in clusters of galaxies

Faraday rotation measurements of background radio sources seen through CLGs have accumulated evidence of $\sim 1\mu G$ magnetic fields in most CLGs (Kronberg 1994). Lack of inverse-Compton hard X-rays from some CLGs, in conjunction with their diffuse synchrotron radio emission, also implies relatively high lower limits for the cluster magnetic fields (e.g. $> 0.1\mu G$; Rephaeli & Gruber 1988). However the implied magnetic pressure is $\sim 2$ orders of magnitude below the plasma pressure, so that the magnetic effect is usually considered to be completely negligible.

Contrary to the popular belief, we expect that the intracluster magnetic field (IMF) can often have significant effects on the ICM dynamics and energetics, because of the following three reasons. Firstly, there are a number of sources of seed magnetic fields, including magnetized jets from AGNs which may exist in CLGs. Secondly the member galaxies in a CLG is making random motions in the gravitational potential to counterbalance the gravitational pull, with a typical velocity of $(3-10) \times 10^7$ cm s$^{-1}$. Since the local ram pressure of the galaxy motion much exceeds the magnetic pressure, the IMF will be violently stretched, entangled, and amplified over the entire cluster volume. Finally the ICM has such a large magnetic Reynolds number that the plasma will be strongly turbulent. Consequently the magnetic pressure will build up, regardless of the initial condition, to a certain level that may not be negligible compared to the plasma pressure.

Observational evidences indicate that the IMF can in fact be quite strong in some cases, particularly in so called 'cooling flow' clusters (see C.4). For example, a giant radio galaxy known as Hydra-A is sitting in the center of a CLG called the Hydra-A cluster. Hydra-A exhibits prominent radio jets and radio lobes, which are thought to be confined by the ICM. The Faraday-rotation measurements of the radio lobes suggest that the central region of the ICM (external to the lobes) of the Hydra-A cluster are magnetized up to $\sim 30\mu G$ (Taylor & Perley 1993), making the field pressure quite comparable to the plasma pressure.

3. A simple two-phase model

We here propose a two-phase ICM model involving co-existing magnetic and non-magnetic phases. The non-magnetic phase is assumed to have a central electron density $n_a$ and zero magnetic field, while the magnetic phase is assumed to have an IMF $B$, an electron density $n_b$ and a plasma pressure $p_b$, all defined at the cluster center. We assume that the magnetic phase has a constant filling factor $f$, and the ratio $k \equiv B^2/(8\pi p_b)$ (inverse of the usual plasma 'beta') for the magnetic phase is constant over the entire cluster. We also assume the two phases to intermix with each other over a sufficiently small scale, to have a common and constant temperature, and to be in a pressure equilibrium so that $n_a = n_b(1+k)$.

Suppose we observe such a CLG and derive an apparent electron density $n_0$, pressure $p_0$, and magnetic field $B_0$, without the knowledge of the two phase nature. While $n_0$ is derived from the observed X-ray surface brightness through the emissivity relation, $B_0$ is estimated with the help of $n_0$ from the radio Faraday rotation of background radio sources. Thus we have $n_0^2 = n_a^2(1-f) + n_b^2 f = n_b^2\left[(1+k)^2(1-f) + f\right]$ and $p_0/p_b = n_0/n_b$. Since the Faraday rotation is proportional to $n_0 B_0 = n_b B f$, we further have $B_0 = Bfn_b/n_0$. Combining these relations, the apparent magnetization parameter $k_0 \equiv B_0^2/(8\pi p_0)$ becomes

$$k_0 = kf^2\left[(1+k)^2(1-f) + f\right]^{-3/2}.$$

Let us for example assume that a considerable fraction $f = 0.25$ is magnetized with $k = 1.5$. If $p_b$ is taken to be $1 \times 10^{-11}$ erg cm$^{-3}$ at the cluster center, $B = 19\mu$ G is implied. Then from the above equation we obtain $k_0 = 0.0085$, $p_0 = 2.2 p_b$, and $B_0 = 2.2\mu$ G. Therefore the ICM appears to have only a microGauss level field, in agreement with the observation. Nevertheless such an ICM is clearly under a strong influence of the IMF.

The two-phase nature would affect the hydrostatic equilibrium of the ICM. Neglecting the buoyance instability, the pressure balance equation becomes

$$\frac{dp_a}{dr} = (1+k)\frac{dp_b}{dr} = -\mu m_p\left[n_a(1-f) + n_b f\right]\frac{d\phi}{dr}.$$

Substituting $n_a = (1+k)n_b$ and using $n_0/n_b = \sqrt{(1+k)^2(1-f)+f} = p_0/p_b$, we obtain

$$\frac{dp_0}{dr} = -\left(1 - \frac{fk}{1+k}\right)\mu m_p n_0 \frac{d\phi}{dr}.$$

Therefore $\beta_i$ ('image beta'), the beta parameter obtained from analysis of the X-ray surface brightness neglecting the two-phase nature, becomes different from $\beta_s = \frac{\mu m_p \sigma^2}{kT}$ ('spectroscopic beta'), obtained by comparing the plasma temperature with the velocity dispersion $\sigma$ of the galaxies. The latter is not subject to the two-phase property. Specifically we have

$$\beta_i = \left(1 - \frac{fk}{1+k}\right)\beta_s.$$

In the above example with $f = 0.25$ and $k = 1.5$, we expect $\beta_i = 0.85\beta_s$. X-ray observations usually yield $\beta_i = (0.5$–$0.7)$ and $\beta_s = (0.7$–$0.9)$. This discrepancy, called the beta-problem, can thus be solved in a natural way by the present model.

4. Cooling flows

The two-phase model introduced above may find its best application to the phenomenon known as cooling flow. In many CLGs in which the giant elliptical galaxy (called cD galaxy) sits at the center, a strong excess in the X-ray surface brightness is observed near the cluster center, and interpreted that the ICM is experiencing a radiative cooling instability (Fabian, Nulsen and Canizares 1991). That is, the ICM near the center is dense enough for the radiative cooling to proceed faster than the Hubble time. The cooling produces a temperature gradient in the ICM, cooler towards the center. The associated pressure drop is compensated by a radial in-flow of the ICM, called cooling flow (CF), which then increases the central ICM density and enhances the cooling.

Although this picture is a theoretically elegant and elaborate one, it is subject to many difficulties. Over the Hubble time a CF would have accumulated a mass comparable to that of the cD galaxy, but little is known about the fate of the CF. There is no evidence of star formation in CF regions. The CF picture takes no, or at least very insufficient, account of the effect of various heating mechanisms, e.g. gravitational compression, heat conduction, drag due to the galaxy motion, supernovae in the cD galaxy, AGN activity within the cluster, magnetic reconnection, and so on. Observationally the CF phenomenon is strongly correlated with the presence of cD galaxy, but their causal relation is not clear either.

Using *Asca*, we have confirmed the presence of cool emission component in the central regions of several CF clusters (Fukazawa *et al.* 1994; Ohashi *et al.* 1994; Makishima 1994). However these results in fact point to a scenario different from that of CF. We have found that the X-ray spectrum taken from the central regions of these CLGs cannot be described by a continuous distribution of plasma temperature, but well by a combination of two discrete plasma temperatures. At least in a few CLGs the two temperatures thus derived are radially quite constant (Fukazawa *et al.* 1994), while the emission measure of the cooler component decreases more rapidly for larger radii than that of the hotter component. The hotter temperature (typically 3–6 keV depending on the clusters) agrees with the plasma temperature observed from outer regions of the same CLG, while the cooler temperature is usually found at $\sim 1$ keV which is typical of gravitational potential of a cD galaxy. Finally in some CLGs, we have observed a dramatic increase in the metal abundance (up to $\sim 1$ solar; Koyama *et al.* 1990; Fukazawa *et al.* 1994) over the region where the cooler component is visible.

While the CF model cannot readily explain any of the above *Asca* results, the two-phase picture we have developed in C.3 can consistently explain all of them. Specifically, we assume that the plasma in the central region of a CF cluster consists of two phases; one is the hotter and metal-poor ICM, while the other is the cooler and metal-rich plasma associated with the cD galaxy. The two phases are separated by the IMF anchored to the cD galaxy. Closed IMF loops may be filled with the cooler cD plasma to form a 'cD corona', whereas open-field regions ('the cD coronal holes') may be dominated by the hotter ICM plasma. The total mass of the cool-phase plasma is estimated from the data to be at most 10–20% of the mass of the cD galaxy, making the overall picture consistent.

The cooler component typically has an X-ray luminosity of $(0.5-5) \times 10^{43}$ ergs s$^{-1}$. The plasma would be subject to the cooling instability unless this luminosity is somehow supplied to the cD galaxy, even though the IMF would strongly suppress the collapse due to cooling. Although the IMF reduces heat conduction from the hotter to the cooler phases, we notice that the IMF also acts as an efficient medium to convert the gravitational energy of the CLG and dissipate it into thermal energy. In fact, the X-ray luminosity of the cool component can be sustained if only a minor fraction ($10^{-3}$–$10^{-4}$) of the total gravitational energy (including that of the dark matter) of the CLG has been released over the Hubble time. Intriguing enough, this fraction is similar to that we invoked to explain the GRXE in B.3. We thus believe that the radiative cooling instability is mostly suppressed, and the phenomenon that has usually been attributed to the CF is in fact due to the presence of a magnetized corona local to the cD galaxy.

4. Future prospects

In the context of observational cosmology, importance of the X-ray observations of CLGs is dramatically increasing. For correct understanding of the CLG plasmas, however, improved knowledge on the IMF is vitally required. We plan to test our view using a series of new data available with *Asca*. We also attempt to update lower limits on the cluster magnetic fields, by reducing upper limits on the inverse-Compton hard X-rays. This will constitute an important objective for the hard X-ray detector which we plan to put onboard the ASTO-E satellite that will succeed *Asca*.

Jets from AGNs are thought to be collimated by magnetic fields, and radio lobes at both ends of the jets are thought to magnetize a large intra-cluster volume very effectively (Kronberg 1994). We have shown that in some AGNs the strong X-ray emission comes from direct synchrotron process at the base of the jets (e.g. Kohmura *et al.* 1994). X-ray observations are thus expected to give magnetic field intensities in these jets. Furthermore, inverse-Compton X-rays from the radio lobes of the radio galaxy NGC1316 (Fornax-A) has been detected with *ROSAT* and *Asca*. A comparison of the synchrotron radio brightness and the inverse-Compton X-ray brightness indicates a magnetic field of 2–4 $\mu$G, which will serve as an important seed field.

Finally we remark on the mysterious cosmic X-ray background (CXB) (Boldt 1987), which has such a uniform surface brightness that the origin is undoubtedly cosmological. The CXB spectrum is well represented with a $kT \sim 40$ keV thermal Bremsstrahlung model, although the possibility of a hot plasma filling the entire universe and producing the CXB has been ruled out by the absence of any distortion in the cosmic microwave background (CMB) spectrum as measured with *COBE*. Thus the currently most popular idea is to regard the CXB as being synthesized with a large number of distant AGNs. However as pointed out by Tajima & Mineshige (a private communication), hot plasmas confined within thin magnetic strings spanning between clusters of galaxies with a very small filling factor can explain the observed X-ray spectrum and surface brightness of the CXB, and the absence of distortion in the CMB spectrum. This possibility provides an interesting case to address the question of cosmological magnetic field.

**REFERENCES**
Boldt, E. A. 1987, Physics Reports **146**, 217.

Bulik, T. et al. 1992, Astrophys. J. **395**, 564.
Chanmugam, G. 1992, Ann. Rev. Astr. Ap **30**, 143.
Clark, G. et al. 1990, Astrophys. J. **353**, 274.
Fabian, A. C., Nulsen, P. E. J. & Canizares, C. R. 1991, Astron. Astrophys. Rev. **2**, 191.
Fukazawa, Y. et al. 1994, Publ. Astr. Soc. Japan **46**, L55.
Iida, K. & Ichimaru, S. 1994, private communication.
Kohmura, Y. et al. 1994, Publ. Astr. Soc. Japan **46**, 131.
Koyama, K. 1989, Publ. Astr. Soc. Japan **41**, 665.
Koyama, K. et al. 1986, Publ. Astr. Soc. Japan **38**, 121.
Koyama, K. et al. 1989a, Publ. Astr. Soc. Japan **41**, 461.
Koyama, K. et al. 1989b, Nature **339**, 603.
Koyama, K., Takano, S. & Tawara, Y. 1990, Nature **350**, 135
Kronberg, P. P. 1994, Rep. Prog. Phys. 325.
Makishima, K. 1986, in *The Physics of Accretion onto Compact Objects*, eds. K. O. Mason, M. G. Watson & N. E. White (Springer-Verlag), p. 249.
Makishima, K. 1992, in *The Structure and Evolution of Neutron Stars*, eds. D. Pines, R. Tamagaki & S. Tsuruta (Adison-Wesley), p. 86.
Makishima, K. et al. 1990, Astrophys. J. **365**, L59.
Makishima, K. & Mihara, T. 1992, in *Frontiers of X-ray Astronomy*, eds. Y. Tanaka & K. Koyama (Tokyo; Universal Academy Press),
Makishima, K. 1994, in *New Horizon of X-ray Astronomy*, eds F. Makino & T. Ohahsi (Tokyo; Universal Academy Press), in press.
Mihara, T. et al. 1990, Nature **346**, 250.
Mihara, T. et al. 1991, Astrophys. J. **379**, L61.
Murakami, T. et al. 1988, Nature **335**, 234.
Nagase, F. 1989, Publ. Astr. Soc. Japan **41**, 1.
Nagase, F. et al. 1992, Astrophys. J. **375** L49.
Ohahsi, T. et al. 1994, in *New Horizon of X-ray Astronomy*, eds F. Makino & T. Ohahsi (Tokyo; Universal Academy Press), in press.
Ohno, H. & Shibata, S. 1993, MNRAS **262**, 953.
Rephaeli Y. & Gruber, D. E. 1988, Astrophys. J. **333**, 133.
Sarazin, C. L. 1988, *X-ray Emissions from Clusters of Galaxies*, (Cambridge University Press).
Tanaka, T., Inoue, H., & Holt, S. S. 1994, Publ. Astr. Soc. Japan **41**, L37.
Taylor, G. B. & Perley, R. A. 1993, Astrophys. J. **416**, 554.
Tsuneta, S. 1994, these proceedings.
Trümper, J. et al. 1978, Astrophys. J. **219**, L105.
Voges, W. et al. 1982, Astrophys. J. **263**, 803.
Wakatsuki, S., Hikita, A., Sato, N & Itoh, N. 1992, Astrophys. J. **392**, 628.
Warwick, R. S., Turner, M. J. L., Watson. M. G., & Willingale, R. 1985, Nature **317**, 218.
Wheaton, Wm. et al. 1979, Nature **282**, 240.
White, N. E., Swank, J. S., & Holt, S. S. 1983, Astrophys. J. **270**, 711.
Worrall, D. M., Marshall, F. E., Boldt, E. A. & Swank, J. H. 1982, Astrophys. J. **255**, 111.
Yamauchi, S. et al. 1990, Astrophys. J. **365**, 532.
Yoshida, A., Murakami, T., Nishimura, J. & Fenimore, E. E. 1992, in *Frontiers of X-ray Astronomy*, eds. Y. Tanaka & K. Koyama (Tokyo; Universal Academy Press), p. 53.

## I.6

# INTENSE MAGNETIC FIELD PHENOMENA

Jon Weisheit

Space Physics and Astronomy Department, Rice University
Houston, TX 77251, U. S. A.

This article surveys three of the many challenging problems involving quantum phenomena in plasmas with magnetic fields $B$ in the range $10^8$–$10^{10}$ Gauss: magnetic white dwarf stars, spectroscopic effects of motional ($\mathbf{v} \times \mathbf{B}$) electric fields, and statistical models of many-electron atoms in strong $B$ fields. It has proved difficult to make progress in this regime of field strengths, where Coulomb and magnetic interactions are comparable.

## I. INTRODUCTION

The behavior of matter in magnetic fields is relevant to many important questions in atomic, plasma and condensed matter physics, and in astrophysics. Through much of this century, investigations tended to explore situations in which magnetic effects were perturbations of the ground or low-lying states of field-free atoms and ions—what one might call the Zeeman regime. Then, in the 1960's pulsars were discovered. It is believed that huge fields, $B > 10^{11}$ Gauss ($10^4$ G = 1 Tesla), exist in neutron stars and underlie the pulsar mechanism [Michel, 1991; Meszaros, 1992]. Largely as a result of pulsars' discovery, research involving magnetic fields has turned to strong-field phenomena, a topic selected by Ginzburg [1978] as one of the "key problems of physics and astrophysics."

Much of what was known about this topic at the time is described in Garstang's [1977] notable review but, of course, there have been many developments since then. In particular, extensive computational efforts have yielded theoretical eigenvalues for hydrogenic systems throughout the range $0 < B < 10^{13}$ G [Muller et al., 1986; Liu and Starace, 1987; Ivanov, 1988; Wunner, 1990], as well as some bound-bound and bound-free radiative transitions strengths for one-electron systems in strong fields [Forster et al., 1984; Alijah et al., 1990; Wang and Greene, 1991; Stancil and Copeland, 1993]. Moreover, self-consistent field methods have been extended to treat multi-electron atoms and molecules in the "Landau regime" of very intense fields ($B > 10^{11}$ G), where an adiabatic approximation can be made to separate (slow) motions parallel to and (fast) motions perpendicular to $\mathbf{B}$ [Neuhauser et al., 1987; Miller and Neuhauser, 1991; Demeur et al., 1994]. There also has been progress, reviewed elsewhere [McDowell and Zarcone, 1985], in understanding elementary collisional processes in strong magnetic fields.

Recall that the Hamiltonian of an atom or ion at rest in a static uniform magnetic field $\mathbf{B}$ is changed from the field-free expression $H_0$ to

$$H = H_0 + H_1 + H_2, \tag{1}$$

where, in atomic units, one has [Bethe and Salpeter, 1957]

$$H_1 = \frac{1}{2}(\alpha \mathbf{B}) \cdot (\mathbf{L} + 2\mathbf{S}) \tag{2}$$

$$H_2 = \frac{1}{8} \Sigma_i (\alpha \mathbf{B} \times \mathbf{r}_i)^2. \tag{3}$$

In Eqs. 2 and 3, $\alpha$ is the fine-structure constant, the $\mathbf{r}_i$ are electron coordinates, $\mathbf{S}$ and $\mathbf{L}$ are total spin and orbital angular momenta, and $\mathbf{B}$ is measured in units of $B_o = (m^2 e^3 c/\hbar^3) = 2.35 \times 10^9$ G. In the quasi-static fields of neutron stars, as well as in the time-harmonic fields that arise during intense ($I > 10^{18}$ W/cm$^2$), but abrupt ($\Delta t < 10^{-12}$ sec), laser irradiation of material targets [Bucksbaum, 1990; Schappert et al., 1990], the magnetic interactions are so strong that Coulomb interactions actually play a subsidiary role in the determination of atomic properties.

Atomic phenomena are more complicated when magnetic and Coulomb interactions are comparable. The truth of this was revealed in the famous experiment of Garton and Tomkins [1969] on the Ba I spectral series limit in kilogauss fields, and it has been reinforced by the discovery of the "chaotic" behavior exhibited by highly excited hydrogen atoms in fields of order $10^4$ G [Kleppner et al., 1983; Holle et al., 1988]. Unfortunately, it is also true that quantal calculations based on perturbation methods generally are unreliable when such situations occur, and that numerical attacks generally are complicated by the different symmetries of $H_0$ and the magnetic terms.

Simple scaling arguments, based on hydrogenic wavefunctions for principal quantum numbers $n$, suggest that non-perturbative treatments are needed for field strengths in the range

$$10^{-2} \frac{Q^2}{n^3} < \frac{B}{B_o} < 10^2 \frac{Q^2}{n^3}, \tag{4}$$

where $Q$ is the effective nuclear charge number. Quasi-static magnetic fields of such intermediate strength arise in certain white dwarf stars [Chanmugam, 1992], in explosive pinch plasmas [Miura and Herlach, 1985], in inertial confinement fusion plasmas [Max, 1982; Wilks et al., 1992], and possibly during the epoch of nucleosynthesis in the early universe [Cheng et al., 1994]. Therefore, the range of applications is wide even though the range of field strengths is not. This article addresses a few aspects of these challenging, but less extreme situations. It should be emphasized that all but the last of the regimes listed above are accessible to spectroscopic measurement, and hence they provide fertile ground for testing various theoretical concepts and models.

## II. MAGNETIC WHITE DWARF STARS

White dwarf stars play a key role in astrophysics, since they serve as testbeds for concepts of stellar and galactic evolution as well as the physics of strongly coupled plasmas [e.g., Weidemann, 1990; Van Horn, 1991]; they even find use as standards for ultraviolet flux calibration [e.g., Davidsen, 1993]. Among the thousand or so catalogued white dwarfs, about 30 of the isolated ones (i.e., ones not in binary systems) have very large surface magnetic fields, $B \geq 10^6$ G [Chanmugam, 1992, and references cited therein]. This knowledge derives mainly from spectroscopic analyses of Balmer lines—their wavelengths, strengths, and shapes—but continuum polarization measurements have proved useful too. Figure 1 shows the positions in the $B$–$T_{eff}$ plane of all known, isolated magnetic white dwarfs, $T_{eff}$ being an effective blackbody temperature for each star (proportional to the fourth root of its luminosity). No trends have been discerned in these data. In particular, since white dwarfs slowly cool as they age, there seems to be no evidence of field decay with time.

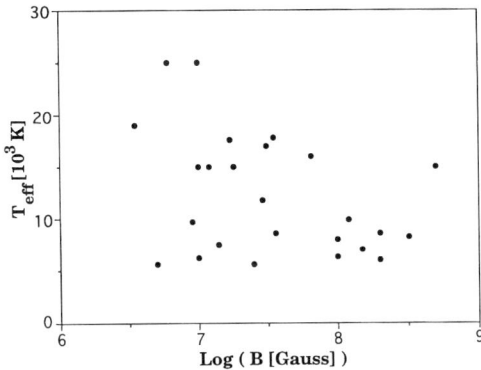

Fig. 1. Magnetic field strengths [in Gauss] and effective temperatures [in $10^3$ K] of isolated magnetic white dwarf stars.

These stars pose several interesting questions, including:
i. Do they arise from a particular kind of progenitor stars, or from a particular set of circumstances ?
ii. To what extent does a strong $B$ field modify the opacity and conductivity in their interiors and, hence, the rate at which magnetic white dwarfs cool ?
iii. Do strong $B$ fields affect the rate at which white dwarfs accrete interstellar matter and, hence, our understanding of their surface compositions ?
iv. Although it is generally believed that a dynamo mechanism generates the fields of most main sequence stars [Moss, 1988], magnetic white dwarf rotation rates exhibit no correlation with field strength, nor do these rates differ much from those of non-magnetic white dwarfs. Why ?

Answers to such questions must await more high-resolution spectroscopic observations, more refined stellar models that include megagauss $B$-field effects on transport coefficients and spectral line features, plus a much better understanding of the role of magnetic fields in star formation.

## III. MOTIONAL ELECTRIC FIELDS

Movement in a strong magnetic field region gives rise to a motional electric field $\mathbf{E}_m$ in an atom's rest frame, and adds to its Hamiltonian a third term,

$$H_3 = -\mathbf{E}_m \cdot \mathbf{d} = -\alpha(\mathbf{v} \times \mathbf{B}) \cdot \mathbf{d}, \tag{5}$$

with $\mathbf{d}$ being the bound electrons' dipole moment and $\mathbf{v}$ ($\ll c$), the atom's center-of-mass velocity. This "external" electric field $\mathbf{E}_m$ augments the "internal" microfield $\mathbf{E}_p$ that results from small-scale (thermal) density fluctuations. If a plasma has a particle number density $\mathcal{N}$ and temperature $T$, then the strengths of these two $E$ fields are approximately in the ratio

$$\frac{E_m}{E_p} \approx \left(\frac{10^{16}\mathrm{cm}^{-3}}{\mathcal{N}}\right)^{2/3} \left(\frac{k_B T}{\mathrm{eV}}\right)^{1/2} \left(\frac{B}{10^6 \mathrm{G}}\right). \tag{6}$$

Conditions in the atmospheres of magnetic white dwarfs, as well as in some laboratory plasmas, are such that these electric fields are comparable.

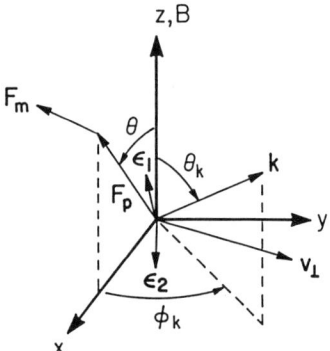

Fig. 2. Physical and geometrical variables relevant to spectral line formation in strongly magnetized plasmas (see text for explanation).

Large motional fields $E_m$ lead to numerous interesting Stark effects. For instance, each radiating atom in a strongly magnetized plasma is perturbed by a field $\mathbf{E} = \mathbf{E}_m + \mathbf{E}_p$ that not only is stochastic, but also depends upon that atom's motion relative to the field direction. This introduces a significant correlation between Doppler and Stark broadening mechanisms, and greatly complicates the calculation of spectral line shapes [Hoe et al., 1981; Mathys, 1990]. The effects of large motional ($\mathbf{v} \times \mathbf{B}$) electric fields are likely to swamp the more familiar ion dynamic effects on spectral line formation [e.g., Stamm et al., 1986]. Figure 2 illustrates the abundance of physical and geometrical variables that are encountered in this more complicated lineshape problem, where polarization (described by unit vectors $\hat{\epsilon}_1$ and $\hat{\epsilon}_2$) of radiation emitted into the line-of-sight direction $\mathbf{k}$ is also relevant for diagnostic purposes. Yet additional complications can arise if the plasma is not only intensely magnetized but also very dense, because then collisional Stark broadening will be affected by the $B$ field too. To date, the problem has only been attacked piecemeal, with various calculations including some of these effects but never all of them.

In the absence of a motional field, plasma ions have their uppermost bound states determined by the strength of electrostatic perturbations that neighboring charges produce. Rogers [1986], Hummer and Mihalas [1988], and Dharma-wardana and Perrot [1992] give alternative descriptions of this so-called continuum lowering phenomenon. A large motional field $E_m \approx E_p$ changes the amount of continuum lowering experienced by ions in a plasma and therefore influences plasma ionization balance. Recently, Pavlov and Meszaros [1993] discussed continuum lowering vis-a-vis just the motional electric field; they also noted that this field will smear out photoionization edges. These various Stark effects all should be manifest in the merging of lines near a spectral series limit (the Inglis-Teller effect), since an intense $B$ field alters line positions, line strengths, line widths and even the number of lines!

Finally, although not discussed here, there are some related, weaker-field phenomena that definitely merit further study. For example, atomic motions in the presence of even modest fields ($B \sim 10^4$ G) lead to Stark effects that can substantially alter rates of dielectronic recombination [Muller et al., 1986; Harmin, 1986], because it proceeds through easily perturbed Rydberg states with $n \gg 1$.

## IV. STATISTICAL ATOMIC MODELS

Kadomtsev [1970] was the first to apply statistical, Thomas-Fermi (TF) theory to the case of an atom in a strong external magnetic field. He assumed that the bound electrons' motions perpendicular to $\mathbf{B} = B\mathbf{z}$ (i.e., in the $\rho$–$\phi$ plane) could be approximated by Landau orbitals characteristic of *free* electrons in that field, and further that only the parallel motion needed to be treated by statistical arguments. He then used certain analytical properties of the Landau orbitals to obtain a new equation relating the number density $N(\mathbf{r})$ of electrons in the atom to its electrostatic potential $\Phi(\mathbf{r})$ (corrected here for a trivial factor of two error),

$$N(\mathbf{r}) = \frac{B}{2\pi^2 B_o} \sqrt{2\Phi(\mathbf{r})}. \tag{7}$$

As in the standard TF development [e.g., Spruch, 1991], he used the Poisson equation to eliminate $N$ and obtain a nonlinear differential equation for the potential $\Phi$. He argued that his approach was valid for a considerable range of field strengths above $B_o$.

In the years since then, this model has been extended to include exchange and density-gradient corrections [see Fushiki et al., 1992, and references therein], and has been used to study material properties of strongly magnetized matter [e.g., Abrahams and Shapiro, 1991]. All of this work has been based on Kadomtsev's original analysis, wherein the usual, spherical TF boundary conditions were applied at large- and small-r values. Of course, as indicated in Eqs. 2 and 3, the atomic Hamiltonian contains terms with axial symmetry, and hence the actual eigenfunctions and the resulting $N(\mathbf{r})$ also must be non-spherical. A simple electrostatics problem can be used to reveal the flaw in Kadomtsev's treatment. Suppose one is asked to find the potential $\Phi(\mathbf{r})$ due to a piece of wire with uniform charge along its length $L$. The answer is conveniently expressed in an infinite series involving Legendre polynomials whose argument is the cosine of the angle between $\mathbf{r}$ and the wire. Only in the limit that $r \gg L$ does $\Phi$ become essentially spherical. Similarly, the atomic potential becomes spherical only far outside the atom (plus, as Kadomtsev correctly assumed, near the nucleus).

Even when this is acknowledged, it is not evident what boundary conditions should be imposed on a modified TF equation for $\Phi(\mathbf{r})$. Therefore, in work recently published by E. P. Lief and the author [1993], a different tactic was employed. First, Poisson's equation was used to eliminate $\Phi(\mathbf{r})$ from Eqn. 7 in favor of $N(\mathbf{r})$:

$$\Delta N^2 = \frac{2(B/B_o)^2 N}{\pi^3}. \tag{8}$$

Then, in the spirit of Kadomtsev's original work, the density of bound electrons was assumed to be factorizable according to the *ansatz*

$$N(\mathbf{r}) = N_z(\rho; z) N_\rho(\rho), \tag{9}$$

with the factor $N_\rho(\rho)$ being determined by the occupied Landau orbitals. This guarantees that the electron density $N(\mathbf{r})$ will have the correct, axial symmetry far from the nucleus. The axial term $N_z(\rho; z)$, which depends only parametrically on $\rho$, must be determined from a numerical solution of the second-order, non-linear partial differential equation which results from the substitution of Eqn. 9 into Eqn. 8.

Lief and Weisheit used an implicit finite-difference scheme to solve that equation for $N_z$. One novel aspect of their approach is that the behavior of $N_z$ at large $z$-values must be adjusted to achieve correct normalization, viz. $\int d^3\mathbf{r}\, N(\mathbf{r}) = Q$. The other unusual aspect pertains to $N_\rho$, since one must specify at the onset which Landau orbitals are occupied. Their calculations have been restricted to atomic ground states, which effectively limits the results to low-temperature plasmas, $k_B T \ll \langle H_1 + H_2 \rangle$. The occupied orbitals then are just those which are consistent with a suitably generalized *Aufbauprinzip*.

Of course, a better model should yield a more tightly bound atom. The atomic binding energy $E_b$ is the difference between the energy $E_{atom}$ that the electrons and nucleus have in the presence of both the field $\mathbf{B}$ and their pair-wise Coulomb interactions (energies $V_{en}$ and $V_{ee}$), and the energy $E_{free}$ that all these charges have when they are in the field but do not interact with each other, i.e., when they are free. The choice of Landau orbitals implies that motion perpendicular to the field is completely regulated by the field itself. Hence, only motion parallel to the field (energy $T_\parallel$) contributes to $E_b$, and the binding energy is the expectation value of the sum of three terms,

$$E_b = E_{atom} - E_{free} = \langle T_\parallel + V_{en} + V_{ee} \rangle. \tag{10}$$

It is necessary to eliminate the parallel kinetic energy $\langle T_\| \rangle$ to obtain a binding energy in terms of just the electrostatic quantities $V_{en}$ and $V_{ee}$ which are straightforward to compute. But, the usual Virial Theorem for spherical atoms cannot be used here. Instead, one needs to appeal to density functional theory to get another relation between these mean energy terms [Banerjee et al., 1974],

$$\langle 3T_\| + V_{en} + 2V_{ee} \rangle = 0, \tag{11}$$

and from these last two equations get the working definition of $E_b$.

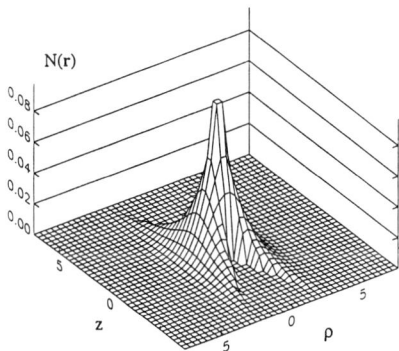

Fig. 3. Contour plot of electron density $N(\mathbf{r})$ for helium in a field of strength $B_o$ directed along the z-axis. All quantities are in atomic units.

Our numerical studies indicate that atoms in large magnetic fields become much more tightly bound when the constraint of spherical symmetry is dropped. Published calculations of Lief and Weisheit [1993] pertain to carbon; another set of computations has since been carried out for helium and is given here in Table I. Although two electrons make for a poor "statistical" atom, these results are encouraging: at the lowest field strengths appropriate to this new statistical method ($B \sim B_o = 2.35 \times 10^9$ G), the binding energy is close to published adiabatic bounds [Park and Starace, 1984]. Moreover, atoms do become much more tightly bound when the (incorrect) constraint of spherical symmetry is dropped. Figure 3 displays a contour plot of the electron density $N(\mathbf{r})$ calculated for He in a field of strength $B = B_o$. Note, in particular, that the butterfly-like shape now computed for this atom in a strong field (and readily apparent in a top-down view) is just what Ruderman predicted long ago [1974] on the basis of very general arguments.

Table 1. Binding energies $E_b$ for He in a strong magnetic field.

| $B/B_0$ | 1 | 10 | 100 |
|---|---|---|---|
| $E_b$ [Kadomtsev] (au) | 1.58 | 3.96 | 9.75 |
| $E_b$ [present work] (au) | 2.83 | 6.93 | 17.6 |

At present, research is underway to study what effect the loss of spherical symmetry in the TF model has on quantum transport coefficients and the plasma's equation of state; earlier work on these topics has been predicated on Kadomtsev's approach [e.g., Fushiki et al., 1989; Abrahams and Shapiro, 1991], or some other model with a spherical atomic charge density [e.g., Hernquist, 1984]. Very recently, there have been some relevant, formal advances regarding anisotropy issues in magnetized plasmas [John and Suttorp, 1994].

## ACKNOWLEDGEMENTS
The author's work on this subject has been supported by the Division of Chemical Sciences, Office of Basic Energy Sciences, Office of Energy Research, U.S. Department of Energy. Past collaborations with E. Lief, E. Avera, and A. Mattingly are happily recorded.

## REFERENCES
Abrahams, A. M., and Shapiro, S. L. 1991, Astrophys. J. **374**, 652.
Alijah, A. Hinze, J., and Broad, J. T., 1990, J. Phys. B: At. Mol. Opt. Phys. **23**, 45.
Banerjee, B., Constantinescu, D. H., and Rehak, P., 1974, Phys. Rev. D **10**, 2384.
Bethe, H, A., and Salpeter, E. E., 1957, *Quantum Mechanics of One- and Two-Electron Atoms* (Springer-Verlag, Berlin).
Bucksbaum, P. H., 1990, in *Atoms in Strong Fields*, edited by C. A. Nicolaides, C. W. Clark, and M. H. Nayfeh (Plenum, New York), p.381.
Chanmugam, G., 1992, Ann. Rev. Astron. Astrophys. **30**, 143.
Cheng, B., Schramm, D. N., and Truran, J. W., 1994, Astrophys. J., in press.
Davidsen, A. F., 1993, Science **259**, 327.
Demeur, M., Heenen, P.-H., and Godefroid, M., 1994, Phys. Rev. A **49**, 176.
Dharma-wardana, M. W. C. and Perrot, F., 1992, Phys. Rev. A **45**, 5883.
Forster, H., Strupat, W., Rosner, W., Wunner, G., Ruder, H., and Herold, H., 1984, J.Phys. B: At. Mol. Phys. **17**, 1301.
Fushiki, I., Gudmundsson, E. H., and Pethick, C. J., 1989, Astrophys. J. **342**, 958.
Fushiki, I., Gudmundsson, E. H., Pethick, C. J., and Yngvason, J. 1992, Annnals Phys. **216**, 29.
Garstang, R. H., 1977, Rep. Prog. Phys. **40**, 105.
Garton, W. R. S., and Tomkins, F. S., 1969, Astrophys. J. **158**, 839.
Ginzburg, V. L., 1978, *Key Problems of Physics and Astrophysics* (MIR, Moscow).

Harmin, D. A., 1986, Phys. Rev. Lett. **57**, 1570.
Hernquist, L., 1984, Astrophys. J. Suppl. **56**, 325.
Hoe, N., Grumberg, J., Canby, M., LeBoucher, E., and Coulaud, G., 1981, Phys. Rev. A **24**, 438.
Holle, A., Main, J., Wiebusch, G., Rottke, H., and Welge, K.H., 1988, Phys. Rev. Lett. **61**, 161.
Hummer, D. G. and Mihalas, D. 1988, Astrophys. J. **331**, 794.
Ivanov, M. V., 1988, J. Phys. B: At. Mol. Opt. Phys. **21**, 447.
John, P., and Suttorp, L. G., 1994, preprint.
Kadomtsev, B. B., 1970, Sov. Phys. JETP **31**, 945.
Kleppner, D. Littman, M. G., and Zimmerman, M. L., 1983, in *Rydberg States of Atoms and Molecules*, edited by R. F. Stebbings and F. B. Dunning (Cambridge University, Cambridge, England), p. 73.
Lief, E. P. and Weisheit, J. C. 1993, Contrib. Plasma Phys. **33**, 471.
Liu, C.-R., and Starace, A. F., 1987, Phys. Rev. A **35**, 647.
Mathys, G., 1990, J. Quant. Spectrosc. Radiat. Transfer **44**, 143.
Max, C. E., 1982, in *Laser-Plasma Interaction*, edited by R. Balain and J.-C. Adam (North-Holland, Amsterdam), p. 301.
McDowell, M. R. C., and Zarcone, M., 1985, Adv. Atom. Molec. Phys. **21**, 255.
Meszaros, P., 1992, *High-Energy Radiation from Magnetized Neutron Stars* (University of Chicago, Chicago).
Michel, F. C., 1991, *Theory of Neutron Star Magnetospheres* (University of Chicago, Chicago).
Miller, M. C., and Neuhauser, D., 1991, Mon. Not. R. Astron. Soc. **253**, 107.
Miura, N., and Herlach, F., 1985, in *Strong and Ultrastrong Magnetic Fields*, edited by F. Herlach (Springer-Verlag, Berlin), p. 247.
Moss, D. 1986, Phys. Repts. **140**, 1.
Muller, A., Belic, D. S., DePaola, B. D., Djuric, N., Dunn, G. H., Mueller, D. W., and Timmer, C., 1986, Phys. Rev. Lett. **56**, 127.
Neuhauser, D., Langanke, K., and Koonin, S. E., 1986, Phys. Rev. A **33**, 2084.
Park, C.-H. and Starace, A. F. 1984, Phys. Rev. A **29**, 442.
Pavlov, G. G., and Meszaros, P., 1993, Astrophys. J. **416**, 752.
Rogers, F. J. 1986, Astrophys. J. **310**, 723.
Ruderman, M.,1974, in *Physics of Dense Matter*, edited by C. J. Hansen (D. Reidel: Dordrecht), p. 117.
Schappert, G. T., Casperson, D. E., Cobble, J. a., Comly, J. C., Jones, L. A., Kyrala, G. A., LaGatutta, K. J., Lee, P. H. Y., Olson, G. L., and Taylor, A. J., 1990, in *Atomic Processes in Plasmas*, edited by Y.-K. Kim and R. C. Elton (A. I. P., New York), p. 217.
Spruch, L., 1991, Rev. Mod. Phys. **63**, 151.
Stamm, R., Talin, B., Pollock, E. L., and Iglesias, C. A., 1986, Phys. Rev. A **34**, 4144.
Stancil, P. C., and Copeland, G. E., 1993, Phys. Rev. A **48**, 516.
Van Horn, H. M., 1991, Science **252**, 384.

Wang, Q., and Greene, C. H., 1991, Phys. Rev. A **44**, 7448.
Weidemann, V., 1990, Ann. Revs. Astron. Astrophys. **28**, 103.
Wilks, S. C., Kruer, W. L., Tabak, M., and Langdon, A. B., 1992, Phys. Rev. Lett. **69**, 1383.
Wunner, G. 1990, in *Spectral Line Shapes*, vol. 6, edited by L. Frommhold and J. W. Keto (A. I. P., New York), p. 563.

# II. Neutrino Astrophysics

II.1

# NEUTRINOS IN COSMOLOGY AND ASTROPHYSICS: AN INTRODUCTORY OVERVIEW OF THEORETICAL ASPECTS

Masataka Fukugita

Yukawa Institute for Theoretical Physics, Kyoto University, Kyoto 606, Japan
Institute for Advanced Study, Princeton, NJ 08540, U. S. A.

An introductory overview is given for current theoretical interests in neutrinos in cosmology and astrophysics.

## I. INTRODUCTION

In this talk I shall try to give an overview as to where neutrinos play a significant role in cosmology and astrophysics. It would be appropriate to start with discussing why neutrinos attract scientists in a variety of fields.

Since its first conjecture, the neutrino has been playing a crucial role in the advancement of our understanding of particle physics, till the completion of the unified theory of electroweak interactions by Glashow, Weinberg and Salam [1]. Its role is not over yet: there are two obvious reasons why it attracts particle physicists' interest: the first is the problem why the neutrino mass is so small, compared with masses of other particles. The most attractive answer to this question involves the physics beyond the standard theory. The other interest in the neutrinos is as a probe for new interactions. The neutrino has no electromagnetic charge nor color charge, and hence it is probably a good place to look for new interactions weaker than the weak interaction.

Astrophysical interest in the neutrino arises from the fact that it is produced copiously in high temperature and/or high density environment and it often dominates the physics of those astrophysical objects. The interaction of the neutrino with matter is so weak that it passes freely through any ordinary matter exiting in the Universe. This makes neutrinos to be a very efficient carrier of energy drain from optically thick objects. At the same time they give a good probe for the interior of such dense objects. Study of astrophysical neutrinos not only brings us new insights for astrophysics, but also provides a new method of exploring particle physics which cannot be reached in laboratories. This also accompanies the interest from nuclear physics and dense plasma physics: neutrinos are produced in very high density environment that cannot be accessible in laboratories, which might lead us hope that we could obtain some information concerning the state of matter in such an unusual environment.

## II. PARTICLE PHYSICS INTERESTS IN THE NEUTRINO

Let us briefly explain the interest in the neutrino as a probe for particle physics beyond the standard theory [2]. The first is the mass issue: mass of the electron neutrino is smaller than $10^{-5}$ the electron mass. One might then suppose that mass of neutrino is exactly vanishing, as usually assumed in the standard theory of electroweak interactions. We must recall that exact vanishing of mass requires some underlying symmetry, just as U(1) gauge symmetry guarantees the masslessness of the photon. No such symmetry is known, nor has been conjectured in a reasonable manner for the

73

neutrino; it is perhaps more natural to suppose finite mass. This leads us to ask the reason why the neutrino mass is much smaller than the electron mass. The fact that the neutrino, unlike other particles, can have a Majorana form of mass term provides an attractive explanation to this problem.

In a modern language the mass term is a term that connects particle of left-handed chirality to that of right-handed chirality, i.e., it is written as $m\overline{\psi}_R\psi_L$. In the Weinberg-Salam theory, masses of particles are generated through the interaction $\mathcal{L} = f\overline{\psi}_R\psi_L\phi$, where $\psi$ is the matter field and $\phi$ the Higgs field [1]. When $\phi$ develops a vacuum expectation value $\langle\phi\rangle \neq 0$, the matter field acquires mass, given by $m = f\langle\phi\rangle$. This mechanism may also apply to the neutrino. We then must assume that $f_\nu/f_e < 10^{-5}$. For the neutrino, however, another form of the mass term is possible: $M_\nu\overline{(\psi_L)^c}\psi_L$, where $(\psi_L)^c$ is a charge conjugation of $\psi_L$ and has right-handed chirality. This is called a Majorana mass term [3]. In the Weinberg-Salam theory, the simplest form of the Majorana interaction is written

$$\mathcal{L} = \frac{g}{M}\overline{(\psi_L)^c}\psi_L\phi\phi, \qquad (1)$$

where $M$ is introduced to make $g$ dimensionless. This term is non-renormalizable, and thus taken to be an effective interaction [4], indicating a new physics at energy scale $M$. With (1), the Majorana mass is given by

$$M_\nu = \frac{g}{M}\langle\phi\rangle^2. \qquad (2)$$

Let us remember the theory of $\beta$ decay, a classical example of the effective interaction. The interaction is given by $\mathcal{L} = G_F(\overline{p}n)(\overline{e}\nu_e)$, where $G_F$ has dimension $M^{-2}$. If one calculates the cross section for $\nu_e + n \to p + e$, it increases with energy $E$ as $\sigma \sim G_F^2 E^2$. On the other hand, unitarity dictates that $\sigma$ be smaller than $2\pi/E^2$. The consistently requires that a new physics appears at the energy scale $E \sim G_F^{-1/2} \sim 100$GeV, and indeed this is realized by W and Z bosons. In the example that concerns us, we see a similar situation: the cross section for $\nu\phi \to \nu^c\phi$ stays constant $\sigma = \pi g^2/M^2$ as energy $E$ increases, while unitarity requires that it be smaller than $2\pi/E^2$. If one demands that this inequality hold up to the energy corresponding to the Planck mass $E = M_{pl} = 1.2 \times 10^{19}$GeV, one obtains that $g/M < 2\pi/M_{pl}$ or $M_\nu < 10^{-5}$eV with the aid of (2). This means that if one would find a neutrino mass, which is larger than $10^{-5}$eV, it indicates a new physics before gravity sets on at the Planck mass[2]. Indeed, in grand unified theories of particle interactions, the neutrino mass usually appears in a manner in agreement with this argument.

Let us mention another interest in neutrinos as a probe of new interactions. One particularly interesting example is the magnetic dipole moment (or transition moment) of the neutrino. The Weinberg-Salam theory predicts that it is $\mu_\nu \simeq 3 \times 10^{-19}(m_\nu/1 \text{ eV})\mu_B$ where $\mu_B$ is the Bohr magneton [5]. On the other hand, the limits obtained from laboratory experiments are of the order of a few $\times 10^{-10}\mu_B$ and those from stellar physics or cosmology are $O(10^{-11}\mu_B)$, an ample gap between the two values. The smallness of the predicted $\mu_\nu$ is a consequence of the chiral-symmetric nature of the standard electroweak theory and the value of $\mu_\nu$ is quite sensitive to chirality-breaking interactions that often appear in various unified theories [6].

## III. NEUTRINOS IN COSMOLOGY

In the early Universe neutrinos are in thermal equilibrium with other particles. When temperature drops to $T \approx 3\text{MeV}$, expansion of the Universe ($\sim T^2/M_{pl}$) becomes faster than the rate of weak interaction ($\sim G_F^2 T^5$), and neutrinos decouple from thermal equilibrium. The neutrino spectrum obeys the Fermi distribution with the effective temperature decreasing as the Universe expands; it is 1.96 K today [7]. The number density of neutrinos is

$$n_{\nu+\bar{\nu}} = \frac{3}{4}\frac{\zeta(3)}{\pi^2}T^3 = 115 \text{ cm}^{-3} \tag{3}$$

for $T = 1.96\text{K}$.

Let us now ask whether or not the neutrino spectrum is exactly the Fermi distribution [8,9]. After neutrinos decouple from other matter fields, $\gamma\gamma \to e^+e^-$ decouples at around 1MeV and $e^+e^- \to \gamma\gamma$ reheats the Universe (this is the origin of the disparity between 2.75K of the cosmic microwave background radiation and 1.96K of neutrino radiation.) Even after decoupling of neutrinos from equilibrium, occasional interactions of neutrinos with electrons that have temperature higher than neutrino temperature heat the neutrinos. This, together with the fact that the neutrino interaction is stronger for a higher energy, causes a distortion of the neutrino spectrum. It is computed by solving a kinetic equation.

$$(\frac{\partial}{\partial t} - Hp\frac{\partial}{\partial p})n_\nu(t,p) = S \tag{4}$$

where $S$ denotes collision terms for $\nu e \to \nu e$, $\nu\bar{\nu} \to e^+e^-$ etc, and $H$ is the expansion rate of the Universe $\dot{a}/a$. Writing the distortion as $n_\nu = n_\nu(\text{Fermi})[1 + \delta(E,t)]$, an approximate solution is given by

$$\delta(E,t) \approx 6 \times 10^{-4}\frac{E}{T}(\frac{11}{4}\frac{E}{T} - 3) \tag{5}$$

for $E \leq 0.7\text{MeV}$ [8]. This is quite a large correction; the energy stored in the distortion amounts to 3%. A consistent result was also obtained in Ref.[9] using a numerical approach. This distortion is compared to the upper limit on the energy stored in a possible distortion of the photon spectrum, $< 0.01\%$ from the COBE satellite [10].

Let us discuss neutrino effects in cosmology in order:

A. Primordial nucleosynthesis

Nucleosynthesis of light elements ($^4\text{He}$, $d$, $^3\text{He}$, $^7\text{Li}$) is a direct consequence of the existence of the cosmic neutrino radiation [11]. After neutrinos decouple from other radiations at $T \approx 3\text{MeV}$, the reactions $n + e^+ \rightleftharpoons \nu + p$, $p + e^- \rightleftharpoons \nu + n$ are still in equilibrium down to $T_\beta \approx 0.7\text{MeV}$, at which the neutron to proton ratio $n/p$ freezes (except for a decrease of $n$ due to its decay.)

When $T$ drops to 0.1 MeV, where photodisintegration of deuterium decouples, light elements are synthesized, the predominant part of neutrons forming $^4\text{He}$. The abundance of $^4\text{He}$ is approximately $Y \simeq 2(n/p)/(1 + n/p)$. The calculation of nucleosynthesis is well converged among the authors and seems very solid (see [12] for a recent calculation).

The distortion of the neutrino spectrum mentioned above modifies the terms that appear in the equation that governs the $n/p$ ratio by 1%, but a cancellation takes place among the increments at around $T \simeq 0.6$ MeV and the net effect in the $n/p$ ratio is very small ($\sim 10^{-4}$) [8,9].

A non-standard scenario of nucleosynthesis [13,14], that considers possible effects of inhomogeneous nucleon density over the space, now lacks the motivation: there is no evidence that QCD phase transition is first order: all lattice QCD simulations indicate the opposite [15]. We have thus no reasons to expect strong inhomogeneity in nucleon density.

While the current standard interpretation is that the standard calculation agrees very well with the observation [12], the observational situation does not seem to be as solid as the calculation. The recent most interesting indication is a possible large deuterium abundance found in a Lyman $\alpha$ cloud [16].

If the neutrino is massless and obeys the standard theory, nucleosynthesis is the only place where neutrinos are important in cosmology.

B. Dark matter in the Universe

If the neutrino has finite mass, we expect from (3) its contribution to mass density of the Universe to be

$$\rho_\nu = 115(m_\nu/\text{eV})\text{eV} \text{cm}^{-3}, \tag{6}$$

which is compared to mass density of baryons $\rho_B \approx 100$ eV cm$^{-3}$ as predicted from nucleosynthesis [12]. This means that relic neutrinos may dominate the Universe gravitationally, if $m_\nu$ exceeds a few eV. By requiring (6) be smaller than the critical mass density of the Universe, $\rho_{\text{crit}} = 10.5 h^2$ keV cm$^{-3}$ ($h$ is the Hubble constant in units of 100 km s$^{-1}$ Mpc$^{-1}$), a limit is obtained on the neutrino mass: $m_\nu < 90 h^2$ eV [17]. For $\nu_\mu$ and $\nu_\tau$ this is much stronger than the limits obtained from laboratory experiments.

If mass of the neutrino lies in between $5 - 90 h^2$ eV, it could be a candidate for the dark matter in the Universe. Such dark matter controls formation of the cosmic structure. A lot of studies were made in 1979 – 1984, but it has been realized that neutrino dark matter scenario is not so successful in understanding the observed structure: in particular galaxy formation, which takes place during collapse of larger scale objects, is too slow in such a universe, if primordial density fluctuations are constrained by anisotropies of the cosmic background radiation. Theorists have, then, invoked a different kind of dark matter, cold dark matter, which was not hot in the early Universe when it decouples from expansion of the Universe, to understand structure formation [18]; in fact this leads to quite a successful scenario of cosmic structure formation and it is even taken as a "standard model" by theorists. Structure formation with neutrinos dark matter has recently been revived as "mixed dark matter scenario", where 60-70% of dark matter is borne by neutrinos and the remaining part by some hypothetical "cold dark matter" [19]. While this scenario successfully explains the connection between the anisotropies discovered by COBE [20] and large-scale structure formation, formation of galaxies seems again too slow to be reconciled with the observation, the most important objection being the ubiquitous existence of galaxy-like objects (damped Lyman $\alpha$ systems) at redshift $z \simeq 3$ [21].

Nevertheless, massive neutrinos (especially $\tau$ neutrinos) are still a candidate for the dark matter. Very interesting neutrino oscillation experiments are now being carried out at CERN that explore the $\tau$ neutrino mass down to 5 eV, the lower limit for the $\tau$ neutrino being the dark matter, and to a quite small mixing angle [22].

C. Baryogenesis

The most widely accepted idea for baryogenesis is that baryons in the Universe were generated in its very early epoch by baryon number non-conservation induced by grand unification (GUT) of strong and electroweak interactions [23]. It has been shown that Sakharov's conditions [24], which are the necessary conditions to generate baryon number, can easily be satisfied with GUT. Unfortunately, we have seen no positive evidence for baryon number non-conservation associated with GUT by now. During the time, 'tHooft [25] pointed out the possibility that baryon number is violated even with the standard electroweak interactions due to a quantum effect called anomaly. Kuzmin, Rubakov and Shaposhnikov [26] then argued that a non-perturbative effect makes this baryon number violation very efficient at high temperatures, $T \approx 100$GeV, say. While this idea had not been received much attention for the first five years after the proposal, a large amount of effort made recently has established this effect to a convincing degree.

This anomalous baryon number non-conservation violates the combination of $B+L$ (baryon number plus lepton number), while it conserves $B-L$. Now, the problem is that it erases all pre-existing baryon numbers that respect $\Delta B = \Delta L$, as is predicted in the standard GUT, whereas it is not easy to generate baryon number with this process. There are by now two promising mechanisms that are considered to generate the baryon number in the Universe: (i) The baryon asymmetry is generated at the time of the electroweak phase transition [27]. For this to happen, this transition must be strong first order so that it provides enough of a departure from thermal equilibrium. The effect of $CP$ violation must also be large. Studies made so far, however, do not seem to give a large enough baryon asymmetry that agrees with the observation. The alternative scenario is that (ii) the baryon asymmetry is the result of a primordial lepton asymmetry generated during some epoch earlier than the electroweak phase transition. This lepton asymmetry is then converted into baryon asymmetry with the action of anomalous baryon number non-conservation at above the electroweak energy scale [28]. If we accept the Majorana mass explanation for the smallness of the neutrino mass, the existence of a heavy Majorana neutrino (at mass scale of the order of $M$ in eq.(1)) is the most likely consequence of the theory [29]. Such a heavy mass Majorana neutrino $\nu_H$ would decay into $\ell + \phi$ and $\bar{\ell} + \bar{\phi}$ [$\ell = (\nu, e)$], and there can be a disparity between the two decay modes due to $CP$ violation, generating net lepton number.

D. Excess uv background

If the neutrino has a mass $5 - 90h^2$ eV and also has a transition magnetic moment $\mu_\nu \simeq 10^{-13} - 10^{-15} \mu_B$, such a massive neutrino decays into a lighter neutrino $\nu$(light) by emitting a photon with lifetime longer than the age of the Universe, and the emitted photons would contribute to a diffuse uv background radiation in the night sky [30]. This large transition moment is not too unreasonable a value in varieties of unified theories of particle interactions [2].

E. Interplay with primordial magnetic fields

If the neutrino has a magnetic moment as large as $\geq 10^{-15} \mu_B$, there appears a possibility that neutrinos in the early Universe interact with a hypothetical primordial magnetic field, causing some observable consequence: if the product of the magnetic moment and the magnitude of the primordial magnetic field is larger than a certain value, a left-handed neutrino, which is in equilibrium with other radiation, flips into a right-handed partner that would contribute to increasing the speed of the expansion of the Universe. The nucleosynthesis argument leads to the limit $(\mu_\nu/10^{-15}\mu_B)$ $(B_0/10^{-11}\text{Gauss}) < 1$, where $B_0$ is the magnitude of the intergalactic magnetic field today. This results from a calculation of the refraction effect of neutrinos due to varieties of physical processes in a relativistic hot plasma [31]; a compete calculation is not yet available, however.

## IV. NEUTRINOS IN STELLAR PHYSICS

The neutrino production rate in hot plasma is generally proportional to a large power of temperature or plasma frequency. Neutrino luminosity becomes comparable to optical luminosity of stars when the central temperature or density reaches $T_c \sim 10^8 K$ or $\rho_c \sim 10^5 \text{g cm}^{-3}$. Therefore, neutrino emission is important only for stars later than helium ignition. In main sequence stars, emission of neutrinos is not energetically important. For the case with the sun, however, interest comes from the fact that solar neutrinos are directly observable, giving information on the interior of the sun and the propagation of neutrinos.

A. Stellar cooling

The importance of neutrinos in stellar cooling was first discussed by Pontecorvo [32]. The basic processes that are important in stellar plasma are [33] (a) photoneutrino production through the Compton-like process, $\gamma + e \to (\nu\bar\nu) + e$, (b) $e^+e^-$ annihilation into a $\nu\bar\nu$ pair, $e^+e^- \to \nu\bar\nu$, (c) plasmon decay $\gamma^* \to \nu\bar\nu$ via virtual $e^+e^-$ production, (d) neutrino bremsstrahlung $e + Z \to (\nu\bar\nu) + e + Z$ [34], and (e) electron capture, $e + (A, Z) \to (A, Z+1) + \nu$. Photoneutrino production is more important than other neutrino production mechanisms, when temperature and density are relatively low as in the core of a main sequence star. In this case, however, neutrino cooling is generally not important. $e^+e^-$ pair annihilation is important only in stars with a core of high ($T \gtrsim 10^{8.5}$K) temperature. In stars with a degenerate core, e.g., low mass helium burning stars and white dwarfs, plasmon decay is the dominant process of stellar cooling. Bremsstrahlung is important only in very dense cores.

The places where neutrino cooling is important are: (i) the critical mass for helium flash. This affects physical characteristics of horizontal branch stars (e.g., RR Lyr stars) for the case of Population II stars, and those of red clumps for Population I stars. (ii) Age of stars with a C, O burning core is basically determined by neutrino cooling. It is, however, difficult to observe this effect mainly due to uncertainties in the model of evolved stars. (iii) Cooling of hot white dwarfs. Neutrino cooling is much more important than photon cooling, and the cooling time scale is set by neutrino emissivity. This effect, which is relatively easy to detect, was used by Stothers (1970) [35] to set an upper and a lower limit on $\nu e$ interactions prior to the detection of $\nu e$ scattering in a laboratory by Reines et al. [36] in 1976. (iv) Stellar core collapse. More than 99% of energies involved in stellar core collapse are liberated by neutrino emission. Core collapse is triggered by process (e); process (b) dominates later phases of collapse. Neutrino transport plays an important role in the explosion; see [37] and also talk by

K. Sato [38]. (v) Cooling of neutron stars [39] is also determined by neutrino emission.

It would be interesting to ask whether some possible corrections to simple formulae used in conventional calculations cause any observable effects. This question is answered best by reviewing the work which attempts to set a limit on the neutrino magnetic moment—another interesting topic concerning neutrino physics as mentioned in Sect. I. If the neutrino has a finite magnetic moment, it enhances decay of plasmons. Therefore, constraints against excess cooling lead to a limit on the magnetic moment [40]. Motivated by the solar neutrino problem [41], a number of quantitative analyses [42-48] have been made to set a limit on the neutrino magnetic moment, as summarized in Table 1. This table shows that the strongest limit, hundred times stronger than the laboratory limit [49,50], is derived from He stars (low luminosity He stars) and cooling white dwarfs. It has been argued that the core mass at the He flash is constrained to as precise as $0.02 M_\odot$ using a stellar evolution model [43]. Using the fact that excess cooling through magnetic moment increases the core mass by $0.015 M_\odot$ per $10^{-12} \mu_B$, a few authors derived the best limit $(1-3) \times 10^{-12} \mu_B$ [43,44]. This limit may probably depend on the detail of models and receive uncertainties in comparing the model with the observation. It seems, however, that $\mu_\nu \simeq 5 \times 10^{-12} \mu_B$ is conservatively ruled out. A similar limit is also obtained from the luminosity function of hot white dwarfs.

It is easy to translate the limits on magnetic moment derived from He stars and white dwarfs into the effective strength of the interaction in units of the standard weak interaction value, since in both cases plasmon decay dominates among the neutrino cooling processes; i.e., the cooling rate due to magnetic moment is related to that of standard plasmon decay as

$$Q_{mag}/Q_{pl} = 2\pi^2 \alpha^2 (\mu_\nu/\mu_B)^2 \, G_F^{-2} \, m_e^{-2} \, \omega_{pl}^{-2} \,, \qquad (7)$$

**Table 1** Constraints on neutrino magnetic moment

| | | | |
|---|---|---|---|
| laboratory ($\nu_e$) | $< 4 \times 10^{-10} \mu_B$ | $\overline{\nu_e}\, e \to \overline{\nu_e}\, e$ | [48] |
| laboratory ($\nu_\mu$) | $< 9 \times 10^{-10} \mu_B$ | $\nu_\mu\, e \to \nu_\mu\, e$ | [49] |
| stellar evolution | | | |
| sun | $< 13 \times 10^{-10} \mu_B$ | age vs. luminosity | [41] |
| He star | $< 0.1 \times 10^{-10} \mu_B$ | red clump HR diagram | [41] |
| | $< 0.03 \times 10^{-10} \mu_B$ | core mass at He flash | [42] |
| | $(< 0.01 \times 10^{-10} \mu_B)$ | core mass at He flash | [43] |
| white dwarf | $< 0.4 \times 10^{-10} \mu_B$ | luminosity function | [41] |
| | $< 0.05 \times 10^{-10} \mu_B$ | luminosity function | [44] |
| neutron star | $< 5000 \times 10^{-10} \mu_B$ | cooling curve | [45] |
| SN1987A | $(< 0.01 \times 10^{-10} \mu_B)$ | neutrino burst (model fixed) | [46] |
| cosmology | | | |
| nucleosynthesis | $< 0.6 \times 10^{-10} \mu_B$ | He abundance | [47] |
| primordial mag. field | $< 10^{-15} \mu_B (B_0/10^{-11})^{-1}$ | He abundance | [30] |

where $G_F$ is the weak decay constant and $\omega_{pl}$ is the plasma frequency [40]. For instance,

$\mu = 0.01 \times 10^{-10} \mu_B$ corresponds to 3 times the standard weak interaction value in a He core, and $\mu = 0.05 \times 10^{-10} \mu_B$ also corresponds to 3 × the standard value in white dwarfs. Our conclusion is : (i) if the standard calculation is modified by a factor of 3, the effect is marginally visible, (ii) if by a factor of 10, there appears an unambiguous observable effect, and (iii) a possibility of modification by a factor of 50–100 is excluded.

A weak dependence of the observable effect on the accuracy of the calculations in core burning stars is a result of a negative feedback effect: if neutrino emissivity is larger than is thought, a slight decrease in temperature leads to a significant suppression in neutrino emissivity. In this sense, an effect in white dwarfs is in principle easier to be detected, since there is no heat source. We cannot even distinguish the Weinberg-Salam theory [51,52] from the old $V - A$ theory [33] in any of the observations of stellar cooling effects.

B. Solar neutrino problem

After the report from the Kamiokande solar neutrino experiment, the long-standing solar neutrinos problem [53] has come into a new stage. The important point is that the Kamiokande and the classical Homestake experiment are sensitive to neutrinos with different energy ranges and yield spectroscopic information on the neutrino flux. The conclusion obtained by combining the two experiments is dramatic: solar neutrino problem is not a problem of astrophysics of the sun, nor a problem with thermonuclear reactions. It is a problem of particle physics [54]. This conclusion is obtained by noting that the Homestake experiment [55], which measures $^{37}Cl(\nu, e^-)^{37}Ar$ and is sensitive to intermediate energy neutrinos from $^7Be$ and $pep$ etc. as well as those form $^8B$, shows the strongest neutrino flux suppression (28%), whereas the Kamiokande experiment, which measures only a higher energy tail of neutrinos from $^8B$, sees only 50% suppression [56]. If the empirical $^8B$ neutrino flux as determined from the Kamiokande experiment is put into the predicted solar neutrino capture rate of the $^{37}Cl$ experiment, we already have $3.1 \pm 0.5$ SNU, marginally consistent with the experimental value $2.2 \pm 0.2$ SNU. Furthermore, $^{37}Cl$ receives a contribution of $1.8 \pm 0.3$ SNU, that depends little on calculations, from $^7Be$, $pep$ and the CNO cycle, giving $4.9 \pm 0.6$ SNU as a sum, which is $4.5\beta P$ away from the experiment. This means that the two experiments cannot be reconciled, unless intermediate energy neutrinos are very strongly suppressed. We note, however, that intermediate energy neutrinos arise from intermediate steps of the thermonuclear reactions in the sun (see Fig. 1). It may not be impossible to reduce the flux from the last step ($^8B$), for example by reducing temperature of the centre of the sun, or by reducing the $^7Be(p,\gamma)^8B$ reaction rate [53], but it seems extremely hard to reduce neutrinos from $^7Be$ to a large extent, while keeping a reduction of $^8B$ neutrinos production to a modest amount.

The results of the long awaited experiment using $^{71}Ga$ by SAGE [57] and GALLEX [58] also corroborate our conclusion: considering the fact that $pp$ neutrino flux is highly correlated with solar luminosity and hence it is difficult to reduce it, the Ga results are consistent with a strong suppression of intermediate energy neutrinos; the observed suppression, 60%, is decomposed into 55% from $pp$ neutrinos (no suppression) and 5% from $^8B$ neutrinos (1/2 suppression); $^7Be$ neutrinos etc. are strongly suppressed.

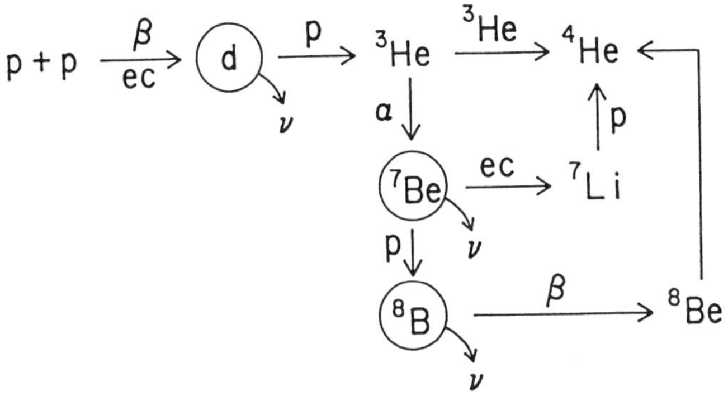

Fig. 1. The main thermonuclear reaction chain (*pp* chain) in the sun.

There are two reasonable particle physics explanations that could account for the strong suppression of intermediate energy neutrinos. The first is to invoke the idea of neutrino oscillation while propagating from the sun to the earth, as has been conjectured for many years [59]. This explanation, however, requires the maximum mixing angle between two neutrino species and a fine tuning of neutrino mass parameters, i.e., a coincidence of the harmonics of the oscillation lengths ($E/\Delta m^2$) with the sun-earth distance for intermediate energy neutrinos ($\Delta m^2$ is the difference of neutrino mass squared). The former seems to be rather unnatural from the viewpoint of particle physics models; for the latter, we must assume that this just happens accidentally. As the second possibility, a very attractive solution has been proposed by Mikheev and Smirnov [60], based on the old idea of Wolfenstein [61]. The solution uses the fact that $\nu_e e \to e \nu_e$ scattering in matter due to the charged current interaction induces an effective mass (refractive index) for electron neutrinos, while this effect is absent for $\nu_\mu$ and $\nu_\tau$. (The neutral current interaction does not cause any effects, since it acts equally for all neutrinos.) Mikheev and Smirnov have pointed out that if a phase from this induced mass cancels that arising from the intrinsic neutrino mass difference, while neutrinos propagate from the centre of the sun to the outer envelope, a complete conversion from $\nu_e$ to $\nu_\mu$ (or $\nu_\tau$) takes place for a wide range of the intrinsic neutrino mixing parameter ($\nu_\mu$ does not interact with $^{37}$Cl, and the cross section of $\nu_\mu e$ scattering is 1/7 that of $\nu_e e$ scattering). This is a level crossing effect, which is familiar to us in two level problems of quantum mechanics, known as the Landau-Zener effect[62]. In this explanation a fine turning is not required for either mixing parameter or neutrino mass difference to cause a large reduction of the neutrino flux. This mechanism explains naturally the suppression of intermediate energy neutrinos, leaving the low energy $pp$ neutrino flux intact and high energy $^8$B neutrinos only loosely suppressed. Neutrino flux reduction observed in the four experiments (Homestake, Kamiokande, GALLEX, SAGE) leads to a solution $\Delta m^2 = 0.3\text{--}1 \times 10^{-5}(\text{eV})^2$ and $\sin^2 \theta_{\nu_e \nu_i} \simeq 0.001$ ($\nu_i = \nu_\mu$ or $\nu_\tau$). (Another solution $\Delta m^2 = 0.4\text{--}3 \times 10^{-5}(\text{eV})^2$ and $\sin^2 \theta_{\nu_e \nu_i} \simeq 0.7$ is also allowed, but it is allowed only thanks to large errors of the experiments.) Since hierarchy is expected among neutrino masses, this indicates $m_{\nu_e} \ll m_{\nu_\mu} = 2\text{--}5 \times 10^{-3}$eV and a new physics energy scale ($M$ in eq. (1)) of $10^{11}$GeV, as the most natural possibility. This scale is what naturally appears in SO(10) Grand Unification [63], and it also explains the Weinberg angle—mixing angle between the electromagnetic and weak interactions, for which a "success" of supersymmetric GUT has been enthusiastically advocated [64]. We are also left with a reasonable chance for $\nu_\tau$ being a dark matter candidate: we see that $\tau$ neutrino could weigh $\sim 10$eV, by scaling mass of $\nu_\mu$ by a factor of the quark mass ratio squared $(m_t/m_c)^2$.

## V. CONCLUSION

The problem with mass of the neutrino is concerned with one of the most fundamental issues in particle physics, in that it would explore new physics beyond the energy scale that is accessible in laboratories. The most significant output to date from the interface of particle physics and astrophysics is the indication for finite neutrino mass from the solar neutrino experiments: in so far as the results from the Homestake and Kamiokande experiments remain unchanged, one of the two explanations with massive neutrinos discussed in B of Sect. IV seems to be compelling. This is a very important hint to understand unification of particle interactions. A promising possibility regarding neutrino mass is that the $\tau$ neutrino may have mass that contributes significantly to mass density of the Universe. While cosmic structure formation with massive neutrino does not provide quite attractive scenario, it is still important to explore the possibility that $\tau$ neutrinos bear all or a significant portion of the dark matter in the Universe. With massive neutrinos the possibility that the baryon asymmetry in the Universe was generated through the lepton number violation that had been taken place in the heavy neutrino sector provides us with an attractive scenario.

Neutrino interactions are also very important for the evolution of stars in their late stage. Unfortunately, the response of the evolutionary effect to a change of neutrino cooling rate is generally small due to negative feedback effects in stars; we do not expect any observable effects even if the cooling rate given by the $V - A$ theory is replaced with that by the Weinberg-Salam theory. That is, while neutrino emissivity depends on the state of matter and it can in principle be a good probe for exploring matter under unusual conditions, it will be a difficult task to find any observable effects, unless a modification induced by environmental effects changes the cooling rate at least by a factor of three.

## REFERENCES

1. S. Weinberg, Phys. Rev. Lett. **19**, 1264 (1967); A. Salam, in *Elementary Particle Theory, Proc. of the 8th Nobel Symposium* Aspenäsgården, 1968, ed. by N. Svartholm (Almqvist and Wiksell, Stockholm, 1968) p. 367.
2. M. Fukugita and T. Yanagida, in *Physics and Astrophysics of Neutrino*, ed. by M. Fukugita and A. Suzuki (Springer, Tokyo, 1994) p. 1.
3. E. Majorana, Nuovo Cim. **14**, 171 (1937).
4. S. Weinberg, in *Proc. 23rd International Conference on High Energy Physics*, Berkeley 1986, ed. by S.C. Loken (World Scientific, Singapore, 1987), vol. I, p. 271.
5. B.W. Lee and R.E. Shrock, Phys. Rev. **D16**, 1444 (1977); W.J. Marciano and A.I. Sanda, Phys. Lett. **67B**, 303 (1977).
6. M. Fukugita and T. Yanagida, Phys. Rev. Lett. **58**, 1807 (1987).
7. e.g., S. Weinberg, *Gravitation and Cosmology*, (John Wiley, New York, 1972); P.J.E.

Peebles, *Physical Cosmology* (Princeton University Press, Princeton, 1971).
8. A.D. Dolgov and M. Fukugita, Phys. Rev. D**46**, 5378 (1992).
9. S. Dodelson and M. S. Turner, Phys. Rev. D**46**, 3372 (1992).
10. J.C. Mather et al., Astrophys. J. **420**, 439 (1994).
11. P.J.E. Peebles, Astrophys. J. **146**, 542 (1966); R.V. Wagoner, W.A. Fowler and F. Hoyle, Astrophys. J. **148**, 3 (1967); H. Sato, Prog. Theor. Phys. **38**, 1083 (1967).
12. T.P. Walker et al., Astrophys. J. **376**, 51 (1991).
13. J.H. Applegate and C.J. Hogan, Phys. Rev. D**30**, 3037 (1985).
14. For a review, R.A. Malaney and G.J. Mathews, Phys. Rep. **229**, 145 (1993).
15. M. Fukugita and C.J. Hogan, Nature **354** 17, (1991).
16. A. Songaila, L.L. Cowie, C.J. Hogan and M. Rugers, Nature **368**, 599 (1994).
17. R. Cowsik and J. McClelland, Phys. Rev. Lett. **29**, 669 (1972); see also S.S. Gershtein and Ya. B. Zeldovich, Pisma Zh. Eksp. Teor. Fiz. **4**, 174 (1966) [JETP Lett. **4**, 120 (1966)].
18. P.J.E. Peebles, Astrophys. J. **263**, L1 (1982); G.R. Blumenthal, S.M. Faber, J.R. Primack, and M.J. Rees, Nature **311**, 517 (1984); Frenk, C. S., White, S.D.M., Davis, M. and G. Efstathiou, Astrophys. J. **327**, 507 (1988).
19. A. Klypin, J. Holtzman, J. Primack and E. Regös, Astrophys. J. **416**, 1 (1993).
20. G.F. Smoot et al., Astrophys. J. **396**, L1 (1992).
21. H.J. Mo and J. Miralda-Escudé, preprint IASSNS-AST-94-7 (1994).
22. K. Niwa, in *Physics and Astrophysics of Neutrinos*, ed. by M. Fukugita and A. Suzuki (Springer, Tokyo, 1994) p. 520.
23. A.Yu. Ignatiev, N.V. Krasnikov, V.A. Kuzmin and A.N. Tavkhelidze, in *Neutrino 77*, ed by M.A. Markov et al. (Nauka, Moscow, 1978) vol. II, p. 293; Phys. Lett. **76B**, 436 (1978); M. Yoshimura, Phys. Rev. Lett. **41**, 281 (1978).
24. A.D. Sakharov, Pisma Zh. Eksp. Teor. Fiz. **5**, 32 (1967) [JETP Lett. **5**, 24 (1967)].
25. G. 'tHooft, Phys. Rev. Lett. **37**, 8 (1976).
26. V.A. Kuzmin, V.A. Rubakov and M.E. Shaposhnikov, Phys. Lett. **155B**, 36 (1985).
27. M.E. Shaposhnikov, Nucl. Phys. B**287**, 757 (1987); L. McLerran, Phys. Rev. Lett. **62**, 1075 (1985); A.G. Cohen, D.B. Kaplan and A.E. Nelson, Nucl. Phys. B**349**, 727 (1991); M. Shaposhnikov, in *Lattice 91*, Nucl. Phys. B (Proc. Suppl.) **26**, 78 (1992) (review); (review); P. Huet and E. Sather, preprint SLAC-6479 (1994).
28. M. Fukugita and T. Yanagida, Phys. Lett. B**174**, 45 (1986).
29. T. Yanagida, in *Proc. of the Workshop on the Unified Theory and Baryon Number in the Universe*, ed. by O. Sawada and A. Sugamoto (KEK report 79-18, 1979), p. 95; M. Gell-Mann, P. Ramond and R. Slansky, in *Supergravity*, ed. by P. van Nieuwenhuizen and D.Z. Freedman (North-Holland, Amsterdam, 1979), p. 315.
30. D.A. Dicus, E.W. Kolb and V.L. Teplitz, Phys. Rev. Lett. **39**, 168 (1977); K. Sato and M. Kobayashi, Prog. Theor. Phys. **58**, 1775 (1977); J.E. Gunn et al., Astrophys. J. **223**, 1015 (1978); A. De Rujúla and S.L. Glashow, Phys. Rev. Lett. **45**, 942 (1980); J. Murthy and R.C. Henry, Phys. Rev. Lett. **58**, 1581 (1987).
31. M. Fukugita, D. Nötzold, G. Raffelt and J. Silk, Phys. Rev. Lett. **60**, 879 (1988); D. Nötzold and G. Raffelt, Nucl. Phys. B**307**, 924 (1988).
32. B.M. Pontecorvo, Zh. Eksp. Teor. Fiz. **36**, 1615 (1959) [Sov. Phys. JETP **9**, 1148 (1959)].
33. G. Beaudet, V. Petrosian and E.E. Salpeter, Astrophys. J. **150**, 979 (1967).
34. G.G. Festa and M.A. Ruderman, Phys. Rev. **180**, 1227 (1969).
35. R.B. Stothers, Phys. Rev. Lett. **24**, 538 (1970).

36. F. Reines, H.S. Gurr and H.W. Sobel, Phys. Rev. Lett. **37**, 315 (1976).
37. H.A. Bethe, in Proc. of Enrico Fermi Summer School, Course XCI, ed. by R.A. Ricci and A. Molinari (North-Holland, Amsterdam, 1986), p. 181.
38. K. Sato, T. Shimizu and S. Yamada, these proceedings.
39. J.N. Bahcall and R.A. Wolf, Phys. Rev. **B140**, 1452 (1965); S. Tsuruta and A.G.W. Cameron, Nature **207**, 364 (1965); S. Tsuruta and K. Nomoto, Astrophys. J. **312**, 711 (1987).
40. J. Bernstein, M. Ruderman and G. Feinberg, Phys. Rev. **132**, 1227 (1963); P. Sutherland et al., Phys. Rev. **D13**, 2700 (1976).
41. M.B. Voloshin, M.I. Vysotski and L.B. Okun, Zh. Eksp. Teor. Fiz. **91**, 754 (1986) [Sov. Phys. JETP **64**, 446 (1986)]; The experimental motivation is given by R. Davis, Jr. in *Proc. of Seventh Workshop on Grand Unification, ICOBAN 86*, ed. by J. Arafune (World Scientific, Singapore, 1987), p. 237.
42. M. Fukugita and S. Yazaki, Phys. Rev. **D36**, 3817 (1987).
43. G.G. Raffelt, Astrophys. J. **365**, 559 (1990).
44. V. Castellani and S. Degl'Innocenti, Astrophys. J. **402**, 574 (1993).
45. S.I. Blinnikov, Moscow preprint ITEP-88-19 (1988).
46. N. Iwamoto, L. Qin, M. Fukugita and S. Tsuruta, Toledo preprint (1994).
47. J.M. Lattimer and J. Cooperstein, Phys. Rev. Lett. **61**, 23 (1988); D. Nötzold, Phys. Rev. **D38**, 1658 (1988).
48. J. Morgan, Phys. Lett. **102B**, 247 (1981).
49. W.J. Marciano, cited in Particle Data Group, Phys. Lett. B **239**, 1 (1990); P. Vogel and J. Engel, Phys. Rev **D39**, 3378 (1989).
50. L.A. Ahrens et al., Phys. Rev. Lett. **D41**, 3297 (1990).
51. D.A. Dicus, Phys. Rev. **D6**, 941 (1972).
52. H. Munakata, Y. Kohyama and N. Itoh, Astrophys. J. **296**, 197 (1985); P.J. Schinder et al., Astrophys. J. **313**, 531 (1987); N. Itoh, these proceedings.
53. J.N. Bahcall, *Neutrino Astrophysics* (Cambridge University Press, Cambridge, 1989); J.N. Bahcall and R. Davis, Jr., in *Essays in Nuclear Astrophysics*, ed. by C.A. Barns et al. (Cambridge University Press, Cambridge, 1982) p.243.
54. J.N. Bahcall and H.A. Bethe, Phys. Rev. Lett. **65**, 2233 (1990); Phys. Rev. **D47**, 1298 (1993). M. Fukugita and T. Yanagida, Mod. Phys. Lett. A**6**, 645 (1991).
55. R. Davis, Jr. in *Frontiers of Neutrino Astrophysics*, ed. by Y. Suzuki and K. Nakamura (Universal Academy Press, Tokyo, 1993), p.47.
56. K.S. Hirata et al., Phys. Rev. Lett. **63**, 16 (1989); Phys. Rev. **D44**, 2241 (1991).
57. A.I. Abazov et al., Phys. Rev. Lett. **67**, 3332 (1991); T. J. Bowles, in *Proc. Neutrino 92*, Granada 1992 (to be published).
58. P. Anselmann et al., Phys. Lett. **B285**, 376 (1992); **B327**, 377 (1994).
59. B. Pontecorvo, Zh. Eksp. Teor. Fiz. **53**, 1717 (1968) [Sov. Phys. JETP **26**, 984 (1968)]; V. Barger, K. Whisnant and R. J. N. Phillips, Phys. Rev. **D24**, 538 (1981).
60. S.P. Mikheev and A.Yu. Smirnov, Yad. Fiz. **42**, 1441 (1985) [Sov. J. Nucl. Phys. **42**, 913 (1985)].
61. L. Wolfenstein, Phys. Rev. **D17**, 2369 (1978).
62. L. Landau, Phys. Zeits. d. Sowjetun. **2**, 46 (1932); C. Zener, Proc. Roy. Soc. London, A**137**, 696 (1932).
63. M. Fukugita and T. Yanagida, ref. 54.
64. P. Langacker and M. Luo, Phys. Rev. **D44**, 817 (1991); U. Amaldi, W. de Boer and H. Fürstenau, Phys. Lett. **B260**, 447 (1991).

II.2

# NEUTRINO EMISSION FROM PROTONEUTRON STARS

H. Suzuki[†] and K. Sumiyoshi[‡]

[†]KEK: National Laboratory for High Energy Physics, Tsukuba, Ibaraki 305, Japan
[‡]RIKEN: The Institute of Physical and Chemical Research
Wako, Saitama 351-01, Japan

Neutrino emission from protoneutron stars and quasistatic evolution of protoneutron stars are investigated by numerical simulations. There are many uncertainties in input physics such as nuclear equation of state and neutrino interaction rates with high density matter. Effects of nuclear symmetric energy upon the protoneutron star cooling are particularly discussed.

## I. INTRODUCTION

It is believed that two types of mechanisms exist for the supernova explosions: one is due to thermonuclear destruction of a white dwarf and the other is driven by core collapse of a massive star. Neutron stars are considered to be formed at the latter. We focus on neutrino emission at the birth of neutron stars.

A core of massive star ($M \gtrsim 8M_\odot$) becomes unstable and begins to collapse. When the central density exceeds nuclear density, the equation of state stiffens suddenly and the collapse is halted. The bounced inner core launches a shock wave into the outer core. Although it is still unsettled whether the neutrino heating of the matter or some instabilities play critical roles for the explosion, blowing off the envelope by the shock wave is the supernova explosion.

At the same time of the shock propagation, the bounced inner core settles into a hydrostatic configuration in the dynamical time scale of millisecond. The matter of the falling outer core is swept and decelerated by the shock wave, accretes onto the unshocked inner core and forms the shocked outer mantle successively. Thus the protoneutron star which consists of the cool (entropy per nucleon $S \sim O(1)$) unshocked inner core and the hot ($S \sim O(10)$) shocked mantle grows until the original stellar core explodes and the matter accretion ceases. Since the inner core of protoneutron star contains many leptons and protons ($\sim 30\%$), we cannot call it a *neutron star*. It is neutrinos that carry out trapped lepton number and thermal energy from the protoneutron star in the neutrino diffusion time scale of $O(10)$ sec. Burrows & Lattimer [1986] and Burrows [1988] investigated this transition phase from a protoneutron star to a normal neutron star with a simple LTEFLD (local thermodynamic equilibrium flux limited diffusion) neutrino transfer method and we [Suzuki, 1990, 1993] performed more detailed numerical simulations using MGFLD (multigroup flux limited diffusion) scheme for neutrino transfer. Fig. 1 shows results of our numerical simulation on the quasistatic evolution of a protoneutron star whose initial model was prepared using numerical results (0.4 sec after bounce) of Mayle & Wilson's dynamical simulations [Wilson and Mayle, 1989]. This cooling stage of a protoneutron star can be divided into two stages; the rapid cooling stage of the shocked outer mantle and the succeeding

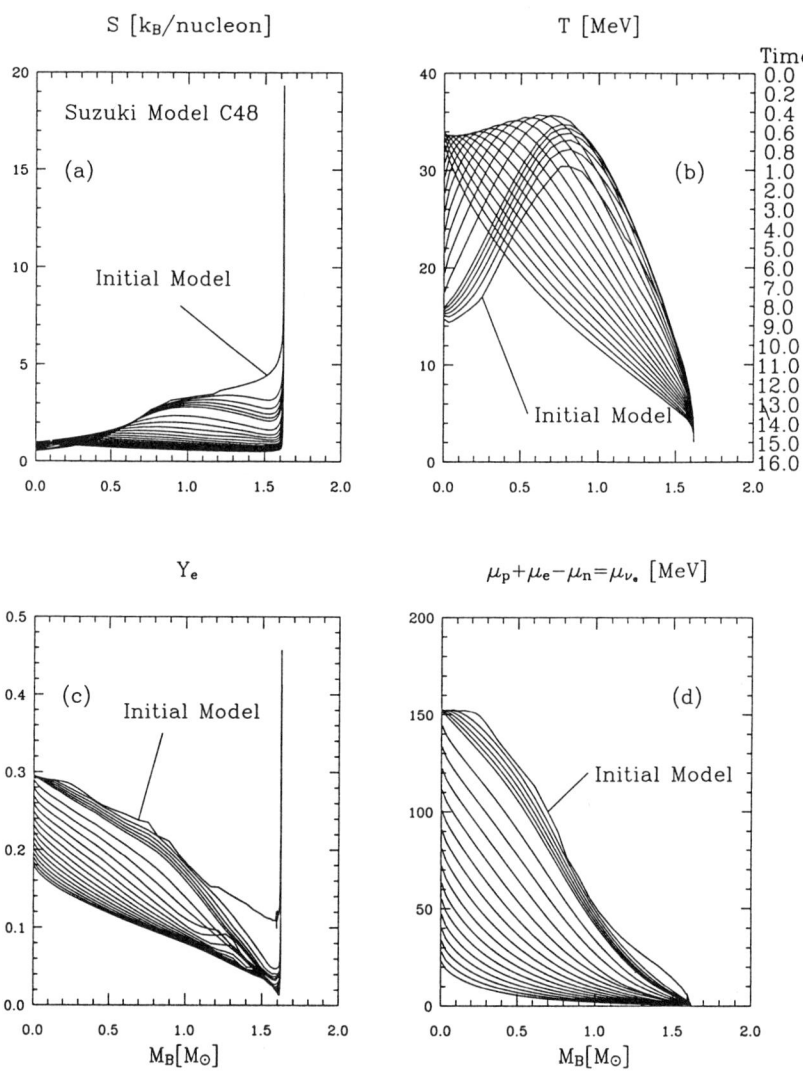

Figure 1. Evolution of the protoneutron star into a 'neutron star'. (a) Snapshots of entropy profile at various times ($t = 0 \sim 16$ sec). Initial profile shows that entropy of shocked outer mantle is much higher than unshocked inner core. (b) Snapshots of temperature profile. Initial peak is not at the center but in the shocked region. (c) Snapshots of electron fraction profile. (d) Snapshots of $\mu_p + \mu_e - \mu_n$.

cooling stage of the inner core. At first, the shocked outer mantle cools down and deleptonizes with emitting neutrinos (Figs. 1a, 1d) because, inside the outer mantle, the density is relatively low and therefore the mean free path of neutrinos is relatively long. The decrease of entropy causes the contraction of the outer mantle with a temperature increase (Fig. 1b). Since, initially, there is a region where the temperature gradient is positive, the heat flux flows inwards there and the entropy of the inner core increases. After about one second, the radius of protoneutron star becomes that of the cold neutron star ($\sim 10$ km) while the initial radius of protoneutron star is about 100 km. Furthermore, after about 10 sec, the temperature profile becomes monotonically decreasing from the center to the surface and the central region also starts to cool down. The whole protoneutron star cools down and deleptonizes gradually to be an ordinary cold 'neutron star' in which neutrino-less $\beta$-equilibrium among neutrons, protons and electrons ($\mu_p + \mu_e = \mu_n$) has been achieved (Fig. 1d). Note that even in the cold 'neutron star' there remain electrons and protons of an amount required for $\beta$-equilibrium.

## II. METHOD OF NUMERICAL SIMULATIONS

Quasistatic evolutions of the spherically symmetric protoneutron stars with neither additional mass accretion nor convection are simulated numerically by solving the Oppenheimer-Volkoff equation with Henyey type method. In these simulations, we use the equation of states (hereafter EOS) for the dense matter at finite temperature calculated with Hartree-Fock method by Wolff [Hillebrandt and Wolff, 1985] (Model C48) and by Sumiyoshi *et al.* [1994]. As for the neutrino transfer, we adopt the multigroup flux limited diffusion scheme [e.g., Bruenn, 1985]. The energy dependence of neutrino transport coefficients can be managed in the multigroup scheme. The flux limiter should be introduced in order to manage the transparent regime with the diffusion approximation. We include the general relativistic effects such as time dilation and red shift of neutrino energy as well. Although there are many types of flux limiters, we adopt Mayle and Wilson's flux limiter [Mayle *et al.*, 1987] in this work. When we use the other flux limiter such as Bruenn's [1985] or Levermore and Pomraning's [1981], the thermal history of protoneutron star and the luminosity curve of the neutrinos are nearly identical to the results with Mayle and Wilson's flux limiter. Although the quantitative values of the mean energy and the peak width of the emergent neutrino flux are affected a bit, the qualitative features discussed below are not influenced by the adopted flux limiter. Three neutrino species, $\nu_e$, $\bar{\nu}_e$ and '$\nu_\mu$', are treated separately, where '$\nu_\mu$' represents the average of $\nu_\mu, \bar{\nu}_\mu, \nu_\tau$ and $\bar{\nu}_\tau$. This is a good approximation in the case that we can neglect the existence of $\mu$ and $\tau$ leptons because of the low temperature ($\lesssim 100$ MeV).

The following neutrino interactions are included as opacity sources or collision terms in the neutrino transfer equations.

$$\begin{array}{llll}
p\, e^- \longleftrightarrow n\, \nu_e, & n\, e^+ \longleftrightarrow p\, \bar{\nu}_e, & A\, e^- \longleftrightarrow A'\, \nu_e \\
e^-\, e^+ \longleftrightarrow \nu\, \bar{\nu}, & \text{plasmon} \longleftrightarrow \nu\, \bar{\nu}, & N\, N' \longleftrightarrow N\, N'\, \nu\, \bar{\nu} \\
N\, \nu \longleftrightarrow N\, \nu, & A\, \nu \longleftrightarrow A\, \nu, & e^\pm\, \nu \longleftrightarrow e^\pm\, \nu
\end{array}$$

where $\nu$ represents all species of neutrinos, A is a representative heavy nucleus, and N is either a proton or a neutron. Most of the interaction rates are taken from Bruenn [1985] with some modifications. At present, many body effects on neutrino opacity are not included except for the multiple scattering suppression effects on nucleon bremsstrahlung process [Raffelt and Seckel, 1991]. Exchange of energy and lepton number between the matter and neutrinos due to the above interactions and the neutrino transport drive the evolution of the protoneutron star.

## III. RESULTS AND DISCUSSION

Uncertainties in the properties of nuclear matter prevent us to construct a precise model of supernova neutrinos. Conversely, however, observations of supernova neutrinos provide us an opportunity to investigate the properties of nuclear matter. Effects of EOS for the nuclear matter upon the supernova neutrinos were studied [e.g., Burrows, 1988]. A soft EOS leads to a large binding energy of neutron star, equivalently large total energy of neutrino burst emitted ( $\int L_\nu dt$ ). It results also in a core with high density and high temperature. Average energy of neutrinos tends to be high and the neutrino diffusion time scale becomes long.

As one of the other properties of nuclear matter than its stiffness, we investigate effects of nuclear symmetric energy upon the protoneutron star cooling. We construct two EOSs with different values of symmetric energy, $W_{sym}$ and compare the numerical results of models using the two EOSs. Both EOSs are calculated using relativistic mean field theory with effective Lagrangian. We use a parameter set for the Lagrangian (TM1) which was obtained by Sugahara and Toki [1993] to fit various experimental data of nuclei including unstable nuclei. In order to construct another EOS, we artificially change the value of coupling constant of $\rho$ meson in the parameter set. For the resultant EOS (TMS), $W_{sym}$ is 28 MeV while the original EOS (TM1) has $W_{sym} = 37$ MeV. Note that stiffness of the two EOSs is similar. In Fig. 2, the initial profiles of the two protoneutron stars with the two EOSs are shown. Density profile and temperature profile are very similar for the same inputs ($S$ and $Y_e$). Difference in $W_{sym}$ can be seen in the differences in the lepton fraction and in the chemical potential of $\nu_e$. Fig. 3 shows the profiles at $t = 15$ sec. Density profiles are still similar because of the similar stiffness, but temperature and electron fraction have different profiles. In particular, lepton fraction for TMS ($W_{sym} = 28$ MeV) is smaller than TM1 ($W_{sym} = 37$ MeV) while it is opposite at $t = 0$ sec. Deleptonization proceeds faster for $W_{sym} = 28$ MeV than for $W_{sym} = 37$ MeV. We find that this is caused by the difference of chemical potential of $\nu_e$. In the central region of protoneutron stars, $\nu_e$'s are degenerate and their diffusion fluxes are roughly proportional to $-\lambda_\nu \partial n_\nu / \partial r$ where $\lambda_\nu$ is the mean free path of $\nu_e$'s and $n_\nu$ is the number density of $\nu_e$'s. Since $\lambda_\nu$ is roughly proportional to the neutrino energy squared, we obtain $\lambda_\nu \propto \mu_\nu^{-2}$ and $n_\nu \propto \mu_\nu^3$ in the degenerate limit. Consequently the diffusion flux in the degenerate limit is proportional to $-\partial \mu_\nu / \partial r$; the smaller the symmetric energy is, the larger the $\nu_e$ flux is and the faster the deleptonization is. Fig. 4 shows time evolution of lepton number flux and neutrino luminosity. It can be seen that lepton number flux for $W_{sym} = 28$ MeV is larger than that for $W_{sym} = 37$ MeV.

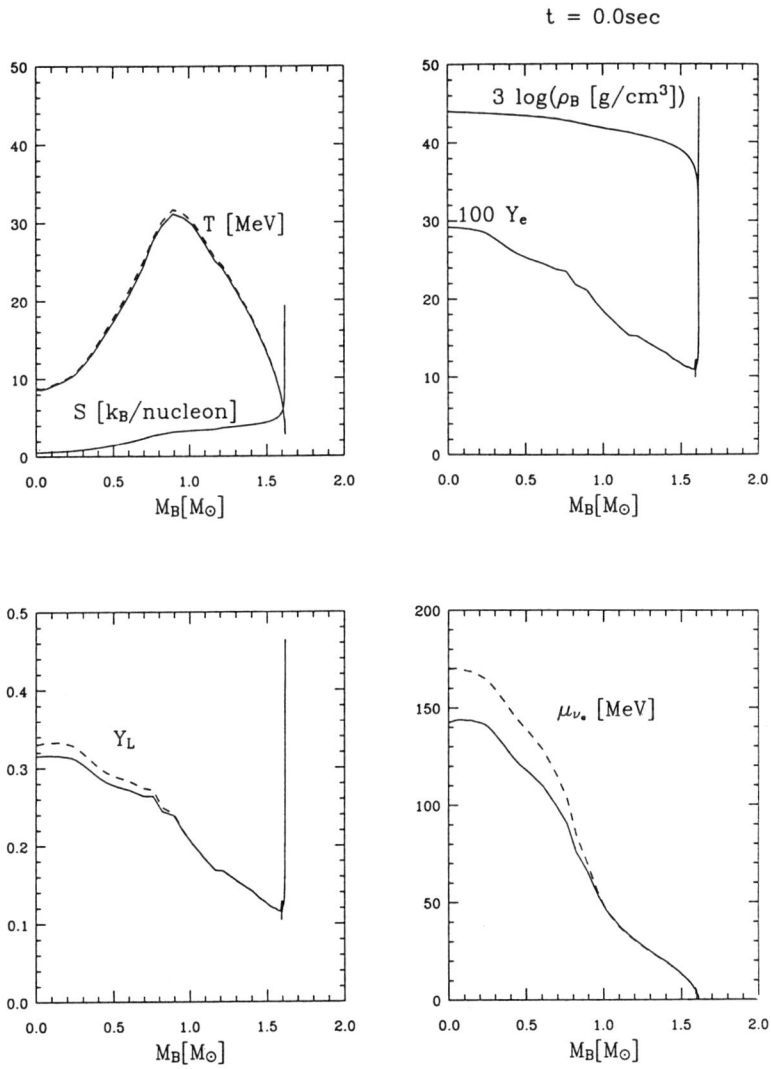

Figure 2. Comparison of models with the two EOSs at $t = 0$ sec. $T$: temperature, $S$: entropy, $\rho_B$: baryon density, $Y_e$: electron fraction, $Y_L$: lepton fraction, $\mu_{\nu_e}$: chemical potential of $\nu_e$. Solid line: TM1, $W_{sym} = 37$ MeV. Dashed line: TMS, $W_{sym} = 28$ MeV.

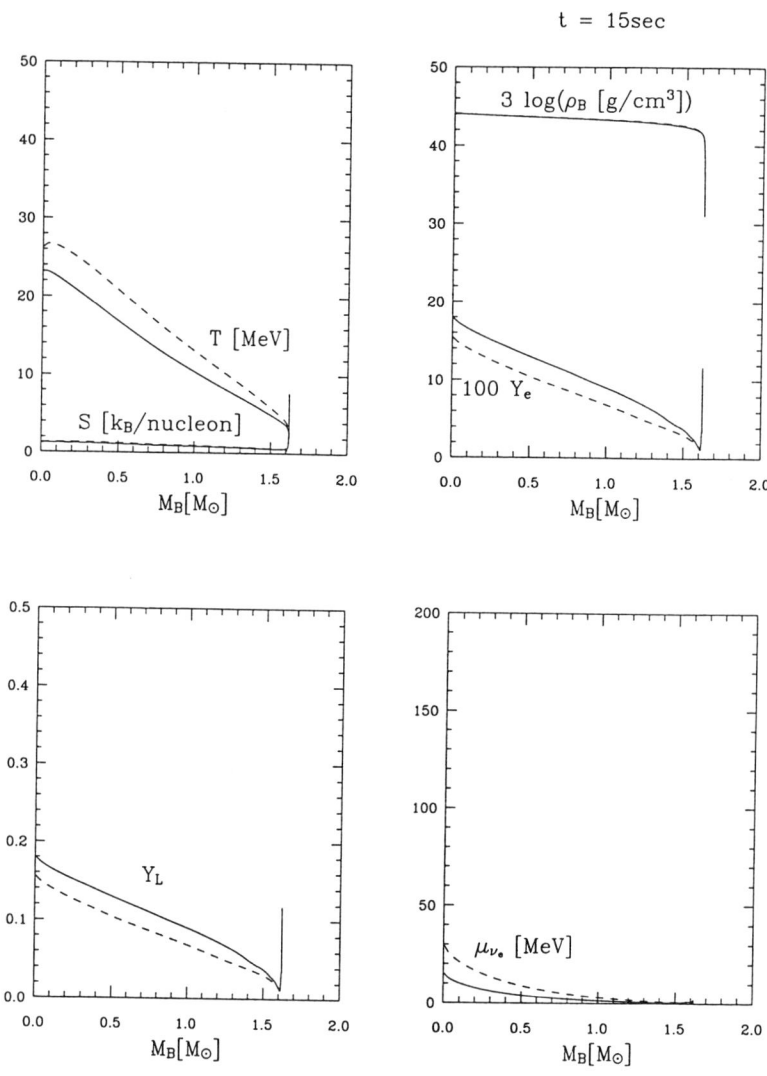

Figure 3. Comparison of models with the two EOSs at $t = 15$ sec. $T$: temperature, $S$: entropy, $\rho_B$: baryon density, $Y_e$: electron fraction, $Y_L$: lepton fraction, $\mu_{\nu_e}$: chemical potential of $\nu_e$. Solid line: TM1, $W_{sym} = 37$ MeV. Dashed line: TMS, $W_{sym} = 28$ MeV.

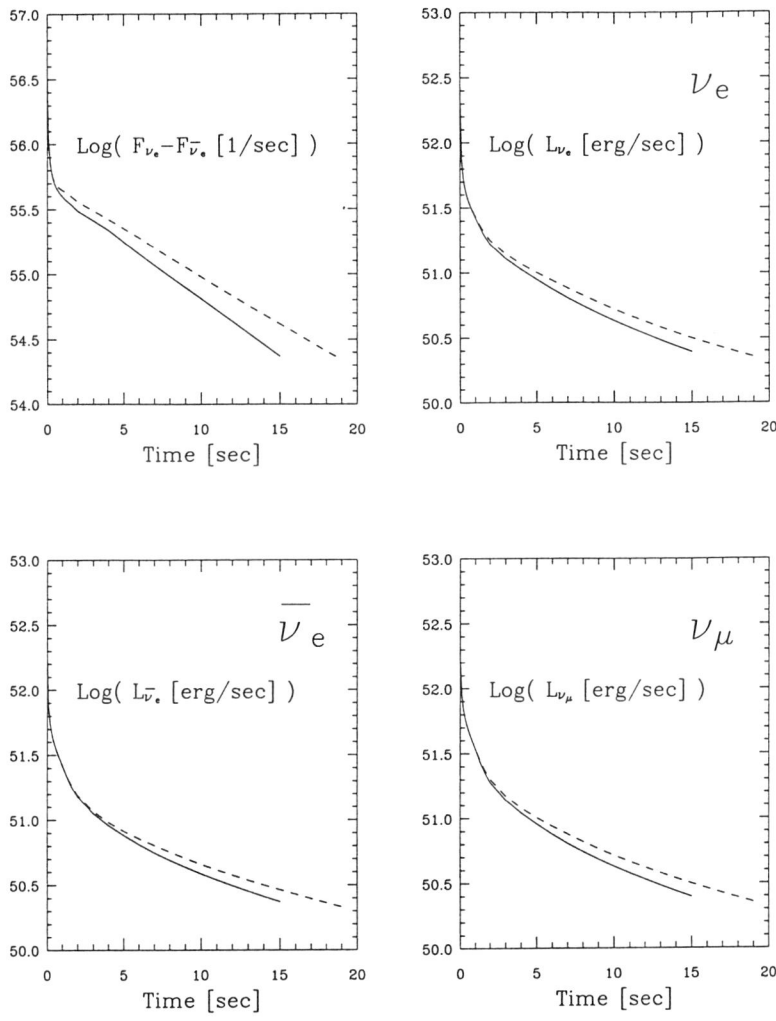

Figure 4. Comparison of models with the two EOSs. Upper left: lepton number flux, others: neutrino luminosity. Solid line: TM1, $W_{sym}$ = 37 MeV. Dashed line: TMS, $W_{sym}$ = 28 MeV.

We also study the effects of uncertainties in neutrino interaction rates. For example, Suzuki [1993] perform numerical simulations of protoneutron star cooling with some different treatments for nucleon bremsstrahlung process. Since there is no detailed calculation for the neutrino energy dependence of the process, we used simple approximations [Ishizuka, 1992] and obtained a result that the average energy of neutrinos emitted from the protoneutron stars might be lowered if one adopts a precise treatment of the nucleon bremsstrahlung process.

Detailed studies on the neutrino interaction rates in the high density matter as well as the equation of state are required for the prediction on the neutrino burst from the next nearby supernova explosion.

## REFERENCES
Bruenn, S. W., 1985, Astrophys. J. Suppl., **58**, 771.
Burrows, A., and J. M. Lattimer, 1986, Astrophys. J., **307**, 178.
Burrows, A., 1988, Astrophys. J., **334**, 891.
Hillebrandt, W., and R. G. Wolff, 1985, in *Nucleosynthesis: Challenges and New Developments*, ed. W. D. Arnett and J. W. Truran (Chicago, The Univ. of Chicago Press), p. 131.
Ishizuka, N., 1992, private communication.
Levermore, C. D., and G. C. Pomraning, 1981, Astrophys. J., **248**, 321.
Mayle, R., J. R. Wilson, and D. N. Schramm, 1987, Astrophys. J., **318**, 288.
Raffelt, G., and D. Seckel, 1991, Phys. Rev. Lett., **67**, 2605.
Sugahara, Y., and H. Toki, 1993, submitted to Nucl. Phys. A.
Sumiyoshi, K., H. Kuwabara, and H. Toki, 1994, preprint.
Suzuki, H., 1990, PhD thesis, Univ. of Tokyo.
Suzuki, H., 1993, in *Proc. of the International Symposium on Neutrino Astrophysics*, ed. Y. Suzuki and K. Nakamura (Tokyo, Universal Academy Press), p. 219.
Wilson, J. R., and R. W. Mayle, 1989, *The Nuclear Equation of State, Part A*, (New York, Plenum Press), p. 731.

# II.3

# SOLAR NEUTRINOS

Y. Suzuki

Institute for Cosmic Ray Research, University of Tokyo
3-2-1 Midori-cho, Tanashi, Tokyo 188, Japan

Status of the solar neutrino experiments and their implications are reviewed. The four on-going solar neutrino experiments all suggest deficits of solar neutrinos. Comparison between the Homestake and the Kamiokande results leads to the conclusion that the $^7$Be-neutrinos are strongly suppressed. If both experiments are correct, new neutrino properties like neutrino masses are needed to explain the solar neutrino problem.

## I. INTRODUCTION

The Sun, our nearest main sequence star, produces neutrinos in its central core. The order of the total neutrino flux ($\sim 6.6 \times 10^{10}/\text{cm}^2/\text{sec}$) is easily estimated through the luminosity of the Sun ($1.37 \times 10^6 \text{erg}/\text{cm}^2/\text{sec}$ at the top of the earth's atmosphere) and the energy released in the net nuclear reaction taken in the core; $4p \rightarrow {}^4\text{He} + 2e^+ + 2\nu_e + 26.2$ MeV. The details of the flux and the spectrum depending mainly on the precise knowledge of the nuclear reaction rates (especially on the nuclear S-factor $S(0)$) and the solar opacity, are calculated by the standard solar models [1,2]. The nuclear reaction chains in the Sun are shown in Fig. 1. The CNO cycle, which becomes much active in higher core temperature than in the Sun, produces less neutrinos than from the pp-chain. The spectra of the solar neutrinos are shown in Fig. 2.

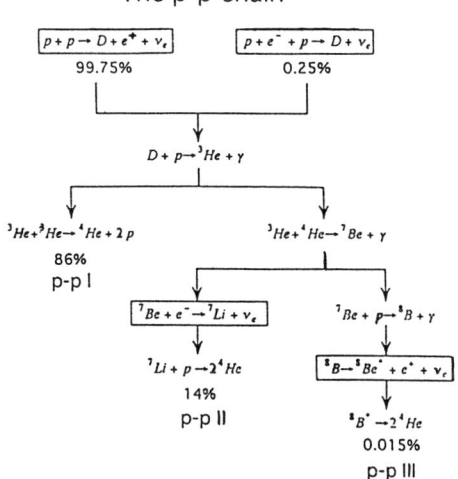

Fig. 1. The pp reaction chain in the Sun.

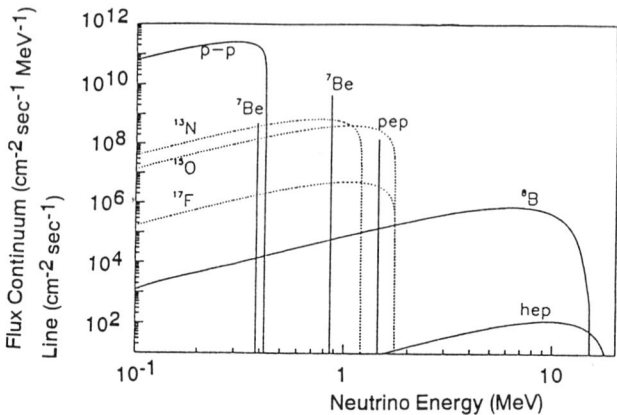

Fig. 2. The solar neutrino spectra.

## II. SOLAR NEUTRINO EXPERIMENTS

There are four on-going solar neutrino experiments: Homestake ($^{37}$Cl), Kamiokande (H$_2$O), Sage ($^{71}$Ga) and Gallex ($^{71}$Ga). The Homestake [3] and the Kamiokande experiment [4] are mainly sensitive to the high energy $^8$B neutrinos (*ppIII*) and Sage [5] and Gallex [6] to the lowest pp-fusion neutrinos (*ppI*). The sensitivities of the $^{37}$Cl and $^{71}$Ga experiments to each neutrino production reaction are listed in table 1. These four experiments are somehow complementary, and the consistency among each experiment is also a clue to the problem.

Table 1. The sensitivity of the solar radiochemical neutrino experiments to neutrinos from each nuclear reaction in the Sun.

|  | Cl | Ga |
|---|---|---|
| pp | — | 70.8 |
| pep | 0.2 | 3.1 |
| $^7$Be | 1.2 | 35.8 |
| $^8$B | 6.2 | 13.8 |
| CNO | 0.4 | 7.9 |
| Total | 8.0 SNU | 132 SNU |

The Homestake experiment [3]—the pioneering solar neutrino experiments started more than 20 years ago—indicated the deficit of solar neutrinos: the solar neutrino problem (SNP). The experiment uses 615 tons of (C$_2$Cl$_4$) and utilizes the reaction, $\nu_e + {}^{37}\text{Cl} \rightarrow e^- + {}^{37}\text{Ar}$ ($E_{th} = 814$ keV) and after collecting the Ar atoms from the tank through He-gas bubbling, they count the number of Ar atoms by measuring the Auger electrons emitted from the k-capture of $^{37}$Ar with a half life of 35 days in small proportional counters. Their observed flux is $2.32 \pm 0.23$ SNU (SNU: $10^{-36}$ capture/atom/sec) for more than 20 years of measurements whereas the expected from the standard solar model (SSM) of Bahcall and Pinsonneault (BP) [1] is $8.0 \pm 3.0$ SNU: they observed only 29% of the expected.

The Kamiokande experiment [4], started its data taking in January 1987, is a 3000 ton water Cherenkov detector—680 tons are used for the solar neutrino measurement—where the event time, the neutrino direction and the recoil electron energy are measurable through the reaction $\nu_e + e \rightarrow \nu_e + e$. These advantages are well demonstrated by the observation of neutrinos from the SN1987A (time and energy) and in Fig. 3—the image of the Sun taken by means of neutrinos. Their energy threshold, 9.3 MeV at the beginning of the experiment, is now down to 7.0 MeV. The experiment is only sensitive to the high energy $^8$B neutrinos. The observed flux obtained from the angular distribution towards the Sun is $0.51 \pm 0.04 \pm 0.06$ of the SSM of BP. If we assume the shape of the $^8$B neutrino spectrum, the flux of the $^8B$ neutrinos observed at Kamiokande is $(2.69^{+0.22}_{-0.21} \pm 0.35) \times 10^6/\text{cm}^2/\text{sec}$.

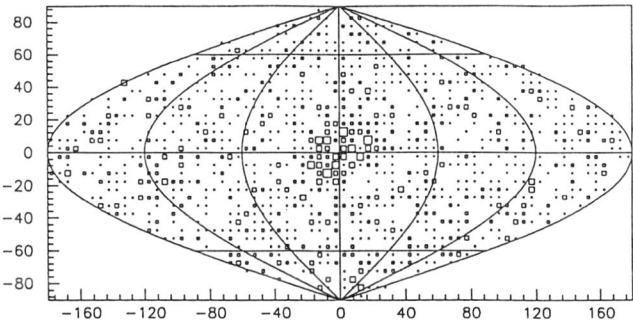

Fig. 3. The image of the Sun taken by neutrinos in the Kamiokande detector shown in the coordinate system where the Sun always sits at the center: the Neutrino Heliograph. The size of each box is proportional to the number of events in 4 deg by 4 deg bins.

The Gallium experiments (Sage and Gallex) use radiochemical technique like the Homestake experiment where they measure the total neutrino flux above 233 keV through the reaction, $\nu_e + ^{71}\text{Ga} \rightarrow e^- + ^{71}\text{Ge}$. Sage uses 60 tons of metallic gallium and Gallex 60 tons of GaCl$_3$, and they measure k and/or L capture of $^{71}$Ge (a half life of 11.4 days) in Xe-proportional counters. The averaged flux of Sage from 1990 to May 1992 is $70 \pm 19$ SNU and that of Gallex for 730 live days from May 1991 to October 1992 is $79 \pm 10 \pm 6$ SNU ($79 \pm 12$ SNU for the combined errors) which can be compare with 132 SNU of the SSM of BP (53 to 60% of the SSM).

The summary of the four solar neutrino experiments is shown in Fig. 4 where the SSM of Turch-Chieze and Lope (TC) [2] as well as of BP are also given as a comparison. It is astonishing that the measured flux of the all four experiments are 30 to 60% of the prediction from the SSM of BP (with the lowest $0.29 \pm 0.029$ for the Homestake experiment).

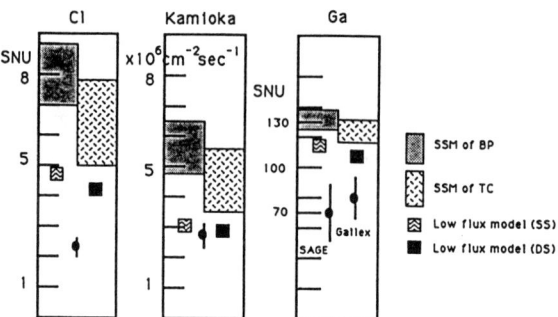

Fig. 4. The summary of the four solar neutrino experiments. The predictions of the SSMs (BP and TC) and the low flux models (see text) are also shown.

First of all, we examine the consistency among those experiments. The Kamiokande experiment measures only $^8$B neutrinos and the Homestake experiment measures both $^8$B and monochromatic $^7$Be neutrinos. For the time being we put aside the discrepancy between the measured Kamiokande flux and the prediction of the SSM (51% of the SSM) and we suppose that the results of the Kamiokande determines the $^8$B flux. The Homestake experiment of which the observed flux is $2.32 \pm 0.23$ SNU, then, must see at least $3.1 \pm 0.3$ SNU for the $^8$B flux (see table 1) and some contribution of $^7$Be (1.2 SNU), pep (0.2 SNU) and CNO (0.4 SNU) neutrinos. Therefore the conclusion we can draw from the comparison is that; (1) $^7$Be neutrinos and part of the $^8$B neutrinos (below Kamiokande threshold) are strongly suppressed; (2) one (or both) of the experiments are wrong; or (3) the cross section of the process $\nu_e + ^{37}Cl \to e^- + ^{37}Ar$ is overestimated. The last case is unrealistic. If these two experiments are correct and do not have unknown large systematic errors around 40 to 50 %, then there must be some mechanism to suppress $^7$Be neutrinos, most likely the new neutrino property, the neutrino mass.

The results of the two gallium experiments are now consistent with each other. The strong suppression of the $^7$Be neutrinos indicated by the comparison of the Homestake and Kamiokande experiments is not inconsistent with the gallium experiments. Suppose that the $^7$Be neutrino flux is zero, then the expected rate for the pp neutrinos for the gallium experiment becomes 76.5 SNU (This is higher than the value shown in table 1 since the constraint from the total luminosity must be satisfied [19]. By adding from the contribution of $^8$B neutrinos of 7 SNU for the Kamiokande measurement, the expected rate for the gallium experiment is 83.5 SNU which must be compare with the measured value of $79 \pm 11.7$ SNU for Gallex and $74 \pm 21.5$ SNU for Sage. The current solar neutrino measurements all suggest that there are strong suppression of the $^7$Be neutrinos and a part of $^8$B neutrinos. It is very difficult to suppress $^7$Be neutrinos more than $^8$B neutrinos in the frame work of the SSMs as is discussed in next section (see the nuclear reaction chains shown in Fig. 1).

## III. ASTROPHYSICAL SOLUTIONS

It is demonstrated by Bahcall [1] that two independent "standard" solar models of BP and TC give identical result if the same input parameters are used. The difference in results of the two SSM's mainly come from the different nuclear reaction rates, especially $S(0)_{17}$ and the opacity.

Recently low flux models, in which at lease $^8$B flux can be adjusted to be close to the observed value, are proposed by two groups: Dar and Shaviv (DS) [7], Shi and Schramm (SS) [8]. The SS model uses $S(0)_{17} = 20.2$ eV·b and $Z = 0.015$ which in effect lower the central temperature; $S(0)_{17} = 22.4$ eV·b and $Z = 0.0196$ for BP. The DS model uses $S(0)_{17} = 17$ eV·b by using the new extrapolation method and the most recent measurement of photon dissociation process of $^8$B beam [9]. They also used $S_{34}(0) = 0.45$ keV·b (0.533 keV·b for BP), and other slightly different treatments which would not effect significantly on the neutrino fluxes. In Fig. 4 the predictions from two such models are also shown. The debate on the validity of these new models are going on [10] and it probably takes time to settle all questions. However, the stress must be on the fact that event these low flux models could not succeed to reproduce the measured value of the Homestake experiment which is still lower than the prediction. The problem can be stated in more general way that the measured ratio of the $^7$Be neutrinos to $^8$B neutrinos ($^7$Be/$^8$B) is significantly lower than the any solar model predictions. The cross section uncertainty on the $p+^7$Be would not effect on the ratio $^7$Be/$^8$B because if you raise the cross section, then the flux of the $^8$B neutrinos becomes larger contradicting with the Kamiokande result; if you lower the cross section, then the ratio becomes larger that contradict with the measured ration of the Homestake and the Kamiokande experiment. This situation is well demonstrated in Fig. 5 which is taken from ref. [11]. Unless you introduce unknown phenomena in the solar modeling, you cannot reproduce those experimental data simultaneously. As is stated in the previous subsection, one or more experiments have to be wrong if the solution of the solar neutrino problem is in the solar modeling.

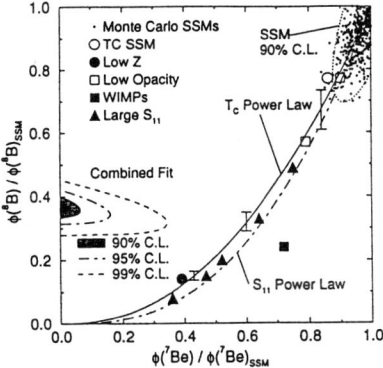

Fig. 5. The $^8$B vs $^7$Be neutrino flux plot from ref. [11]. The SSMs and their modifications cannot explain both Kamiokande and Homestake results at the same time.

## IV. NEUTRINO PHYSICS SOLUTIONS

It is a common belief that the most natural solutions of the SNP is the matter enhanced neutrino oscillations: the MSW effect (following the names of, and Wolfenstein) [12]. The MSW solution of the SNP is shown in Fig. 6 where the SSM of BP is used. There are two allowed regions: one in the large angle region and the other in the small angle region. The size and position of these regions changes if other SSMs are used, but the qualitative argument would not change. The vacuum neutrino oscillation, especially the "Just so" neutrino oscillation [13], is another choice of the solution, but the statistical significance is now very small [14].

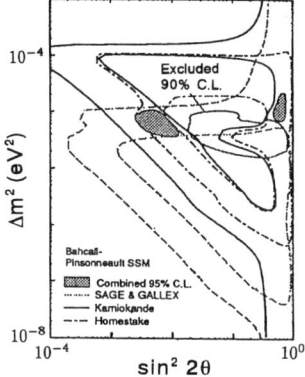

Fig. 6. The MSW solutions for the current solar neutrino experiments with the standard solar model from ref. [11]

## V. HOW TO RESOLVE THE PROBLEM IN FUTURE EXPERIMENTS

There are many criticism both on the experimental results and the solar model calculations, partly because those experiments are extremely difficult and very low statistics and some of the input parameters for the calculation have large uncertainties. The next generation experiments must be high statistics and should derive the conclusion independent of the model calculations.

The Superkamiokande [15], one of the next generation experiments currently under construction, is a high statistics and high resolution experiment—a scale up of the current kamiokande detector—therefore it enables us to measure the detailed spectrum of the recoil electron and to study the time variations, especially the day/night flux difference that is very powerful to discriminate the MSW solutions.

In Fig. 7 we show the day/night flux difference due to the MSW effect through the earth: the data are shown in terms of the path length of neutrinos through the earth. The current data from the Kamiokande experiment cannot discriminate the large and the small angle solutions, but only one month of the Superkamiokande data can discriminate them. This is independent of the SSM neutrino flux. The shape of the recoil electron shown in Fig. 8 has another power that they can discriminate among the answer. With the three years of the data accumulation in Superkamiokande, it is possible that they can pin down the small angle solution. The Superkamiokande is expected to start to take data in April, 1996.

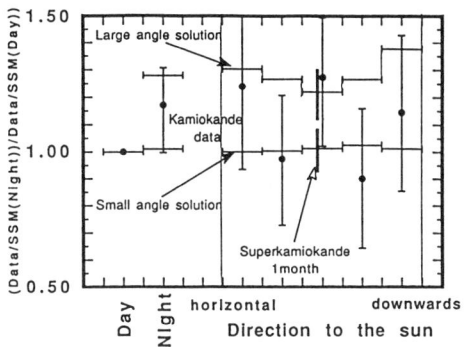

Fig. 7. The Day/Night flux differences. The kamiokande results, the MSW distortions for the small and large angle solutions, the expected results of the Superkamiokande are shown. Only one month measurement can resolve the large and small angle solutions.

The SNO experiment [16], supposed to start in early 1996, uses heavy water as a target material that they are able to measure neutral current interactions as well as charged current interactions. The neutral current interactions are independent of the neutrino species and therefore it serves as a normalization of the total neutrino flux, therefore the SNO experiment is also able to study the SNP independent of the SSMs.

Borexino [17], which is not yet fully approved, has an aim to measure $^7$Be neutrinos with a high event rate of $\sim 50/$day. It is very interesting since the current experimental result suggests strong suppression of the $^7$Be neutrinos.

Those experiments will start within two years, we can expect that some conclusions (solutions) will be obtained in near future.

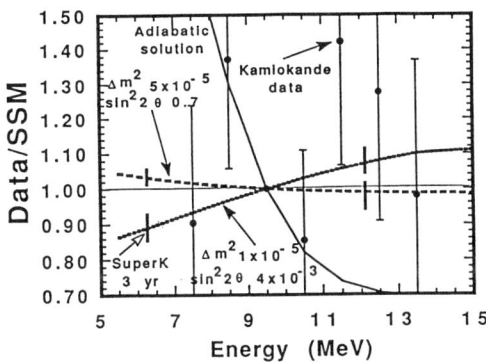

Fig. 8. The recoil electron energy spectrum. The Kamiokande result, the Monte Carlo simulations for the various MSW parameters and the expected results from the Superkamiokande are shown.

## REFERENCES

[1] J. N. Bahcall and M. H. Pinsonneault, Rev. Mod. Phys. **64** (1992) 885.
[2] S. Turck-Chieze and I. Lopes, Astrophys. J. **408** (1993) 347.
[3] K. Lande, in *Proc. of the XVI International Conference on Neutrino Physics and Astrophysics*, Eilat, Israel, 1994.
[4] Y. Suzuki, in *Proceedings of the International symposium on Neutrino Astrophysics*, Takayama, Kamioka, 1992.
[5] J. N. Abdurashitov *et al.*, Phys. Lett. B **328** (1994) 234.
[6] P. Anselmann *et al.*, Phys. Lett. B **327** (1994) 377.
[7] A. Dar and G. Shaviv, Technion preprint PH-94-5 (1994).
[8] X. Shi and D. N. Schramm, *Particle World*, **3**, No 4 (1993) 149.
[9] T. Motobayashi *et al.*, Rikko RUP 94-2 (1994).
[10] J. N. Bahcall *et al.*, Princeton preprint IASSN-AST 94/13.
[11] See for example, N. Hata, UPR-0612T, and references therein.
[12] S.P. Mikheyev and A. Yu. Smirnov, Sov. J. Nucl. Phys. **42** (1985) 913; Nuovo Cimento **9C** (1986) 17; L. Wolfenstein, Phys. Rec. D **17** (1978) 2369.
[13] S.L. Glashow and L.M. Krauss, Phys. Lett. **90** (1987) 199.
[14] K. Inoue, Ph. D. Thesis, University of Tokyo, 1994; P.I. Krastev and S.T. Petkov, IFP-480-UNC.
[15] Y. Suzuki, Nucl. Phys. B (Proc. Suppl.) **35** (1994) 273.
[16] A.B. McDonald, Nucl. Phys. B (Proc. Suppl.) **35** (1994) 345.
[17] M.G. Giammarchi, Nucl. Phys. B (Proc. Suppl.) **35** (1994) 433.
[18] G. Barr, T.K. Gaisser, and T. Stanev, Phys. Rev. D **39** (1989) 3532.
[19] V. Berezinsky, LNGS-94/101.

# II.4

# NEUTRINO EMISSION PROCESSES IN STARS

Naoki Itoh

Department of physics, Sophia University, Chiyoda-ku, Tokyo 102, Japan

Neutrino emission processes play a vital role in the late stages of stellar evolution. Thanks to the firm establishment of the Weinberg-Salam theory, we are now in a position to be able to provide accurate numerical results of the neutrino energy-loss rates. The present author and his collaborators have systematically investigated various neutrino emission processes in stars in the past decade. Significant improvements in the numerical results have been thereby achieved. In this paper we will review the recent progress in the calculation of the neutrino energy-loss rates in the stellar interiors.

## I. INTRODUCTION

In the past decade the present author together with his collaborators has systematically investigated the neutrino energy-loss in stellar interiors based on the Weinberg-Salam theory [Itoh and Kohyama 1983; Itoh et al. 1984a,b,c; Munakata, Kohyama, and Itoh 1985; Kohyama, Itoh, and Munakata 1986; Munakata, Kohyama, and Itoh 1987; Itoh et al. 1989; Itoh et al. 1992; Kohyama et al. 1993; Kohyama et al. 1994]. They have dealt with the pair, photo-, plasma, bremsstrahlung, and recombination neutrino processes. Dicus [1973], Dicus et al. [1976], Schinder et al. [1987], Braaten [1991], and Braaten and Segel [1993] also calculated the neutrino energy-loss rates using the Weinberg-Salam theory. The calculations of the neutrino energy-loss rates based on the Feynman-Gell-Mann theory were summarized by Beaudet, Petrosian, and Salpeter [1967].

In this paper we review the most recent developments in the calculations of the neutrino energy-loss rates in the stellar interiors. The present paper is organized as follows. The plasma neutrino process is reviewed in Sect. II. The axial-vector contribution to the plasma neutrino process is reviewed in Sect. III. The recombination neutrino process is reviewed in Sect. IV. Concluding remarks are given in Sect. V.

## II. PLASMA NEUTRINO PROCESS

In 1991 Braaten pointed out that the plasmon dispersion relations used originally by Beaudet, Petrosian, and Salpeter [1967] for the plasma neutrino calculation and followed by the later references were not accurate enough for ultrarelativistic electrons. He has given the expressions for the plasma neutrino energy-loss rates which are valid for ultrarelativistic electrons. However, his expressions were not practical enough for astrophysical applications, because the ultrarelativistic limit is not always realized in the interior of the real stars. Therefore, Itoh et al. [1992] calculated the neutrino energy-loss rates due to the plasma neutrino process for strongly degenerate electrons by using the accurate relativistic dispersion relations for the longitudinal and transverse plasmons originally calculated by Jancovici [1962]. The calculated plasma neutrino energy-loss rates are generally valid for relativistically degenerate electrons as well as for nonrelativistically degenerate electrons. The ratio of the result of their calculation to that of the previous calculations ranges between 0.39 and 3.22 depending upon densities and temperatures. They expressed the results of the calculation by an analytical fitting formula.

In Figures 1-3 we show the plasmon dispersion relations for the electron Fermi monenta $f = 1.0 m_e$, $f = 10.0 m_e$, and $f = 100 m_e$. The plasmon frequency is denoted by $\omega_0$. The plasma dispersion relations are solved numerically by using the Jancovici dielectric functions.

The energy-loss rate due to the plasma neutrino process is written as

$$Q_{plasma} = \left(C_V^2 + nC_V'^2\right) Q_V, \tag{1}$$

$$C_V = \frac{1}{2} + 2\sin^2\theta_W, \tag{2}$$

$$C_V' = 1 - C_V, \tag{3}$$

$$\sin^2\theta_W = 0.2325 \pm 0.0008. \tag{4}$$

In the above $\theta_W$ is the Weinberg angle and $n$ is the number of the neutrino flavors other than the electron neutrino whose masses can be neglected compared with $k_B T$. The vector contribution $Q_V$ in equation (1) consists of two parts: the contribution of the longitudinal plasmon $Q_L$ and that of the transverse plasmon $Q_T$,

$$Q_V = Q_L + Q_T. \tag{5}$$

In Figures 4-8 we show the results of the calculations of $Q_L$ and $Q_T$.

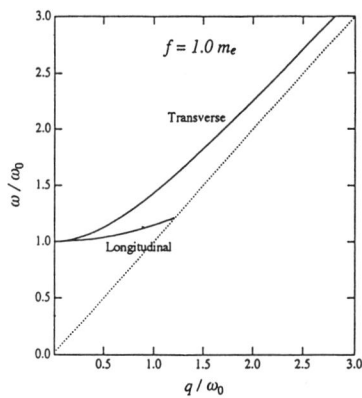

Fig. 1 – Plasmon dispersion curves for $f = 1.0 m_e$.

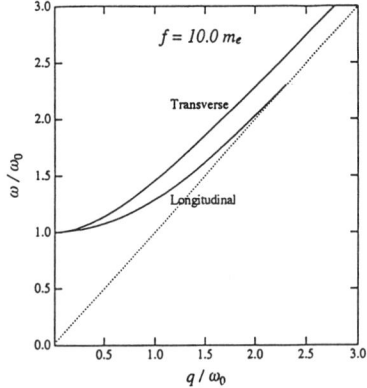

Fig. 2 – Plasmon dispersion curves for $f = 10.0 m_e$.

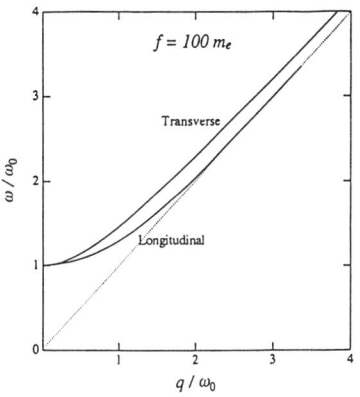

Fig. 3 – Plasmon dispersion curves for $f = 100\,m_e$.

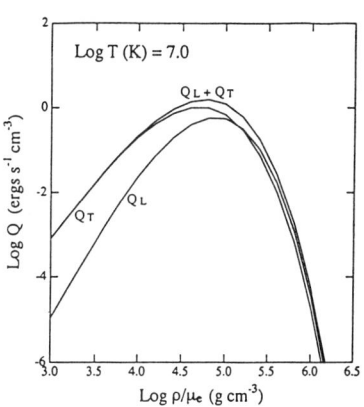

Fig. 4 – $Q_L$ and $Q_T$ for $T = 10^{7.0}$ K.

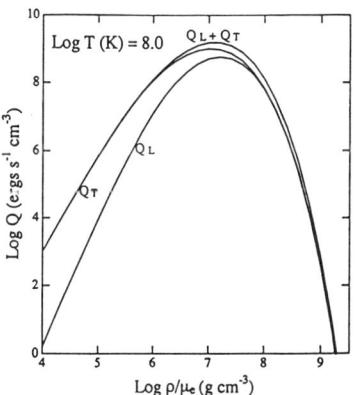

Fig. 5 – $Q_L$ and $Q_T$ for $T = 10^{8.0}$ K.

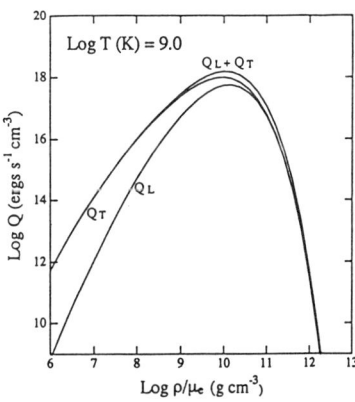

Fig. 6 – $Q_L$ and $Q_T$ for $T = 10^{9.0}$ K.

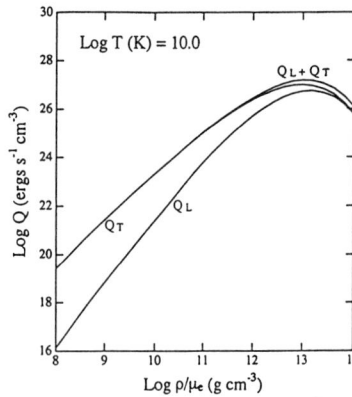
Fig. 7 – $Q_L$ and $Q_T$ for $T = 10^{10.0}$K.

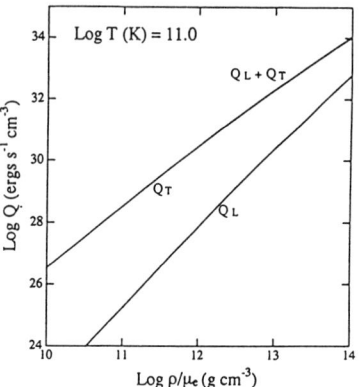
Fig. 8 – $Q_L$ and $Q_T$ for $T = 10^{11.0}$K.

## III. AXIAL-VECTOR CONTRIBUTION TO THE PLASMA NEUTRINO ENERGY LOSS RATE

We discussed the plasma neutrino process in the previous section. Strictly speaking, they are vector contributions. Kohyama et al. [1986] have shown that the axial-vector contribution to the plasma neutrino energy-loss rate is at most on the order of $10^{-4}$ of the vector contribution for $T \leq 10^{11}$K. However, Braaten and Segel [1993] presented results of their numerical calculation in which the axial-vector contribution becomes on the order of $10^{-1}$ for $T = 10^{11}$K and $\rho/\mu_e = 10^{14}$gcm$^{-3}$. Kohyama et al. [1994] recalculated the axial-vector contribution to the plasma neutrino energy-loss rate for strongly degenerate electrons using Jancovici's dielectric functions and confirmed that the calculation of Kohyama et al. [1986] was essentially correct. Braaten and Segel's [1993] calculation was found to be in numerical error.

In Figures 9-13 we compare the axial-vector contribution $Q_A$ with the vector contribution $Q_V$. We find that the axial-vector contribution is at most on the order of $10^{-4}$ of the vector contribution for $T \leq 10^{11}$K and $\rho/\mu_e \leq 10^{14}$gcm$^{-3}$.

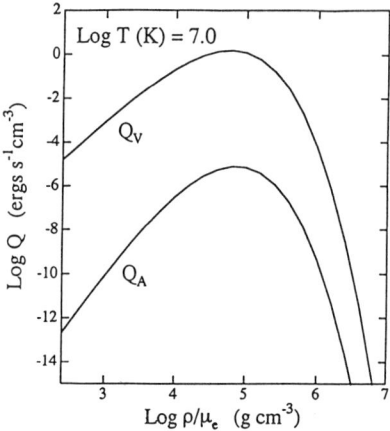
Fig. 9 – The vector and axial-vector contributions to the plasma neutrino energy loss rate at $T = 10^7$K; $\mu_e$ is the electron mean molecular weight.

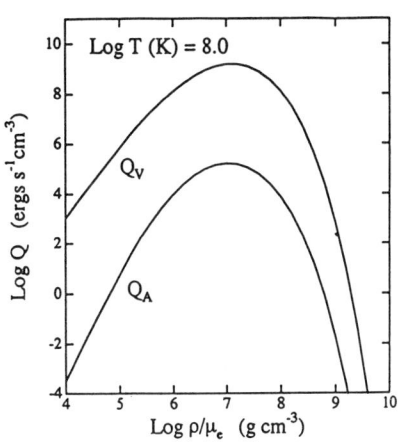

Fig. 10 – Same as Fig. 9, for $T = 10^8$ K.

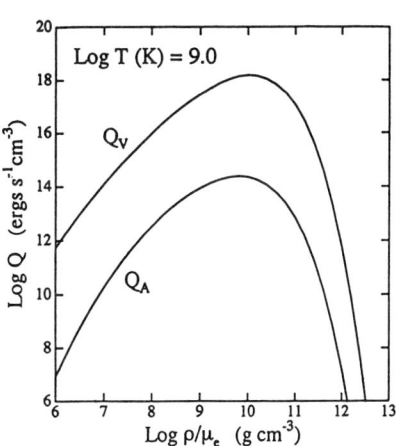

Fig. 11 – Same as Fig. 9, for $T = 10^9$ K.

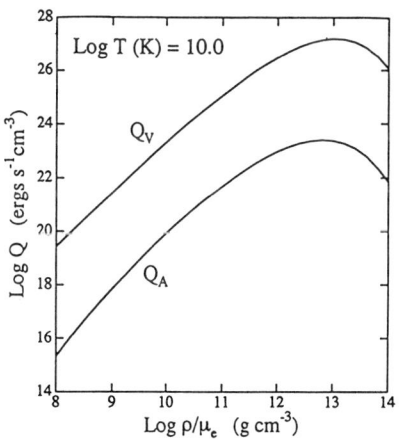

Fig. 12 – Same as Fig. 9, for $T = 10^{10}$ K.

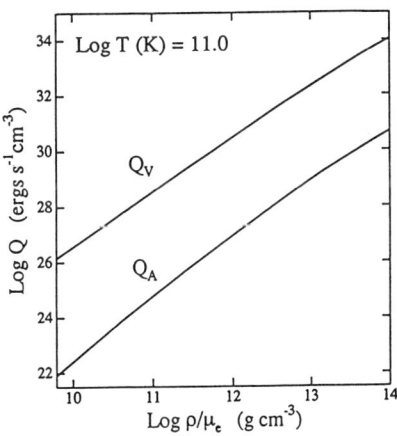

Fig. 13 – Same as Fig. 9, for $T = 10^{11}$ K.

## IV. RECOMBINATION NEUTRINO PROCESS

Beaudet, Petrosian, and Salpeter [1967] in their neutrino energy-loss calculation based on the Feynman-Gell-Mann theory noted possible importance of the recombination neutrino process [Pinaev 1964]. This is the neutrino process in which an electron in the continuum state makes a transition to a bound state, thereby emitting a neutrino pair. Therefore, one might alternatively call this process a "free-bound neutrino process".

Kohyama et al. [1993] calculated the energy-loss rate due to the recombination neutrino process for nonrelativistic electrons in the framework of the Weinberg-Salam theory. The Coulomb distortion effects for the electrons in the continuum states were accurately taken into account. These effects have been found to reduce the neutrino energy-loss rates drastically, by more than 1.5 orders of magnitude. Comparison with the other neutrino processes was made. It has been found that the recombination neutrino process is a dominant neutrino process for relatively large $Z$-values at relatively low densities and low temperatures. They expressed the results of their calculation by an analytic fitting formula.

The results of their calculation are shown in Figures 14-23. From Figures 14-23 we conclude that the recombination neutrino process is not a dominant process for the $^1$H and $^4$He matter. For the $^{12}$C and $^{16}$O matter it can be a dominant process at relatively low temperatures. However, the neutrino energy-loss rates are small at these low temperatures. For the pure $^{56}$Fe matter the recombination neutrino process is a dominant process at $T \leq 10^{8.2}$K. However, it is unlikely that the pure $^{56}$Fe matter state is realized in stars of relatively low densities and temperatures. Therefore, it is not likely that the recombination neutrino process plays a major role in stellar evolution.

The conclusion in the above is in sharp contrast to the conclusion reached by Beaudet, Petrosian and Salpeter [1967]. This is because they adopted the plane-wave approximation for the initial electron (continuum) state, thereby overestimating the neutrino energy-loss rates by more than 1.5 orders of magnitude. In Figures 24-25 we compare the results of Kohyama et al. [1993] with those of Beaudet, Petrosian, and Salpeter [1967]. It is readily seen that the plane-wave approximation leads to overestimation of the neutrino energy-loss rates by more than 1.5 orders of magnitude.

## V. CONCLUDING REMARKS

In this paper we have reviewed the most recent developments in the calculation of the neutrino energy-loss rates. Because of the limited space, we were not able to report on the other neutrino processes. We wish to make a fully comprehensive review of the neutrino energy-loss rates elsewhere. We believe it is fair to say that the investigation of the neutrino energy-loss rates in stellar interiors is approaching its completion. Therefore, we shall be in a position to be able to discuss the late stages of stellar evolution with great confidence in the neutrino energy-loss rates.

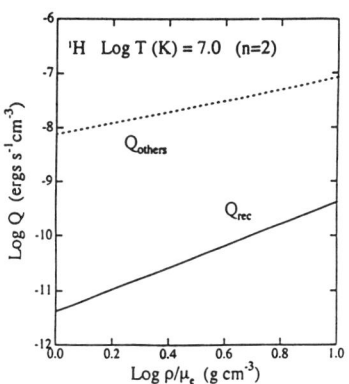

Fig. 14 – Comparison with the neutrino processes, $Q_{others} = Q_{pair} + Q_{photo} + Q_{plasma} + Q_{brems}$ for $^1H$ matter with $n = 2$ at $T = 10^{7.0}$K.

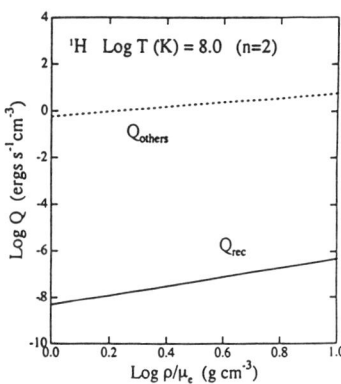

Fig. 15 – Same as Fig. 14, for $^1H$, $T = 10^{8.0}$K.

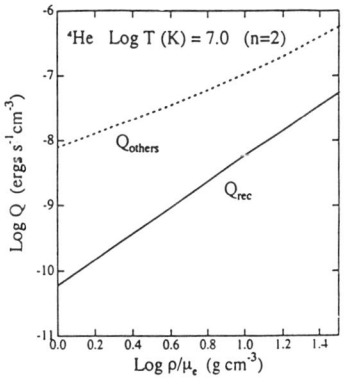

Fig. 16 – Same as Fig. 14, for $^4He$, $T = 10^{7.0}$K.

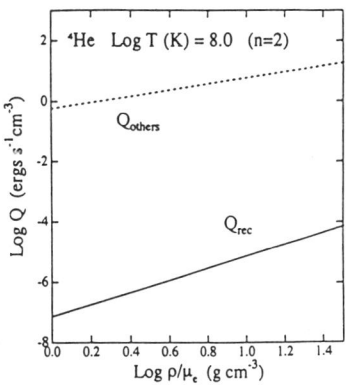

Fig. 17 – Same as Fig. 14, for $^4He$, $T = 10^{8.0}$K.

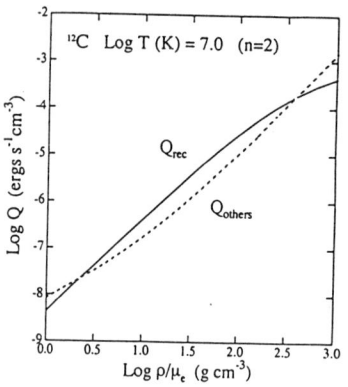

Fig. 18 – Same as Fig. 14, for $^{12}C$, $T = 10^{7.0}$K.

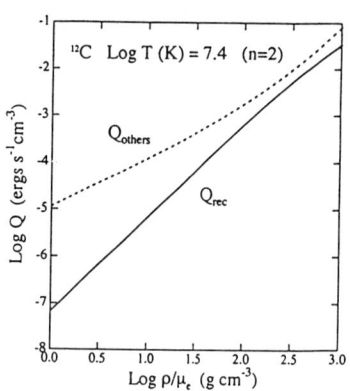

Fig. 19 – Same as Fig. 14, for $^{12}C$, $T = 10^{7.4}$K.

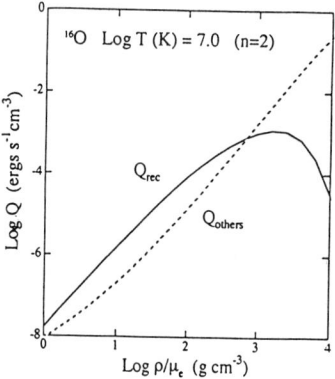

Fig. 20 – Same as Fig. 14, for $^{16}O$, $T = 10^{7.0}$K.

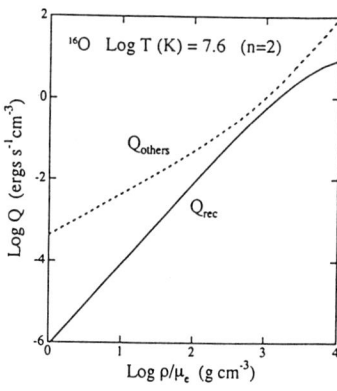

Fig. 21 – Same as Fig. 14, for $^{16}O$, $T = 10^{7.6}$K.

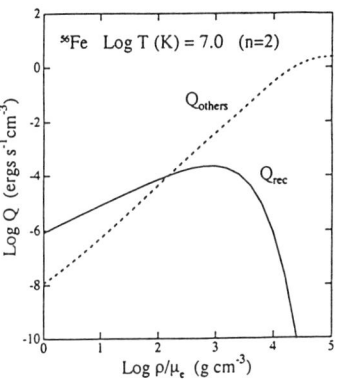

Fig. 22 – Same as Fig. 14, for $^{56}$Fe, $T = 10^{7.0}$ K.

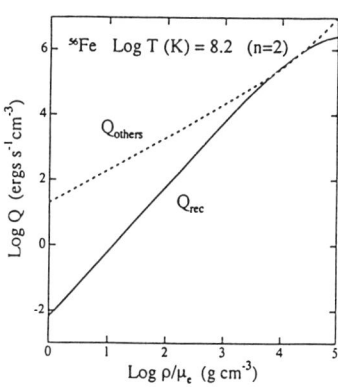

Fig. 23 – Same as Fig. 14, for $^{56}$Fe, $T = 10^{8.2}$ K.

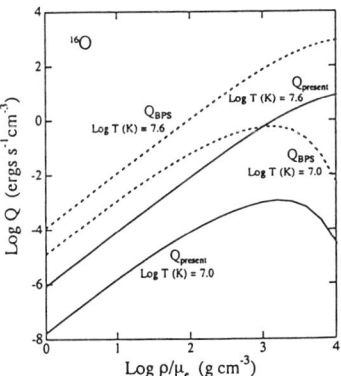

Fig. 24 – Comparison of the results of the present calculation with those of Beaudet, Petrosian, and Salpeter (1967) for the case of $^{16}$O.

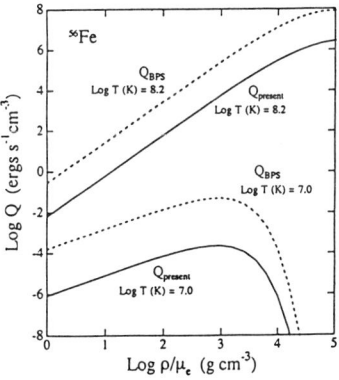

Fig. 25 – Same as Fig. 24, for $^{56}$Fe.

## REFERENCES

Beaudet, G., Petrosian, V., and Salpeter, E. E. 1967, Astrophys. J., **150**, 979.
Braaten, E. 1991, Phys. Rev. Lett., **66**, 1655.
Braaten, E., and Segel, D. 1993, Phys. Rev., **D48**, 1478.
Dicus, D. A., 1973, Phys. Rev., **D6**, 941.
Dicus, D. A., Kolb, E. W., Schramm, D. N., and Tubbs, D. L. 1976, Astrophys. J., **210**, 481.
Itoh, N., Adachi, T., Nakagawa, M., Kohyama, Y., and Munakata, H. 1989, Astrophys. J., **339**, 354; erratum **360**, 741 (1990).
Itoh, N.,and Kohyama, Y. 1983, Astrophys. J., **275**, 858.
Itoh, N., Kohyama, Y., Matsumoto, N., and Seki, M. 1984a, Astrophys. J., **280**, 787; erratum **404**, 418 (1993).
Itoh, N., Kohyama, Y., Matsumoto, N., and Seki, M. 1984b, Astrophys. J., **285**, 304; erratum **322**, 584 (1987).
Itoh, N., Matsumoto, N., Seki, M., and Kohyama, Y. 1984c, Astrophys. J., **279**, 413.
Itoh, N., Mutoh, H., Hikita, A., and Kohyama, Y. 1992, Astrophys. J., **395**, 622; erratum **404**, 418 (1993).
Jancovici, B. 1962, Nuov. Cimento, **25**, 428.
Kohyama, Y., Itoh, N., and Munakata, H. 1986, Astrophys. J., **310**, 815.
Kohyama, Y., Itoh, N., Obama, A., and Hayashi, H. 1994, Astrophys. J., in press.
Kohyama, Y., Itoh, N., Obama, A., and Mutoh, H. 1993, Astrophys. J., **415**, 267.
Munakata, H., Kohyama, Y., and Itoh, N. 1985, Astrophys. J., **296**, 197; erratum **304**, 580 (1986).
Munakata, H., Kohyama, Y., and Itoh, N. 1987, Atrophys. J., **316**, 708.
Pinaev, V. S. 1964, Soviet Phys. – JETP, **18**, 377.
Schinder, P. J., Schramm, D. N., Wiita, P. J., Margolis, S. H., and Tubbs, D. L. 1987, Astrophys. J., **313**, 531.

# III. Nuclear Processes

# III.1

# NUCLEAR REACTIONS IN DENSE MATTER

S. Ichimaru and H. Kitamura

Department of Physics, University of Tokyo, Bunkyo, Tokyo 113, Japan

The rates of nuclear fusion reactions in dense matter in astrophysical and/or terrestrial settings are reviewed from a condensed-matter-theoretic point of view. Quantum-statistical treatments of the effects of electronic screening and of internuclear correlations on the contact probabilities between reacting nuclei in dense matter are shown to yield *density* enhancement factors for thermonuclear and fluid-pycnonuclear rates over those in a dilute (vacuum) state, and *thermal* enhancement factors for solid-pycnonuclear rates over those in the ground state. Combining the concepts of pycnonuclear reactions at low temperatures and their enhancement due to strong internuclear Coulombic correlations, both applicable in ultrahigh pressure metals, we present concrete cases of a "supernova on the earth" scheme for fusion researches, based on the reactions: $^2$H$(p, \gamma)^3$He, $^3$H$(d, n)^4$He, and $^7$Li$(p, \alpha)^4$He.

## I. INTRODUCTION

Astrophysical condensed matter includes that found in the interiors, surfaces, and outer envelopes of such astronomical objects as neutron stars, white dwarfs, the sun, brown dwarfs, and giant planets. Among the nuclear processes of interest are the resonant and nonresonant fusion reactions, $\beta$ captures, and neutrino processes. The rates of these processes in condensed matter depend sensitively on the density and on the nature of the quantum state which characterizes it; one encounters changes in these states associated with a freezing transition and/or chemical separation, which may take place in the interior of a dense star. Mainly due to progress in condensed matter physics, there have been a number of recent improvements in theoretical calculations which incorporate the interplay of these states of matter and the calculations of the rates of those nuclear processes. Thus, an effort stemming from a recent collaboration between the University of Tokyo and the University of Rochester has advanced condensed-matter-theoretic calculations of the pycnonuclear reaction rates in dense Coulombic solids, accounting for the possible involvement of multiple nuclear species and thermal enhancement of the pycnonuclear rates near the melting transition [Ichimaru, Ogata, and Van Horn, 1992; Ichimaru, Kitamura, and Ogata, 1994]. A second example involves the theory of neutrino processes, where it has been demonstrated that band-structure effects suppress bremsstrahlung of neutrino pairs by electrons in the crusts of neutron stars at lower temperatures [Pethick and Thorsson, 1994]. These developments are expected to influence significantly the predictions of the internal structure and thermal evolution of various astronomical objects.

In this article, we review recent developments in the theories [Ichimaru, 1993, 1994] on the rates of nuclear fusion reactions, applied to dense matter in astrophysical as well as terrestrial settings. Attention will be paid to involvement of multiple nuclear species and of freezing transition. A specific proposal is made for an enhanced pycnonuclear fusion research in ultrahigh pressure metals, which may be called a "supernova on the earth" scheme [Ichimaru and Kitamura, 1994].

## II. NUCLEAR CROSS-SECTION FACTORS

We consider dense binary-ionic substances with mass density $\rho_m$, pressure $P$, and temperature $T$, which consist of nuclear species with charge number $Z_i$, mass number $A_i$, and molar fraction $x_i$ ($i = 1, 2$). The number density of nuclei and that of electrons are then

$$n_i = \frac{x_i \rho_m}{(x_1 A_1 + x_2 A_2) m_N} \;, \quad n_e = \frac{(x_1 Z_1 + x_2 Z_2)\rho_m}{(x_1 A_1 + x_2 A_2) m_N} \;, \quad (1)$$

where $m_N = 1.6605 \times 10^{-24}$g is an average mass per nucleon. We expect that those electrons are all in metallic (conductive) states if their Fermi energy is greater than the largest atomic binding energy of either nuclei; that is, $\hbar^2 (3\pi^2 n_e)^{2/3}/2m > m Z_i^2 e^4 / 2\hbar^2$.

The cross section factor $S_{ij}(E)$ of nuclear reactions [Salpeter, 1952; Barnes, 1971; Harris et al., 1983; Fowler, 1984] between i and j are defined by

$$S_{ij}(E) \equiv \sigma_{ij}(E) E \exp\left(\frac{2\pi e^2 Z_i Z_j}{\hbar v}\right) \;, \quad (2)$$

where $\sigma_{ij}(E)$ is the nuclear reaction cross section; $E$ and $v$ refer to the center-of-momentum energy and velocity, respectively. The cross section factors for reactions involving nuclei such as protons, deuterons, $\alpha$-particles, Li, Be, C, O, Ne, ... are essential quantities in predicting the rates of energy production and nucleosyntheses in various astrophysical circumstances. At lower energies of astrophysical interest, the experimental values of some of those cross section factors have been found to take on anomalously large magnitude [Engstler et al., 1988, 1992]. The physical origin of such anomalies may be examined through investigation of the influence of electronic screening effects on internuclear collisions at the stellar energies; this subject is, however, outside the scope of the present review.

## III. THERMONUCLEAR REACTIONS

Consider scattering between nuclei, i and j, with relative velocity $v$ and reduced mass

$$\mu_{ij} = \frac{A_i A_j}{A_i + A_j} m_N \quad (3)$$

via the bare Coulomb potential, $W_{ij}^{(0)}(r) = Z_i Z_j e^2 / r$. Events of scattering are described by the wave function $\Psi_{ij}(\mathbf{r})$ for the colliding pairs at internuclear separation $\mathbf{r}$. An important parameter characterizing such a Coulomb scattering in the short range is the *nuclear Bohr radius*

$$r_{ij}^* = \frac{\hbar^2}{2\mu_{ij} Z_i Z_j e^2} \;. \quad (4)$$

The controlling factor in the analysis of such a scattering event is the effective potential between the nuclei in the short-range domain, where the potential may be regarded as isotropic. A calculation of reaction rates may then be facilitated by the observation that the major contributions to the contact probabilities arise from the s-wave scattering acts between the reacting nuclei. Such an observation stems from the fact that the wave function of scattering in a spherically symmetric potential with the azimuthal quantum number $l$ is proportional to $r^l$ in short ranges. Since one can generally assume that the nuclear reaction radius $r_N < r_{ij}^*$, the s-wave scattering gives the major contribution to the reaction rate; hence $r_N \approx 0$ may be taken for the calculations of the contact probabilities $|\Psi_{ij}(r_N)|^2$.

The rates of thermonuclear reactions stemming from those binary Coulomb-scattering processes may be calculated by the thermal average, $\langle \sigma_{ij}(E)v \rangle$, over the Boltzmann distribution at temperature $T$,

$$f_B(E) = \frac{2}{k_B T} \left( \frac{E}{\pi k_B T} \right)^{1/2} \exp\left( -\frac{E}{k_B T} \right). \tag{5}$$

The result yields the Gamow rate [Gamow and Teller, 1938; Thompson, 1957] of the thermonuclear reactions (in units of power per unit volume)

$$P_G(\rho_m, T) = \frac{16 Q_{ij} S_{ij} r_{ij}^* \tau_{ij}^2}{3^{5/2} \pi (1 + \delta_{ij}) \hbar} n_i n_j \exp(-\tau_{ij}). \tag{6}$$

Here $Q_{ij}$ is the nuclear energy released by the reaction, $S_{ij}$ refers to a thermal average of the nuclear cross-section S-factor, $\delta_{ij}$ denotes Kronecker's delta, and

$$\tau_{ij} \cong 33.70 (Z_i Z_j)^{2/3} \left( \frac{2\mu_{ij}}{m_N} \right)^{1/3} \left( \frac{T}{10^6 \text{ K}} \right)^{-1/3}. \tag{7}$$

The reaction rate (6) contains a factor $\exp(-\tau_{ij})$ that decreases very steeply at lower temperatures. The magnitude of $\tau_{ij}$ increases with the charge numbers and/or with the reduced mass.

The integration leading to the Gamow rate contains in its integrand a product between a steeply rising term $\sigma_{ij}(E)$ of $E$ and a steeply decreasing Boltzmann factor (5). The product thus exhibits a *Gamow peak* at the energy, $E_{GP} = \frac{1}{3}\tau_{ij} k_B T$. The radius $r_{TP}$ of the classical turning point for a colliding pair with the Gamow peak energy is given by

$$r_{TP} = \frac{3 Z_i Z_j e^2}{k_B T \tau_{ij}}. \tag{8}$$

## IV. FLUID-PYCNONUCLEAR REACTIONS

A condensed matter, an example of which is a metal, contains many electrons, which act to reduce or screen the Coulombic repulsion between atomic nuclei. In a metallic substance under ultrahigh pressure, effects of such screening become so effective that the rates of nuclear reactions at relatively low temperatures take on values independent of the temperature [Salpeter and Van Horn, 1969; Ichimaru, 1993]. A.G.W. Cameron [1959] coined the term *"pycnonuclear"* reactions for such nuclear processes, thought to be applicable in a white-dwarf progenitor of supernova.

Short-range screening distance $D_s$ of the electrons responsible for the pycnonuclear reactions may be calculated as

$$\frac{1}{D_s} = \frac{2}{\pi} \int_0^\infty dk \left[ 1 - \frac{1}{\epsilon_e(k,0)} \right], \tag{9}$$

where $\epsilon_e(k,\omega)$ refers to the wave-number and frequency dependent dielectric function for the electrons with local-field corrections appropriately taken into account [Ichimaru and Utsumi, 1981; Ichimaru, 1994]. We then define the screening temperature as [Ichimaru, 1993]

$$T_s \text{ (K)} \equiv 5.7 \times 10^4 \text{ (K)} \sqrt{Z_i Z_j} \left(\frac{2\mu_{ij}}{m_N}\right)^{-1/2} \left(\frac{D_s}{10^{-9} \text{ cm}}\right)^{-3/2}. \tag{10}$$

In the weak screening regime such that $T > T_s$, the rate of release in fusion energy is calculated through a perturbative modification of Gamow's thermonuclear rate [Ichimaru, 1993].

In the strong screening regime $T < T_s$, the pycnonuclear counterpart to Eq. (6), is given by [Ichimaru, 1993]

$$P_s(\rho_m, T) = \frac{2Q_{ij}S_{ij}r_{ij}^*}{(1+\delta_{ij})\hbar} n_i n_j \sqrt{\frac{D_s}{r_{ij}^*}} \exp\left(-\pi\sqrt{\frac{D_s}{r_{ij}^*}}\right), \tag{11}$$

in which case the classical turning point takes place at $r_{TP} = D_s$, a quantity almost independent of $T$ in a condensed matter. Contrary to the Gamow rate (6), which changes sharply with the temperature via $\tau_{ij}$, the fluid-pycnonuclear rate (11) is practically independent of the temperature, and decreases, as $D_s$, the charge product $Z_i Z_j$, and/or the reduced mass increase.

Table 1 lists some of the parameters pertinent to the effects of electron screening in examples of dense astrophysical and laboratory plasmas. We remark that the first three astrophysical examples have turned out to be the cases of weak electron screening, while the last terrestrial example corresponds to the case of strong electron screening. This is somewhat ironic, since Cameron's original idea of pycnonuclear processes was for the interiors of degenerate stars such as white dwarfs. In these stars, however, the actual temperatures $T$ are usually higher than the critical temperatures $T_s$ of electron screening; hence the effect of electron screening on reaction rates is relatively weak. A huge enhancement in the reaction rates expected in those degenerate stars stems principally from the screening potentials [Ichimaru, 1993] produced by the internuclear many-particle correlations, which we shall treat in the subsequent section.

TABLE 1. Electron screening effects in nuclear reactions. Mass densities $\rho_m$ and temperatures $T$ are those assumed for the reacting nuclei.

| Case | white dwarf | brown dwarf | giant planet | pressurized metal |
|---|---|---|---|---|
| Reaction | $^{12}$C-$^{12}$C | p-p | d-p | $^7$Li-p |
| $\rho_m$ (g/cm$^3$) | $2 \times 10^9$ | $1 \times 10^3$ | 5 | 30 |
| $Z_1, Z_2$ | 6, 6 | 1, 1 | 1, 1 | 3, 1 |
| $A_1, A_2$ | 12, 12 | 1, 1 | 2, 1 | 7, 1 |
| $D_s$ ($10^{-9}$cm) | 0.042 | 1.3 | 3.2 | 2.1 |
| $T_s$ (K) | $1.1 \times 10^7$ | $4.0 \times 10^4$ | $8.6 \times 10^3$ | $2.4 \times 10^4$ |
| $T$ (K) | $5 \times 10^7$ | $3 \times 10^6$ | $2 \times 10^4$ | $1 \times 10^3$ |

## V. DENSITY ENHANCEMENT FACTORS

Rates of nuclear reactions in dense matter depend on quantum-statistical correlations between reacting nuclei at extremely short distances. Multiparticle correlations act to enhance the reaction rates through the screening potentials, which are closely related to the thermodynamic functions of dense plasmas [Salpeter and Van Horn, 1969; Ogata, Iyetomi, and Ichimaru, 1991; Ichimaru, 1993; Ichimaru, Ogata, and Tsuruta, 1994]. These relationships are summarized in this section.

Enhancement factor for the thermonuclear or the fluid-pycnonuclear rate is expressed compactly as [Ichimaru, 1993]

$$A_{ij}(\rho_m, T) = \exp(\xi_{ij}) \tag{12}$$

where

$$\xi_{ij} = \frac{H_{ij}(0)}{k_B T} - \frac{5}{32} \Gamma_s \left(\frac{r_{TP}}{a_{ij}}\right)^2 \left[1 + (C_1 + C_2 \ln \Gamma_s)\frac{r_{TP}}{a_{ij}} + C_3 \left(\frac{r_{TP}}{a_{ij}}\right)^2\right] \tag{13}$$

with $C_1 = 1.1858$, $C_2 = -0.2472$, and $C_3 = -0.07009$ [Ogata, Iyetomi, and Ichimaru, 1991]. Here

$$a_{ij} = \frac{1}{2}\left[\left(\frac{3Z_i}{4\pi n_e}\right)^{1/3} + \left(\frac{3Z_j}{4\pi n_e}\right)^{1/3}\right], \tag{14}$$

$$\Gamma_s = \frac{Z_i Z_j e^2}{k_B T a_{ij}} \exp\left(-\frac{a_{ij}}{D_s}\right), \tag{15}$$

and the first term on the right-hand side of Eq. (13) may be calculated as

$$H_{ij}(0)/k_B T = 1.148\Gamma_s - 0.00944 \Gamma_s \ln \Gamma_s - 0.000168 \Gamma_s (\ln \Gamma_s)^2. \tag{16}$$

The expression (13) is applicable for $1 \leq \Gamma_s \leq 170$.

Note that the enhancement factor (12) depends sensitively on $\rho_m$ and $T$; it increases with $\rho_m$ and decreases sharply as $T$ increases. Consequently, the total power production rate,

$$P_{ij}(\rho_m, T) = A_{ij}(\rho_m, T) P_{G(s)}(\rho_m, T), \tag{17}$$

is also a sensitive function of $\rho_m$ and $T$.

## VI. SOLID-PYCNONUCLEAR RATES IN THE GROUND STATE

The body-centered cubic (bcc) crystalline structures are known to have the lowest values in the Madelung energy for the Coulombic crystals [e.g., Ichimaru, 1994]; hence a bcc structure is usually assumed for a Coulomb solid. The principal problem then is the evaluation of contact probabilities $|\Psi_{ij}(0)|^2$ arising from s-wave scattering between nearest neighbor particles. The factor that crucially controls such a contact probability is the effective potential between the nuclei in the short-range domain. Such a potential is called the *lattice potential*; it may be approximated as isotropic for the s-wave scattering. Pycnonuclear rates for Coulomb solids are thus obtained from a solution to the relevant Schrödinger equation with inclusion of the lattice potential.

Salpeter and Van Horn [1969] originally evaluated the lattice potential, by choosing a pair of nearest-neighbor particles in a bcc crystal and then calculating the electrostatic energy as a function of the interparticle separation $r$ with the center of mass fixed. In their fully relaxed approximation, these authors adjusted the resultant screening potential near $r = 0$ in accordance with the ion sphere model of Salpeter [1954].

The screening potentials can be analyzed through a Monte Carlo (MC) method, which samples the joint probability densities $g_{ij}(r)$ between those pairs of particles located around the nearest-neighbor sites of the bcc lattice [Ogata, Iyetomi, and Ichimaru, 1991; Ogata et al., 1993]. The exact MC nearest-neighbor separation $r_{m,ij}$ may be determined from the observed peak position of $g_{ij}(r)$; the results are then expressed as the sum of the ion-sphere scaling contribution and a deviation therefrom:

$$r_{m,ij} = 1.76 a_{ij} + \Delta r_{m,ij} . \tag{18}$$

The deviation $\Delta r_{m,ij}$, characterizing extra distortion in the particle configurations due to the charge disparities ($R_Z \equiv Z_2/Z_1$) in the binary-ionic-mixture (BIM) solids, have been measured in the MC data; the results can be summarized in the following parametrized forms for $0 \leq x(= x_2) < 1$ and $1 \leq R_Z < 4.5$:

$$\frac{\Delta r_{m,11}}{a_{11}} = 0.44 \frac{(R_Z - 1)(2.3 - R_Z)}{R_Z^2} x^{1.3} , \tag{19a}$$

$$\frac{\Delta r_{m,12}}{a_{12}} = -0.043 \frac{\sqrt{R_Z - 1}}{1 + 100 x^{1.3}} , \tag{19b}$$

$$\frac{\Delta r_{m,22}}{a_{22}} = -0.17 \frac{(R_Z - 1)(2.3 - R_Z)}{R_Z^2} (1 - x)^{1.3} . \tag{19c}$$

Pycnonuclear rates may be obtained from the solution to a Schrödinger equation with a lattice potential for the s-wave scattering. The solution $\Psi_{ij}(r)$ in the ground state should correspond to the zero-point vibration of the nuclei and exhibits a peak at $r_{m,ij}$ given by Eq. (18). Short-range values of the solution, which determine the contact probabilities, are governed by the cusp condition,

$$\lim_{r \to 0} \frac{d \ln \Psi_{ij}(r)}{dr} = \frac{1}{2 r_{ij}^*} . \tag{20}$$

Hence the contact probabilities should scale with the two length parameters, $r_{m,ij}$ and $r_{ij}^*$.

The pycnonuclear rates for BIM solids in the ground state are thus obtained as [Ichimaru, Ogata, and Van Horn, 1992]

$$P_{ij}^{(0)}(\rho_m, T)(\text{W/cm}^3) = \frac{1.34 \times 10^{32}}{1 + \delta_{ij}} \frac{X_i X_j (A_i + A_j)}{Z_i Z_j (A_i A_j)^2}$$
$$\times \rho_m^2 Q_{ij} S_{ij} \lambda_{ij}^{-1.809} \exp(-2.460 \lambda_{ij}^{-1/2}) . \tag{21}$$

Here $X_i$ denotes the mass fraction, $Q_{ij}$ is the nuclear energy released by a reaction expressed in Joule, $S_{ij}$ is the cross-section factor in MeV·barns, and the quantity $\lambda_{ij}$ is defined by

$$\lambda_{ij} \equiv \left( \frac{3}{4} \right)^{1/2} \frac{r_{ij}^*}{r_{m,ij}} . \tag{22}$$

We remark that the exponent in the important exponential factor of Eq. (21) is proportional to $\rho_m^{-1/6}$. Thus, the pycnonuclear rate in the ground state increases steeply with the density.

## VII. THERMAL ENHANCEMENT FACTORS

The nuclei forming a quantum solid perform zero-point vibrations about their equilibrium sites in the ground state. Their reaction rates proportional to the contact probabilities between nearest-neighbor pairs depend only on the density but are independent of the temperature.

At an elevated temperature, the nuclei may occupy excited states of lattice vibrations and the reaction rates may take on significantly enhanced values over those in the ground state, especially near the melting conditions. Enhancement factors arising from such thermal excitations may likewise be evaluated through solutions to the Schrödinger equation for excited states with the cusp boundary condition,

$$\lim_{r \to 0} \frac{d \ln \Psi_\nu(r)}{dr} = \frac{1}{2r^*} . \tag{23}$$

Here $\Psi_\nu(r)$ refers to the wave function of the s-wave scattering for the $\nu$-th excited state of nuclei with energy eigenvalue $E_\nu$. In this section we confine ourselves to the cases of one-component-plasma (OCP) solid, for which subscripts such as i and j are to be deleted.

The states of a quantum OCP solid [Iyetomi, Ogata, and Ichimaru, 1993] may be characterized by two dimensionless parameters, $R_s$ and $Y$,

$$R_s = \frac{0.4924}{\lambda} = 0.4924 A Z^2 \left( \frac{1}{A} \frac{\rho_m}{1.3574 \times 10^{11} \text{ g/cm}^3} \right)^{-1/3} , \tag{24a}$$

$$Y = 14.30 \frac{Z}{A} \left( \frac{\rho_m}{10^7 \text{ g/cm}^3} \right)^{1/2} \left( \frac{T}{10^6 \text{ K}} \right)^{-1} , \tag{24b}$$

where $\lambda$ is the OCP counterpart to Eq. (22) related to the ratio of $r^*$ to $r_m$. The parameters in Eqs. (24) are related to the OCP Coulomb-coupling parameter [Ichimaru, 1994] via $\Gamma = Y\sqrt{R_s}$.

Schrödinger equation with a lattice potential has been solved [Ichimaru, Kitamura, and Ogata, 1994] for the excited states in the ranges of density-temperature combinations:

$$10^3 \leq R_s \leq 5 \times 10^4, \ Y \leq 20, \text{ and } \Gamma \geq 172.$$

The temperature-dependent enhancement factor $A(\rho_m, T)$ of the pycnonuclear rates is thus calculated as

$$A(\rho_m, T) = \frac{\sum_{\nu=0}^{\infty} \exp[-\beta(E_\nu - E_0)] |\Psi_\nu(0)|^2}{\sum_{\nu=0}^{\infty} \exp[-\beta(E_\nu - E_0)] |\Psi_0(0)|^2} .$$

The values of $A(\rho_\mathrm{m}, T)$ so computed are then fitted in an analytic formula with

$$A(\rho_\mathrm{m}, T) = \exp[F(Y) R_\mathrm{s}^{1/2}] \,, \tag{25a}$$

where

$$F(Y) = 0.613 \frac{1.678 + Y^2}{0.779 + Y^2} \exp(-0.780 Y) \,. \tag{25b}$$

Errors in fitting by this formula are confined within 0.5 in absolute magnitude. In this formula, we see that enhancement factor can be as large as $10^{54}$ near the melting conditions when $R_\mathrm{s} = 5 \times 10^4$. This does not mean a large value of the reaction rate itself, however, since the basic pycnonuclear rate at $T = 0$, Eq. (21), takes on a minuscule magnitude at a relatively low density such as $R_\mathrm{s} = 5 \times 10^4$.

A principal effect of the huge enhancement predicted in Eqs. (25) is to make the total reaction rate in the solid regime *join smoothly across a thermal melting line* into that in the fluid regime, which is given by a product between the Gamow rate (6) and the enhancement (12). The smooth connection has been achieved in terms of both physical understanding and numerical results. Since evolution lines for a considerable class of degenerate cores [e.g., Nomoto, 1982] follow close to the melting line, it is a significant issue to elucidate pycnonuclear enhancement in the vicinity of the melting conditions.

Finally, combining Eq. (25a) with the pycnonuclear rate (21) for OCP, we obtain the pycnonuclear energy-production rate at finite temperatures expressed as

$$\begin{aligned} P_\mathrm{PN}(\rho_\mathrm{m}, T) \,(\mathrm{W/cm^3}) &= A(\rho_\mathrm{m}, T) P^{(0)}(\rho_\mathrm{m}, T) \\ &= 4.83 \times 10^{32} \left[\frac{\rho_\mathrm{m}(\mathrm{g/cm^3})}{A}\right]^2 \frac{Q(\mathrm{Joule}) S(\mathrm{MeV \cdot barns})}{A Z^2} \\ &\quad \times R_\mathrm{s}^{1.809} \exp\{[F(Y) - 3.506] R_\mathrm{s}^{1/2}\} \,. \end{aligned} \tag{26}$$

As we have noted in conjunction with the pycnonuclear rates (21), the density dependence of the exponential factor reflects the features germane in solids in that particle motions are confined around their lattice sites. Since $F(Y)$ and consequently the exponent of Eq. (25a) increase rapidly with the temperature for $Y < 1$, the thermal effects in the pycnonuclear rates become very pronounced near the classical (i.e., thermal) melting conditions. The pycnonuclear rate (26) is applicable over the entire parameter regime of OCP solids.

## VIII. ENHANCED PYCNONUCLEAR REACTIONS IN ULTRAHIGH PRESSURE METALS

In a ultradense material, as we have elaborated in Sec. V, we may expect *enhancement* of the thermonuclear or the fluid-pycnonuclear rate due to strong internuclear correlations. In a white-dwarf progenitor of supernova, enhancement by a factor of 20 to 30 orders of magnitude is considered in some cases; virtually no enhancement is expected, however, in the cases of the solar interior and an inertial-confinement fusion (ICF) plasma [Ichimaru, 1993]. Possibilities of utilizing those enhanced pycnonuclear reactions in ultrahigh pressure liquid metals for power production were first predicted and investigated by one of the present authors [Ichimaru, 1991, 1993, 1994].

For a conclusion of the present article, combining the concepts of pycnonuclear reactions at low temperatures and their enhancement due to strong internuclear Coulombic correlations, both applicable in ultrahigh pressure metals, we investigate in this section the possibilities of a "supernova on the earth" scheme for nuclear fusion researches. As concrete examples, we consider temporal evolution of density and temperature for the cases with the reactions—$^2$H$(p, \gamma)^3$He, $^3$H$(d, n)^4$He, and $^7$Li$(p, \alpha)^4$He—and compare their merits as well as demerits quantitatively from the point of view of technical accessibility. We will thereby find that a scheme using $^2$H$(p, \gamma)^3$He should be most advantageous among the three, and predict typically that a pulsed compression with 1 kJ input to initial conditions—mass density $\approx$ 20 g/cm$^3$, temperature $\approx$ 1400 K, pressure $\approx$ 490 Mbar, and radius $\approx$ 0.017 cm—will yield a fusion energy of approximately 33 kJ in an interval of 0.03 fs. The state of matter is expected to be in the regime of ultradense liquid-metallic hydrogen near the freezing conditions, which should be far more *stable* thermohydrodynamically than one in an extremely high-temperature, gaseous state of an ICF scheme. For a feasibility study, we also propose a scaled-down experiment, which may be attainable through extrapolation of the current ultrahigh pressure metal technologies [Nellis et al., 1992; Meyer-ter-Vehn and Oparin, 1994].

Once nuclear reactions take place, charged fusion products deposit their energies in the substance, heat the remaining fuel, and thereby quench the reactions. Temporal evolution of fuel mass density and temperature is described by

$$\rho_{\rm m}(t) = \rho_{\rm m}(0) - \frac{(A_i + A_j)m_{\rm N}}{Q_{ij}} \int_0^t {\rm d}t'\, P_{ij}[(\rho_{\rm m}(t'), T(t')]\,, \qquad (27{\rm a})$$

$$T(t) = T(0) + \frac{\eta_{ij}w_{ij}}{Q_{ij}} \int_0^t \frac{{\rm d}t'}{c_{\rm P} t_{\rm H}} \int_0^{t'} {\rm d}t''\, P_{ij}[(\rho_{\rm m}(t''), T(t'')]\,. \qquad (27{\rm b})$$

Here $c_{\rm P}$ is the specific heat per unit volume at constant pressure, $w_{ij}$ denotes the kinetic energy carried by a charged fusion product, $\eta_{ij}$ is the number of such products per reaction, and the heating time $t_{\rm H} = w_{ij}/({\rm d}w_{ij}/{\rm d}t)$ where the stopping power ${\rm d}w_{ij}/{\rm d}t$ is calculated in the usual way [Bethe, 1930; Yan et al., 1985] by taking into account scattering with both the electrons and the nuclei. The total fusion output is thus evaluated as

$$W = \int_0^\infty {\rm d}t\, V(t) P_{ij}[\rho_{\rm m}(t), T(t)]\,, \qquad (28)$$

where $V(t)$ is the fuel volume. Since $PV$ gives a measure of energy necessary for the initial compression, we define and calculate the gain factor as $G = W/PV$, and the average duration of reactions as

$$\Delta t = \frac{W}{V(0)P_{ij}[\rho_{\rm m}(0), T(0)]}\,. \qquad (29)$$

In the calculation of (28), we observe that the reactions once initiated will be quenched by fuel heating (27b), since exponent (13) of the enhancement factor (12) is inversely proportional to $T$.

TABLE 2. Three cases of the nuclear reactions treated and typical parameters considered for pressurized metals and ICF.

|  | pressurized metals | | | ICF |
| --- | --- | --- | --- | --- |
| reactions | $^2$H$(p, \gamma)^3$He | $^3$H$(d, n)^4$He | $^7$Li$(p, \alpha)^4$He | $^3$H$(d, n)^4$He |
| $S_{ij}$ (keV b) | $2.5 \times 10^{-4}$ | $1.7 \times 10^4$ | 52 | $1.7 \times 10^4$ |
| $Q_{ij}$ (MeV) | 5.494 | 17.6 | 17.347 | 17.6 |
| $\rho_\mathrm{m}$ (g/cm$^3$) | 20 | $2 \times 10^4$ | $2 \times 10^5$ | 30 |
| $T$ (K) | $1.4 \times 10^3$ | $2.2 \times 10^4$ | $1.3 \times 10^5$ | $5 \times 10^7$ |
| $P$ (Mbar) | $4.9 \times 10^2$ | $3 \times 10^7$ | $2.1 \times 10^9$ | $1 \times 10^5$ |
| $D_\mathrm{s}$ ($10^{-9}$ cm) | 3.13 | 0.96 | 0.62 | 11.9 |

We have examined the possibilities of energy release by the enhanced pycnonuclear processes for three cases of ultrahigh pressure metals listed in Table 2; for comparison, a case of ICF is also included. In all cases, $x_1 = x_2 = 1/2$ is assumed; $\rho_\mathrm{m}$ and $T$ denote the assumed initial density and temperature, at which a gain of approximately 10 is expected.

Though the S-factor takes on a value far larger in $d$-$t$ than in $p$-$d$, substantially larger values of $\rho_\mathrm{m}$, $T$, and $P$ need to be attained for a similar gain (and hence far more difficult to realize experimentally) with $d$-$t$ than with $p$-$d$, due mainly to the difference in the reduced masses. In $p$-$d$, bulk of $Q_{ij}(= 5.494$ MeV$)$ goes into the $\gamma$-ray (5.489 MeV), while $^3$He carries an energy of 5.40 keV. In $d$-$t$, neutrons carry away bulk of the fusion energy.

Though the S-factor still takes on a value larger in $p$-$^7$Li than in $p$-$d$, again substantially larger values of $\rho_\mathrm{m}$, $T$, and $P$ are required for a similar gain for $p$-$^7$Li than for $p$-$d$, due mainly to the difference in the products between nuclear charges. Merits with $p$-$^7$Li rest on the facts that a lithium hydride is a stable solid substance and that fusion yields consist all in $\alpha$-particles.

Figure 1 shows constant-gain and constant-$\Gamma_\mathrm{s}$ contours on $\rho_\mathrm{m}$-$T$ plane for $p$-$d$ cases. The narrow strip between $G = 1$ and $\Gamma_\mathrm{s} = 170$, approximately the estimated criterion for solidification, signifies the power range. Given the input energy $PV$ for pulsed compression, we then calculate the radius of compressed fuel as $R = (3V/4\pi)^{1/3}$; the mass and energy confinement times are then estimated as $t_\mathrm{S} = R/c_\mathrm{S}$ and $t_\mathrm{E} = c_\mathrm{P} R^2/\kappa$, respectively, where $c_\mathrm{S}$ refers to the adiabatic sound velocity and $\kappa$ is the thermal conductivity [Ichimaru, 1994]. Assuming $PV = 1$ kJ, we plot and compare various physical parameters in Figs. 2 and 3 for the $p$-$d$ cases.

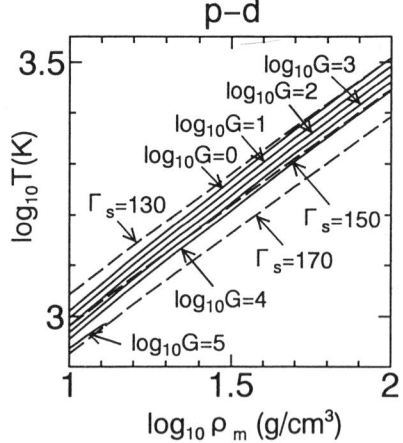

Fig. 1. Constant-$G$ and constant-$\Gamma_\mathrm{s}$ contours on $\rho_\mathrm{m}$-$T$ plane for the $p$-$d$ cases.

In all the cases, the confinement times remain far greater than the heating time, which basically determines the duration of nuclear reactions; the reaction time $t_R$—the inverse of the reaction rate—on the other hand, depends on the initial conditions, which set the gain.

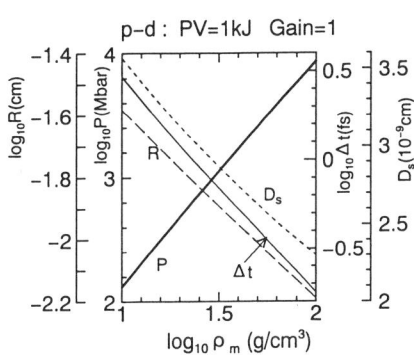

Fig. 2. Pressure, short-range screening distance, confinement radius, and average duration of reactions for the $p$-$d$ cases with $G=1$ and $PV=1$kJ.

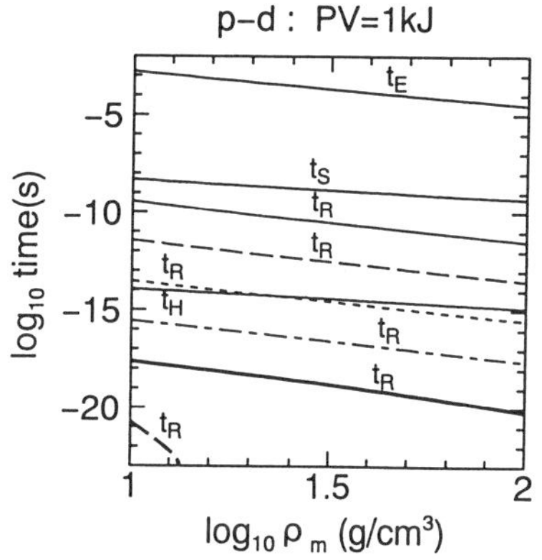

Fig. 3. Various time scales with the $p$-$d$ reactions. Solid lines: $G=1$; dashed line: $G=10$; dotted line: $G=10^2$; chain line: $G=10^3$; thick solid line: $G=10^4$; thick dashed line: $G=10^5$.

In Table 3, we list a set of physical parameters at which experimental feasibility study for the enhanced pycnonuclear reactions may be undertaken with scaled-down initial conditions (Case I) and another under power-producing conditions (Case II). The former conditions may be attainable through extrapolation of the current ultrahigh pressure metal technologies [Nellis et al., 1992; Meyer-ter-Vehn, and Oparin, 1994].

TABLE 3. Physical parameters on a scaled-down experiment for feasibility study (Case I) and a power-production experiment (Case II) both for $p$-$d$ reactions.

| Case | I | II |
|---|---|---|
| $\rho_m$ (g/cm$^3$) | 6 | 20 |
| $T$ (K) | 770 | 1400 |
| $P$ (Mbar) | 49 | 490 |
| $\log_{10} \Delta t$ (s) | $-14.53$ | $-16.46$ |
| $\log_{10} t_R$ (s) | $-9.48$ | $-13.11$ |
| $\log_{10} t_H$ (s) | $-13.71$ | $-14.20$ |
| $\log_{10} t_E$ (s) | $-6.34$ | $-1.29$ |
| $\log_{10} t_S$ (s) | $-10.04$ | $-7.60$ |
| $PV$ (Joule) | $1 \times 10^{-3}$ | $1 \times 10^6$ |
| $R$ (cm) | $3.6 \times 10^{-4}$ | $0.17$ |
| $W$ (Joule) | $1.9 \times 10^{-3}$ | $3.3 \times 10^7$ |
| reactions | $2.2 \times 10^9$ | $3.7 \times 10^{19}$ |

We have thus shown experimental possibilities of utilizing the enhanced pycnonuclear reactions, such as $^2$H$(p, \gamma)^3$He, $^3$H$(d, n)^4$He, and $^7$Li$(p, \alpha)^4$He, in ultrahigh pressure metals. Experimental feasibility is assessed as most promising with the $p$-$d$ reactions, but manageability of fuel material and fusion yields in the $p$-Li cases should not be dismissed.

## ACKNOWLEDGMENTS

The work reported here is an outcome of Japan-US Cooperative Science Programs on astrophysical dense plasmas, and we wish to thank Dr. H. Iyetomi, Dr. S. Ogata, and Professor H.M. Van Horn for collaboration in these years through the Programs. We acknowledge hospitality of Aspen Center for Physics, where a part of this work was carried out.

## REFERENCES

Barnes, C.A., 1971, Adv. Nucl. Phys., **4**, 133.
Bethe, H., 1930, Ann. Phys. (Leipzig), **5**, 325.
Cameron, A.G.W., 1959, Astrophys. J., **130**, 916.
Engstler, S., A. Krauss, K. Neldner, C. Rolfs, U. Schröder, and K. Langanke, 1988, Phys. Lett. B, **202**, 179.
Engstler, S., G. Raimann, C. Angulo, U. Greif, C. Rolfs, U. Schröder, E. Somorjai, B. Kirch, and K. Langanke, 1992, Phys. Lett. B, **279**, 20.
Fowler, W.A., 1984, Rev. Mod. Phys., **56**, 149.
Gamow, G., and E. Teller, 1938, Phys. Rev., **53**, 608.
Harris, M.J., W.A. Fowler, G.R. Caughlan, and B.A. Zimmerman, 1983, Ann. Rev. Astron. Astrophys., **21**, 165.
Ichimaru, S., 1991, J. Phys. Soc. Jpn., **60**, 1437.
Ichimaru, S., 1993, Rev. Mod. Phys., **65**, 255.
Ichimaru, S., 1994, *Statistical Plasma Physics II. Condensed Plasmas*, (Addison-Wesley, Reading, MS).
Ichimaru, S., H. Kitamura, and S. Ogata, 1994, Publ. Astron. Soc. Japan, **46**, 285.
Ichimaru, S., and H. Kitamura, 1994, preprint.
Ichimaru, S., S. Ogata, and K. Tsuruta, 1994, Phys. Rev. E, **50**, 2977.
Ichimaru, S., S. Ogata, and H.M. Van Horn, 1992, Astrophys. J. Lett., **401**, L35.
Ichimaru, S., and K. Utsumi, 1981, Phys. Rev. B, **24**, 7385.
Iyetomi, H., S. Ogata, and S. Ichimaru, 1993, Phys. Rev. B, **47**, 11703.
Meyer-ter-Vehn, J., and A. Oparin, 1994, in these proceedings.
Nellis, W.J., A.C. Mitchell, P.C. McCandless, D.J. Erskine, and S.T. Weir, 1992, Phys. Rev. Lett., **68**, 2937.
Nomoto, K., 1980, Astophys. J., **253**, 798; **257**, 780.
Ogata, S., H. Iyetomi, and S. Ichimaru, 1991, Astrophys. J., **372**, 259.
Ogata, S., H. Iyetomi, S. Ichimaru, and H.M. Van Horn, 1993, Phys. Rev. E, **48**, 1344.
Pethick, C.J., and V. Thorsson, 1994, Phys. Rev. Lett., **72**, 1964.
Salpeter, E.E., 1952, Phys. Rev., **88**, 547.
Salpeter, E.E., 1954, Aust. J. Phys., **7**, 373.
Salpeter, E.E., and H.M. Van Horn, 1969, Astrophys. J., **155**, 183.
Thompson, W.B., 1957, Proc. Phys. Soc. (London) B, **70**, 1.
Yan, X.-Z., S. Tanaka, S. Mitake, and S. Ichimaru, 1985, Phys. Rev. A, **32**, 1785.

# III.2

# EXPERIMENTAL APPROACH TO NUCLEOSYNTHESIS IN THE UNIVERSE

S. Kubono

Institute for Nuclear Study (INS), University of Tokyo
3-2-1 Midori-cho, Tanashi, Tokyo 188, Japan

Recent experimental developments in studying thermonuclear reactions relevant to the nucleosynthesis in the universe are surveyed. Specifically, a new field using radioactive nuclear beams is discussed, which is a powerful tool for investigating the mechanism of nucleosynthesis in explosive phenomena in high-density high-temperature stellar sites. The scope of the field is also discussed.

## I. WHY AND WHERE ARE NUCLEAR ASTROPHYSICAL EXPERIMENTS NEEDED?

Although nuclear reactions are known to play a crucial role in the evolution of the universe, nuclear physics information is often scarce for important processes. These are due to partly the difficulty of the experiments in the laboratories, and partly the lack of development in nuclear physics. There are many nuclear reactions which are so difficult to investigate by the current experimental nuclear physics technology. These, for instance, can be the case for the charged-particle induced nuclear reactions just above the threshold because of the small cross sections due to the Coulomb barrier. The second point can be seen in the case of investigating the explosive nucleosynthesis, which involves nuclear reactions of unstable nuclei [1,2], which were not accessible except for a few nuclides a decade ago. A new development has been made in the study of structure of unstable nuclei and their reactions in nuclear physics since then. Figure 1 depicts nucleosynthesis scenarios on the nuclear chart. For instance, the explosive hydrogen-burning process, which is called the rapid-proton (rp) process [3], is considered to run through the proton-rich unstable nuclear region where experimental studies just have started. Nucleosynthesis of the rp-process goes more or less through resonant states. However, the level density near the threshold is not high as the threshold comes down in energy as one approaches to the nuclei near the proton drip line. Thus, experimental investigation is really required to study the rp-process.

One typical example that indicates clearly the necessity of experiment is a discovery of the first excited state above the proton threshold in $^{20}$Na [4]. This nucleus is important to understand the onset mechanism of the rp-process through the $^{19}$Ne$(p,\gamma)^{20}$Na thermonuclear reaction. In the old data, there was only one broad level known at 2.9 MeV above the threshold (2.199 MeV). The new data, however, shows a lowest state at 2.637 MeV, which is much lower in energy. Actually, the reaction rate estimated based on the new data is more than two orders of magnitude enhanced at the temperature region of $1 \sim 10 \times 10^8$ K, which is a typical temperature for explosive hydrogen burning in novae and supernovae. This reduces to a considerable lowering of the onset temperature of the process. As we discuss later, this reaction rate is not well determine yet, and is still one of the hottest subjects under investigation.

More difficult situation can be seen in the investigation of the rapid-process (r-process) which should have produced very heavy elements, although the site of the r-process is not identified yet. It is still quite difficult in the laboratories to produce the nuclei on the possible r-process path even by the modern high-energy heavy-ion machine or high-energy, high-intensity proton machines except for a few nuclei near the stability line. Because of this reason, the r-process path itself is not well known yet.

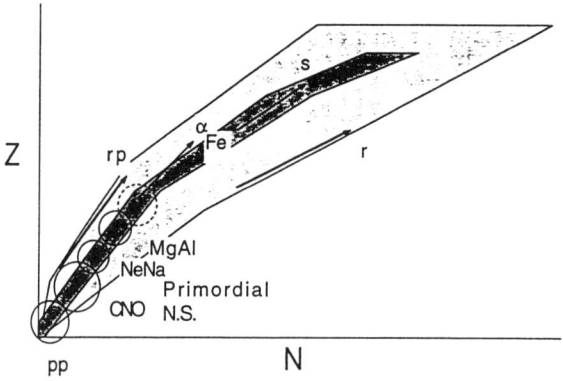

Fig. 1 Nucleosynthesis scenarios on the nuclear chart (N-Z plane). The nucleosynthesis at high temperature locates in unstable nuclear region.

## II. THE CLUSTER NUCLEOSYNTHESIS DIAGRAM AND $^{16}$O SYNTHESIS IN HE-BURNING

The nucleosynthesis at low temperature involves more or less stable and/or long-lived nuclides because of the slow processes involved. There are still some critical reactions not well investigated. Among them, the most crucial and long-standing problem is $^{16}$O production rate by the $^{12}$C$(\alpha,\gamma)^{16}$O reaction in He burning [5]. The reaction rate of this process plays a decisive role in the later half of the stellar evolution. The core/shell size of C and O should be changed according to this rate, which finally brings to a different scenario for the remnant of supernovae.

Let me first discuss here a little on the relation of stellar evolution and the energy generation. Figure 2 shows the Cluster-Nucleosynthesis (CN) diagram that we have proposed recently [6]. This explains stellar evolution from He burning to Si burning which leads to Fe, and the relation to the energy generation in nucleosynthesis point of view. In the universe, $4n$ nuclei, e.g., $^4$He, $^{12}$C, and $^{16}$O, are the most abundant nuclides except for hydrogen, and often participate in the nucleosynthesis, where cluster states above the cluster threshold play a crucial role. The nuclear clustering of nuclei exists in nature, and it is favored to minimize the energy of the nuclear systems because of the large binding energies of these nuclei. The nucleosynthesis in nature of course goes through a synthesis flow-path that obeys the same principle under the site condition of interest. This CN-diagram is similar to the so called cluster diagram in nuclei [7], but includes the dynamics of nucleosynthesis and the typical scenarios. This diagram explains very well the nature of the main nucleosynthesis leading to Fe. The ordinate measures the released energy (or binding energy in nuclear physics) per nucleon, and thus more energy will be released as the nucleosynthesis develops by going to the right-lower level until Fe. One should see the diagram from the a cluster threshold to the ground states of B (= $\alpha$+ A), where the reaction goes through a resonant cluster state of $\alpha$+ A and/or continuum state to the ground state B. In nuclear physics on the contrary, one would discuss the development of clustering in nucleus B by putting excitation energies to the nuclear system B, although the basic physics is the same. Subsequent scenarios till Fe, i.e., C-burning, O-burning, and Si-burning, can be also understood well on the CN-diagram.

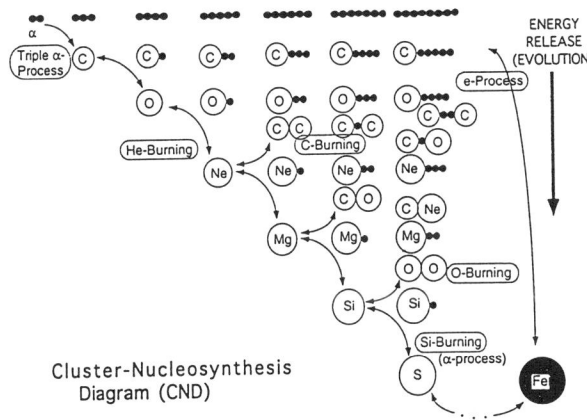

Fig. 2  The Cluster Nucleosynthesis (CN) diagram [6]. Nucleosynthesis of heavy elements goes in the direction from top-left to bottom-right till Fe, releasing the energy to the stellar system.

After hydrogen burning in the main sequence, He, the ash of the pp-chain, can be the main fuel for the next nucleosynthesis, which is called He-burning. The first step is the synthesis of $^{12}$C by the triple $\alpha$-process, and then the $^{12}$C$(\alpha,\gamma)$ reaction to produce $^{16}$O, but it would not go much beyond because the next possible step, $^{16}$O$(\alpha,\gamma)$, has a small reaction rate due to the nuclear structure just above the a threshold in $^{20}$Ne. Although the $^{12}$C$(\alpha,\gamma)$ reaction should be very important in the He-burning stage as discussed above, the reaction rate is not determined yet at the temperature region of interest [8-17], i.e., $\sim 2 \times 10^8$ K or 300 keV above the threshold. The problem here is that the cross sections will be extremely small ($\sim 10^{-8}$ nb/sr estimated at 300 keV), and there will be an interference of the tails of the sub-threshold $1^-$ state at 7.117 MeV and the broad $1^-$ resonance at 9.585 MeV. The gamma decay is considered to be predominantly E1 together with some E2 strength.

Table 1 summarizes the astrophysical $S$-factors estimated from the experimental reaction rate of $^{12}$C$(\alpha,\gamma)^{16}$O determined at $E_{c.m.} \geq 0.94$ MeV by measuring the singles gamma rays [9-11] or the gamma rays and the recoil product $^{16}$O in coincidence [12]. Here, the E1 and E2 strengths were separated by measuring angular distributions. The extrapolation procedures have a quite large ambiguity depending on the theories used, as can be seen in table 1.

Recently, a very small $S_{E1}(300)$ value was reported by Ouellet et al. [14], as shown in table 1. This is much smaller than those previously reported. Extensive measurements were also performed for the b-delayed a decay of $^{16}$N. The measurement has an advantage [8] for the $\beta$-decay to the lower excitation energies and the decay is also sensitive to the E1 strength. Buchmann et al. [15] and Zhao et al. [16] reported S-factors estimated by analyzing the $\alpha$ spectrum from $^{16}$N with $K$-matrix and $R$-matrix theories. The results are consistent each other and much larger than that in ref. 14. These are, however, roughly a factor of two smaller than those currently used for the stellar models [5,18]. It should be concluded that the estimated values are scattered in a wide range, and thus it is poorly determined in both experiment and theory.

Table 1 Astrophysical $S$-factors of $^{12}C(\alpha,\gamma)^{16}O$ estimated theoretically and experimentally in unit of keV·b.

| $S_{E1}(300keV)$ | $S_{E2}(300)$ | $S_{tot}(300)$ | Reference |
|---|---|---|---|
| 140 (three-level R.) | | | Dyer (74) |
| 80 (R. + optical M.) | | | |
| 250 | 180 | 430 | Kettner (82) |
| 200 | 96 | 296 | Redder (87) |
| ≤ 140 | | | Kremer (88) |
| 1 | 40 | | Ouellet (92) |
| 57 (K-matrix) | | | Buchmann (93) |
| 79 (Global R-matrix) | | | |
| 95 (R-matrix) | | | Zhao (93) |
| 69 (Global R-matrix) | | | |

There are many on-going experiments on this subject. One possibility is to use the Coulomb dissociation method [19], which will be discussed in sec. III. This method gives a quite large yield enhancement because of a large phase volume for the inverse kinematical condition as well as of a thicker target usable. One might go down to the region of $E_{c.m.} = 0.5$ MeV, although this method is sensitive to the E2 part in this reaction. One also has to deal carefully with final-state interaction effects and the nuclear-force contribution in a continuum region. These require both theoretical and experimental developments.

Of course one could go down in energy with the direct measurement of $^{12}C(\alpha,\gamma)^{16}O$, or with $\beta$-delayed $\alpha$ decay measurement using a higher-intensity beam with an improved detector system, such as the one in development at the Institute for Nuclear Study (INS), University of Tokyo. This subject still deserves much intense works in both experiment and theory.

## III. COULOMB DISSOCIATION METHOD FOR THE SOLAR NEUTRINO PROBLEM

Another crucial nucleosynthesis, which is not well determined, is the $^7Be(p,\gamma)^8B$ reaction which is directly related to the solar neutrino problem. Before going into the problem of the $^7\text{Be}(p,\gamma)^8\text{B}$ reaction study, I will discuss on the Coulomb dissociation method, which is a new method [19] developed recently for nuclear astrophysics, and could be applied to this problem. For a study of $A(p,\gamma)B$ thermonuclear reaction, one can measure two particles $p$ and $A$ in the inverse reaction, $B(\gamma,p)A$, and then obtain the cross section of $A(p,\gamma)B$ using the principle of detailed balance. The favorable condition may be realized if the inverse reaction is dominated only by Coulomb interaction. A successful application was made to the study of the $^{13}\text{N}(p,\gamma)^{14}\text{O}$ reaction [20], where an unstable $^{14}\text{O}$ beam from the projectile fragment separator at RIKEN was bombarded on a $^{208}\text{Pb}$ target. The virtual photons, produced by the scattering in the large Coulomb field of the target, dissociate $^{14}\text{O}$ into $p+^{13}\text{N}$, giving the gamma width of $3.1\pm0.6$ eV for the resonant state at 5.17 MeV just above the proton threshold in $^{14}\text{O}$. Here, the two dissociated fragments, $^{13}\text{N}$ and $p$, were detected at small relative energies which corresponded to the stellar temperatures of interest. This result is consistent with that measured directly by the $^{13}\text{N}(p,\gamma)^{14}\text{O}$ reaction using a radioactive beam of $^{13}\text{N}$ [21], and has given a definite conclusion on the onset of the Hot-CNO cycle. There are some cautions when applying this method; a) the interaction should be nearly Coulomb force, and b) the nuclear state of interest should be coupled directly to the ground state in the target system.

Recently, the $^7\text{Be}(p,\gamma)^8\text{B}$ thermonuclear reaction [22] was studied by the Coulomb dissociation method, where a $^8\text{B}$ beam was obtained from a secondary beam line at RIKEN, and dissociated on a $^{208}\text{Pb}$ target. The result is still preliminary, but it seems promising to give a clear result for the astrophysical S-factor. A refined experiment is in progress.

## IV. THE EXPLOSIVE HYDROGEN BURNING PROCESS

Explosive hydrogen-burning, which is the rp-process [3] discussed earlier, takes place in high-temperature high-density proton-rich stellar or primordial sites. Figure 3 depicts some cycles, and the breakout processes from one cycle to the next. The rp-process supposedly begins by breaking out from the Hot-CNO cycle to the NeNa cycle through a reaction chain of $^{15}\text{O}(\alpha,\gamma)^{19}\text{Ne}(p,\gamma)^{20}\text{Na}(p,\gamma)^{21}\text{Mg}\cdots$ [1-4,23]. Specifically, the $^{19}\text{Ne}(p,\gamma)^{20}\text{Na}$ reaction has been investigated extensively [24-30], but the reaction rate is not well determined yet experimentally. Here, note that unstable proton-rich nuclei are inevitably involved in these nuclear reactions [1,2], as can be seen in the figure, but nuclear physics information is very scarce for these unstable nuclei. The nuclear reactions which involve unstable nuclei can be investigated either by learning the level property associated, or by directly measuring the production cross sections using the unstable nuclear beams.

Figure 4 displays three reaction rates [1,2] so far estimated for the breakout process from the Hot-CNO cycle. The most important finding in the last decade was the discovery of the first excited state at 2.637 MeV just above the proton threshold in $^{20}\text{Na}$ [4,25], as discussed in sec. I. However, none of the rates in the figure is determined definitely, because the critical physical parameters [32] for the reaction rates of the capture processes are not measured yet. They are more or less assumed by taking the values from the analog states in the mirror nuclei.

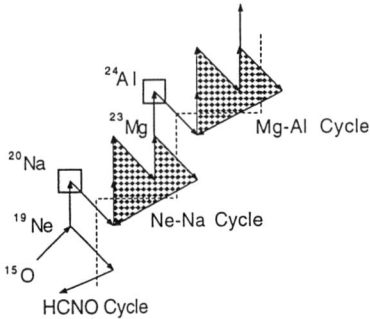

Fig. 3 A schematic nucleosynthesis flow diagram for the leakout from the Hot-CNO cycle to the NeNa cycle, and so on. The left-hand side of the dashed lines is proton-rich unstable nuclei.

There are several experimental efforts made for $^{19}$Ne$(p,\gamma)^{20}$Na to determine the gamma width since then. The gamma decays from this state were searched in the beta delayed gamma spectra of $^{20}$Mg that was produced from the fragmentation process of $^{24}$Mg at RIKEN [26], but no feeding was observed to the 2.637 MeV state in $^{20}$Na. Same type of experiments were also pursued later at MSU [27] and GANIL [28], resulting in mostly the same conclusion. The charge-symmetric reaction $^{20}$Ne$(t,^{3}$He$)^{20}$F [29] did not excite the postulated analog state in $^{20}$F, but excited the 2.966 MeV $3^+$ state with a reasonable cross section. An experiment to measure directly the gamma width is being undertaken at Louvain-la-Neuve [12] by using a radioactive nuclear beam of $^{19}$Ne accelerated up to the energy of the resonance of $p+^{19}$Ne. Further experimental works on this subject are also being prepared [2] at the radioactive beam facilities, e.g., the facility at INS, University of Tokyo, which is under construction.

There are many other experiments made for the study of the early stage of the rp-process and related topics. One is the breakout process of the NeNa cycle to the next MgAl cycle [33]. This problem is related to the onset of the rp-process as this breakout as well as that of the HCNO cycle determines the first stage of the process. One important progress is made by investigating the nuclear levels of $^{24}$Al to learn about the breakout process, $^{23}$Mg$(p,\gamma)^{24}$Al [34]. The experimental result suggests a temperature reduction of about 12 % for the ignition of this breakout.

Another interesting and important problem of the rp-process is where the explosive process would terminate on the nuclear chart. A theoretical calculation suggests that the process will run up to somewhere mass 100 region [35] under a typical condition of x-ray burst ($T = 10^9$ K and $\rho = 10^6$ g/cm$^3$). Here, important is again that almost none is known about the proton-rich unstable nuclei and their reactions, even the existence of the nuclei in the intermediate mass region. An extensive effort is being made on this subject at Michigan State University [36].

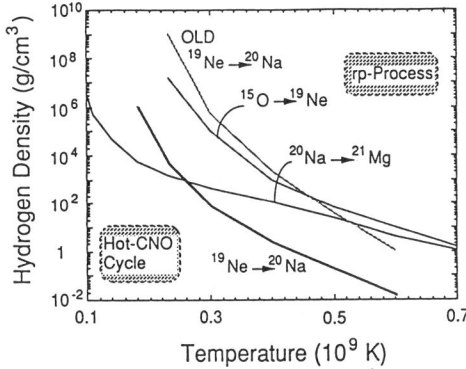

Fig. 4 The estimated thermonuclear reaction rates of the three successive reactions for the onset of the explosive hydrogen burning process [1,2].

Isotopic anomalies in meteorites are also very useful in investigating the stellar burning processes. One of the interesting subjects is so called the Ne-E problem [37], where very high $^{22}$Ne enrichments were observed in presolar grains in meteorites. The most plausible explanation is that the nucleus $^{22}$Na, which was produced in some stellar event, would have beta decayed to $^{22}$Ne afterward in the meteorites as $^{22}$Na has a long half-life of 2.6 y. Thus, the nuclear astrophysical problem here is to know the reaction rates associated with $^{22}$Na [38-42]. At relatively low temperatures, $^{22}$Na would have been produced in the NeNa cycle. The destruction process, $^{22}$Na$(p,\gamma)^{23}$Mg, was recently investigated, and a new level was discovered [41] at the important temperature region of $0.1 \sim 1.0 \times 10^8$ K in $^{23}$Mg. The result suggests that more flux will be carried away by the $^{22}$Na$(p,\gamma)^{23}$Mg reaction than thought before.

## V. BETA DECAY OF IONIZED ATOMIC NUCLEI

Beta decay rates are also of great concern in nuclear astrophysics. As we discussed in the preceding sections, nucleosynthesis problem is always a competition of the nuclear reaction process of interest and the beta decay rate in each step when the target nucleus is unstable. Beta decays to the excited states or that from the excited states also enhance the rate considerably in high temperature sites. Beta decay life time also changes depending on the ionization state of the atom, and also by the electron density of the site nearby [43]. Thus, beta stability of nuclei changes inside cores of massive developed stars. A high ionization might also alter the conclusion of cosmochronology, the estimate of age of the galaxy, if one uses a chronometer of $^{187}_{75}$Re–$^{187}_{76}$Os, for instance [43]. In a dramatic case, the beta stability changes, and a highly ionized atomic nucleus, which is stable when it is neutral, will beta decay, if the $Q$-value is small enough to be negative for a bound-state beta decay. Such situation is expected in $^{163}$Dy.

The abundance of $^{164}$Er is about an order of magnitude larger than the other p-process nuclei nearby in the solar abundance. This anomaly could be explained if the inverse beta decay occurs for $^{163}_{66}$Dy, followed by $^{163}_{67}$Ho$(n,\gamma)^{164}$Ho$(\beta-)^{164}_{68}$Er. The nucleus $^{163}$Dy is stable when it is neutral in charge, but it can beta-decay to $^{163}$Ho when it is fully ionized, which would be realized in the s-process condition. It was difficult to check such process experimentally because full ionization of such heavy element was not possible before.

A beautiful experiment [44] was reported recently on this subject. A high energy heavy ion accelerator made it possible to ionize fully $^{163}$Dy at GSI, Germany. The fully ionized $^{163}_{66}$Dy$^{66+}$ was stored in a storage ring ESR, and the beta decay product $^{163}_{67}$Ho$^{66+}$ was detected. Here, $e^-$ from the beta decay is captured in a bound state of $^{163}_{67}$Ho. The measurement gives $T_{1/2} = 47^{+5}_{-4}$ days for the half life of $^{163}_{66}$Dy$^{66+}$. This value should be compared to a theoretical estimate, $T_{1/2} = 36.7$ days, which is no so off, but indicates clearly an importance of experiment.

## VI. SCOPE OF EXPERIMENTAL NUCLEAR ASTROPHYSICS

As we sketched in the preceding sections, unstable nuclei are inevitably involved in explosive burning. On the other hand, unstable nuclear beams become now available from very low to high energies in the laboratories. Although the quality of the secondary beams from the projectile fragmentation process of heavy ion beams is limited, there are many interesting experiments to be made for nuclear astrophysics, as we discussed here.

A promising method for direct simulation of stellar reactions is to use unstable nuclear beams accelerated. This requires an Isotope Separator On-Line (ISOL) system to produce intense unstable nuclei and an accelerator to boost the kinetic energy up to the energy region of astrophysical interest. The beam quality to be provided is as good as ordinary stable nuclear beams. This method should provide more intense beams than the projectile fragment method, since here one can use much thicker target for high energy proton beams for unstable-nuclei production. The major concern here may be the absolute efficiency of ion sources, which is under development. This type factory of unstable nuclear beams is under construction at the Oak Ridge National Laboratory in U.S.A., and GANIL in France as well as at INS, University of Tokyo as mentioned earlier. There are similar, other plans in Europe and Russia.

Nuclear astrophysics program is of course one of the major research subjects in these facilities. Experimental research there will open a new step to nuclear astrophysics. This new phase in nuclear astrophysics is strongly promoted by the research activities in nuclear physics. Nuclear astrophysical problems have attracted many nuclear physicists, and now it becomes one of the major subjects in nuclear physics. This of course will bring fruitful results for both nuclear astrophysics and nuclear physics.

## REFERENCES

[1] S. Kubono, Proc. Fourth Int. Conf. Nucleus-Nucleus Collisions (Kanazawa, Japan, 1991), Nucl. Phys. A **538** (1992) 505c.
[2] S. Kubono, Comm. on Astrophys. **16** (1993) 287.
[3] R. K. Wallace and S. E. Woosley, Astrophys. J. Suppl. **45** (1981) 389.
[4] S. Kubono, H. Orihara, S. Kato, and T. Kajino, Astrophys. J. **344**, (1989) 460.
[5] T. A. Weaver, and S. E. Woosley, Phys. Rep. **227** (1993) 65.

[6] S. Kubono, Z. Phys. A, (1994) in press.
[7] H. Morinaga, private communication, and K. Ikeda, et al., Prog. Teor. Phys. Suppl. Extra Number (1968) 464.
[8] F. Barker, Aust. J. Phys. **24** (1971) 777.
[9] Dyer and C. A. Barnes, Nucl. Phys. A **233** (1974) 495.
[10] K. U. Kettner, et al., Z. Phys. A **308** (1982) 73.
[11] A. Redder, et al., Nucl. Phys. A **462** (1987) 385.
[12] R. M. Kremer, et al., Phys. Rev. Lett. **60** (1988)1475.
[13] X. Ji, B. M. Filippone, J. Humblet, and S. E. Koonin, Phys. Rev. C **41** (1990) 1736.
[14] J. M. L. Ouellet, et al., Phys. Rev. Lett. **69** (1992) 1896.
[15] L. Buchmann, et al., Phys. Rev. Lett. **70** (1993) 726.
[16] Z. Zhao, et al., Phys. Rev. Lett. **70** (1993) 2066.
[17] L. Buchmann and R. E. Azuma, (1993), private communication.
[18] F.-K. Thielemann and T. Rauscher, Proc. of the International Symposium on Origin and Evolution of the Elements (Tokyo,1992), ed. S. Kubono and T. Kajino, World Scientific, 1993, 254.
[19] G. Baur and M. Weber, Nucl. Phys. A **504** (1989) 352.
[20] T. Motobayashi, et al., Phys. Lett. **264B** (1991) 259.
[21] P. Decrock, et al., Phys. Rev. Lett. **67** (1991) 808.
[22] T. Motobayashi, et al., preprint RUP 94-2 (Rikkyo Univ.)
[23] K. Langanke, M. Wiescher, and W. A. Fowler, Astrophys. J. **301** (1986) 629.
[24] S. Kubono, et al., Z. Phys. A **331** (1988) 359.
[25] L. O. Lamm, et al., Nucl. Phys. A **510** (1990) 503.
[26] S. Kubono, et al., Phys. Rev. C **46** (1992) 361.
[27] J. Görres, et al., Phys. Rev. C **46** (1992) R833.
[28] A. Piechaczek, et al., Proc. Int. Conf. Nuclei Far from Stability/ Atomic Masses and Fundamental Constants (Bernkastel-Kues, 1992), Inst. Phys. Pub., 1993, IPC Series **132**, p. 851.
[29] N. M. Clarke, S. Roman, C. N. Pindor, and P. R. Hayes, J. Phys. G. **19** (1993) 1411.
[30] R. D. Page, et al., Paper presented at the 4th Int. Conf. of Radioactive Nuclear Beams (East Lansing, 1993)
[31] F. Ajzenberg-Selove, Nucl. Phys. A **475** (1987) 1.
[32]When the stellar reaction goes through a resonance, the reaction rate is proportional to the resonance strength, $w\gamma = (2J + 1)/(2J_T + 1)(2J_p + 1) \cdot \Gamma_p \Gamma_g / \Gamma_{tot}$, where $\Gamma_{tot}$ is the total width, $\Gamma_p$ is the proton decay width, and $\Gamma_g$ is the gamma decay width of the resonant state.
[33] M. Wiescher, J. Görres, F. -K. Thielemann, and H. Ritter, Astr. Astrophys. **160** (1986) 56.
[34] S. Kubono, et al., Proc. Int. Nucl. Phys. Conf. (Wiesbaden, 1992), ed. R. Bock, et al.; Nucl. Phys. A **553** (1993) 481c.
[35] S. E. Wooseley, Proc. Accelerated Radioactive Beams Workshop, TRI-85-1 (TRI-UMF), 1986, p. 4.
[36] M. F. Mohar, et al., Phys. Rev. Lett. **66** (1991) 1571, and B. M. Sherrill, private communication.

[37] M. Arnould and H. Nørgaard, Astron. Astrophys. **64** (1978) 195, and the references therein.
[38] J. Görres, et al., Phys. Rev. C **39** (1989) 8.
[39] S. Seuthe, et al., Nucl. Phys. A **514** (1990) 471.
[40] M. Wiescher and K. Langanke, Z. Phys. A **325** (1986) 309.
[41] S. Kubono, et al., Z. Phys. A **348** (1994) 59.
[42] P. Schmalbrock, et al., AIP Conf. **125** (1985) 785.
[43] K. Takahashi and K. Yokoi, Nucl. Phys. A **404** (1983) 578.
[44] M. Jung, et al., Phys. Rev. Lett. **69** (1992) 2164.

III.3

# COSMOLOGICAL PHASE TRANSITION AND NUCLEOSYNTHESIS

T. Kajino

Division of Theoretical Astrophysics, National Astronomical Observatory
Mitaka, Tokyo 181, Japan

Large amplitude fluctuations of inhomogeneous baryon-number density distribution, which are presumed to originate from several cosmological phase transitions, allow primordial nucleosynthesis of heavy elements, as well as the light elements $A \leq 7$, in the $0.01 \leq \Omega_B \leq 0.15$ Universe models. We propose that the astronomical observation of heavy elements from unprocessed materials in the halo stars or inter-galactic medium would constrain the epoch of making the inhomogeneous baryon-number density distribution in the early Universe.

## I. INTRODUCTION

The mass of the Universe consists of luminous baryons and dark matter. The observed luminous component takes less than one percent of the critical mass to marginally close the Universe, i.e. $\Omega_B^{(\text{LUM})}(\text{OBS}) \leq 0.01$. Total dynamical mass including dark matter amounts to $\Omega_{\text{DYN}}(\text{OBS}) = \rho/\rho_C = 0.1$–$0.3$. In the standard big-bang model (hereafter SBBM) (Wagoner et al. 1967; Yang et al. 1981) for primordial nucleosynthesis, this $\Omega_{\text{DYN}}$ value arises no longer from the baryonic dark matter because this model (Walker et al. 1991) infers

$$0.01 \leq \Omega_B h_{100}^2 \leq 0.015 , \qquad (1)$$

from light element abundance constraints, where current estimates of the Hubble constant range between $0.4 < h_{100} < 1.0$. Recent detection (Songaila et al. 1994; Carswell et al. 1994) of Lyman-$\alpha$ deuterium absorption line at high redshift $z = 3.32$ along the sight to a quasar shows extremely high deuterium abundance $1.9 \times 10^{-4} \leq \text{D/H} \leq 2.5 \times 10^{-4}$. If we take this abundance constraint in SBBM, the allowed $\Omega_B$ changes (Mathews, Kajino and Orito 1994) to even smaller value

$$0.005 \leq \Omega_B h_{100}^2 \leq 0.007 . \qquad (2)$$

However, it is known that the baryonic mass in the form of hot X-ray gas (White et al. 1993), a kind of dark matter by definition, in dense galactic clusters makes more contribution to $\Omega_B$ as large as $\approx 0.15$. In addition, if recently detected MACHO's (Alcock et al. 1993; Aubourg et al. 1993) are made of brown dwarfs or small mass stars, their expected contribution to $\Omega_B$ is thought of $\approx 0.07$ (Freeman 1994). Both observations show one order of magnitude larger baryonic mass in the Universe from the SBBM prediction.

There has been particular interest in the possibility that the primordial nucleosynthesis may have occurred in an environment of inhomogeneous baryon distribution

(Wagoner 1973; Zel'dovich 1975; Applegate, Hogan and Scherrer 1987; Alcock, Fuller and Mathews 1987; Fuller, Mathews and Alcock 1988; Malaney and Fowler 1988; Terasawa and Sato 1989, 1990; Kurki-Suonio et al. 1990; Boyd and Kajino 1989; Kajino and Boyd 1990; Kajino, Mathews and Fuller 1990; Mathews et al. 1990) because high baryonic contribution as high as $\Omega_B \leq 0.15$ is possible in this model (Mathews, Kajino and Orito 1994). Such fluctuations in baryon density distribution are thought to have been produced by a number of processes operating in the early Universe; electroweak baryogenesis (Jedamzik et al. 1994a), inflation generated isocurvature fluctuations (Dogonov and Silk 1993), kaon condensation (Nelson 1990), magnetic fields from superconducting cosmic strings (Malaney and Butler 1989), and others. A first order QCD phase transition is one possibility, although the order of the phase transition is still not convincing. In this paper I would discuss a possible mechanism of creating strong baryon inhomogeneity in the cosmological QCD phase transition, based on the phenomenological model (Sumiyoshi et al. 1990; Kajino 1991a). I would then present observable signature of primordial production of heavy elements in the baryon inhomogeneous big-bang nucleosynthesis models (hereafter IBBM) (Mathews, Kajino and Orito 1994; Jedamzik, Fuller, Mathews and Kajino 1994).

## II. COSMOLOGICAL PHASE TRANSITION AND CREATION OF BARYON INHOMOGENEITY

If the QCD phase transition is first order, the dynamics of the phase transition is phenomenologically described in terms of a few fundamental parameters in QCD. They are the critical temperature of the phase transition, $T_C$ ($\approx$ 100–200 MeV), intrinsic surface tension of the phase boundary between quark-gluon-plasma (QGP) and hadron gas, $\sigma$ ($\approx T_C^3$), and baryon permeability through the boundary, $\lambda$ and $\lambda'$. The last parameters $\lambda$ and $\lambda'$ were calculated in the chromoelectric flux tube model (Sumiyoshi et al. 1989) which is an effective model of quark confinement.

When the expanding Universe was first cooled to the critical temperature $T_C$ at $t \approx 10^{-4}$ sec, hadronic bubbles were nucleated in QGP. The liberated latent heat then reheated the Universe to $T = T_C$. During this epoch of coexisting two phases, the baryon number which was originally carried by only quarks inside QGP was transferred between the two phases, leading eventually to inhomogeneous baryon density distribution.

Probability of finding the nucleation sites and distribution of their distances are given by the isothermal fluctuation theory (Fuller et al. 1988). Time variation of the baryon number densities, $n_B^q$ for QGP and $n_B^h$ for hadron gas, is given by

$$\frac{dn_B^q}{dt} = -\lambda n_B^q + \lambda' n_B^h - \left(\frac{\dot{V}}{V} + \frac{\dot{f_V}}{f_V}\right) n_B^q, \qquad (3)$$

$$\frac{dn_B^h}{dt} = \frac{f_V}{1-f_V}\left(-\lambda' n_B^h + \lambda n_B^q + \frac{\dot{f_V}}{f_V} n_B^h\right) - \frac{\dot{V}}{V} n_B^h, \qquad (4)$$

where $V$ is the horizon volume and $f_V$ is the volume fraction of QGP, which are obtained by solving the Einstein equation in weak supercooling.

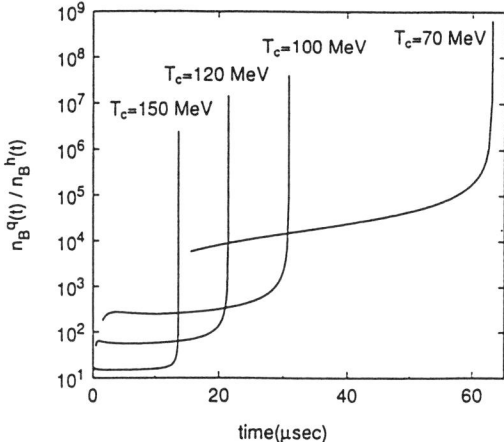

Fig. 1. Calculated baryon-number density ratio, $R = n_B^q/n_B^h$, versus time. Time = 0 refers to the epoch when $T_C$ is first reached. Calculated curves start from the end of supercooling.

Solutions of eqs. (3) and (4) are displayed in Fig. 1; $R = n_B^q/n_B^h$, for $T_C$ = 70, 100, 120 and 150 MeV and the fixed $\sigma = 10^6$ MeV$^3$. If the whole system is in the statistical equilibrium, both $n_B^q$ and $n_B^h$ stay at the initial values in the chemical equilibrium. However, the calculation (Sumiyoshi et al. 1990) shows clearly that the baryon number transfer occurs under a statistically non-equilibrium condition until the end of the phase transition, at which the speed of the phase boundary approaches sound velocity of relativistic plasma $v = c/\sqrt{3}$. After this time the QGP droplets lose energy source and dive into the supercooling phase, from which the high baryon density spots emerge promptly. The obtained density contrast can reach $10^6 < R$.

The final result depends upon the two QCD parameters. Repeating numerical calculations by varying $T_C$ and $\sigma$ in a wide range, we found that the strong baryon-gas fluctuations are created for $T_C \leq 140$ MeV and $10^5 \leq \sigma \leq 10^7$ MeV$^3$ (Sumiyoshi et al. 1990; Kajino 1991a), which affect strongly the primordial nucleosynthesis to be discussed later. For another parameter space 70 MeV $\leq T_C \leq$ 140 MeV and $\sigma \leq 5 \times 10^5$ MeV$^3$, $n_B^q$ reaches nuclear matter density or more, suggesting the creation of quark matter nugget (Kajino 1991a, b).

## III. NUCLEOSYNTHESIS

The amplitude for baryon density fluctuation, $R$, has long been assumed in many papers to be of order $R = R_{eq} \approx 10$–100, which is simply given by the thermodynamic ratio of the equilibrium baryon densities of the two phases (that is shown in Fig. 1 at the beginning of each curve). The actual density contrast, however, could be as large as $R \geq 10^6$ as discussed in the previous section in an example calculation of phenomenological QCD.

The baryon density fluctuation formed long before the nucleosynthesis epoch of $t \approx 100$ sec is affected by neutrino-photo-induced damping in radiation dominated era. Recent theoretical calculation including this effect (Jedamzik et al. 1994b, and references therein) has indicated that any high density zones are inflated to $R \approx 10^5$–$10^6$ before QCD era at $t \approx 10^{-4}$ sec. After that, the density contrast stays there until the beginning of primordial nucleosynthesis. We therefore adopted here $R \approx 10^6$ at the $T = T_C$ (QCD) era. The other geometry parameters in IBBM, the volume fraction of high density zones, $f_V$, and the mean separation distance between the fluctuations, $2r$, were varied. Note however that the following discussions of primordial nucleosynthesis have nothing to do with the QCD phase transition specifically. We will actually discuss later how to know the epoch of making the baryon inhomogeneity in the early Universe, including EW era as well as QCD.

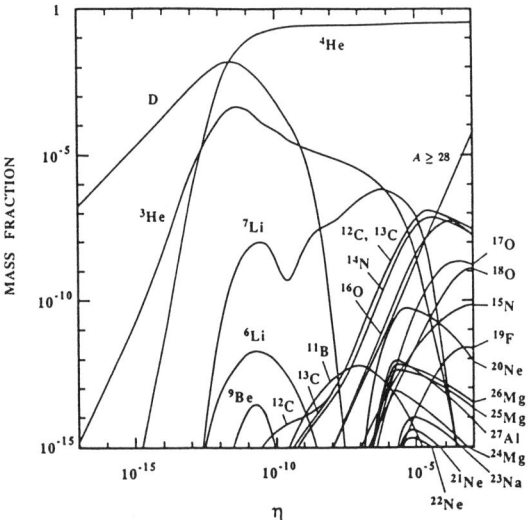

Fig. 2. Standard homogeneous big-bang nucleosynthesis for wide range of $\eta = n_B/n_\gamma$.

Numerical calculations have been done in 16 zones of variable width as in Mathews et al. 1990, including neutron and proton diffusion before, during and after the nucleosynthesis. Three light neutrino family is assumed, and neutron life 889.8 sec was used.

A. Nucleosynthesis in dilute Universe model

The calculated results of SBBM for various densities for $\eta = 10^{-17}$–$10^{-3}$ are displayed in Fig. 2 (Jedamzik, Fuller, Mathews and Kajino 1994), where $\eta$ is the baryon-to-photon number ratio satisfying $\eta = n_B/n_\gamma = 2.68 \times 10^{-8} \Omega_B h_{100}^2$. Given the standard model value $\eta(\text{SBBM}) \approx 3 \times 10^{-10}$, only D, $^3$He, $^4$He and small amount of

$^7$Li are produced. It is clear from this figure that the baryon density fluctuation of order $R_{\text{eq}} = 10$–$100$ around $\eta(\text{SBBM})$ causes difficulty in the observed abundance constraints, particularly for $^7$Li whose abundance is overproduced because the observed abundance hits the theoretical value of local minimum at $\eta \approx \eta(\text{SBBM})$. However, if the fraction of high density zones $f_V$ is smaller than 0.02 as constrained from the $^4$He abundance and the fluctuation amplitude is as large as $R \approx 10^6$ around $\eta \approx 3 \times 10^{-10}$, the abundance constraints on the light elements are still satisfied. This is because at high density zones having $\eta_{\text{high}} = \eta(\text{SBBM}) \times R \approx 3 \times 10^{-4}$ is produced practically no D, $^3$He and $^7$Li.

Baryon diffusion modifies these results. As far as $f_V \ll 1$, however, the modification in D and $^3$He abundances is as small as 4% and it is not larger than the ambiguity arising from the observed abundances. We found that the calculated abundance for heavy elements $A \geq 12$ is most sensitive to the diffusion effect. Fig. 3 shows this sensitivity. For smaller proper separation distances $r < 6$ m (at $T = T_C = 100$ MeV), both protons and neutrons diffuse out of the fluctuations, and the result is very similar to the homogeneous nucleosynthesis. For larger distances 6 m $< r$, however, the heavy element abundances quickly increase due to proton and alpha induced reactions. Baryon diffusion does not disturb the process of heavy element production in high density zones.

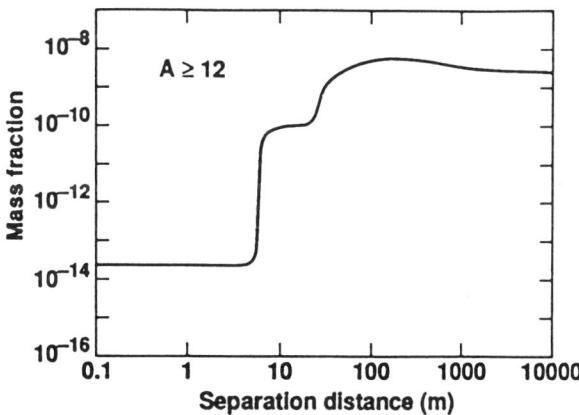

Fig. 3. Sensitivity of baryon diffusion to heavy element abundances. Average yield for heavy elements is shown as a function of proper separation distance (in meter) at $T = 100$ MeV.

B. Constraints on the epoch of producing the baryon inhomogeneity

Figure 4 identifies the region in the $M_B$ vs. $f_V$ plane in which heavy element nucleosynthesis may imply a signature of high amplitude baryon fluctuations, where $M_B$ is the total baryon mass of a single high-density fluctuation zone. Total baryon mass within the horizon for the EW and the QCD phase transition era is also shown in this figure. For $M_B/M_{\text{solar}} \geq 10^{-18}$ of EW baryon mass scale and $f_V \leq 0.02$, heavy elements are formed in amount $[Z] = \log[N(A \geq 12)/N(\text{H})] - \log[N(A \geq 12)/N(\text{H})]_{\text{solar}} \geq -7$ without overproducing helium and lithium. For $M_B/M_{\text{solar}} < 10^{-21}$ it is also possible to satisfy the helium and lithium constraints, but the baryon diffusion erases the inhomogeneities before the nucleosynthesis begins and there is no essential difference between the standard homogeneous Universe model.

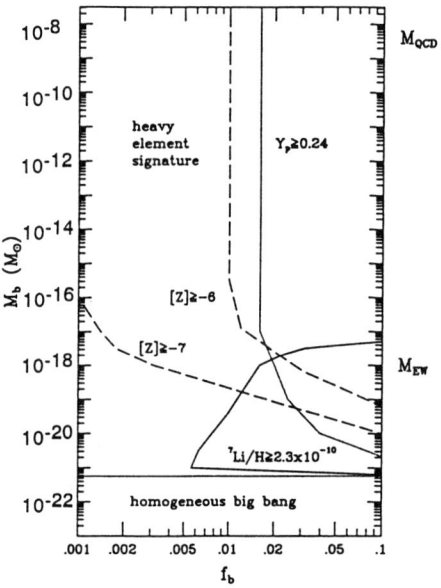

Fig. 4. Heavy element nucleosynthesis signatures of high amplitude fluctuations $\eta_i \geq 10^{-4}$ and $\eta_{\text{low}} = 2.8 \times 10^{-10}$ (Jedamzik, Fuller, Mathews and Kajino 1994). $M_b = M_B$, and $f_b = f_V$.

C. Inhomogeneous nucleosynthesis in baryon dominated Universe models

The discussions in the previous section lead us to seek carefully for more baryon dominated Universe models $\Omega_B \approx 0.2$ in IBBM (Mathews, Kajino and Orito 1994). In the condensed sphere fluctuation geometry the constraints from the light element abundances, $Y_p \leq 0.24$, D/H $\geq 1.8 \times 10^{-5}$, (D+$^3$He)/H $\leq 10^{-4}$ and Li/H $\leq 2.3 \times 10^{-10}$ (Smith et al. 1993) are satisfied in the parameter region of $r \approx 10$–$50$ m (at $T = 100$ MeV), $f_V^{1/3} \approx 0.25$ and $R = 10^6$, allowing $0.015 \leq \Omega_B h_{100}^2 \leq 0.07$.

Recent theoretical studies of stellar evolution of main-sequence metal-poor dwarfs (Deliyannis et al. 1990; Pinsonneault et al. 1992) have indicated that the rotation-driven mixing and diffusion of materials in the stellar atmosphere could destroy lithium by one order of magnitude, and thus the primordial Li abundance in the plateau could be as large as $^7$Li/H $\leq 1.3 \times 10^{-9}$. Adopting this constraint, we can get larger $\Omega_B$ value; $\Omega_B h_{100}^2 \leq 0.15$. We also obtain $\Omega_B h_{50}^2 \leq 0.5$, taking the smallest possible value of Hubble constant $H_0 = 50$ km/s/Mpc.

Finally adopting high D/H abundance as observed quite recently (Songaila et al. 1994; Carswell et al. 1994), $1.9 \times 10^{-4} \leq \text{D/H} \leq 2.5 \times 10^{-4}$, we can estimate $(D+^3\text{He})/\text{H} \leq 3.0 \times 10^{-4}$ together with $1.9 \times 10^{-4} \leq \text{D/H}$ in a simple galactic chemical evolution model (Mathews, Kajino and Orito 1994; Yoshii, Mathews and Kajino 1994). These constraints can still allow $\Omega_B h_{50}^2 \leq 0.15$.

## IV. SUMMARY

Baryon inhomogeneous big-bang models have a potential possibility to resolve the dark matter problem without any hypothetical mass from non-baryonic dark matter. Provided that the high amplitude baryon fluctuations $R \approx n_B^{\text{high}}/n_B^{\text{low}} \approx 10^6$ with small volume fraction of high density zones $f_V \leq 0.02$ and large proper mean separation distance $r \geq 10$ m (at $T \approx 100$ MeV) are created at the era of EW or QCD phase transition, the primordial heavy element abundances are enhanced extremely, still satisfying the light element abundance constraints. Astronomical observation of heavy elements in halo stars or old unprocessed inter-galactic medium is highly desired.

## REFERENCES

Alcock, C. R., Fuller, G. M. and Mathews, G. J. 1987, Astrophys. J. **320**, 439.
Alcock, C. R. et al. 1993, Nature **365**, 621.
Applegate, J. H., Hogan, C. J. and Scherrer, R. J. 1987, Phys. Rev. **D35**, 1151.
Aubourg, E. et al. 1993, Nature **365**, 623.
Boyd, R. N. and Kajino, T. 1989, Astrophys. J. **336**, L55.
Carswell, R. F. et al. 1994, MNRAS **268**, L1.
Dolgov, A. and Silk, J. 1993, Phys. Rev. **D47**, 4244.
Deliyannis, C. P., Demarque, P. and Kawaler, S. D. 1990, Astrophys. J. Suppl. **73**, 21.
Fuller, G. M., Mathews, G. J. and Alcock, C. R. 1988, Phys. Rev. **D37**, 1380.
Freeman, K. C. 1994, Proc. 6th Asia Pacific Physics Conference and 11th Australian Institute Physics Congress.
Jedamzik, K., Fuller, G. M., Mathews, G. J. and Olinto, A. 1994a, Phys. Lett. B, in press.
Jedamzik, K., Fuller, G. M., Mathews, G. J. and Kajino, T. 1994b, Astrophys. J. **422**, 423.
Kajino, T. and Boyd, R. N. 1990, Astrophys. J. **359**, 267.
Kajino, T., Mathews, G. J. and Fuller, G. M. 1990, Astrophys. J. **364**, 7.
Kajino, T. 1991a, Nucl. Phys. (Proc. Suppl.) **B24**, 74.
Kajino, T. 1991b, Phys. Rev. Lett. **66**, 125.
Kurki-Suonio, H. et al. 1990, Astrophys. J. **353**, 406.
Malaney, R. A. and Fowler, W. A. 1988, Astrophys. J. **333**, 14.
Malaney, R. A. and Butler, M. N. 1989, Phys. Rev. Lett. **62**, 117.
Mathews, G. J., Kajino, T. and Orito, M. 1994, Astrophys. J., submitted.
Mathews, G. M., Meyer, B. S., Alcock, C. R. and Fuller, G. M. 1990, Astrophys. J. **358**, 36.
Nelson, A. 1990, Phys. Lett. **240B**, 179.
Pinsonneault, M. H., Deliyannis, C. P. and Demarque, P. 1992, Astrophys. J. Suppl. **78**, 179.
Smith, M. S., Kawano, L. H. and Malaney, R. A. 1993, Astrophys. J. Suppl. **85**, 219.
Songaila, et al. 1994, Nature **368**, 599.
Sumiyoshi, K., Kusaka, K., Kamio, K. and Kajino, T. 1989, Phys. Lett. **225B**, 10.

Sumiyoshi, K., Kajino, T., Alcock, C. R. and Mathews, G. J. 1990, Phys. Rev. **D42**, 3963.
Terasawa, N. and Sato, K. 1989, Phys. Rev. **D39**, 2893.
Terasawa, N. and Sato, K. 1990, Astrophys. J. **362**, L47.
Wagoner, R. V., Fowler, W. A. and Hoyle, F. 1967, Astrophys. J. **148**, 3.
Wagoner, R. V. 1973, Astrophys. J. **197**, 343.
Walker, T. P., Steigmann, G., Schramm, D. N., Olive, K. A. and Kang, H. 1991, Astrophys. J. **376**, 51.
White, S. D. M. et al. 1993, Nature **366**, 429.
Yang, J., Turner, M. S., Steigman, G., Schramm, D. N. and Olive, K. A. 1984, Astrophys. J. **281**, 493.
Yoshii, Y., Mathews, G. J. and Kajino, T. 1994, Astrophys. J., submitted.
Zel'dovich, Ya. B. 1975, Sov. Astr. Lett. **1**, 5.

# IV. Atomic and Optical Processes in Stellar Matter

# IV.1

# EQUATION OF STATE, METALLIZATION, AND ENERGY TRANSPORT IN DENSE STELLAR MATTER

S. Ogata, H. Kitamura, and S. Ichimaru

Department of Physics, University of Tokyo, Bunkyo, Tokyo 113, Japan

Equations of state and phase diagrams for partially to fully ionized dense stellar matter are investigated by taking into account physical effects of strong coupling between atoms and plasma particles, such as decrease and elimination of the atomic levels, multiple scattering of electrons by the atoms, and dispersion force at finite temperatures. Phase diagrams for $^4$He and $^{12}$C matter are exhibited as examples. Electric and thermal resistivities are calculated and expressed in parametrized forms through explicit account of the phase shifts in electron-atom scattering. A novel formulation of the free-free opacity is applied for examination of the many-particle correlation effects represented by the dynamic dielectric function of electrons, the complex Gaunt factor, and the static structure factor of the ions.

## I. INTRODUCTION

Dense stellar matter [Van Horn, 1990, 1991; Ichimaru, 1994] expected in the interiors of giant planets (e.g., Jupiter) and brown dwarfs, and in the envelopes of white dwarfs and neutron stars, exhibits various types of phase transition including metallization [Mott, 1974] and change in the degrees of ionization [Ichimaru, 1994]. The rate coefficients of energy transport in dense matter, such as conductive and radiative opacities, may depend sensitively on the nature of the quantum states which characterize it. Equation of state (EOS) and transport properties of dense matter affect crucially the predictions on the internal structure and thermal evolution of various astronomical objects [Van Horn, 1990, 1991; Clayton, 1983].

In Sec. II, we briefly summarize thermodynamic properties for various realizations of dense matter and plasmas. A new scheme of evaluating EOS and thereby constructing the phase diagram for partially ionized dense matter is described in Sec. III. Parametrized expressions of the electric and thermal resistivities for such a dense matter are obtained in Sec. IV. A novel formulation of the photo-absorption cross section is applied in Sec. V for examination of various effects of interparticle correlation on the free-free opacity.

## II. THERMODYNAMIC FUNCTIONS FOR DENSE MATTER AND PLASMAS

Thermodynamic functions for various realizations of dense plasma materials have been described in recent monographs by one of the present authors [Ichimaru, 1993, 1994]. Here, without repeating the fundamental formulas contained therein, we confine ourselves to presentation of the features improved over or additional to these formulas, including involvement of the partially ionized atoms.

A. Electron liquids

Thermodynamic properties of the electron liquids with number density $n_e$ ($= N_e/V$) are characterized by the dimensionless parameters,

$$r_s = a_e/a_B \quad \text{and} \quad \Theta = k_B T/E_F , \qquad (1)$$

where $a_e = (4\pi n_e/3)^{-1/3}$, $a_B = \hbar^2/me^2$, and $E_F$ is the Fermi energy. The parameter $r_s$ characterizes the strength of Coulomb coupling, and $\Theta$ measures the degree of Fermi degeneracy.

The Helmholtz free energy of an electron liquid, normalized by $N_e k_B T$, is given by

$$f_e(n_e, T) = f_e^{\text{id}} + f_e^{\text{x}} + f_e^{\text{c}} . \qquad (2)$$

Here $f_e^{\text{id}}$ and $f_e^{\text{x}}$ denote the ideal electron-gas term and the first-order exchange term, respectively; useful fitting-formulas for these energies applicable at any combinations of $r_s$ and $\Theta$ exist [e.g., Ichimaru, 1994; Clayton, 1983].

Effects of the Coulomb correlation are represented in $f_e^{\text{c}}$. The correlation energy $f_e^{\text{c}}$ in the ground state has been accurately evaluated by Green's function Monte Carlo method [Ceperley and Alder, 1980]; this quantity in the classical limit likewise has been well investigated [e.g., Ogata and Ichimaru, 1987]. In the regime of intermediate degeneracy ($\Theta = 0.1, 1, 5$), Tanaka and Ichimaru [1986] computed the correlation energy in Singwi-Tosi-Land-Sjölander [1968] scheme of the dielectric formulation [e.g., Ichimaru, 1994]. Accounting for all of those evaluations, fitting formula for $f_e^{\text{xc}} = f_e^{\text{x}} + f_e^{\text{c}}$ applicable at any $r_s$ and $\Theta$ has been obtained in Ichimaru et al. [1987] as Eq. (3.83) combined with Eqs. (3.64) and (3.79)–(3.82).

B. Ionic liquids

One-component plasma (OCP) is an assembly of charged point-particles in a uniform background of neutralizing charges, which obeys the classical statistics [Ichimaru, 1982]. Degrees of Coulomb correlation in such an OCP are measured by

$$\Gamma = \frac{(Ze)^2}{a_{\text{ion}} k_B T} \quad \text{with} \quad a_{\text{ion}} = (4\pi n_{\text{ion}}/3)^{-1/3} . \qquad (3)$$

Abe [1959] derived an analytic formula for the normalized interaction energies applicable for $\Gamma < 1$ by a giant-cluster expansion technique.

$$u_{\text{ex}}^{\text{ABE}}(\Gamma) = -\frac{\sqrt{3}}{2}\Gamma^{3/2} - 3\Gamma^3 \left[\frac{3}{8}\ln(3\Gamma) + \frac{\gamma}{2} - \frac{1}{3}\right] \qquad (4a)$$

where $\gamma = 0.57721\cdots$ is Euler's constant. In the strong-coupling liquid regime, $1 \leq \Gamma < 180$, values of the interaction energy have been calculated accurately by Monte Carlo (MC) simulations [Brush et al., 1966; Slattery et al., 1980, 1982; Ogata and Ichimaru, 1987]. Thus we express for $0 < \Gamma < 180$ [Ichimaru, 1994]

$$u_{\text{OCP}}^{\text{int}}(\Gamma) = \frac{u_{\text{ex}}^{\text{ABE}}(\Gamma) + (3 \times 10^3)\Gamma^{5.7} u_{\text{ex}}(\Gamma)}{1 + (3 \times 10^3)\Gamma^{5.7}} \qquad (4b)$$

where
$$u_{\text{ex}}(\Gamma) = -0.898004\Gamma + 0.96786\Gamma^{1/4} - 0.86097 + 0.220703\Gamma^{-1/4} . \qquad (4c)$$

A coupling-constant integration yields the expression for the total free energy normalized by $N_{\text{ion}}k_{\text{B}}T$ as [e.g., Ichimaru, 1994]

$$f_{\text{OCP}}(n_{\text{ion}}, T, Z) = f_{\text{ion}}^{\text{id}} + f_{\text{OCP}}^{\text{int}} \quad \text{with} \quad f_{\text{OCP}}^{\text{int}}(\Gamma) = \int_0^\Gamma d\Gamma' \frac{u_{\text{OCP}}^{\text{int}}(\Gamma')}{\Gamma'} . \qquad (4d)$$

In an electron-screened OCP, the electron-screened Coulomb-coupling parameter,

$$\Gamma_{\text{s}} = \Gamma S_{\text{c}}(a_{\text{ion}}) \qquad (5)$$

with the screening function $S_{\text{c}}(r)$ defined for example in Eq. (A84) of Ichimaru [1993], may replace the $\Gamma$ of Eq. (3) in the expression of the interaction energy.

In the cases of multi-ionic mixtures composed of $N_i$ ions of charge $Z_i e$ ($i$: species) in a volume $V$, interaction free-energy $F_{\text{MIM}}^{\text{int}}$ is well described by a linear mixing [Ichimaru et al., 1987] of the OCP formula,

$$F_{\text{MIM}}^{\text{LM}} = \sum_i^{\{\text{species}\}} N_i f_{\text{OCP}}^{\text{int}}(\Gamma_i) , \qquad (6)$$

where $\Gamma_i = (Z_i e)^2/a_i k_{\text{B}} T$ with $a_i = (4\pi \sum_j N_j Z_j / 3VZ_i)^{-1/3}$. Extensive MC simulations have been performed [Ogata et al., 1993] for binary and ternary mixtures, and have shown that the the possible deviations of $F_{\text{MIM}}^{\text{LM}}$ from $F_{\text{MIM}}^{\text{int}}$ are less than 1% near the freezing conditions. It should be remarked, however, that such deviations, albeit small, do play an essential part in the analysis of chemical separation associated with a freezing transition in a multi-ionic mixture plasma [Ogata et al., 1993].

Interaction free energy for charged hard spheres [Hansen and McDonald, 1986] is given by

$$f_{\text{CHS}}^{\text{int}} = f_{\text{OCP}}^{\text{int}}(\Gamma) + \zeta \frac{\eta(4 - 3\eta)}{(1 - \eta)^2} \quad \text{with} \quad \zeta = \frac{\eta}{\eta + 0.00065\Gamma^{1.5}} . \qquad (7)$$

Here $\eta = \pi n_{\text{ion}} d^3/6$ is the packing fraction and $d$ is the hard sphere diameter. Function $\zeta$ in Eq. (7) is determined by using the results [Hansen and Weis, 1977] of MC simulation in charged hard-sphere systems at parametric combinations of $\Gamma = 20 \sim 70$ and $\eta = 0.343 \sim 0.4$. Since Coulomb repulsion acts to lower the contact probabilities, $\zeta$ takes on a smaller value as $\Gamma$ increases.

C. Strong electron-ion correlations—Incipient Rydberg states

Tanaka et al. [1990] analyzed strong interparticle correlation in fully ionized, dense hydrogen plasmas near a metal-insulator transition using an integral equation approach. They thereby found emergence of an "incipient Rydberg state (IRS)" as the kinetic energy of an electron approaches the binding energy between an electron and an ion, that is, Rydberg energy. Ichimaru [1993] advanced interaction energy formulas in terms of such an IRS model, which accurately reproduce the computed values of the interaction energy. Here we present improvement of these formulas by newly introducing a low-density correction term (Eq. 15 below), to ensure a correct boundary condition in the limit of low densities.

We consider a two-component plasma of electrons and ions with charge $Ze$. In the IRS model, conduction electrons are regarded as being a mixture of "free" and IRS electrons. Fraction of electrons in IRS is characterized by the IRS parameter

$$X = \frac{x_b}{1+x_b} \quad \text{with} \quad x_b^2 = r_s \tanh\left(\hbar\sqrt{\frac{2\pi}{mk_BT}} n_e^{1/3}\right) . \quad (8)$$

Screening parameters associated with "free" and IRS electrons, respectively, are

$$k_f = \sqrt{\frac{4\pi(1-X)n_e e^2}{[\partial P_F(n_e)/\partial n_e]_T}} \quad \text{and} \quad k_b = \frac{2}{a_B} \frac{8 + 12a_B k_f}{(2+a_B k_f)^3} . \quad (9)$$

In Eq. (9), $P_F$ is the pressure of an ideal Fermi gas. The interaction energy between electrons and ions in units of $N_{\text{ion}} k_B T$ is given by

$$u_{ei} = (1-X)u_{ei}^f + X u_{ei}^b \quad (10)$$

with

$$u_{ei}^f = -\left[0.95 + 0.05\exp\left(-\Theta^{-1/2}\right)\right]\frac{(Ze)^2 k_f}{k_B T} \quad (11)$$

and

$$u_{ei}^b = -\tanh\left(\hbar\sqrt{\frac{2\pi}{mk_BT}} n_e^{1/3}\right)\frac{(Ze)^2 k_b}{2k_B T} . \quad (12)$$

We obtain the normalized free energy due to electron-ion correlation as

$$f_{ei}(n_e, T, Z) = \int_0^{e^2} \frac{de'^2}{e'^2}\left[u_{ei}(e'^2) + \delta u_{ei}^{(1)}(e'^2) + \delta u_{ei}^{(2)}(e'^2)\right] . \quad (13)$$

To ensure boundary conditions in the weak coupling regime, we have introduced in Eq. (13) a high-temperature correction term

$$\delta u_{ei}^{(1)} = \left[(\sqrt{2}-1)u_{\text{ex}}^{\text{ABE}}(\Gamma_s(Z)) - (\sqrt{2}-1)\frac{\sqrt{3}}{2}\Gamma_e^{3/2}\right]\exp\left(-p_1\Gamma^{3/2} - p_2/\Theta\right) , \quad (14)$$

and a low-density correction term

$$\delta u_{ei}^{(2)} = -u_{ei}\exp\left(-p_3\frac{E_F}{e^2/a_e}\right) . \quad (15)$$

Parameters $p_1$, $p_2$, $p_3$ in Eqs. (14) and (15) are found to be $p_1 = 0.5$, $p_2 = 0.1$, and $p_3 = 6$ through comparison with the total internal energies calculated by Tanaka et al. [1990].

## III. PHASE DIAGRAMS ASSOCIATED WITH IONIZATION TRANSITION

We consider a dense matter composed of single species of atoms with the mass number $A$ and the charge number $Z$. For a given set of mass density $\rho_m$, temperature $T$, and fractional populations $x(j)$ for $j$-th ionized atoms, the density of conduction electrons is calculated as

$$n_e = \sum_{j=1}^{Z} j x(j) n_A \quad \text{with} \quad n_A = \frac{\rho_m}{A m_u}, \qquad (16)$$

where $m_u = 1.661 \times 10^{-24}$g is the unit of nuclear mass.

### A. Free-energy formulas

The Helmholtz free energy $F_{tot}$ at a given set of $(n_A, T, \{x(j)\})$ may be expressed as

$$\frac{F_{tot}}{N_A k_B T} \equiv f_{tot}(n_A, T, \{x(j)\}) = \frac{N_e}{N_A} f_e + f_{eA} + f_A + f_{attr} + \sum_j x(j) f_A^{intra}(j). \qquad (17)$$

Here the first three terms on the right-hand side denote those contributions arising from electron-electron, electron-atom, and atom-atom correlations; $f_{attr}$, from the attractive dispersion force between neutral atoms; and $f_A^{intra}(j)$, from the atomic levels. Ionization equilibrium for the matter at a given combination of $n_A$ and $T$ is determined by minimizing $f_{tot}$ in Eq. (17) with respect to $\{x(j)\}$. Once a set $\{x(j)\}$ is determined, we may calculate thermodynamic quantities in a standard procedure. The Gibbs free energy, for instance, is calculated as $G(P,T) = F_{tot} + PV$ with $P = -\frac{\partial F_{tot}}{\partial V}\Big]_{T,\{x(j)\}}$.

Physical effects considered in the present analyses include excluded volume of the atoms, decrease ("shallowing") and disappearance of atomic levels due to plasma screening, EOS of strongly coupled plasmas, and thermal effects in the dispersion forces. Explicit expression for each term in Eq. (17) is given as follows.

#### 1. Excluded volume

Electrons bound in atoms act to reduce the effective volume for plasma particles. Such an effect may be approximately described by replacing $n (= n_e, n_{ion})$ for

$$\tilde{n} = \frac{n}{1 - \delta V/V} \quad \text{with} \quad \frac{\delta V}{V} = \sum_{j=0}^{Z-1} \frac{\pi n_A x(j)}{6} [d_{HS0}(j)]^3. \qquad (18)$$

Here $d_{HS0}(j)$ is the diameter of a $j$-th ionized atom in vacuum. We take experimental values [Allen, 1973] for $d_{HS0}(0)$. For ionized atoms ($j = 1, 2, \cdots, Z-1$), we estimate them as

$$d_{HS0}(j) = d_{HS0}(0) / \left[1 + 0.13\sqrt{\delta E_{j+1}^{B0}(\text{eV})}\right], \qquad (19)$$

where $\delta E_{j+1}^{B0} \equiv E_{j+1}^{B0} - E_j^{B0}$ is the $(j+1)$st ionization energy [Allen, 1973] in vacuum. Equation (19) can well reproduce the diameters [Allen, 1973] of ionized atoms.

2. Modification of the atomic levels

Atomic levels become shallow or disappear due to screening effects of the plasma particles; extent of such a modification depends on $\rho_m$ and $T$. We evaluate the screening parameter $D_s(j)$ for a $j$-th ionized atom by using $S_c(r)$ in Eq. (5) as

$$S_c(a_j) = \exp\left(-\frac{a_j}{D_s(j)}\right) \quad \text{with} \quad a_j = j^{1/3} a_e . \tag{20}$$

By defining "Bohr radius" $R_B(j) = (j+1)e^2/(2\delta E_{j+1}^{B0})$, the $(j+1)$st ionization energy $\delta E_{j+1}^B$ in a dense plasma may be estimated as

$$\delta E_{j+1}^B = f\left(\frac{R_B(j)}{D_s(j)}\right) \delta E_{j+1}^{B0} . \tag{21}$$

Here $f(x)$ is determined from a 1s solution to Schrödinger equation with a potential $V(r) = -Ze^2 \exp(-r/D_s)/r$ and takes the form [Fushimi et al., 1994] $f(x) = 1 - 1.9585x + 1.2172x^2 - 0.24900x^4 + 0.012973x^4$. A level is eliminated when $x \geq 1.17$ (i.e., $f(x) \leq 0$).

Normalized free energy associated with the bound electrons is given by

$$f_A^{\text{intra}}(j) = \begin{cases} E_j^B/k_B T & \text{for } j \geq 1 , \\ -\ln\left[\sum_l g_l \exp(-E_l/k_B T)\right] & \text{for } j = 0 , \end{cases} \tag{22}$$

where $E_j^B = E_{j+1}^B - \delta E_{j+1}^B$ with $E_Z^B = 0$, $g_l$ is the statistical weight, and $E_l$ is the atomic level in vacuum.

3. EOS of the strongly-coupled plasmas

The contributions $f_e$, $f_{eA}$, $f_A$ in Eq. (17) are calculated as

$$f_e = f_e(\tilde{n}_e, T) , \quad f_{eA} = \sum_{j=1}^{Z} x(j) f_{ei}(\tilde{n}_e, T, j) ,$$

$$f_A = f_{\text{ion}}^{\text{id}}(\tilde{n}_{\text{ion}}, T) + \sum_{j=0}^{Z} x(j) \int_0^{e^2} \frac{de'^2}{e'^2} u_{\text{OCP}}^{\text{int}}(\Gamma_s'(j)) \tag{23}$$

$$+ \Upsilon \left[\frac{\Pi(3s + t - 3s\Pi)}{(1-\Pi)^2} + (t-1)\ln(1-\Pi)\right] .$$

The term with [ ] corresponds to the interaction free-energy [Lebowitz et al., 1964] for multi-component hard spheres, where $\Upsilon = \Pi/(\Pi + 0.00065\langle\Gamma_s\rangle^{1.5})$, $\Pi \equiv \sum_{j=0}^{Z} \pi n_A x(j) [d_{\text{HS}}(j)]^3/6$, and $d_{\text{HS}}(j)$ represents the atomic diameter in a plasma. In Eq. (23), $s \equiv \langle d_{\text{HS}}^2\rangle\langle d_{\text{HS}}\rangle/\langle d_{\text{HS}}^3\rangle$; $t \equiv \langle d_{\text{HS}}^2\rangle^3/\langle d_{\text{HS}}^3\rangle^2$; and the average $\langle \ \rangle$ is defined as $\langle B^k\rangle \equiv \sum_j B(j)^k x(j)$. The diameter $d_{\text{HS}}(j)$ is calculated by

$$\frac{d_{\text{HS}}(j)}{d_{\text{HS0}}(j)} = 1 + 0.09\left(\frac{d_{\text{HS0}}(j)}{2D_s(j)}\right)^2 . \tag{24}$$

Formula (24) derives from an expansion of the 1s-orbital in the potential $V(r) = -Ze^2 \exp(-r/D_s)/r$.

4. Thermal effects in dispersion forces

The dipole-dipole attractive potential between neutral atoms is calculated as [e.g., Kihara, 1978]

$$V_{\text{attr}}(r) = -\frac{\mu(T)}{r^6} \quad \text{with} \quad \mu(T) = \frac{3\hbar}{\pi} \int_0^\infty \alpha^2(i\omega) d\omega , \quad (25)$$

where the dynamic dipole-polarizability $\alpha(\omega)$ is given by

$$\alpha(i\omega) = \sum_l^{\{\text{bound}\}} \sum_n^{n \neq l} \frac{e^2}{m} \frac{f_{ln}}{(E_l - E_n)^2/\hbar^2 + \omega^2} \frac{g_l \exp(-E_l/k_B T)}{\sum_p^{\{\text{bound}\}} g_p \exp(-E_p/k_B T)} \quad (26)$$

with the oscillator strength $f_{ln}$, the atomic levels $E_l$, and the statistical weight $g_l$; suffix $n$ runs over all the states including continuum. Coefficient $\mu(T)$ may depend sensitively on the detailed structure of the levels as well as their occupation probabilities at finite temperatures. Energy intervals between adjacent levels are generally smaller at higher levels. When $T$ is raised, transition probability between higher levels increases, and so does the $\mu(T)$.

We calculate $\mu(T)$ between He-He and C-C pairs using experimental data [Moore, 1971; Wiese and Glennon, 1972] for $g_l$, $E_l$, and $f_{ln}$; total numbers of the atomic levels used are 14 for He and 15 for C, and those of the transitions are 74 for He and 58 for C. The $f$-sum rule has been evoked for evaluation of the oscillator strengths including continuum transitions.

Figure 1 shows the values of $\mu$ so calculated as a function of $T$. In the ground state, $\mu(0) = 0.71$ eV·Å$^6$ for He, in good agreement with the experimental value [Dalgarno, 1967] 0.87 eV·Å$^6$ at $T \sim 4$K. In both cases of He and C, we find a substantial increase in $\mu(T)$ when $k_B T$ takes on a few eV, a magnitude of astrophysical significance.

Contribution of the attractive potential to the total free energy may be calculated as

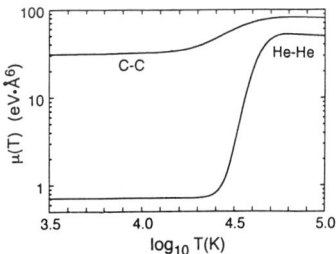

Fig. 1. Temperature dependence of $\mu$ for He-He and C-C pairs.

$$f_{\text{attr}} = \frac{n_A x(0)}{2} \int_{d_{\text{HS0}}(0)}^\infty \{1 - \exp[-V_{\text{attr}}(r)/k_B T]\} 4\pi r^2 dr . \quad (27)$$

B. Phase diagrams

For a construction of the phase diagram associated with ionization transition, we rely on experimental date for the ionization energies [Allen, 1973] $\delta E_j^{B0}$, the diameter of a neutral atom [Allen, 1973] $d_{\text{HS0}}(0)$, and the atomic levels $E_l$ and their statistical weight $g_l$ for a neutral atom [Moore, 1971]. In the case of $^4$He atom, $\delta E_j^{B0} = 24.59$eV ($j = 1$), 54.42eV (2), and $d_{\mathrm{HS0}}(0) = 2.1$Å; for an $^{12}$C atom, $\delta E_j^{B0} = 11.26$eV ($j = 0$), 24.38eV (1), 47.86eV (2), 64.48eV (3), 392.0eV (4), 490.0eV (5), and $d_{\mathrm{HS0}}(0) = 1.4$Å. We have used 14 atomic levels for He, and 15 for C, consistently with the calculations of $\mu(T)$ in Eq. (25).

Figure 2 plots contours of constant $P$ and $\langle Z \rangle \equiv \sum_j j x(j)$ for $^4$He matter. In the hatched region, the matter at constant $P$ and $T$ separates into low- and high-$\rho_m$ phases. Figure 3 depicts variations of $x(j)$ as a function of $\rho_m$ at $T = 10^{4.8}$K, $10^5$K, and $10^{5.2}$K. Dotted curves represent $x(0)$; dashed curves, $x(1)$; and solid curves, $x(2)$.

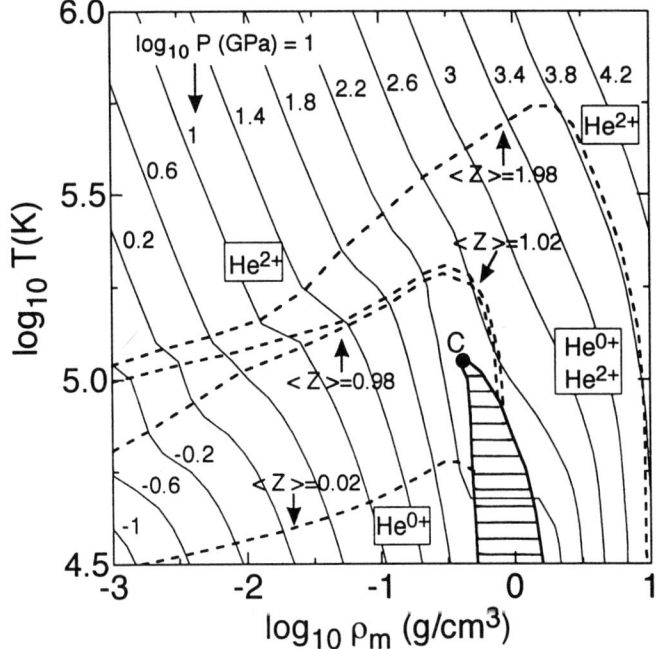

Fig. 2. Contour plots of $P$ and $\langle Z \rangle$ for $^4$He matter.

In a low $\rho_m$ regime ($< 10^{-1}$g/cm$^3$), ionization proceeds from He$^{0+}$ to He$^{1+}$ then to He$^{2+}$ as $T$ increases. Slow variation of $\langle Z \rangle \approx 1$, observed at $T \sim 10^5$K results from large difference in the ionization energies; that is, $\delta E_1^B \approx 14$eV and $\delta E_2^B \approx 54$eV at $(\rho_m, T) = (10^{-3}$g/cm$^3, 10^5$K$)$. Mean charge number $\langle Z \rangle$ decreases as we increase $\rho_m$ at a fixed value of $T$; it is due to the elevation of chemical potential for electrons.

At $\rho_m \sim 10^0$g/cm$^3$, a first-order phase transition [Ichimaru, 1994] associated with metallization occurs at low $T$ as a result of the strong Coulomb correlation effects contained in $f_{ee}$, $f_{eA}$, and $f_{AA}$. We note that the transition proceeds from He$^{0+}$ directly to He$^{2+}$. We find the critical point for such a transition at $(\rho_c, T_c) = (10^{-0.4}\text{g/cm}^3, 10^{5.1}\text{K})$. Since thermal effects enhance ionization, the value of $\rho_m$ at which such a transition occurs decreases as $T$ increases. Table I lists $\rho_m$ dependence of various parameters at $T = T_c$. Shallowing of the level for He$^{1+}$ is significant at $\rho_m \sim \rho_c$, leading to a negligibly small fraction of He$^{1+}$ at the transition.

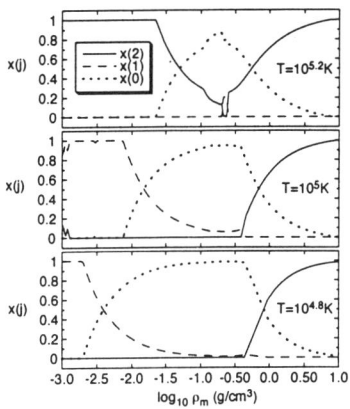

Fig. 3. Variation of $x(j)$ for $^4$He matter.

TABLE I. Density dependence of various parameters at $T = 10^{5.1}$K for $^4$He matter.

| $\rho_m$ (g/cm$^3$) | $P$ (GPa) | $\langle Z \rangle$ | $r_s$ | $D_s(1)/a_B$ | $\delta E_2^B / \delta E_2^{B0}$ | $\langle \Gamma_s \rangle$ | $\Pi$ |
|---|---|---|---|---|---|---|---|
| $10^{-1}$ | $10^{1.6}$ | 0.15 | 8.7 | 5.8 | 0.84 | 0.0094 | 0.064 |
| $10^{-0.4}$ | $10^{2.3}$ | 0.23 | 4.5 | 2.6 | 0.67 | 0.027 | 0.21 |
| $10^{0}$ | $10^{2.7}$ | 1.38 | 1.8 | 0.94 | 0.26 | 0.26 | 0.23 |

Figure 4 shows the results of analogous calculations for $^{12}$C matter. At $T < 10^{4.9}$K, we find three stages of the first-order transition as $\rho_m$ increases from the value $10^{-5}$g/cm$^3$. At the first stage, the atomic states change from C$^{2+}$ to C$^{0+}$; at the second, from a mixture of C$^{0+}$ and C$^{2+}$ to C$^{4+}$; at the third, from C$^{4+}$ to C$^{6+}$. The critical temperatures for these transitions are: $T = 10^{4.9}$K, $10^{5.7}$K, $10^{5.5}$K, respectively. Predominance of the atomic states with C$^{2+}$ and C$^{4+}$ is observed, reflecting the nature of electronic level structures in terms of the ionization energies. Physical parameters at the critical points are listed in Table II. Since the atomic levels significantly decrease or even disappear in the vicinity of the critical point c$_3$, one finds an inequality, $T_{c_3} < T_{c_2}$.

The thermal enhancement of $\mu(T)$ in Eq. (25) acts to increase the fraction of neutral atoms. If we set $\mu(T)$ at a constant level $\mu(0)$ for $^{12}$C matter, the critical temperature related to the transition between C$^{2+}$ and C$^{0+}$ would decrease significantly and take on a value $T_{c_1} = 10^{4.5}$K.

The phase diagram for $^4$He matter obtained in the present calculation (Fig. 2) differs from that obtained by Förster et al. [1992] in a number of ways. In particular the latter authors derived a phase diagram with two critical points: one for the transition between He$^{0+}$ and He$^{1+}$ with the critical point $(\rho_m, T) = (10^{0.3}\text{g/cm}^3, 10^{4.5}\text{K})$, and the other for He$^{1+}$ to He$^{2+}$ with the critical point $(10^{0.9}\text{g/cm}^3, 10^{5.1}\text{K})$. In these calculations, various high-density effects have been ignored, however, including the effects of level shallowing and disappearance due to plasma screening and the thermal enhancement of the dispersion forces. At a density $\rho_m \sim 10^{0.5} \text{g/cm}^3$, the $r_s$ value for the electrons is less than unity. It is thus clear that an atomic level in $\text{He}^{1+}$ should be unstable at such a high density.

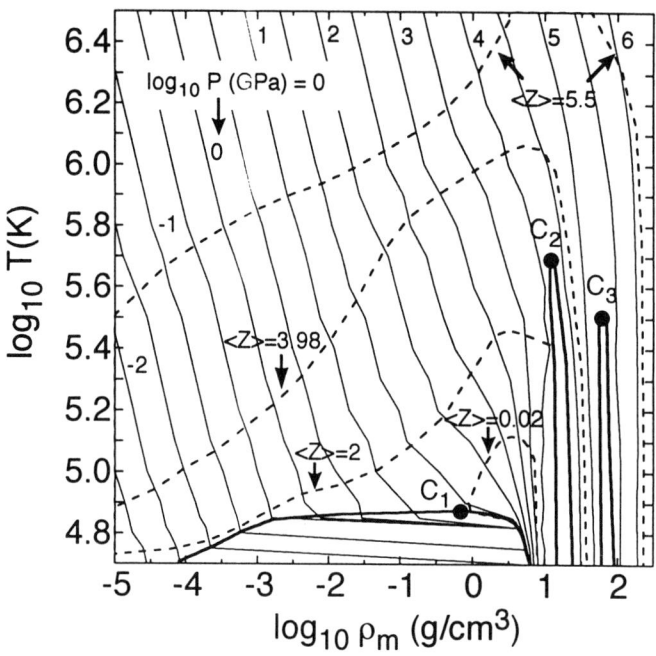

Fig. 4. Contour plots of $P$ and $\langle Z \rangle$ for $^{12}\text{C}$ matter.

TABLE II. Various parameters at the critical points for $^{12}\text{C}$ matter.

| point | $T$ (K) | $\rho_m$ (g/cm$^3$) | $P$ (GPa) | $\langle Z \rangle$ | $r_s$ | $\Theta$ | $\langle \Gamma_s \rangle$ | $\Pi$ |
|---|---|---|---|---|---|---|---|---|
| $c_1$ | $10^{4.9}$ | $10^{-0.2}$ | $10^{1.4}$ | 0.037 | 11 | 17 | 0.0026 | 0.043 |
| $c_2$ | $10^{5.7}$ | $10^{1.1}$ | $10^{4.2}$ | 2.4 | 0.84 | 0.61 | 0.98 | 0.21 |
| $c_3$ | $10^{5.5}$ | $10^{1.8}$ | $10^{5.3}$ | 5.1 | 0.39 | 0.085 | 7.7 | 0.37 |

## IV. CONDUCTIVE TRANSPORT: ELECTRIC AND THERMAL RESISTIVITIES

In an earlier investigation [Ichimaru, 1994], parametrized expressions for the electric and thermal resistivities of electron-ion two-component plasmas (TCPs) were derived on the basis of the microscopic calculations [Tanaka, et al., 1990] obtained through the solution to quantum-mechanical transport equations for the electrons. In this section, we extend those results to the cases of high-$Z$ TCPs, guided by an additional calculation of the $Z$-dependent effects in the scattering cross sections of the ion spheres [Kitamura and Ichimaru, 1994]. The results are then applied to evaluation of the resistivities in partially ionized plasmas, in which the degrees of ionization formulated in Sec. III are incorporated. We take into account the effects of IRS, core scattering, and heat capacities in the resistivities.

A. $Z$-dependent electric and thermal resistivities

To begin, we revisit the derivation of parametrized formulas for the resistivities in a fully ionized TCP with the ionic charge number $Z$. Electric and thermal resistivities, $\rho_E$ and $\rho_T$, are expressed as [Ichimaru, 1994]

$$\rho_E = \frac{8}{3}\left(\frac{\pi}{2}\right)^{1/2} \frac{Z^2 e^2 m^{1/2} n_{\text{ion}}}{n_e (k_\text{B} T)^{3/2}} L_E(Z) , \qquad (28\text{a})$$

$$\rho_T = \frac{c_P^{(0)}}{c_P} \frac{52(2\pi)^{1/2}}{75} \frac{Z^2 e^4 m^{1/2} n_{\text{ion}}}{n_e k_\text{B} (k_\text{B} T)^{5/2}} L_T(Z) . \qquad (28\text{b})$$

Here $c_P$ and $c_P^{(0)}$ refer to the specific heat (per unit volume) at constant pressure for the plasma and that for the ideal-gas electrons, respectively. The factor $c_P^{(0)}/c_P$ has been introduced in Eq. (28b) under the assumption that the plasma as a whole couples strongly with its partial component of the electrons; under these circumstances, we expect that the electrons may transport energy by the amount $c_P T$ per unit volume. In the usual treatment based on a quantum transport equation [e.g., Ichimaru, 1992], however, only the ideal-gas part for the electrons, $c_P^{(0)} T$, would be taken into consideration.

Formulas (28) define the generalized Coulomb logarithms $L_E(Z)$ and $L_T(Z)$. In the classical ($\Theta \gg 1$) and weak-coupling ($\Gamma \ll 1$) limit, $L_E(Z)$ and $L_T(Z)$ approach the Debye-Hückel result,

$$L_0(Z) = -\frac{1}{2}\ln \zeta - \frac{1}{2}\left[\gamma + \frac{1}{Z}\ln(Z+1)\right] + O(\zeta) , \qquad (29)$$

where $\zeta = \hbar^2 (Z+1) k_{\text{De}}^2 / 8 m k_\text{B} T$, $k_{\text{De}} = (4\pi n_e e^2 / k_\text{B} T)^{1/2}$, and $\gamma = 0.57721\cdots$ is Euler's constant. Equation (29) was derived by Kivelson and DuBois [1964] with the aid of the quantum-mechanical version of Balescu-Guernsey-Lenard equation [e.g., Ichimaru, 1992].

In the degenerate ($\Theta \ll 1$) and strong-coupling ($\Gamma > 1$) regime, the interparticle correlations are described by the ion-sphere model [e.g., Ichimaru, 1982], in which one regards an ion as being surrounded by a uniform electronic charge sphere of a radius $a = (3Z/4\pi n_e)^{1/3}$. In this model, the potential of electron-ion scattering is written as

$$U(r) = \begin{cases} Ze^2\left(-\dfrac{1}{r} + \dfrac{3}{2a} - \dfrac{r^2}{2a^3}\right) & \text{for } r \leq a , \\ 0 & \text{for } r > a , \end{cases} \qquad (30)$$

where we have set the position of the ion at the origin $\mathbf{r} = 0$.

The resistivities are proportional to the transport cross section $Q_m(k_F)$ for the electrons with wave number $k = k_F = (3\pi^2 n_e)^{1/3}$. We have evaluated $Q_m(k_F)$ from the phase shifts, which have been calculated through a numerical solution to Schrödinger equations with the scattering potential $U(r)$.

The Born approximation is applicable for $E_F > Z^2 e^4 m/2\hbar^2$. In this regime, values of $Q_m(k_F)$ obtained in the phase shift analyses in fact exhibit good agreement with the results in the Born approximation; the results can be expressed in a fitting formula,

$$Q_m^{\text{Born}}(k_F) = 1.14 a^2 r_s^2 \, Z^{8/3} \exp(-1.47 Z^{1/3}) \,. \tag{31}$$

This formula is applicable for $Z \leq 26$.

On the basis of the results, Eqs. (29) and (31), in the two limiting cases, we present parametrized formulas of the Coulomb logarithms $L_E(Z)$ and $L_T(Z)$ as

$$L_{E(T)}(Z) = \frac{1}{2} \ln\left[1 + \alpha_{E(T)} \left(\frac{1}{\zeta_{\text{DH}}} + \tanh \frac{1}{\zeta_{\text{Born}}}\right)\right]$$
$$\times \left\{1 + A_{E(T)} x_b^2 \exp(-C r_s^D) + B_{E(T)} x_b^{10} [\exp(-C r_s^D)]^5\right\} \,, \tag{32}$$

where $\alpha_E = 1$, $\alpha_T = 75/13\pi^2$, and

$$\zeta_{\text{DH}} = \frac{(Z+1)^{1+\frac{1}{Z}} \exp \gamma}{(12\pi^2)^{1/3}} \frac{\Gamma_e}{\Theta} \,, \quad \zeta_{\text{Born}} = \frac{1}{K \Theta^{3/2} Z^{4/3} \exp(-1.47 Z^{1/3})} \,. \tag{33}$$

In those formulas, $A_E = 0.42$, $B_E = 0.063$, $A_T = 0.38$, $B_T = 0.049$, $C = 6 \times 10^{-4}$, $D = 2$, and $K = 2.5$, which have been determined through fit to the numerical values of Coulomb logarithms for hydrogen plasmas [Tanaka et al., 1990] computed in the ranges of $0.01 \leq \Theta \leq 10$, $0.05 \leq \Gamma \leq 43.441$, and $x_b \leq 1.5$.

The term inside the braces in formula (32), which is a steeply increasing function of the IRS fractional parameter $x_b$, represents the enhancement of scattering due to the strong electron-ion Coulomb coupling. In the low-density ($r_s \gg 1$) limit, however, IRS should vanish since the probability of finding electrons within a Bohr radius of an atom is extremely small; the factor $\exp(-C r_s^D)$ accounts for such a consideration.

Those analytic formulas retain the following features.
(i) In the classical ($\Theta \gg 1$) and weak-coupling ($\Gamma \ll 1$) case, since $\zeta_{\text{DH}} \ll 1$ and $\zeta_{\text{Born}} \ll 1$, Eq. (32) reproduces the Debye-Hückel result, Eq. (29).
(ii) In the degenerate ($\Theta \ll 1$) and strong-coupling ($\Gamma \gg 1$) case, since $1 \ll \zeta_{\text{Born}} < \zeta_{\text{DH}}$, we find $L_E \approx \alpha_E/2\zeta_{\text{Born}}$ and $L_T \approx \alpha_T/2\zeta_{\text{Born}}$. The transport cross section $Q_m(k_F)$ obtained from $\rho_E$ via the Drude formula, $Q_m(k_F) = n_e e^2 \rho_E/\hbar n_{\text{ion}} k_F$, is thus proportional to $Q_m^{\text{Born}}(k_F)$ of Eq. (31). Furthermore the Wiedemann-Frantz relation [e.g., Ziman, 1972], $\rho_E/\rho_T = \pi^2 k_B^2 T/3 e^2$, is satisfied.

In Figs. 5 and 6, values of $\rho_T$ for a fully ionized $^4$He TCP at $T = 3.2 \times 10^5$ K and a fully ionized $^{12}$C TCP at $T = 1 \times 10^6$ K are shown and compared with other theoretical predictions.

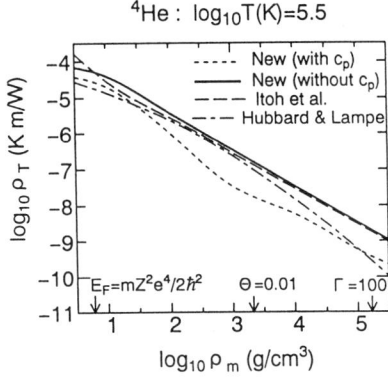

Fig. 5. Thermal resistivity of fully ionized $^4$He plasmas in various theories.

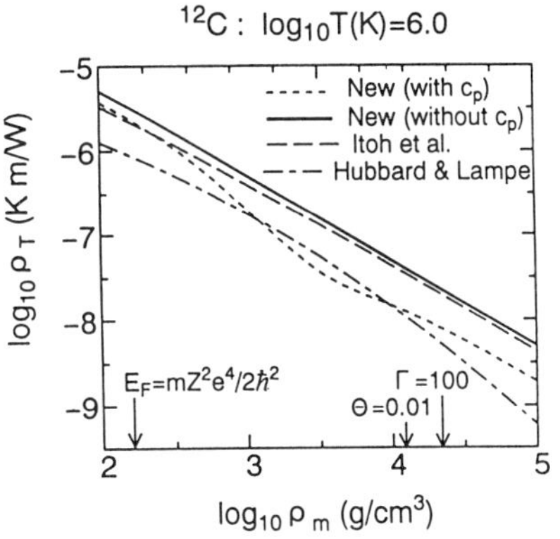

Fig. 6. Thermal resistivity of fully ionized $^{12}$C plasmas in various theories.

Hubbard and Lampe [1969] computed thermal resistivities in dense stellar matter through the Chapman-Enskog method in the weak-coupling regime. In the strong-coupling regime $10 \lesssim \Gamma \lesssim 100$, however, those authors used an interpolation formula between fluid and solid state. The resistivities are thus underestimated, as Figs. 5 and 6 illustrate.

Itoh et al. [1983] calculated the resistivities using the Ziman formula with the RPA dielectric function of the relativistic electrons and the ionic structure factor for the classical OCP. The relativistic term in the resistivity is ignored for the comparison in Figs. 5 and 6. Without the effect of heat capacity, the results agree well with the present calculations in the limit of $\Theta \ll 1$, despite the difference in the evaluations of the correlation functions.

We remark that in the classical limit $\Theta \gg 1$, Eqs. (28) take on values somewhat larger than the Spitzer values [Spitzer and Härm, 1953], because the effects of electron-electron interaction are not appropriately taken into account in the deformation of the electronic distribution function [e.g., Ichimaru, 1992].

B. Application to partially ionized plasmas

The degrees of ionization $\{x(j)\}$ in partially ionized plasmas can be determined through the procedure described in Sec. III. Since an atom in the $j$-th stage of ionization has a net charge $je$, the Coulomb resistivities may be calculated as

$$\rho_E^{\text{Coulomb}} = \frac{8}{3}\left(\frac{\pi}{2}\right)^{1/2} \frac{e^2 m^{1/2} n_A}{n_e (k_B T)^{3/2}} \sum_{j=1}^{Z} j^2 x(j) L_E(j) , \qquad (34\text{a})$$

$$\rho_T^{\text{Coulomb}} = \frac{52(2\pi)^{1/2}}{75} \frac{e^4 m^{1/2} n_A}{n_e k_B (k_B T)^{5/2}} \sum_{j=1}^{Z} j^2 x(j) L_T(j) , \qquad (34\text{b})$$

where $L_E(j)$ and $L_T(j)$ are given by Eq. (32) in which $Z$ is replaced by $j$. As far as the electronic transport is concerned, we need not take into account the excluded volume effect for the electrons; the partially ionized atoms may be regarded as the sources of external potential field which scatter the carrier electrons.

In addition to those Coulomb resistivities, contributions stemming from scattering of electrons against the neutral atoms should be taken into account. Denoting the transport cross section for such a scattering by $Q_m^{\text{core}}$, we calculate the core resistivities through solution to Boltzmann equation for the electrons in the Lorentz-gas approximation as

$$\rho_E^{\text{core}} = \frac{3\pi^2 \hbar^3 n_A x(0) Q_m^{\text{core}}}{2e^2 m k_B T I_0(\alpha)} , \qquad \rho_T^{\text{core}} = \frac{3\pi^2 \hbar^3 n_A x(0) Q_m^{\text{core}}}{2m k_B^3 T^2 F(\alpha)} , \qquad (35)$$

where

$$I_\nu(\alpha) = \int_0^\infty dt \frac{t^\nu}{\exp(t - \alpha) + 1} \qquad (36)$$

are the Fermi integrals with the normalized electronic chemical potential $\alpha = \mu/k_B T$, and $F(\alpha) = 3I_2(\alpha) - 4I_1^2(\alpha)/I_0(\alpha)$.

The total electric and thermal resistivities are thus given by

$$\rho_E = \rho_E^{\text{Coulomb}} + \rho_E^{\text{core}} , \qquad \rho_T = \frac{c_P^{(0)}}{c_P} (\rho_T^{\text{Coulomb}} + \rho_T^{\text{core}}) . \qquad (37)$$

We evaluate $c_P$ with the ideal and Coulombic parts of the EOS, Eq. (23). Here we assume that the neutral atoms do not contribute to the heat transport through $c_P$; those contributions are therefore excluded in the evaluation of $c_P$.

Figures 7 and 8 illustrate the thermal resistivities so computed for $^4$He and $^{12}$C matters, respectively, at $T = 1 \times 10^5$ K. We adopt $Q_m^{\text{core}} = 4.6 a_B^2$ for helium atom [Massey and Burhop, 1952] and $Q_m^{\text{core}} = 11 a_B^2$ for carbon atom [Cooper and Martin, 1962], for incident electrons of energy $k_B T$.

In the low density region where the electrons are classical, we observe large enhancement of the resistivities due to the large IRS fraction, $x_b \approx 1.6$. We see that the core resistivities may become comparable to the Coulomb resistivities in the region where the plasmas are largely dominated by neutral atoms. The factor $c_P^{(0)}/c_P$ significantly reduces the resistivities especially in the high-density strong-coupling region. In Fig. 8, we note that for $10^{0.15}$g/cm$^3 \lesssim \rho_m \lesssim 10^{0.85}$g/cm$^3$, all the atoms are neutral and there remain no carrier electrons.

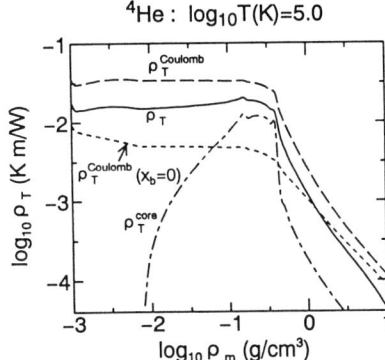
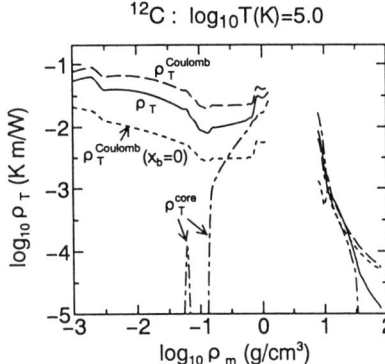

Fig. 7. Thermal resistivity of $^4$He matter.

Fig. 8. Thermal resistivity of $^{12}$C matter.

## V. RADIATIVE TRANSPORT: FREE-FREE OPACITY

In dense matter, the rate of photo-absorption depends sensitively on the electronic states and the interparticle correlation; the dispersion relation of photons is affected by the plasma. We here consider the photo-absorption process due to the conduction electrons.

A. Formulation of the free-free absorption cross section

Dispersion relation of photon with angular frequency $\omega$ is given by $\omega^2 \epsilon_\perp(k,\omega) = c^2 k^2$ [e.g., Pines, 1963], where $\epsilon_\perp(k,\omega)$ is the transverse dielectric function, $k$ is the wave number, and $c$ is the velocity of light in vacuum. Since $k \ll n_e^{1/3}$, we may assume $k = 0$ in $\epsilon_\perp(k,\omega)$, which we shall denote as $\epsilon_\perp(\omega)$. For $\text{Im}\,\epsilon_\perp(\omega) \ll 1$, photo-absorption cross section $\sigma(\omega)$ per atom is thus calculated as

$$\sigma(\omega) n_A = 2\,\text{Im}\,k = \frac{\sqrt{2}\omega\,\text{Im}\,\epsilon_\perp(\omega)}{c\sqrt{\text{Re}\,\epsilon_\perp(\omega) + |\epsilon_\perp(\omega)|}} \approx \frac{\omega^2}{c\sqrt{\omega^2 - \omega_p^2}} \text{Im}\,\epsilon_\perp(\omega). \quad (38)$$

The function $\epsilon_\perp(\omega)$ is generally related to the current-current correlation function [Pines, 1963; Mahan, 1981]. In calculating such a correlation function, we express Fourier component of the electron density fluctuation, $\delta n_e(\vec{q},\omega)$, in the dielectric formulation [e.g., Ichimaru, 1994] as

$$\delta n_e(\vec{q},\omega) = \chi_0(\vec{q},\omega) v_e(q)$$
$$\times \left\{ -[1 - G_{ei}(\vec{q},\omega)] \sum_j j \delta n_j(\vec{q},\omega) + [1 - G_{ee}(\vec{q},\omega)] \delta n_e(\vec{q},\omega) \right\}. \quad (39)$$

Here $\chi_0(\vec{q},\omega)$ is the free-electron polarizability, $v_e(q) = 4\pi e^2/q^2$, and $\delta n_j(\vec{q},\omega)$ refers to the density fluctuation of $j$-th ionized atoms. The local field corrections, $G_{ee}(\vec{q},\omega)$ and $G_{ei}(\vec{q},\omega)$, take into account strong interparticle correlation beyond a mean-field description and describe scattering between particles beyond the Born approximation.

In terms of these functions, $\epsilon_\perp(\omega)$ is now calculated as [Hasegawa and Watabe, 1969; Mahan, 1981]

$$\epsilon_\perp(\omega) = 1 - \frac{\omega_p^2}{\omega^2} - \frac{n_A}{m^2 \omega^4} \sum_{\vec{q}} (\vec{q}\cdot\hat{\eta})^2 q^2 \tag{40}$$
$$\times \left|\frac{v_e(q)}{\tilde{\epsilon}(q)}\right|^2 \langle Z^2\rangle S_Z(q) \left[\frac{1 - G_{ei}(q,\omega)}{\epsilon_e(q,\omega)} - \frac{1 - G_{ei}(q,0)}{\epsilon_e(q,0)}\right].$$

Here $\omega_p = \sqrt{4\pi n_e e^2/m}$, $\hat{\eta}$ denotes a unit vector,

$$\tilde{\epsilon}(q) = 1 - v_e(q)[1 - G_{ee}(q,0)]\chi_0(q,0), \tag{41a}$$

$$\frac{1}{\epsilon_e(q,\omega)} = \frac{1 + v_e(q)G_{ee}(q,\omega)\chi_0(q,\omega)}{1 - v_e(q)[1 - G_{ee}(q,\omega)]\chi_0(q,\omega)}, \tag{41b}$$

$$S_Z(q) = \frac{1}{N_A \langle Z^2\rangle}\left\langle \left|\sum_j j n_j(\vec{q})\right|^2\right\rangle, \quad \langle Z^2\rangle = \sum_j j^2 x(j), \tag{41c}$$

and $\langle\ \rangle$ refers to a statistical average. Since $m_u \gg m$, we have adopted the static approximation for the ion density fluctuations in Eq. (40); that is,

$$\left\langle\left|\sum_j j n_j(\vec{q},\omega)\right|^2\right\rangle \approx \left\langle\left|\sum_j j n_j(\vec{q},0)\right|^2\right\rangle \delta(\omega).$$

Combining Eqs. (38) and (40), we thus obtain

$$\sigma(\omega) = -\frac{1}{m^2\omega^2\sqrt{\omega^2 - \omega_p^2}cV}\sum_{\vec{q}}(\vec{q}\cdot\hat{\eta})^2 q^2 \left|\frac{v_e(q)}{\tilde{\epsilon}(q)}\right|^2 \tag{42}$$
$$\times \langle Z^2\rangle S_Z(q)\,\text{Im}\left[\frac{1 - G_{ei}(q,\omega)}{\epsilon_e(q,\omega)}\right].$$

We note that Eq. (42) provides a rigorous framework in which different approximations may be compared. Accuracy of a given approximation may be assessed through numerical comparison of the physics contents in $\tilde{\epsilon}(q)$, $S_Z(q)$, $G_{ei}(q,\omega)$, and $\epsilon_e(q,\omega)$.

In the free-free processes, absorption of a photon is accompanied by acceleration of an electron in the field produced by a nucleus or an ion; such an acceleration is manifested in the imaginary part of $G_{ei}(q,\omega)$ in Eq. (42). In the ideal-gas limit of high-$T$ and low-$\rho_m$, the cross section $\sigma(\omega)$ in Eq. (42) should reduce to $\sigma_0(\omega) \equiv \langle \sigma_S(\omega; v, v')f(v)[1-f(v')]\rangle_{\vec{v},\vec{v}'}$ where $\sigma_S$ is the Sommerfeld [1953] cross-section obtained by the use of the exact Coulomb wave functions with the electron velocities, $v$ and $v'$, before and after the absorption [Karzas and Latter, 1961]; $f$ denotes the Fermi distribution function. Comparing between $\sigma_0(\omega)$ and $\sigma(\omega)$ with $\tilde{\epsilon} = S_Z = 1$, Re $G_{ei} = $

1, and the Hartree-Fock (HF) values for $\epsilon_e$, we obtain a limiting formula for Im $G_{ei}$ at $\Theta \gg 1$ as

$$\text{Im } G_{ei}(q,\omega) = \frac{4}{3\sqrt{\pi}} \sqrt{\frac{\text{Ry}}{k_B T}} \frac{1}{Y^3} \exp\left[-\frac{(\Omega + Y^2)^2}{4\Theta Y^2}\right], \quad (43a)$$

where $\text{Ry} = me^4/2\hbar^2$, $Y = q/k_F$, and $\Omega = \hbar\omega/E_F$. At $\Theta \gg 1$, Im $G_{ei} \propto 1/(\sqrt{T}q^3)$, which implies significant effects of acceleration for low-velocity electrons.

We may then identify the Born-Elwert expression [e.g., Bethe and Salpeter, 1957] for the Gaunt factor as Re $G_{ei}(q,\omega)$ averaged over the distribution of $\vec{k}$; i.e.,

$$\langle \text{Re } G_{ei}(q,\omega)\rangle_{\vec{k}} = G_{ei}^{\text{BE}}(\omega) = 1 - \left\langle \frac{1 - \exp(-2\pi\xi_{\vec{k}+\vec{q}})}{2\pi\xi_{\vec{k}+\vec{q}}} \frac{2\pi\xi_{\vec{k}}}{1 - \exp(-2\pi\xi_{\vec{k}})}\right\rangle_{\vec{k}}. \quad (43b)$$

Here $\xi_{\vec{k}} = \frac{Ze^2 m}{\hbar^2 k}$, $\xi_{\vec{k}+\vec{q}} = \frac{Ze^2 m}{\hbar^2 |\vec{k}+\vec{q}|}$, and $\frac{\hbar^2 |\vec{k}+\vec{q}|^2}{2m} = \frac{\hbar^2 k^2}{2m} + \hbar\omega$.

Correlation between electrons and ions contained in $\tilde{\epsilon}$, $S_Z$, and $\epsilon_e$ affects $\sigma(\omega)$ in the following manner: First, $\tilde{\epsilon}(q) \ (> 1)$ acts to reduce the rate of inelastic scattering. Figure 8 compares the HF and RPA values of $-\text{Im}[1/\epsilon_e(q,\omega)]$ at $\omega = 1.05\omega_p$ with $r_s = 1$ and $\Theta = 0.1, 10$. For RPA, we observe that (i) plasmon-peak appears at $q \ll k_F$, (ii) probabilities of electron-hole excitations at $q \sim k_F$ are suppressed, and (iii) thermal effects act to broaden the width of a plasmon peak. Since $S_Z(q) < 1$ at $q < n_{\text{ion}}^{-1/3} \sim k_F$, combination of $S_Z$ and $-\text{Im}[1/\epsilon_e^{\text{RPA}}]$ thus reduces $\sigma(\omega)$.

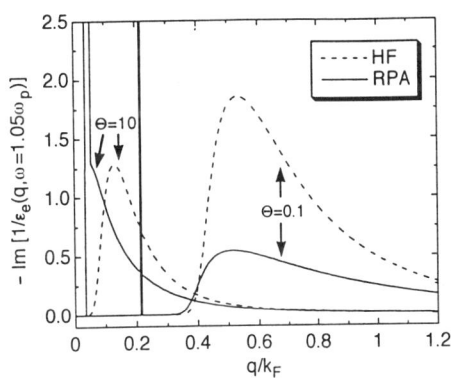

Fig. 9. Comparison of $-\text{Im }[1/\epsilon_e(q, \omega = 1.05\omega_p)]$ between HF and RPA calculations at $r_s = 1$.

B. Free-free opacity

Rosseland-mean opacity $\kappa$ is defined by [e.g., Clayton, 1983]

$$H = \frac{16\sigma_{\text{SB}}}{3\rho_m} \frac{1}{\kappa} T^3 \frac{dT(z)}{dz}, \quad (44)$$

where $H$ is the heat flux, $\sigma_{\text{SB}} \ (= 5.67 \times 10^{-8}$ W·m$^{-2}$·K$^{-4})$ is the Stefan-Boltzmann constant, and $z$ is the spatial coordinate in the direction of temperature gradient. In light of Kirchhoff's law, the heat flux in the plasma is given by

$$H = \frac{d}{dz}\left[\int_{\omega_p}^{\infty} d\omega \frac{v_g(\omega;\omega_p)u(\omega;\omega_p,T)}{n_A(1 - e^{-\hbar\omega/k_B T})\sigma(\omega) + n_e\sigma_T}\right] \quad (45)$$

where $v_g(\omega;\omega_p)$ is the group velocity of photon at a plasma density $\omega_p$, $u(\omega;\omega_p,T)$ refers to the frequency distribution of photon energy-density, and $\sigma_T = \frac{8}{3}\pi(e^2/mc^2)^2$ is the cross section for Thomson scattering. Since the dispersion relation of photon may be approximated as $\omega^2 = \omega_p^2 + c^2k^2$, we obtain

$$\frac{1}{\kappa(\text{cm}^2/\text{g})} = \frac{15}{4\pi^4} Am_u \int_{y_p}^{\infty} \frac{W(y,y_p)dy}{\sigma(\omega)[1-e^{-y}] + \langle Z \rangle \sigma_T}, \quad (46)$$

where $W(y,y_p) = y^4(1-y_p^2/y^2)e^y/(e^y-1)^2$ with $y = \hbar\omega/k_B T$ and $y_p = \hbar\omega_p/k_B T$.

We calculate $\kappa$ in the following four schemes: Scheme 0 takes $\tilde{\epsilon} = S_Z = 1$ and $\epsilon_e$ given by HF evaluation. Scheme 1 includes $e$-$e$ correlation by adopting the RPA for $\epsilon_e(q,\omega)$. In scheme 2, we include $i$-$i$ correlation by adopting the OCP structure factors $S_{\text{OCP}}(q)$ at $\Gamma = \langle\Gamma_s\rangle$ for $S_Z$; $\tilde{\epsilon}(q)$ is calculated by using $G_{ee}(q)$ due to Ichimaru and Utsumi [1981]. Debye-Hückel values are used for $S_{\text{OCP}}(q)$ at $\Gamma \leq 1$; improved HNC values [Iyetomi and Ichimaru, 1982; 1983] at $\Gamma > 1$. In the schemes 0, 1, 2, we have set $\omega_p = 0$ in the calculations of Eq. (46). Finally, scheme 3 takes into account the photon dispersion relation, $\omega^2 = \omega_p^2 + c^2k^2$, in addition to the $e$-$e$ and $i$-$i$ correlations of schemes 1 and 2.

Table III compares $\kappa(i)$ calculated in the scheme $i$ for He matter at various $\rho_m$ and $T$; the degrees of ionization considered in Sec. III are incorporated in these calculations. Reduction in $\kappa$ due to $\tilde{\epsilon}$, $S_Z$, and $\epsilon_e^{\text{RPA}}$ is considerable in magnitude at high densities, as $\kappa(1)$ and $\kappa(2)$ illustrate.

The effect of plasmon cut-off in the photon dispersion acts to decrease both the energy density and group velocity in thermal equilibrium, leading to enhancement of the opacities. Since the heat flux is dominated by photons with $\hbar\omega \sim 5k_B T$ in vacuum, such an enhancement is significant for the cases with $\hbar\omega_p \geq 5k_B T$, as we observe in $\kappa(3)$

The OPAL opacities, $\kappa(\text{OPAL})$ in Table III, have been taken from Table 12 in Rogers and Iglesias [1992]. In general, the present opacity, $\kappa(3)$, is close to OPAL opacity with discrepancies less than $\sim 15\%$ for $T \geq 10^{6.5}$K and $\rho_m \leq 1$g/cm$^3$. We note, however, that $\kappa(3)$ takes on values smaller than $\kappa(\text{OPAL})$ by a factor of two for $(T,\rho_m) = (10^6 \text{K}, 10^{-2}\text{g/cm}^3)$ and $(10^6 \text{K}, 10^{-1}\text{g/cm}^3)$, where helium matter is expected to be fully ionized in the light of Fig. 2. We remark at $(T,\rho_m) = (10^6 \text{K}, 10^{-1}\text{g/cm}^3)$, $\kappa(0) = 5.1$cm$^2$/g and Kramers free-free opacity = $4.0$cm$^2$/g [e.g., Eq. (3-170) in Clayton (1983)]. In fact, Kramers's semiclassical cross section is known to be larger slightly than Sommerfeld's value by a factor of $2\sqrt{3}/\pi \sim 1.1$ in the limit of high temperatures [Karzas and Latter, 1961]. Many-particle correlation effects contained in $\tilde{\epsilon}$, $S_Z$, and $\epsilon_e$, reduce $\kappa$ slightly as shown in Table III. Since these effects cannot account for the discrepancy between $\kappa(3)$ and $\kappa(\text{OPAL})$ at these conditions, the source of the discrepancy may possibly be attributed to different assessment in the degrees of ionization.

TABLE III. Comparison of $\kappa$ calculated in the different approximation schemes for He matter.

| $T$(K) | $\rho_m$(g/cm$^3$) | $\langle Z \rangle$ | $r_s$ | $\Theta$ | $\langle \Gamma_s \rangle$ | $\frac{\hbar\omega_p}{k_BT}$ | $\kappa(0)$(cm$^2$/g) | $\frac{\kappa(1)}{\kappa(0)}$ | $\frac{\kappa(2)}{\kappa(0)}$ | $\frac{\kappa(3)}{\kappa(0)}$ | $\frac{\kappa(\mathrm{OPAL})}{\kappa(0)}$ |
|---|---|---|---|---|---|---|---|---|---|---|---|
| $10^{6.5}$ | $10^{-2}$ | 2.0 | 8.1 | 360 | 0.02 | 0.008 | $2.33\times10^{-1}$ | 1.000 | 1.000 | 1.000 | 1.02 |
| $10^{6.5}$ | $10^{-1}$ | 2.0 | 3.8 | 77 | 0.04 | 0.02 | $3.42\times10^{-1}$ | 1.000 | 1.000 | 1.000 | 1.09 |
| $10^{6.5}$ | 1 | 2.0 | 1.7 | 17 | 0.07 | 0.08 | $9.24\times10^{-1}$ | 1.000 | 1.000 | 1.000 | 1.15 |
| $10^{6.5}$ | $10^1$ | 2.0 | 0.8 | 3.6 | 0.17 | 0.24 | 5.41 | 1.000 | 0.995 | 0.997 | |
| $10^{6.5}$ | $10^2$ | 2.0 | 0.4 | 0.8 | 0.37 | 0.76 | $4.47\times10^1$ | 1.002 | 0.946 | 0.961 | |
| $10^6$ | $10^{-2}$ | 2.0 | 8.1 | 110 | 0.05 | 0.02 | $7.31\times10^{-1}$ | 1.000 | 1.000 | 1.000 | 2.26 |
| $10^6$ | $10^{-1}$ | 2.0 | 3.8 | 24 | 0.09 | 0.08 | 3.89 | 1.000 | 1.000 | 1.000 | 2.67 |
| $10^6$ | 1 | 2.0 | 1.7 | 5.3 | 0.16 | 0.24 | $3.37\times10^1$ | 1.000 | 0.996 | 0.998 | |
| $10^6$ | $10^1$ | 2.0 | 0.8 | 1.1 | 0.37 | 0.75 | $3.07\times10^2$ | 1.002 | 0.960 | 0.976 | |
| $10^6$ | $10^2$ | 2.0 | 0.4 | 0.2 | 0.90 | 2.4 | $2.45\times10^3$ | 0.996 | 0.668 | 0.752 | |
| $10^{5.5}$ | $10^{-2}$ | 2.0 | 8.1 | 36 | 0.10 | 0.08 | $2.22\times10^1$ | 1.000 | 1.000 | 1.000 | |
| $10^{5.5}$ | $10^{-1}$ | 2.0 | 3.8 | 7.7 | 0.18 | 0.24 | $2.14\times10^2$ | 1.000 | 0.997 | 0.999 | |
| $10^{5.5}$ | 1 | 1.7 | 1.9 | 1.9 | 0.23 | 0.68 | $1.38\times10^3$ | 1.002 | 0.980 | 0.994 | |
| $10^{5.5}$ | $10^1$ | 2.0 | 0.8 | 0.4 | 0.80 | 2.4 | $1.73\times10^4$ | 1.003 | 0.734 | 0.831 | |
| $10^{5.5}$ | $10^2$ | 2.0 | 0.4 | 0.08 | 2.6 | 7.5 | $1.05\times10^5$ | 0.474 | 0.247 | 2.66 | |
| $10^5$ | $10^{-2}$ | 0.7 | 12 | 23 | 0.04 | 0.14 | $6.79\times10^1$ | 1.000 | 1.000 | 1.000 | |
| $10^5$ | $10^{-1}$ | 0.08 | 11 | 21 | 0.005 | 0.15 | 9.48 | 1.000 | 1.000 | 1.000 | |
| $10^5$ | 1 | 1.3 | 2.0 | 0.7 | 0.24 | 1.9 | $4.93\times10^4$ | 1.007 | 0.891 | 0.977 | |
| $10^5$ | $10^1$ | 2.0 | 0.8 | 0.1 | 2.1 | 7.5 | $8.86\times10^5$ | 0.439 | 0.298 | 2.62 | |
| $10^5$ | $10^2$ | 2.0 | 0.4 | 0.02 | 8.3 | 24 | $8.37\times10^5$ | 0.197 | 0.061 | $6.48\times10^5$ | |

## ACKNOWLEDGMENTS

The work reported here is an outcome of Japan-US Cooperative Science Programs on astrophysical dense plasmas and we wish to thank Dr. H. Iyetomi and Professor H.M. Van Horn for collaboration in these years through the Programs. We acknowledge hospitality of Aspen Center for Physics, where this work was initiated.

## REFERENCES

Abe, R., 1959, Prog. Theor. Phys., **21**, 475.
Allen, C. W., 1973, *Astrophysical Quantities*, 3rd ed., (Athlone, London).
Bethe, H. A., and E. E. Salpeter, 1957, *Quantum Mechanics of One- and Two-electron Atoms*, (Springer, Berlin).
Brush, S. G., H. L. Sahlin, and E. J. Teller, 1966, J, Chem. Phys., **45**, 2102.
Ceperley, D.M., and B.J. Alder, 1980, Phys. Rev. Lett., **45**, 566.
Clayton, D. D., 1983, *Principles of Stellar Evolution and Nucleosynthesis*, (Univ. of Chicago Press, Chicago).
Cooper, J. W., and J. B. Martin, 1962, Phys. Rev., **126**, 1482.
Cox, A. N., and J. N. Stewart, 1965, Astrophys. J. Suppl., **11**, 22,
Cox, A. N., and J. N. Stewart, 1969, Astrophys. J. Suppl., **19**, 243,
Dalgarno, A., 1967, Adv. Chem. Phys., **12**, 1.

Förster, A., T. Kahlbaum, and W. Ebeling, 1992, Laser and Particle Beams, **10**, 253.
Fushimi, A., H. Iyetomi, and S. Ichimaru, 1994, unpublished.
Hansen, J.-P., and I. R. McDonald, 1986, *Theory of Simple Liquids*, 2nd ed., (London, Academic).
Hansen, J.-P., and J. J. Weis, 1977, Mol. Phys., **33**, 1379.
Hasegawa, M., and M. Watabe, 1969, J. Phys. Soc. Jpn., **27**, 1393.
Hubbard, W. B., and M. Lampe, 1969, Astrophys. J. Suppl., **18**, 297.
Ichimaru, S., 1982, Rev. Mod. Phys., **54**, 1017.
Ichimaru, S., 1992, *Statistical Plasma Physics, Vol. I: Basic Principles*, (Addison-Wesley, Redwood City, CA).
Ichimaru, S., 1993, Rev. Mod. Phys., **65**, 255.
Ichimaru, S., 1994, *Statistical Plasma Physics, Vol. II: Condensed Plasmas*, (Addison-Wesley, New York).
Ichimaru, S., H. Iyetomi, and S. Tanaka, 1987, Phys. Rep., **149**, 92.
Ichimaru, S., and K. Utsumi, 1981, Phys. Rev. B, **24**, 7385.
Itoh, N., S. Mitake, H. Iyetomi, and S. Ichimaru, 1983, Astrophys. J., **273**, 774.
Iyetomi, H., and S. Ichimaru, 1982, Phys. Rev. A, **25**, 2434.
Iyetomi, H., and S. Ichimaru, 1983, Phys. Rev. A, **27**, 3241.
Karzas, W. J., and R. Latter, 1961, Astrophys. J. Suppl., **6**, 167.
Kihara, T., 1978, *Intermolecular Forces*, translated by S. Ichimaru (Wiley & Sons, New York).
Kitamura, H., and S. Ichimaru, 1994, preprint.
Kivelson, M. G., and D. F. DuBois, 1964, Phys. Fluids, **7**, 1578.
Lebowitz, J. L., and J. S. Rowlinson, 1964, J. Chem. Phys., **41**, 133.
Mahan, G. D., 1981, *Many-Particle Physics*, (Plenum, New York), Chap. 8.
Massey, H. S. W., and E. H. S. Burhop, 1952, *Electronic and Ionic Impact Phenomena*, (Clarendon, Oxford), Chap. I.
Moore, C. B., 1971, *Atomic Energy Levels as Derived from the Analysis of Optical Spectra*, Vol. 1, (NSRDS-NBS 35).
Mott, N. F., 1974, *Metal-Insulator Transitions*, (Taylor & Fracis, London).
Ogata, S., and S. Ichimaru, 1987, Phys. Rev. A, **36**, 5451.
Ogata, S., H. Iyetomi, S. Ichimaru, and H. M. Van Horn, 1993, Phys. Rev. E, **48**, 1344.
Pines, D., 1963, *Elementary Excitations in Solids*, (Benjamin, New York).
Rogers, F. J., and C. A. Iglesias, 1992, Astrophys. J. Suppl., **79**, 597.
Singwi, K. S., M. P. Tosi, R. H. Land, and A. Sjölander, 1968, Phys. Rev., **48**, 589.
Slattery, W. L., G. D. Doolen, and H. E. DeWitt, 1980, Phys. Rev. A, **21**, 2087.
Slattery, W. L., G. D. Doolen, and H. E. DeWitt, 1982, Phys. Rev. A, **26**, 2255.
Sommerfeld, A. J. F., 1953, *Atombau und Spektrallinien*, (Ungar, New York), Vol. 2, Chap. 7.
Spitzer, L., and R. Härm, 1953, Phys. Rev., **89**, 977.
Tanaka, S., X.-Z. Yan, and S. Ichimaru, 1990, Phys. Rev. A, **41**, 5616.
Tanaka, S., and S. Ichimaru, 1986, J. Phys. Soc. Jpn., **55**, 2278.
Van Horn, H. M., 1990, in *Strongly Coupled Plasma Physics*, edited by S. Ichimaru (North-Holland/Yamada Science Foundation, Amsterdam), p. 3.
Van Horn, H. M., 1991, Science, **252**, 384.
Wiese, W. L., and B. M. Glennon, 1972, in *American Institute of Physics Handbook*, edited by D. E. Gray (McGraw-Hill, New York), Chap. 7.
Ziman, J. M., 1972, *Principles of the Theory of Solids*, (Cambridge Univ. Press, London), Chap 7.

# IV.2

# THERMODYNAMICS, KINETICS, AND PHASE TRANSITIONS OF DENSE PLASMAS

Werner Ebeling and Andreas Förster

Institut für Physik, Humboldt-Universität zu Berlin
Invalidenstrasse 110, D-10099 Berlin, Germany

We study analytical representations of the thermodynamic functions for quantum plasmas. In the low density limit, the contributions to the free energy density are calculated in the chemical picture up to the order of $\mathcal{O}(n^{5/2})$. The theory is extended to isothermal and isentropic nonequilibrium conditions. Stochastic equations for the chemical kinetics of in the occupation number space are formulated. By using the method of Jacobi-Padé approximations a systematic approach to thermodynamics of strongly coupled plasmas is given. The effects of bound shell electrons are taken into account by the concepts of excluded volume and charged hard spheres. The theory is applied to pure and mixed plasmas of helium, carbon, and hydrogen. Plasma phase transitions are predicted and compared with results from other theories.

## I. INTRODUCTION

The modern theory of quantum plasmas is based on a variety of methods which were developed in parallel (DeWitt, 1961, 1962, 1966; Ebeling, 1966, 1967, 1968, 1969; Ebeling *et al.*, 1970, 1976; Ichimaru, 1973, 1982, 1993a, 1994; Kraeft *et al.*, 1986; Alastuey and Martin, 1989; Alastuey and Perez, 1992; Perez, 1994; Alastuey *et al.*, 1994). The chemical picture of quantum plasmas with bound states, which is of special relevance for applications, is used often in a rather intuitive way leaving open the question of its correct formulation which is consistent with statistical physics. In spite of the varieties of the different theories and their success in concrete applications (see, e.g., Ebeling *et al.*, 1992) one must insist on the internal consistency of these approaches as far as they are based on correct methods of statistical physics. It is the aim of this paper to construct formulae for practical applications to real plasmas which are based on one hand on available analytical results and are formulated on the other hand in the powerful chemical picture. The theory is sketched for isothermal and isentropic situations as well and is based on Jacobi-Padé approximations for the interaction contributions. The special need for tractable analytical formulae arises at present time in connection with the development of new experimental approaches to dense plasmas and with new astrophysical observations of dense plasma objects (see, e.g., Däppen, 1992; Christensen-Dalsgaard and Däppen, 1992; Chabrier and Schatzman, 1994). We believe that the formulation of tractable analytical theories based on the chemical picture is of principal importance for the interaction between theory and experiment. Due to limited space we must restrict ourselves to a survey of results without giving all details of the derivations.

## II. DISCUSSION OF THE MAIN EFFECTS

The simplest theory of nonideal plasmas with bound states may be based on the Debye-Hückel-Bjerrum (DHB) model for classical charged spheres (Falkenhagen, 1971; Ebeling, 1971a,b). In the framework of this model the chemical potential of free charged particles $a$ with the density $n_a^*$ is given by

$$\mu_a = k_B T \ln\left(n_a^* \Lambda_a^3\right) - \frac{e_a^2 \kappa}{8\pi\epsilon_0(1 + \kappa d)} . \tag{1}$$

Here, $e_a$ and $m_a$ are the charge and the mass of the particles, respectively, $d$ is the mean diameter, and

$$\Lambda_a = h\left(2\pi m_a k_B T\right)^{-1/2} \tag{2}$$

is the thermal de Broglie wave length. The inverse Debye length $\kappa$ is defined by

$$\kappa^2 = \beta \sum n_a^* e_a^2/\epsilon_0 , \quad \beta = 1/k_B T . \tag{3a,b}$$

The ideal part of the chemical potential of bound state consisting of a pair (ab) is

$$\mu_{(ab)} = k_B T \ln\left(n_{(ab)}^* \Lambda_{(ab)}^3 / \sigma_{(ab)}(T)\right) , \tag{4}$$

where $\sigma$ is the partition function of the bound state which is in the classical theory given by the Bjerrum expression (Falkenhagen, 1971). The chemical equilibrium is described by the condition

$$\mu_a + \mu_b = \mu_{(ab)} . \tag{5}$$

As shown already some time ago (Ebeling, 1971a,b; Ebeling et al., 1976), the classical DHB-theory yields a first order phase transition at $T < T_c$ where the critical temperature and the critical values of the reciprocal Debye radius are given by

$$T_c = \frac{e^2}{64\pi\epsilon_0 k_B d} , \quad \kappa_c = 1/d . \tag{6a,b}$$

Recently this classical plasma phase transition was studied in detail by Fisher and Levin (1993).

Quantum effects may be introduced in an elementary way based on Heisenberg's uncertainty relation. The thermal momenta of the electrons lead to an uncertainty of the position of the charge. From the uncertainty relation follows an effective thermal diameter of electrons (Ebeling et al., 1976)

$$d_e(T) = \Lambda_e/8 . \tag{7}$$

In this way we may conclude from the classical result that a similar phase transition may be predicted for dense plasmas. Based on Eqs. (6-7), the expected critical temperatures and densities of the plasma phase transition are

$$T_c = \frac{e^2}{32\pi\epsilon_0 k_B a_B} \approx 13000\text{K} , \quad n_{e,c} = \frac{1}{(4\pi)^4 a_B^3} \approx 2.7 \times 10^{20} \text{cm}^{-3} . \tag{8a,b}$$

Obviously, these numbers may be considered only as a rough estimate.

The question of a correct treatment of bound states remains open in the elementary theory. The divergence of the atomic partition function was known already to Bohr, Herzfeld, Fermi, and Planck. One procedure to avoid divergencies is based on the idea to omit the responsible terms in the series expansion of the exponent. This leads to the Planck-Brillouin-Larkin partition function,

$$\sigma(T) = \sum_{s=1}^{\infty} s^2 \left[\exp(-\beta E_{(ab),s}) - 1 + \beta E_{(ab),s}\right] . \tag{9}$$

So far we did not discuss the effect of external fields which lead to considerable difficulties in the statistical theory. Let us mention for example the work of Lehmann and March (1994) devoted to the effects of external electric fields which give rise to the well known ambiguity considering the bound states. On the one hand it can be argued quantum mechanically that there are no bound states; for every energy there is a finite ionization probability and, once ionized, the electron is accelerated infinitely along the field. On the other hand, many experimental findings prove it useful to retain the concept of bound states, especially for low fields. By comparing the number of states in a gas of noninteracting electrons in an electric field with and without a nucleus Lehmann and March observe a difference which has to be understood as binding. The number of states $N(\epsilon)$ per energy $\epsilon$ is defined as the all-space integral of the local density of states which can be carried out very conveniently in $k$-space:

$$N(\epsilon) = \int\int dr dr' \nu(r, r', \epsilon) = \bar{\nu}(0, 0, \epsilon) . \tag{10}$$

For an electron gas in an $E$-field alone ($E$ denoting the field strength along the $z$-axis) one has $N_\epsilon = 1$. For the presence of a nucleus (denoted by the index $Z$ for the nuclear charge) an integral equation for $\bar{\nu}_{EZ}(k, k', \epsilon)$ can be derived

$$\bar{\nu}_{EZ}(k, k', \epsilon) = \bar{\nu}_E(k, k', \epsilon) - 2\int dr \tilde{\nu}_E(k, r, \epsilon) V_Z(r) \tilde{\nu}_{EZ}(r, k', \epsilon) . \tag{11}$$

This equation can be solved in statistical mechanical perturbation theory, taking the electric field case as the starting point. For a one-dimensional model atom (i.e., keeping $1/r$ as the nuclear potential) this perturbation series can be summed up exactly, giving $N_{EZ} = L(\epsilon, Z)$ with

$$L = \left(1 - 2ZE^{-1/3}(2\pi)^{-3} I_0(\epsilon E^{-2/3})\right)^{-1} , \tag{12a}$$

$$-\frac{\partial^2 I_0}{\partial \epsilon^2} = \left(2^{4/3} E^{-2/3} \pi^2 \text{Ai}(-2^{1/3}\epsilon E^{-2/3})\right)^2 . \tag{12b}$$

Since now always $L > 1$, the positive number

$$N_b = N_E - N_{EZ} - 1 - L(\epsilon, Z) \tag{13}$$

describes the vanishing of states out of the electron gas due to the presence of a nucleus. Physically, this has to be interpreted as binding. I seem remarkable that in this statistical mechanical approach the above mentioned ambiguity can, at least in principle, be reconciled. For any quantitative analysis, however, further knowledge of the atomic region is required. In the work of Lehmann and March (1994), first order results for the 3d case and a series solution for $L(\epsilon, Z)$ are obtained.

## III. QUANTUM-STATISTICAL COULOMB CONTRIBUTION IN THE OCP

For simplicity let us first discuss the one-component plasmas (OCP) of point charges in thermal equilibrium. The interaction part of the free energy density is given by the binary correlation function $F_2$ in the form

$$f_{\text{int}} = \frac{n^2}{2} \int_0^1 d\lambda \int dr' V(r,r') F_2(r,r',\lambda) , \qquad (14)$$

with $\lambda$ being the charging parameter. The binary correlations as well as the corresponding contributions to the free energy density may be represented as a sum of classical and quantum-mechanical contributions

$$F_2(r) = F_2^{\text{cl}}(r) + F_2^{\text{qu}}(r) , \quad f = f_{\text{cl}} + f_{\text{qu}} . \qquad (15\text{a,b})$$

Explicit calculations for the classical contribution were given already in the pioneering work of Abe (1959). The result for the low-density limit of the free energy density is

$$\begin{aligned} f_{\text{cl}} = f_0 - k_{\text{B}} T n & \left\{ \frac{\beta e^2 \kappa}{12\pi\epsilon_0} + \frac{1}{12} \left( \frac{\beta e^2 \kappa}{4\pi\epsilon_0} \right)^2 \left[ \ln\left( \frac{\beta e^2 \kappa}{4\pi\epsilon_0} \right) + C_E \right] \right. \\ & \left. + \frac{1}{12} \left( \frac{\beta e^2 \kappa}{4\pi\epsilon_0} \right)^3 \ln\left( \frac{\beta e^2 \kappa}{4\pi\epsilon_0} \right) + \mathcal{O}\left[ \left( \frac{\beta e^2 \kappa}{4\pi\epsilon_0} \right)^3 \right] \right\} , \end{aligned} \qquad (16)$$

where $C_E$ is the Euler number. An exact expression for the quantum mechanical contribution can be found by using the method of Jost functions from quantum scattering theory (Ebeling et al., 1970, 1976). In order to reproduce the exact result we study not the free energy itself but consider instead the difference between the free energy density of a quantum plasma and that of a classical plasma. In lowest order with respect to the density, we expect terms of the order $\mathcal{O}(n^2)$. Assuming that at lease in this order the divergencies are of entirely classical origin we postulate

$$\delta f_{\text{qu}} = -k_{\text{B}} T n^2 \left( 1 + \frac{\beta e^2 \kappa}{4\pi\epsilon_0} \right) \delta B_2(T) , \qquad (17\text{a})$$

$$\delta B_2(T) = B_2(T) - \lim_{\hbar \to 0} B_2(T) . \qquad (17\text{b})$$

We express now the quantum-mechanical trace in the virial coefficient $B_2$ by a complex integral (resolvent representation)

$$\delta B_2(T) = \text{const.} \int dz e^{-\beta z} F(z) - \lim_{\hbar \to 0} \text{const.} \int dz e^{-\beta z} F(z) . \qquad (18)$$

Here the trace of the resolvent is defined by

$$F(z) = \mathrm{Tr}\left((H_2 - z) - (H_2^0 - z)\right), \qquad (19)$$

which is closely related to the Jost functions of the quantum scattering theory. The virial coefficient may be calculated by exploiting the fact that the Jost functions of the Coulomb scattering problem are exactly known (Ebeling et al., 1976). We get for Fermi particles with spin $s$ the expression

$$f_{\mathrm{qu}} = -2\pi k_B T n^2 \lambda^3 \left(1 + \frac{\beta e^2 \kappa}{4\pi\epsilon_0}\right) \left[Q(-\xi) - \frac{1}{2s+1} E(-\xi) - \frac{\xi^3}{6}(\ln\xi - C_E)\right] \qquad (20)$$

with

$$\xi = -\frac{e^2}{4\pi\epsilon_0 k_B T \lambda}, \qquad \lambda = \frac{\hbar}{\sqrt{2mk_B T}}. \qquad (21\mathrm{a,b})$$

The virial functions $Q(x)$ and $E(x)$ are given by infinite Tayler series

$$Q(x) = \sum_{n=1}^{\infty} q_n x^n, \qquad E(x) = \sum_{n=0}^{\infty} e_n x^n; \qquad (22\mathrm{a,b})$$

$$q_1 = -\frac{1}{6}, \quad q_2 = -\frac{\sqrt{\pi}}{8}, \quad q_3 = -\frac{1}{6}\left(\frac{1}{2}C_E + \ln 3 - \frac{1}{2}\right),$$

$$q_n = \sqrt{\pi}\zeta(n-2)2^{-n}/\Gamma(1+n/2), \quad \text{if } n \geq 4;$$

$$e_0 = \frac{\sqrt{\pi}}{4}, \quad e_1 = \frac{1}{2}, \quad e_2 = \frac{\sqrt{\pi}}{4}\ln 2, \quad e_3 = \frac{\pi^2}{72},$$

$$e_n = \sqrt{\pi}\left(1 - 2^{2-n}\right)\zeta(n-1)2^{-n}/\Gamma(1+n/2), \quad \text{if } n \geq 4. \qquad (23\mathrm{a\text{-}i})$$

These convergent series which define analytic functions were introduced and discussed already in earlier work (Ebeling, 1966, 1967, 1968, 1969; Ebeling et al., 1970).

A new theory developed recently by Alastuey and Martin (1989), Alastuey and Perez (1992), Perez (1994), and Alastuey et al. (1994) confirmed our findings with respect to the quantum virial functions. In this way the full quantum contribution up to the order $\mathcal{O}(n^2)$ is exactly known. The term of order $\mathcal{O}(n^{5/2})$ is still under discussion since here discrepancies between different versions of the quantum statistical theory are observed (Ebeling, 1993; Schlanges et al., 1993). For examples, the result given above does not fully agree with the term of the order $\mathcal{O}(n^{5/2})$ given by Alastuey et al. (1994). In difference to these authors another prefactor of the direct contribution to the term $\mathcal{O}(\kappa^5)$ is found. Let us summarize the result of a detailed comparison (Ebeling, 1993) of 4 different analytical theories. The 4 theories which were compared are:

1) The Slater sum method, the low order Green's function method and the Jost function method described above (Kraeft et al., 1986).
2) The ring summations according to Montroll-Ward and DeWitt (1961, 1962, 1966).
3) The Feynman-Kac technique according to Alastuey et al. (1994).
4) The Green's function technique including higher order diagrams (Schlanges et al., 1993; DeWitt et al., 1994).

The comparison showed that the 4 approaches mentioned above agree in all terms up to $\mathcal{O}(n^2)$. The contribution to the order $\mathcal{O}(n^{5/2})$ are:

$$\text{1) } \delta f_{5/2} = -\frac{1}{12\pi}\kappa^5\lambda^2\left(\frac{1}{4} + \frac{3}{8s+4}\right),$$

$$\text{2) } \delta f_{5/2} = -\frac{1}{12\pi}\kappa^5\lambda^2\left(\frac{1}{8} + \frac{3}{4s+2}\right),$$

$$\text{3) } \delta f_{5/2} = -\frac{1}{12\pi}\kappa^5\lambda^2\left(\frac{1}{8} + \frac{3}{8s+4}\right), \quad\quad\quad (24\text{a-d})$$

$$\text{4) } \delta f_{5/2} = -\frac{1}{12\pi}\kappa^5\lambda^2\left(\frac{1}{8} + \frac{3}{4s+2} - \frac{4\pi^{3/2}n\lambda}{\kappa^2(2s+1)}\right).$$

We see that expressions 2), 3), and 4) agree in the direct contributions and the expressions 1) and 3) in the exchange contribution. Taking agreement of at least two independent groups as a criterion for correctness, one would like to conclude that the expression 3) is the best one. However, the expression 4) contains higher order diagrams which are not included into 1), 2), and 3). The missing factor of 1/2 in the direct term of 1) is probably a consequence of the quasi-classical screening procedure.

## IV. MULTICOMPONENT PLASMAS AND CHEMICAL PICTURE

Instead of extending in detail the method given above to the multi-component case, we use the more simple procedure of analytical continuations. This method is based on the analytical properties of the virial functions, which were obtained above only for the OCP, i.e. for the case of pure scattering states. Instead of perturbation parameter $\xi$ we use in the multicomponent case the interaction parameter for a pair of species $a$ and $b$ of point charges:

$$\xi_{ab} = -\frac{e_a e_b}{4\pi\epsilon_0 k_B T \lambda_{ab}}, \quad \lambda_{ab} = \frac{\hbar}{\sqrt{2m_{ab}k_B T}}. \quad\quad (25\text{a,b})$$

The new parameter $\xi_{ab}$ is negative for charges with equal sign and positive for charges with opposite sign. For two electrons we find $\xi_{ee} = -\xi$. Therefore, $Q(-\xi)$ and $E(-\xi)$ are to be replaced by $Q(\xi_{ab})$ and $E(\xi_{ab})$. In this way, the virial contribution from a pair of two arbitrary charges $e_a$ and $e_b$ is given by

$$Q(\xi_{ab}) + D_{ab}E(\xi_{ab}), \quad \text{with} \quad D_{ab} = (-1)^{2S_a}\delta_{ab}/(2S_a+1). \quad\quad (26\text{a,b})$$

Only pairs of identical particles may contribute to the exchange term $E$; all pairs contribute to the correlation term $Q$.

The term $Q$ contains also the contribution of the bound states. This follows from an identity which holds for positive values of $\xi$ (i.e. opposite charges):

$$Q(\xi) = 2\sqrt{\pi}\sigma(T) - \sqrt{\pi}\xi^2/4 - Q(-\xi), \quad\quad (27)$$

where $\sigma$ is the PBL-partition function (Planck-Brillouin-Larkin)

$$\sigma(T) = \sum_{s=1}^{\infty} \sigma_{(ab),s}(T) , \qquad (28a)$$

$$\sigma_{(ab),s} = s^2 \left[\exp(-\beta E_{(ab),s}) - 1 + \beta E_{(ab),s}\right] , \qquad (28b)$$

$$E_{(ab),s} = -\frac{e_a^2 e_b^2 m_{ab}}{32\pi^2 \epsilon_0^2 \hbar^2 s^2} . \qquad (28c)$$

In this way we see that the PBL procedure follows in natural way from the exact quantum virial functions. Other cut-off procedures are possible but they look less naturally than the quantum statistical results. The transition to the chemical picture is based on the idea, that the term containing $\sigma(T)$ is identified as a contribution of free atoms to the Helmholtz energy. Let us introduce now $n_a^*$ as the density of free elementary particles of kind $a$ and $n_{(ab),s}$ as the density of bound states of two opposite charges $a$ and $b$ with principal quantum number $s$. In the new representation the free energy density of low-density multicomponent quantum plasmas is given by

$$f(n_a^*, n_{(ab),s}, T) = f_{\text{id}} + f_{\text{int}} . \qquad (29)$$

The Boltzmann contribution to the free energy density is given by

$$\begin{aligned}f_{\text{id}} =& k_B T \sum_a n_a^* \left(\ln(n_a^* \Lambda_a^3) - 1\right) \\ &+ k_B T \sum_{(ab)} \sum_s n_{(ab),s} \left(\ln(n_{(ab),s} \Lambda_{(ab)}^3 / \sigma_{(ab),s}) - 1\right) .\end{aligned} \qquad (30)$$

The interaction part of the free energy density including all terms up to the order $\mathcal{O}(n^{5/2})$ is given by:

$$\begin{aligned}f_{\text{int}} = &-\frac{k_B T}{12\pi} \kappa^3 + k_B T 2\pi \sum_a \sum_b n_a^* n_b^* \left[\left(1 + \frac{\beta e_a e_b \kappa}{4\pi\epsilon_0}\right) \lambda_{ab}^3 \right. \\ &\left. \times \left(\sqrt{\pi}\xi^2/4 + \frac{1}{6}\xi_{ab}^3 \ln(\kappa\lambda_{ab}) + Q(-|\xi_{ab}|) - D_{ab}E(\xi_{ab})\right) + A_0 \left(\frac{\beta e_a e_b}{4\pi\epsilon_0}\right)^4 \kappa \right] \\ &+ k_B T A_1 \sum_a \sum_b \sum_c n_a^* n_b^* n_c^* \frac{\beta^5 e_a^4 e_b^3 e_c^3}{(4\pi\epsilon_0)^5 \kappa} \\ &- k_B T A_2 \sum_a \sum_b \sum_c \sum_d n_a^* n_b^* n_c^* n_d^* \frac{\beta^6 e_a^3 e_b^3 e_c^3 e_d^3}{(4\pi\epsilon_0)^6 \kappa^3} ,\end{aligned} \qquad (31)$$

with the constants

$$A_0 = \pi[1 - \ln(4/3)]/3 \approx 0.7459 , \quad A_1 \approx 10.72 , \quad A_2 \approx 10.13 . \qquad (32\text{a-c})$$

We note that these classical coefficients were first calculated by Czerwon (1972). We may state again agreement with the expressions given by Alastuey and Perez except an additional term in the AP theory

$$\delta f_{\text{int}} = k_\text{B} T \sum_a n_a^* \frac{h^2 \beta e_a^2 \kappa^3}{96\pi\epsilon_0 m_a} , \qquad (33)$$

which is missing in our calculations so far, probably due to the incomplete screening procedure.

The independent parameters in the chemical picture are the densities of the free states $n_a^*$ of the elementary particles and the densities of the bound states $n_{(ab),s}$ which are considered as composite particles. Constraints are given by the conservation of the total number of elementary particles

$$n_a = n_a^* + \sum_b \sum_s n_{(ab),s} . \qquad (34)$$

The free particle densities in the chemical picture are found by a minimization of the free energy density

$$f(n_a^*, n_{(ab),s}, \beta) = \min . \qquad (35)$$

observing the constraints of constant total number of elementary particles. The construction of the minimum may be replaced by conditions for the chemical potentials of the species, which are also known as mass-action laws or Saha equations.

## V. NONEQUILIBRIUM KINETICS

We consider first nonequilibrium plasmas under isentropic conditions. Let us define the entropy of the system in the chemical picture by

$$S(V, T, N_a^*, N_{(ab),s}) = -V \frac{\partial f(n_a^*, n_{(ab),s}, T)}{\partial T} . \qquad (36)$$

Nonequilibrium processes are called isentropic if the entropy is a constant of the macroscopic motion:

$$S(V(t), T(t), N_a^*(t), N_{(ab),s}(t)) = S(V_0, T_0, N_{a,0}^*, N_{(ab),s,0}) = \text{const.} \qquad (37)$$

This is a relation between the variables which may be used to eliminate the time variation for one of them, e.g.

$$T(t) = g(V(t), N_a^*(t), N_{(ab),s}(t), V_0, T_0, N_{a,0}^*, N_{(ab),s,0}) . \qquad (38)$$

This relation is well known for the case of ideal gases, but may be easily calculated now for nonideal plasmas.

The chemical equilibrium under isentropic conditions is found as the minimum of the internal energy $U(S, V, T, N_a^*, N_{(ab),s})$ under the constraints given by the balance equations (34). In this way we find the isentropic path in the temperature-density plane.

Strictly speaking the isentropic condition is an idealization and there are no real nonequilibrium processes which are totally isentropic. However there are many examples of nonequilibrium plasmas which follow in good approximation the isentropic path. Examples are the adiabatically expanding plasmas produced by laser pulses or explosions (Meyer-ter-Vehn and Oparin, 1994). The theory of the thermodynamics and kinetics of isentropic plasmas is still at its infancy but, as we believe, it may find many applications.

Let us consider now the chemical kinetics of nonequilibrium plasmas. We shall assume in the following, that the plasma is strictly uniform and in equilibrium with respect to the velocity distributions of all species. The kinetics of the transitions in the chemical picture which finally lead to the minimum of the free energy may be formulated in a stochastic theory. The idea is the following: We first number all possible states of the electrons by $i$ and associate occupation number $N_i$ with these state. An example of numbering is $i = 0$ for the free states of the electrons, $i = 1, 2, 3, \ldots, i_{(ab)}, \ldots$ for the bound states of the pairs $(ab)$ etc. Then we introduce a probability distribution $P(N_1, N_2, \ldots, t)$ and a master equation

$$\partial_t P(N_1, N_2, \ldots, t) = \sum{}' \left[ W(N_1, N_2, \ldots | N_1', N_2', \ldots) P(N_1', N_2', \ldots, t) \right. \\ \left. - W(N_1', N_2', \ldots | N_1, N_2, \ldots) P(N_1, N_2, \ldots, t) \right] . \quad (39)$$

Under isothermal conditions the transition probability are constructed on the basis of the free energy by using a Monte Carlo procedure:

$$W(i \to j) = \begin{cases} g(i \to j), & \text{if } \Delta f(i \to j) < 0 \\ g(i \to j) \exp[\beta V \Delta f(i \to j)], & \text{if } \Delta f(i \to j) \geq 0 \end{cases} . \quad (40)$$

Here $\Delta f(i \to j)$ are the changes in the free energy density according to the transition $i$ to state $j$. The $g(i \to j)$ are chosen proportional to the atomic transition rates.

The process defined above converges to the minimum of the free energy in a stochastic way. In average the deterministic theory of transitions in nonideal plasmas formulated earlier (Ebeling et al., 1991) is obtained.

Under isentropic conditions the transition probabilities are defined by means of the internal energy densities $u = U(V, S, N_a^*, N_{(ab),s})/V$ by using again Monte Carlo type assumptions:

$$W(i \to j) = \begin{cases} g(i \to j), & \text{if } \Delta u(i \to j) < 0 \\ g(i \to j) \exp[\beta V \Delta u(i \to j)], & \text{if } \Delta u(i \to j) \geq 0 \end{cases} . \quad (41)$$

Now the kinetics will converge to the minimum of the internal energy, i.e. to the equilibrium distributions on the adiabatic path.

## VI. APPROXIMATIONS INCLUDING HIGHER ORDERS IN THE DENSITIES

The quantum-mechanical results discussed in section IV are valid in the region of weak interaction and week degeneracy. The Coulomb contribution to the thermodynamic functions are analytically known also in other limiting cases as, e.g., in the Gellmann-Brueckner or in the Wigner region. For an interpolation between the limit cases we used Jacobi-Padé formulae. The idea of these approximations is based on the construction of rational fractions which reflect the available knowledge on the analytical behavior of the function under consideration. For the Coulomb part of the free energy density we have constructed approximation formulae which are applicable to many-component plasmas including multiply charged ions (Ebeling et al. 1991, section 3.3). Generally, we use the following partition of the free energy

$$f_{\text{Coul}} = f_e + f_i + f_{ie} \quad (42)$$

representing the electron gas, the ion gas, and the ion-electron interaction contributions, respectively. They yield the quantum-corrected Debye law and expressions for the classical screened ion lattice for in the low- and high-density limit, respectively.

So far we considered mainly systems of point charges. However, at small distances between the particles the bound-electron shells appear to be essential and, due to the Pauli exclusion principle, lead to a strong repulsion between all ions and atoms. Generally, serious problems arise for a strong theoretical treatment of the short-range interaction between the finite-size plasma particles. The theoretical determination of the structure factor is a tedious task and has been carried out only for a small number of ions. As an alternative, we suggest a rather crude approach by modelling the ions as charged hard spheres utilizing the concept of the mean spherical approximation (MSA) (Förster and Ebeling, 1992, 1993). One of the convenient features of the MSA is that it gives explicit, yet general results for the thermodynamic functions. On the other hand it reduces to the Debye law for a plasma of point charges what makes it inapplicable for the description of pressure ionization in light elements. In the present paper, we use a combination of thermodynamic results for charged hard spheres and for point charges utilizing a semi-empirical linear mixing (LM) rule. Furthermore, we pay attention to the reduction of the effective volume for free electrons (Förster et al., 1992; Kaulbaum and Förster, 1992). A consequent application of the reduced-volume hypothesis requires modification of the MSA contribution (Förster and Ebeling, 1993).

## VII. APPLICATIONS TO PLASMAS OF LIGHT ELEMENTS

In the last part of the present paper we present a few explicit results which were obtained for the model discussed above. The main features may be characterized by the following composition of the free energy density

$$f = f_{\text{id},i} + f^*_{\text{id},e} + f^*_{\text{LM}} + f_{\text{hs}} + f^*_{\text{xc},e} + f_{e-i} \ . \quad (43)$$

The first two terms denote the usual ideal contributions of atoms and ions and of electrons (e) following from Boltzmann and Fermi-Dirac statistics respectively. The third term gives the Coulomb contribution of the ionic subsystem within linear mixing. The excess contribution of the uncharged hard-sphere system (hs) is modeled by a widely used quasi-exact expression. The fifth term stands for the exchange-correlation contribution of the electron gas (xc,e). Here, we take advantage again a Padé-fit formula (Tanaka et al., 1985) which reproduces numerical data obtained from the Singwi-Tosi-Land-Sjölander theory, The screening contributions ($e-i$) are taken from our earlier work (Ebeling et al., 1991, Eq. (3.3.5)). That provides the complete Debye law for ions and classical electrons at low density and the electronic screening contributions to the OCP at high densities (full ionization). The asterisk * means here that the expression is rescaled according to the concept of the reduced volume.

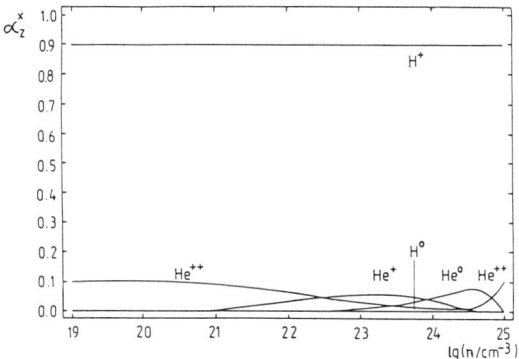

Figure 1: Relative abundances of helium and hydrogen ions $\alpha_Z^x$ versus total heavy particle density $n$ for a temperature of $T=200000K$. The relative fractions of helium and hydrogen are: $\nu_{He}=0.1$, $\nu_H=0.9$.

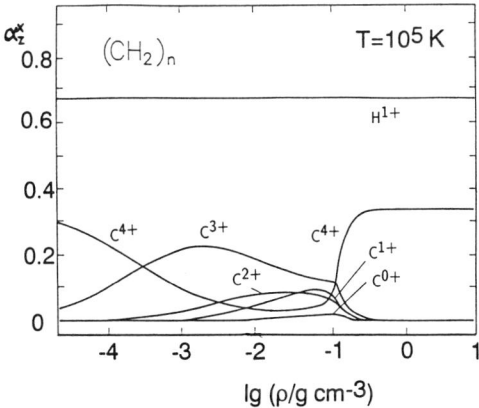

Figure 2: Relative abundances of carbon and hydrogen ions $\alpha_Z^x$ versus mass density $\rho$ for a temperature of $T=100000K$. The relative fractions of carbon and hydrogen are: $\nu_C=1/3$, $\nu_H=2/3$.

The equilibrium state was determined by solving the coupled set of nonideal Saha equations, supplemented by the electro-neutrality condition. For the consideration of first-order phase transition the introduction of the combined chemical potential $M$ for multicomponent plasmas is of utility:

$$M = \sum_x \nu_x M^x , \quad M^x = \mu_z^x + z\mu_e , \quad Z - \text{arbitrary} . \tag{44}$$

The index $x$ runs over the chemical elements which are present in the mixture. The $\nu_x$ are the relative fractions of the chemical elements and $z$ is the charge number of the ions. The $M^x$ is the combined chemical potentials of the chemical element $x$ and could be understood as the chemical potential of a simple neutral combination of plasma particles of the element $x$, namely one $z$-fold charged ion and $z$ electrons. (Förster et al., 1991, 1992). Thus, the combined chemical potential $M$ appears as the natural generalization of the "plasma potential" $\mu_1 + \mu_e$ for hydrogen. Let $n_\text{I}(T)$ and $n_\text{II}(T)$ be the total densities of the coexisting phases I and II of one Plasma-Phase Transition (PPT) at temperature $T$. Then the Maxwell construction for the coexisting area reads

$$p(n_\text{I}, T) = p(n_\text{II}, T) \quad \text{and} \quad M(n_\text{I}, T) = M(n_\text{II}, T) . \tag{45}$$

Our model has been applied to pure and mixed plasmas of helium, carbon, and hydrogen. Fig. 1 shows the detailed plasma composition for a mixture of 10% helium and 90% hydrogen, which is typical for the sun, for a fixed temperature of 200000K. At this temperature hydrogen is fully ionized. The full lines is the lower part of the picture demonstrate the ionization process of helium with increasing density. Due to the hydrogen "background" the $He^+$ ions are suppressed at pressure ionization. Fig. 2 is devoted to a carbon-hydrogen mixture which is of importance for dynamical high-pressure experiments with an organic material like polyethylene (Conrads et al., 1995). The influence of the hydrogen background for the ionization process of carbon is illustrated on Fig. 3: due to the additional free electrons, the Saha equilibrium of carbon is shifted to states of lower ionization. The plasma phase transition appears only in the pure-carbon system, since the charged averages $\langle z^p \rangle$ which enter the Coulomb contributions are generally higher for pure carbon plasma than for the carbon-hydrogen mixture and, hence, the Coulomb effects are stronger. Fig. 4 gives a comparison of phase diagrams of pure helium on the pressure-temperature plane as published by different authors (FKE - Förster et al., 1991, 1992; SBT - Schlanges et al., 1992; FE - Förster and Ebeling, 1993; OI - see Ichimaru, 1993b). Some typical features of the underlying theories are: FKE - Padé approximations for the Coulomb contributions, SBT - same Padé approximations plus second virial coefficient for the charge-neutral interaction, neglection of $He^{2+}$ ions, FE - linear mixing between Mean Spherical Approximation and Padé formulae, OI - HNC-MCA calculations and shallowing of bound states. On Fig. 5 we give a comparison of our theoretical equation of state with the date from Nellis et al. (1984) which are based on shock wave experiments in pure helium. Our theoretical prediction is within the error bars of the experiment. Finally on Fig. 6 we present a phase diagram of pure carbon. Due to the strongly bound 1s-electrons a plateau characterized by the dominance of $C^{4+}$ is formed. Once more we observe a plasma phase transition which is now connected with the ionization step leading to $C^{4+}$. This transition was already found by Ogata and Ichimaru. Our Fig. 6 may be directly compared with Fig. 4 of Ichimaru (1993b) (see also Ichimaru (1994)). Both calculations for the equation of state perfectly coincide at high temperatures. The so far published data of Ogata and Ichimaru show two additional plasma phase transitions for carbon, one in the region of almost ideal carbon plasma at low densities and another in the region of pressure ionization of neutral carbon, which we can not observe.

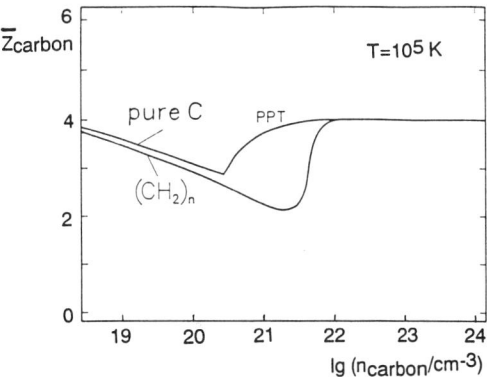

Figure 3: Mean charge $\bar{Z}$ of carbon ions versus total density of carbon ions for a fixed temperature of 100000K. The upper curve corresponds to pure carbon. The lower curve corresponds to carbon imbedded in fully ionized hydrogen (conditions of figure 2).

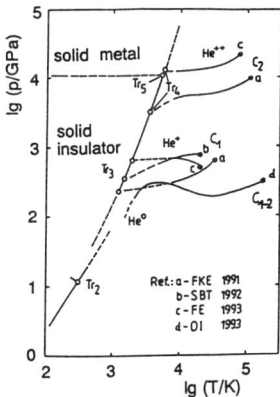

Figure 4: Phase diagrams of pure helium on the pressure-temperature plane: comparison of different results for the PPTs. For details of the models see text. Tr - Triple points: 2 - fcc/hcp(?)/fluid; 3 - solid/$He^+$/$He^0$ (PPT 1); 4 - solid/$He^{++}$/$He^+$ (PPT 2); 5 - solid insulator/solid metal/plasma. C - Critical points: 1 - PPT 1, 2 - PPT 2, 1-2 - 1 and 2 merged into one. Thick lines - first order phase transitions (extrapolations are dashed).

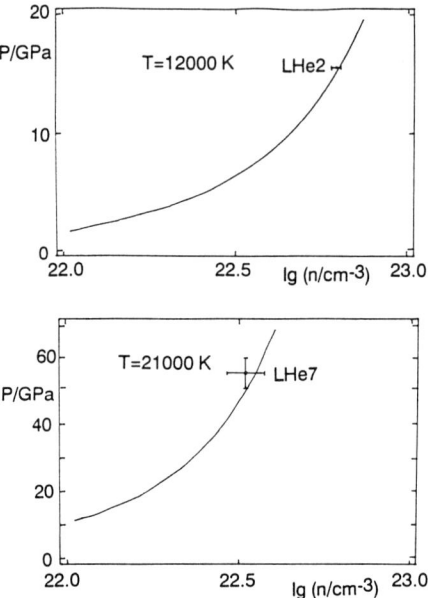

Figures 5a,b: Comparison of our theoretical equation of state with shock-wave data from Livermore (Nellis et al., 1984) for pure helium. LHe 2 - single shock experiment, LHe7 - double shock experiment.

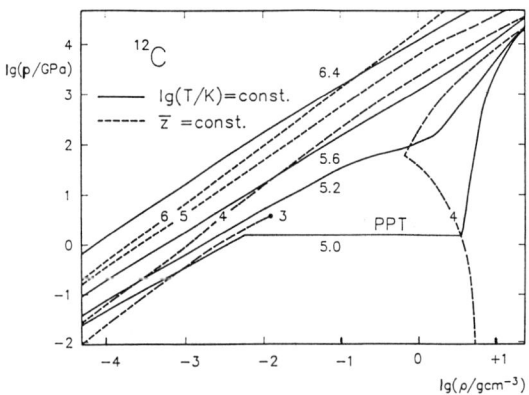

Figure 6: Equation of state $p$, mean ionization state $\bar{z}$, and plasma phase transition of carbon.

## VIII. CONCLUSIONS

The results given above may be summarized as follows:
(i) With respect to the thermodynamic functions the considered theories for the free energy density in the low density region are in full agreement up to the second

order in the density. In the order $\mathcal{O}(n^{5/2})$ there are still divergencies between the different theories which are not fully explored.

(ii) The contributions up to $\mathcal{O}(n^2)$ contain the bound states in the form of Plank-Brillouin-Larkin (PBL) partition functions and may be reformulated into the chemical picture. The minimization procedure contained in the chemical picture, is the appropriate representation in the presence of bound states.

(iii) Under isothermal and isentropic conditions, the kinetic theory of uniform plasmas may be formulated by means of a Monte Carlo procedure.

(iv) The most relevant higher order interaction contributions may be condensed into Padé approximations and by the introduction of a effective free volume for the electrons.

(v) Dense plasmas of the light elements show characteristic Coulomb phase transitions.

The authors thank H. Lehmann for a discussion on the effects of external fields and W.D. Kraeft and M. Schlanges for discussions on the higher order quantum corrections of the Debye law.

## REFERENCES

Abe, R., 1959, Progr. Theor. Phys. **22**, 213.
Alastuey, A., and Martin, Ph.A., 1989, Phys. Rev. A. **40**, 6485.
Alastuey, A., and Perez, A., 1992, Europhys. Lett. **20**, 19.
Alastuey, A., Cornu, F., and Perez, A., 1994, Phys. Rev. E. **49**, 1077.
Chabrier, G., and Schatzman, E., editors, 1994, *The Equation of State in Astrophysics*, (Cambridge, Cambridge University Press).
Czerwon, H.J., 1972, Diplom Thesis, (Rostock, Wilhelm-Pieck-Universität).
Christensen-Dalsgaard, J., and Däppen, W., 1992, Astron. Astrophys. Rev. **4**, 267.
Conrads, H., Ebeling, W., and Förster, A., 1995, in preparation.
Däppen, W., 1992, Rev. Mexicanna Astron. Astrof. **23**, 141.
DeWitt, H.E., 1961, Nuclear Energy C: Plasma Physics **2**, 27.
DeWitt, H.E., 1962, J. Math. Phys. **3**, 1216.
DeWitt, H.E., 1966, J. Math. Phys. **7**, 616.
DeWitt, H.E., Schlanges, M., and Kraeft, W.D., 1994, personal communication.
Ebeling, W., 1966, Ann. Physik (Leipzig) **17**, 415.
Ebeling, W., 1967, Ann. Physik (Leipzig) **19**, 104; **21**, 315; **22**, 33, 383, 392.
Ebeling, W., 1968, Physica **38**, 373; **40**, 290.
Ebeling, W., 1969, Physica **43**, 293.
Ebeling, W., 1971a, Z. Physik. Chem. (Leipzig) **247**, 340.
Ebeling, W., 1971b, Phys. Stat. Sol. (b) **46**, 243.
Ebeling, W., 1993, Contr. Plasma Phys. **33**, 492.
Ebeling, W., Förster, A., Fortov, V.E., Gryaznov, V.K., and Polishchuk, A.Ya., 1991, *Thermophysical Properties of Hot Dense Plasmas* (Stuttgard/Leipzig, Teubner).
Ebeling, W., Förster, A., and Radtke, R., editors, 1992, *Physics of Nonideal Plasmas* (Stuttgard/Leipzig, Teubner).
Ebeling, W., Kraeft, W.D., and Kremp, D., 1970, Beitr. Plasmaphysik **10**, 237.

Ebeling, W., Kraeft, W.-D., and Kremp, D., 1976, *Theory of Bound States and Ionization Equilibrium in Plasmas and Solids* (Berlin, Akademie-Verlag).
Falkenhagen, H., 1971, *Theorie der Elektrolyte* (Leipzig, Hirzel-Verlag).
Fisher, M.E., and Levin, Y., 1993, Phys. Rev. Lett. **71**, 3826.
Förster, A., and Ebeling, W., 1992, in *Physics of Nonideal Plasmas*, edited by W. Ebeling, A. Förster, and R. Radtke (Stuttgard and Leipzig, Teuber), p. 95.
Förster, A., and Ebeling, W., 1993, in *Strongly Coupled Plasmas*, edited by H.M. Van Horn and S. Ichimaru, (Rochester, University of Rochester Press), p. 347.
Förster, A., Kahlbaum, T., and Ebeling, W., 1991, High Press. Res. **7**, 375.
Förster, A., Kahlbaum, T., and Ebeling, W., 1992, Laser Part. Beams **10**, 253.
Ichimaru, S., 1973, *Basic Principles of Plasma Physics: A Statistical Approach* (Ner York, W.A. Benjamin).
Ichimaru, S., 1982, Rev. Mod. Phys. **54**, 1017.
Ichimaru, S., 1993a, Rev. Mod. Phys. **65**, 255.
Ichimaru, S., 1993b, Contrib. Plasma Phys. **33**, 345.
Ichimaru, S., 1994, *Statistical Plasma Physics, Vol. II: Condensed Plasmas* (Reading, Massachusetts, Addison-Wesley).
Kahlbaum, T., and Förster, A., 1992, Fluid Phase Equil. **76**, 71.
Kraeft, W.-D., Kremp, D., Ebeling, W., and Röpke, G., 1986, *Quantum Statistics of Charged Particle Systems* (Berlin, Akademie-Verlag).
Lehmann, H., and March, N.H., 1994, J. Stat. Phys., submitted.
Meyer-ter-Vehn, J., and A. Oparin, 1994, in this volume.
Nellis, W.J., Holmes, N.C., Mitchell, A.C., Trainor, R.J., Governo, G.K., Ross, M., and Young, D.A., 1984, Phys. Rev. Lett. **53**, 1248.
Perez, A., 1994, *Developpments diagrammatiques pour un plasma quantique dans la représentation de Feynmann-Kac* (PhD Thesis) (Lyon, Université Claude Bernard-Lyon I).
Schlanges, M., Bonitz, M., and Tschttschjan, A., 1992, Poster at STATPHYS 18, Berlin.
Schlanges, M., Kraeft, W.D., and DeWitt, H.E., 1993, Contr. Plasma Physics **33**, 437.
Tanaka, S., Mitake, S., and Ichimaru, S., 1985, Phys. Rev. A **32**, 1896.

# IV.3

# NEW ASTROPHYSICAL OPACITIES AND THEIR EFFECT ON STELLAR MODELS

F.J. Rogers and C.A. Iglesias

Lawrence Livermore National Laboratory, Livermore, CA 94550, U. S. A.

Recent advances in astrophysical opacity calculations are discussed. A description and comparison of results from two new independent efforts is given. The large increases in the calculated opacity in the few hundred-thousand degree range have led to improved agreement with a wide range of observations. Some directions for future work are suggested.

## I. INTRODUCTION

The rapid increase in the breadth and quality of stellar observations in recent years has presented a challenge to stellar models. The most widely known example has been attempts to explain the solar neutrino production rate. This, plus a number of other observed properties of stars that could not be adequately explained by theoretical models were known to be sensitive to the opacity. It was noticed that arbitrarily increasing the opacity in the few hundred-thousand degree temperature range by factors of 2–3 would improve some properties of variable stars (Fricke, Stobbie, and Strittmatter 1971; Peterson 1974; Stellingswerf 1978; Simon 1982), increasing the opacity by about 40% at slightly higher temperatures would improve the calculated Li abundance in the Hyades cluster (Swenson, Stringfellow, and Faulkner 1990), and increasing the opacity by 10–20% in the few million degree range would improve the predicted solar p-mode frequencies (Christensen-Dalsgaard, et al. 1985; Korzenik and Ulrich 1989; Cox et al. 1989).

The primary source of opacity data from the early 1960's until recently has been the Los Alamos National Laboratory (Cox and Stewart 1962; Cox and Tabor 1976; Huebner et al. 1977). St Andrews University has also provided opacity data (Carson, Mayers, and Stibbs 1968; Carson 1976). In response to a plea by Simon (1982) to reinvestigate the opacity, Magee et al. (1984) concluded that errors in the existing Los Alamos data were around 20% and rejected the possibility of large increases in the theoretical opacity.

Due to the importance of radiation transport in opacity for a wide range of stellar properties, the Los Alamos study notwithstanding, two groups undertook completely new, independent efforts to calculate astrophysical opacities. One of these, known as the Opacity Project (OP), is an international collaboration concerned with stellar envelopes. The other, known as OPAL and located at Lawrence Livermore National Laboratory, is concerned with the full range of interior opacities needed in stellar modeling. Contrary to the conclusions of Magee et al. (1984) preliminary OPAL results showed substantial increases in opacity (Iglesias, Rogers, and Wilson 1987) and up to factors of four have been reported in more complete calculations (Iglesias and Rogers 1991; Rogers and Iglesias 1992; Seaton et al. 1994). These increases are largely due to the rich spectrum of transitions originating in M-shell iron with an important contribution from the intra-M-shell transitions neglected by Los Alamos.

The format of the paper is as follows. Section II gives a brief description f the OP and OPAL efforts and Section III discusses the equation of state approaches that underlie the opacity calculations. New calculations of astrophysical opacities that include very heavy elements are presented in Section IV and Section V suggests some future directions for opacity work.

## II. OPACITY

The radiation transfer equation describes the transport of energy by photons and is equivalent to the Boltzmann equation in the kinetic theory of particle transport (Mihalas 1978). For steady-state conditions it is given by

$$\frac{dI_\nu(s,\mathbf{n})}{ds} = -\kappa_\nu I_\nu(s,\mathbf{n}) + j_\nu \quad (1)$$

where $I_\nu(s,\mathbf{n})$ is the intensity of radiation of frequency $\nu$ in the direction $\mathbf{n}$ as a function of distance $s$, $j_\nu$ is the emissivity, and $i$ is the elemental type. The monochromatic opacity, $\kappa_\nu$, is given by

$$\kappa_\nu = \sum_i \chi_i \left(\kappa_i^{bb} + \kappa_i^{bf} + \kappa_i^{ff}\right)\left(1 - e^{-h\nu/kT}\right) + \kappa^s, \quad (2)$$

where $\chi_i$ is the number fraction, $\kappa_i^{bb}$, $\kappa_i^{bf}$, and $\kappa_i^{ff}$, are the bound-bound, bound-free, and free-free absorption cross-sections, respectively, $\kappa^s$ is the scattering cross-section and the factor $(1 - \exp(-h\nu/kT))$ is the correction factor for stimulated emission.

Conditions inside a star are close to local thermodynamic equilibrium so that according to Kirchhoff's law,

$$j_\nu = \kappa_\nu B_\nu(T), \quad (3)$$

where,

$$B_\nu = (2\pi^2 \nu^3/c^2)\left(e^{h\nu/kT} - 1\right)^{-1} \quad (4)$$

is the Planck function. In addition, conditions change slowly over many photon mean-free paths and the radiation transfer equation greatly simplifies. In this limit, known as the diffusion approximation, $\kappa_\nu$ can be replaced with a flux weighted harmonic mean according to

$$\frac{1}{\kappa_R} = \int_0^\infty d\nu \frac{1}{\kappa_\nu}\frac{dB_\nu}{dT} \bigg/ \int_0^\infty \frac{dB_\nu}{dT}. \quad (5)$$

In Eq. (5) $\kappa_R$ is the Rosseland mean opacity or simply the opacity. A large value of the opacity indicates strong absorption from a beam of photons, whereas a small value indicates that the beam loses very little energy as it passes through the medium.

It was apparent from the beginning that the most intensive part of the opacity calculations would be the vast amount of atomic data needed to calculate bound-bound and bound-free absorption cross-sections. Several possible levels of detail for including atomic data are possible. For example, consider transitions that connect two electrons in an sp configuration to a $p^2$ configuration through a one electron jump. In the simplest approximation the transitions are degenerate. Opacity calculations that carry out a sum over a large number of configurations in this approximation are referred to as detailed configuration accounting (DCA) methods. However, the DCA approach neglects non-spherical interactions that remove the degeneracy and lead to configuration term structure. In light elements the dominant non-spherical term is the Coulomb interaction between the electrons, which leads to pure LS coupling. In LS coupling the single spectral line of the DCA method is split into three distinct lines corresponding to a triplet and two singlet terms. With heavier elements, such as those in the iron group, the interaction between the electron spin and the magnetic field resulting from the electron orbital motion is no longer negligible and requires intermediate coupling (Cowan 1981). In intermediate coupling the three lines of the LS coupling scheme in our example split into eight lines having no net total spin change, $\Delta S = 0$, and 6 intercombination lines having $\Delta S = \pm 1$. In more complicated configurations the increase in the number of spectral lines can be more dramatic.

To calculate the required atomic data the OPAL group developed a parametric potential method that is fast enough to allow on-line calculations, while achieving accuracy comparable to single configuration Dirac-Fock results (Rogers, Iglesias, and Wilson 1988). This on-line capability also provides flexibility to study easily the effects of atomic physics approximations such as various angular momentum couplings or data averaging methods. By contrast, the OP group uses first principle methods to construct detailed atomic databases that can be used in other types of investigations (Seaton 1987; Seaton et al. 1994). The large increase in the iron opacity obtained with the LS coupling scheme compared to calculations that neglect term splitting were an indication that fine structure should also be included (Rogers and Iglesias 1992). The OPAL group responded with calculations that include the spin-orbit effects in full intermediate coupling (Iglesias, Rogers, and Wilson 1992), while the OP group uses an approximate method that does not include spin changing transitions (Seaton et al. 1994). The OPAL calculation assumes single configurations, while OP includes configuration-interaction effects in both the bound-bound and bound-free calculations. Configuration-interaction is most important for neutral and near neutral transitions. The available Los Alamos opacity data is mostly in the DCA approximation, but more detailed calculations are in progress (Magee 1993).

The OPAL calculations include degeneracy and plasma collective effects in the free-free absorption using a screened form of the parametric potentials, whereas, these effects are not included in OP. In both OPAL and OP collective effects on the Thomson scattering are obtained from the method of Boercker (1987). In OPAL spectral line broadening for one, two, and three electrons ions are obtained from a suite of codes provided by R.W. Lee (1988) that include linear Stark theory. For all other transitions the OPAL calculations use Voigt profiles where the Gaussian width is due to Doppler broadening and the Lorentz width is due to natural plus electron impact collision broadening (Dimitrievic and Konjevic 1980). The OP line broadening follows similar lines, except broadening by ions is included only for hydrogenic systems and HeI (Seaton 1990).

The source of the largest opacity increases obtained by the OPAL and OP groups has been improved atomic physics, but equation of state and line-broadening improvements have also been important.

## III. EQUATION OF STATE

The equation of state plays an important role in opacity calculations by providing the occupation numbers needed to evaluate Eq. (5). The OP and OPAL equation of state approaches are quite different, but for typical stellar conditions give very similar thermodynamic results (Däppen 1992). However, the predicted occupation numbers are different for highly excited states that can affect the opacity at high density; the OPAL values being somewhat larger. The OP calculation is based on a free energy minimization, or chemical picture method, whereas the OPAL calculation is based on activity expansions of the grand canonical ensemble, or physical picture method.

A major difficulty that arises in existing free energy minimization methods is obtaining a convergent partition function. To treat this problem the OP work uses an occupation probability formalism that is thermodynamically consistent and produces continuous free energies (Däppen, Anderson, and Mihalas 1987; Hummer and Mihalas; Mihalas, Däppen, and Hummer 1988). It is commonly referred to as the MHD equation of state. Based on experimental measurements of level shifts (Goldsmith, Griem and Cohen 1984), they assume that the bound states of atoms and ions are unshifted by the plasma environment.

The MHD approach is based on the observation that if a configurational free energy, $f(V, T, \{n_i\})$, that depends explicitly on the occupation numbers of the individual states is added to the ideal free energy terms, then the ratio of the occupation of a state $i$ of a given ion to the total occupation is given by

$$n_i/n = \exp\left[-\beta(E_i + \partial f/\partial n_i)\right]/\tilde{Z}_{\text{int}} \qquad (6)$$

where

$$\tilde{Z}_{\text{int}} = \sum_i \exp\left[-\beta(E_i + \partial f/\partial n_i)\right] \qquad (7)$$

plays the role of the internal partition function,

$$\omega_i = \exp\left[-\beta \partial f/\partial n_i\right] \qquad (8)$$

is the occupation probability, $n_i$ is the occupation number for state $i$ and $n$ is the total occupation for a given species. The occupation probability is a measure of the number of bound states of type $i$ that are available to be occupied. The quantity $1 - \omega_i$ is, thus, a measure of the fraction of total states that have been severely affected by plasma perturbations and no longer act like localized states. In order to make progress it is necessary to have either a good estimate for $f$ or a good estimate for $\omega_i$, whichever is easier. The MHD approach mixes the two possibilities. In the case of neutral particle interactions the free energy of a parameterized hard sphere gas is used to determine the occupation probability. For ion-ion interactions the occupation probabilities are determined from the electric microfield (Stark-ionization theory). For ion-neutral interactions a product of the two forms is used. The method is thus phenomenological, but uses experimental data to fit free parameters in the occupation probability function. The MHD method has been used in numerous stellar modeling sensitivity studies (Christensen-Dalsgaard and Däppen 1993; Dziembowski, Pamyatnykh, and Sienkiewicz 1992). A method similar to MHD has also been developed by Sevastyanenko (1985).

The OPAL approach treats the system in terms of its fundamental constituents, so that bound complexes arise naturally from the theory. The procedure is to expand the pressure as a sum of two body, three body terms etc., i.e., a cluster expansion. The long range of the Coulomb potential introduces substantial complications. In addition, the quantum nature of electrons introduces degeneracy and exchange corrections. The attractive electron-ion interaction leads to short distance divergences in classical cluster coefficients, so that the use of quantum mechanical methods is essential. The activity expansion method uses graphical resummation procedures to remove the long-range divergences occurring in all cluster coefficients of plasmas. Composite particles, i.e., ions, atoms, and molecules, arise naturally in the physical picture, such that plasma screening effects on the bound states are determined from theory (Rogers 1994; 1989; 1986). This is a definite advantage over the chemical picture methods in current use which must introduce intuitive models to obtain these effects. A detailed description of the procedure is given elsewhere (Ebeling et al. 1976; Kraeft et al. 1986; Rogers 1981).

The simple case of a hydrogen plasma can be used to illustrate results from the physical picture (Rogers 1994; 1989; 1986). To relate the results to the chemical picture we display the low density free energy obtained in the physical picture, as described in Rogers (1994),

$$\frac{F}{kT} = -N_l \ln\left(\frac{eg_e}{\rho_e \lambda_e^3}\right) - N_p \ln\left(\frac{eg_p}{\rho_p \lambda_p^3}\right) - N_H \ln\left(\frac{eg_H}{\rho_H \lambda_H^3} Z_{\text{int}}^{pl}\right) - \frac{F_{\text{DH}}}{kT} \quad (9)$$

where

$$Z_{\text{int}}^{pl} = \sum_{nl}(2l+1)\left(e^{-\beta E_{nl}} - 1 + \beta E_{nl}\right) \quad (10)$$

is the so called Planck-Larkin partition function and

$$\frac{F_{\text{DH}}}{kT} = -\frac{V}{12\pi}\left(\frac{kT}{4\pi e^2 \sum_i Z^2 \rho_i}\right)^{3/2} \quad (11)$$

is the Debye-Hückel free energy, and $\rho_i$ is the number density of ions of type $i$ including electrons. The sum in $Z_{\text{int}}^{pl}$ ranges over the allowed states in a screened potential that approaches the Debye-Hückel potential at very low density. However as described in Rogers (1986; 1981) the energy levels appearing in $Z_{\text{int}}^{pl}$ are unscreened except for high lying states near the plasma continuum. The states that are screened change with plasma conditions. As a result $Z_{\text{int}}^{pl}$ is both finite and a continuous function of temperature and density; although the density dependence is very slight for normal stellar conditions. The MHD equation of state displays a similar property through the use of the occupation probability formalism. The OPAL approach obtains systematic higher order Coulomb corrections not considered in the OP work and is expected to produce more realistic results at higher density.

## IV. NEW RESULTS

Accurate models of stellar structure require detailed computer calculations. However, in regions where radiation pressure dominates, the ratio of the matter pressure to the radiation pressure is approximately constant (Bohm-Vitense 1992). Using the non-relativistic ideal gas pressure in combination with the Stefan-Boltzmann law, one obtains with the above assumption a constant value for density/(temperature)$^3$. Thus, it is convenient to tabulate the Rosseland mean opacity at constant values of $R$ against temperature where $R = \rho/T_6^3$, $\rho$ is the material density in g/cm$^3$, and $T_6$ is the temperature in units of $10^6$ Kelvin.

In order to calculate the opacity it is necessary to specify the chemical composition. Estimated elemental abundances have changed appreciably over the years and have been a major source of opacity uncertainties. For example, improvements in measuring techniques have brought the solar photospheric abundances into close agreement with the meterioritic abundances. Of particular importance to the opacity calculations is the recent 30% reduction in the photospheric iron abundance (Biemont et al. 1991). Although, abundance estimates are available for all naturally occurring elements, those above zinc have not been included in any of the existing astrophysical opacity calculations. The abundance of a number of elements of odd atomic number and a few of even atomic number below zinc are also very low. It has been computationally expedient to eliminate these elements from the calculations by combining their abundance with their neighbors. The most recent changes in solar abundances have been small (Grevesse and Noels 1993).

Figure 1 displays the opacity as a function of temperature for $\log R = -3$ for the abundances used by Cox-Tabor (1976). The mixture in this example assumes a hydrogen mass fraction $X = 0.7$. OPAL results are displayed for a mixture having no metals and for a mixture that is composed of 2% metals by mass. In the astrophysical literature metals refers to all elements heavier than helium and their total fractional mass is indicated by the symbol $Z$. The helium mass contents of the two mixtures are, thus, $Y = 0.3$ and 0.28, respectively. The results for a commonly used set of Los Alamos opacities (Cox and Tabor 1976) with $Z = 0.02$ are also plotted. The curves show a series of bumps in $\log k_R$ that are due to strong absorption features in the spectrum of the indicated species. The H, He and He$^+$ bumps long have been associated with pulsational driving in variable stars (Bohm-Vitense 1992). However, the substantial bump in the new opacities around $\log T = 5.4$ is missing in the old results. This new Z bump has been shown to resolve several long-standing pulsation problems. The bump near $\log T = 6.3$ is important to helioseismology as well as Li depletion. A major reason that large deficiencies in the astrophysical opacities persisted over such a long time was due to the lack of laboratory experiments. A few experiments that measure opacity for conditions relevant to stellar envelopes have recently appeared (DaSilva et al. 1992; Springer et al. 1992).

The OPAL calculations were carried out with a 14 element mixture; reduced first from the Anders-Grevesse (1989) abundances and later the Grevesse (1991) abundances. As a consequence of the large increase in opacity resulting from a careful treatment of transitions originating from the M-shell, even lowly abundant elements from the iron group make contributions to the total opacity (Rogers and Iglesias 1992).

More recent OP data has been calculated including Cr, Mn, and Ni (Seaton et al. 1994). OPAL calculations that include these elements in the opacity tables are underway. In Fig. 2 we compare OP and OPAL opacities for the same reduced 14 element Grevess 1991 mixture having $X = 0.7$ and $Z = 0.02$. The overall agreement is good. The largest differences (30–40%) occur for $\log R \geq -2$ and $\log T_6 \geq 5.8$ and for $\log R < -3$ and $\log T_6 \approx 5.2$. In both cases the OP opacity is lower than OPAL. A comparison of the same OPAL data with the 17 element OP data reduced from Noels and Grevesse (1992) that includes Cr, Mn, Ni is given in Fig. 18 of Seaton et al. (1994). This brings the OP results closer to OPAL, which are in this situation slightly lower than OP in some places. Similar increases in the OPAL data can be expected when the same 17 element mixture is used.

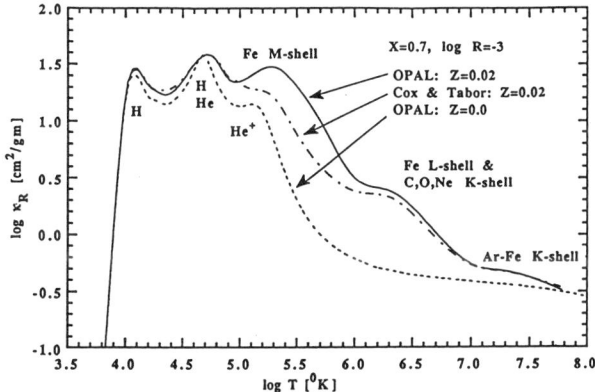

FIG. 1. Opacity versus temperature for a 14 element mixture of solar composition, having $X = 0.7$, along a track of constant $R = 10^{-3}$. OPAL results for $Z = 0$ and $0.02$ are shown. Results for a commonly used set of Los Alamos opacities with $Z = 0.02$ is also plotted.

A large number of calculations have been reported that use the new opacities; e.g., Rogers and Iglesias (1994). For example: pulsational instability for b-Cephei stars, as actually observed, is now predicted (Cox et al. 1992: Kiriakidis, El Eid, and Glatzel 1992; Moskalik and Dziembowski 1992) and calculated period ratios for double mode classical Cepheids and RR-Lyrae stars agree closely with observation at evolutionary masses (Kovacs, Buchler and Marom 1992; Cox 1991; Moskalik, Buchler and Marom 1992). In addition the location of stellar models in the observational planes relating mass, luminosity, effective temperature, and radius is substantially improved for main-sequence stars (Stothers and Chin 1991); the observed lithium abundance of stars in the Hyades cluster can now be modeled without invoking exotic theories (Swenson et al. 1994); the solar seismic frequencies as well as the calculated radius of the inner boundary of the solar convection zone are in close agreement with observations (Dziembowski, Pamyatnykh, and Sienkiewicz 1992; Guenther et al. 1992; Bahcall and Pinsonnealt 1992; Cox and Guzik 1993); and the calculated light curves for the decay phase of classical novae agree better with observation (Kato 1994).

## V. EFFECTS OF VERY HEAVY ELEMENTS ON OPACITY

The iron number fraction in the sun is down by almost 5 orders of magnitude compared to hydrogen and represents only about 2% of the number fraction of elements heavier than helium. Even so the major reason for the large increase in opacity in the few hundred-thousand degree temperature range with OPAL and OP is due to a better treatment of the atomic physics of iron. The combined number abundance of Cr and Ni is more than an order of magnitude less than iron, but still can increase the total opacity by as much as 20%. Consequently, even though the abundance of still heavier elements is down another couple orders of magnitude, the aggregate effect of these elements needs to be examined. Especially since they will display N shell (and higher) spectra that have still greater photoabsorption than the M-shell spectra.

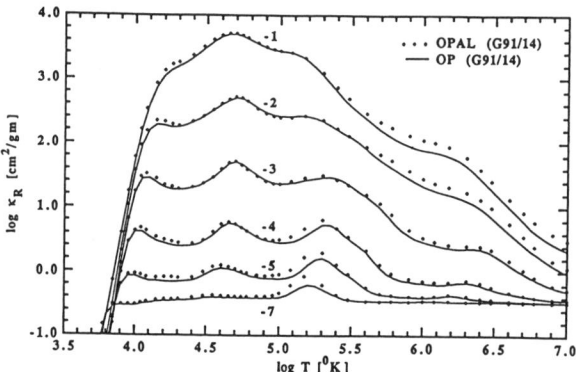

FIG. 2. Comparison of OPAL and OP for the reduced G91 abundances. The calculations assume $X = 0.7$ and $Z = 0.02$. The OPAL results are represented by solid lines, OP results by dashed lines. Plots are shown for five values of $\log R$.

Preliminary calculations that approximately include the effects of heavy elements up to neodymium have recently been carried out by Iglesias *et al.* (1994). The procedure was to augment the 14 element OPAL calculations for the Grevesse-Noels (1993) mixture with the calculations of very heavy elements from the Super Transition Array (STA) code (Bar-Shalom *et al.* 1994). The STA code uses statistical methods to treat the very large number of lines in heavy elements, where detailed accounting of individual lines becomes intractable. It includes in an approximate way all possible transitions between electronic states of the various ions in the plasma. The STA code uses parametric potentials, similar to OPAL, to obtain reasonably accurate atomic data (Klapisch 1971; Klapisch *et al.* 1977) and includes configuration interactions between neighboring j-j configurations. The statistical approach makes it possible to address cases where the number of relevant configurations is immense.

A comparison between OPAL and STA shown in Fig. 3 helps clarify the STA method. The calculations are for Ga at a temperature and density near the peak of the absorption bump due to transitions originating in the M-shell. The STA accurately reproduces the envelope of the OPAL results, but since it is a statistical approach it does not resolve the details of the spectrum. Consequently the STA method tends to overestimate the Rosseland mean opacity (Iglesias, Rogers, and Wilson 1990). The advantage is that the STA approach is computationally fast and can be applied to the heavier elements where the detailed line accounting methods are not practical.

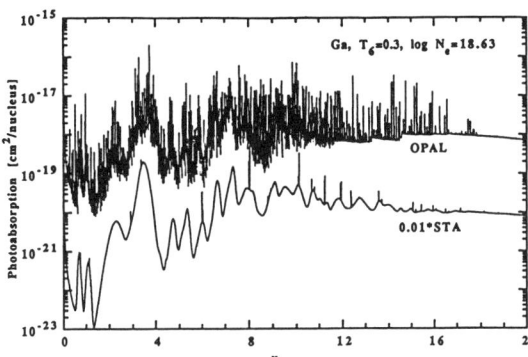

FIG. 3. Monochromatic opacity vs. $u$ =photon energy/$kT$ for gallium. The STA results have been multiplied by 0.01.

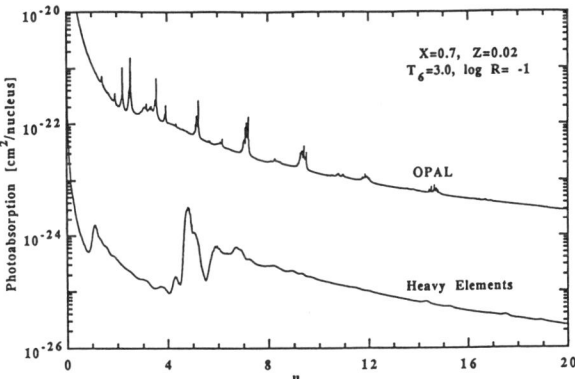

FIG. 4. Contribution of heavy elements for conditions near the base of the solar convection zone. The OPAL 14 element reduced mixture results for elements below zinc are also shown.

The Iglesias et al. (1994) calculations treated the heavy elements in groups of five, placing the total abundance with the central element of the group. They found that for solar element abundances that the effect is insignificant for $T_6 > 1$. An example is given in Fig. 4 for conditions near the bottom of the solar convection zone ($T_6 = 3$). The individual element absorption is greater for the heavy elements at these high temperatures, but that increase is not enough to overcome the greatly decreased abundances. However, for stars with peculiar element abundances, such as those of type Am and Ap, that have significant increases in some heavy elements, there may be important contributions to the opacity. For normal solar composition the heavy elements make the largest contribution in the $0.1 < T_6 < 1$ temperature region similar to the iron group elements.

## VI. NEW DIRECTIONS

The applications of the new opacities have so far concentrated on main sequence and near main sequence stars. It remains to be seen what the effects of the new opacity will be on more evolved stars that have exhausted most of their core hydrogen and perhaps even their helium. Depending on their initial mass and metallicity, these stars can display a wide range of phenomena ending up as exploding supernova in one extreme or as cooling white dwarfs in the other.

The current OPAL opacity tables only extend far enough to model white dwarf stars whose surface temperatures are greater than about 15000 K (even higher with OP). However, the coolest white dwarfs ($T_{\text{eff}} \approx 4000$ K) are remnants of the first generation of stars in the local disk of the Galaxy. Opacities that allow for the study of these white dwarfs would provide an independent estimate of the age of the local disk and could shed some light on the formation of spiral galaxies in general. This will require extending the equation of state and opacity calculations several orders of magnitude in density, presenting new challenges to the calculations.

There has been speculation that observed stellar abundance anomalies in the envelopes of some A and B stars may be produced by radiative levitation of highly absorptive heavy elements and diffusion (Richer, Michaud, and Proffitt 1992; Proffitt 1994). The radiation force acting on different species in a stellar mixture is given by a formula very similar to Eq. 1 (Michaud 1976) and can be calculated using slight modifications to the OP and OPAL codes.

There are a number of recent neutron star flux detections with the Hubble telescope UV and especially the ROSAT soft X-ray satellites. These stars have dense iron atmospheres and calculations of the emergent spectrum are sensitive to the opacity (Romani 1987). Modeling of the atmospheres of neutron stars is also sensitive to non-ideal Coulomb corrections in the thermodynamic properties. In addition the presence of large magnetic fields in some neutron stars may affect the opacity and thermodynamic properties (Van Riper 1988).

Cool stars with surface temperatures below 5000 K can form a wide variety of molecules in their atmospheres. Molecular opacities in these stars affect the surface temperature and are important for determining their masses. A substantial part of the mass in the galaxy may be locked-up in low mass, cool stars and reliable opacities are needed to answer this question. Several groups are in the process of improving molecular opacities and new results are becoming available (Alexander and Ferguson 1994; Allard *et al.* 1994; Sharp 1993).

## ACKNOWLEDGEMENTS

We thank M. J. Seaton for useful discussions and for providing OP data. Work performed under the auspices of the U.S. Department of Energy by the Lawrence Livermore National Laboratory under contract W-7405-Eng-48.

## REFERENCES

Allard, F., P. H. Hauschildt, R. L. Miller, and S. Tennyson 1994, Astrophys. J. (in press).
Alexander, D. J., and J. W. Ferguson, Astrophys. J. (in press).
Anders, E. and N. Grevesse, 1989, Geochim. Cosmochim. Acta, **53**, 197.
Bar-Shalom, A., J. Oreg, and W. Goldstein, 1994, J. Quant. Spectrosc. Radiat. Transfer, **51**, 27.
Bahcall, J. N., and M.H. Pinsonneault, 1992, Rev. Mod. Phys., **64**, 885.
Biemont, E., M. Baudoux, R. L. Kurutz, W. Ansbacker, E. H. Pinnington, 1991, Astron. Astrophys., **249**, 510.
Boercker, D.B., 1987, Astrophys. J., **316**, L95.
Bohm-Vitense, E., *Introduction to Stellar Astrophysics*, (Cambridge Univ. Press, Cambridge, 1992), vols. 2 and 3.
Carson, T. R., 1976, Astron. Astrophys., **14**, 95.
Carson, T. R., D. F. Mayers and D. W. N. Stibbs, 1968, MNRAS, **140**.
Christensen-Dalsgaard, J. and W. Däppen, 1993, Astron. Astrophys. Rev., **4**, 267.
Christensen-Dalsgaard, J., T. L. Duvall, D. O. Gough, J. W. Harvey, and E. J. Rhodes, 1985, Nature, **315**, 378.
Cowan, R. D., 1981, *The Theory of Atomic Structure* (Univ. of California Press, Berkeley).
Cox, A. N. and J. N. Stewart, 1962, Astrophys. J. Suppl., **67**, 113.
Cox, A. N. and J. E. Tabor, 1976, Astrophys. J. Suppl., **31**, 271.
Cox, A. N., 1991, Astrophys. J., **381**, L71.
Cox, A. N., S. B. Morgan, F. J. Rogers, and C. A. Iglesias, 1992, Astrophys. J., **393**, 272.
Cox, A. N., J. A. Guzik, and R. B. Kidman, 1989, Astrophys. J., **342**, 1187.
Cox, A. N. and J. A. Guzik, 1993, Astrophys. J., **411**, 394.
Däppen, W., L. S., Anderson, and D. Mihalas, 1987, Astrophys. J., **319**, 195.
Däppen, W., 1992, Rev. Mexicana Astron. and Astrof., **23**, 1144.
Da Silva, L. B., *et al.*, 1992, Phys. Rev. Lett., **69**, 438.
Dimitrievic, M.S. and N. Konjevic, 1980, J. Quant. Spectrosc. Rad. Transf., **24**, 451.
Dziembowski, W. A., A. A. Pamyatnykh, and R. Sienkiewicz, 1992, Acta Astronomica, **14**, 5.
Ebeling, W., W. D. Kraeft, and D. Kremp, 1977, *Theory of Bound States and Ionization Equilibrium in Plasmas and Solids* (Berlin, Akademie-Verlag; New York, Plenum).
Fricke, K., R. S. Stobie, and P. A. Strittmatter, 1971, MNRAS, **154**, 23.
Glatzel, W., and M. Kiriakidis, 1993, MNRAS, **262**, 85.
Goldsmith, S., H. E. Griem, and L. Cohen, 1984, Phys. Rev. A, **30**, 2775.
Grevesse, N., 1991, private communication; see Iglesias, C.A., F.J. Rogers, and B. G. Wilson, 1992, for details.
Grevesse, N. and A. Noels, 1993, N. Prantzo, E. Vangioni-Flam, and M. Casse eds. in *Origin and Evolution of the Elements* (Cambridge Univ. Press, Cambridge).
Guenther, D. B., P. Demarque, Y.-C. Kim, and M. H. Pinsonneault, 1992, Astrophys. J., **387**, 372.
Huebner, W. F., A. L. Merts, N. H. Magee, and M. F. Argo, 1977, Los Alamos Scientific Report La-6760-M.
Hummer, D. G., and D. Mihalas, 1988, Astrophys. J., **331**, 794.
Iglesias, C. A., B. G. Wilson, F. J. Rogers, A. Bar-Shalom, and J. Oreg, 1994, in preparation.
Iglesias, C. A., F. J. Rogers, and B. G. Wilson, 1987, Astrophys. J., **322**, l45.

Iglesias, C. A., F. J. Rogers, and B. G. Wilson, 1992, Astrophys. J., **397**, 717.
Iglesias, C. A. and F. J. Rogers, 1991, Astrophys. J., **371**, 408.
Iglesias, C. A., and F. J. Rogers, 1993, Astrophys. J., **412**, 752.
Kato, M., 1994, Astron. and Astrophys., **281**, L49.
Kiriakidis, M., M. F. El Eid, and W. Glatzel, 1992, MNRAS, **255**, 1.
Klapisch, M., 1971, Computer Phys. Comm., **2**, 239.
Klapish, M., J. L. Schwob, B. S. Fraenhel, and J. Oreg, 1977, JOSA, **61**, 148.
Korzennik, S. G. and R. K. Ulrich, 1989, Astrophys. J., **339**, 1144.
Kovacs, G., J. R. Buchler, and A. Marom, 1992, Astron. Astrophys. J., **252**, L27.
Kraeft, W. D., D. Kremp, W. Ebeling, and G. Röpke, 1986, *Quantum Statistics of Charged Particle Systems*, (Plenum Press, New York).
Lee, R.W., 1988, J. Quant. Spectrosc. Rad. Transf., **40**, 561.
Magee, N. H., 1993, *New Opacity Library and SESAME Opacity Data*, Los Alamos National Laboratory Report, LA-UR-93-1248.
Magee, N. H., A. L. Merts, and W. F. Huebner, 1984, Astrophys. J., **283**, 264.
Michaud, G., Y. Garland, S. Vauclair, and G. Vauclair, 1976, Astrophys. J., **210**, 447.
Mihalas, D., 1978, Stellar Atmospheres (W. H. Freeman & Co., San Francisco).
Mihalas, D., W. Däppen, and D. G. Hummer, 1988, Astrophys. J., **331**, 815.
Moskalik, P., J. R. Buchler, and A. Marom, 1992, Astrophys. J., **385**, 685.
Moskalik, P. and W. A. Dziembowski, 1992, Astron. Astrophys., **256**, L5.
Petersen, J. O., 1974, Astron. Astrophys., **34**, 309.
Petersen, J. O., 1992, Astron. Astrophys., **265**, 555.
Proffitt, C. R., 1994, Astrophys. J., **425**, 849.
Ricger, J., G. Michaud, and C. R. Proffitt, 1992, Astrophys. J. Suppl., **82**, 329.
Rogers, F. J., 1981, Phys. Rev. A, **24**, 1531.
Rogers, F. J., 1986, Astrophys. J., **310**, 723.
Rogers, F. J., 1990, Astrophys. J., **352**, 689.
Rogers, F. J., 1994, in *The Equation of Sate in Astrophysics* IAU Colllloquiun **147**, edited by G. Chabrier and E. Schatzman (Cambridge University Press).
Rogers, F. J., and C. A. Iglesias, 1992, Astrophys. J. Suppl., **79**, 507.
Rogers, F. J. and C. A. Iglesias, 1994, Science, **263**, 50.
Rogers, F. J., C. A. Iglesias, and B. G. Wilson, 1988, Phys. Rev. A, **38**, 5007.
Romani, R. W., 1987, Astrophys. J., **313**, 718.
Seaton, M.J., Y. Yan, D. Mihalas, and A.K. Pradhan, 1994, MNRAS, **266**, 805.
Sevastyanenko, V., 1985, Beitr Plasmaphys. **25**, 151.
Stothers, R. B., 1992, Astrophys. J., **392**, 706; R. B. Stothers and C.-w. Chin, 1993, Astrophys. J., **408**, L85.
Sharp, C. M., 1993, *Inside The Stars*, eds. W. W. Weis and A. Baglin, ASP Conf. Series **40**, 263.
Seaton, M. J., 1987, J. Phys. B, **20**, 6363.
Seaton, M. J., 1990, J. Phys. B., **23**, 3255.
Simon, N. R., 1982, Astrophys. J., **260**, L87.
Springer, P., *et al.*, 1992, Phys. Rev. Lett., **69**, 3735.
Stellingwerf, W. F., 1978, Astrophys. J., **83**, 1184.
Stothers, R. B. and C-w Chin, 1991, Astrophys. J.. **381**, L67.
Swenson, F. J., G. S. Stringfellow, and J. Faulkner, 1990, Astrophys. J., **348**, L33.
Swenson, F. J., J. Faulkner, C. A. Iglesias, F. J. Rogers, and D. R. Alexander, 1994, Astrophys. J., **422**, L79.
Van Riper, K. A., 1988, Astrophys. J., **329**, 339.

# IV.4

# ATMOSPHERIC STRUCTURE OF VERY LOW MASS STARS: M DWARFS, SUBDWARFS, AND BROWN DWARFS

T. Tsuji and K. Ohnaka

Institute of Astronomy, University of Tokyo, Mitaka, Tokyo 181, Japan

Atmospheric structures of very low mass stars (M dwarfs/subdwarfs) and substellar objects (brown dwarfs) are discussed on the basis of the model-atmosphere method. For this purpose, radiative-convective model atmospheres have been extended to substellar regime with $T_{\text{eff}}$ as low as 1000 K. The major source of molecular opacities is $H_2O$ in M (sub)dwarfs, and $CH_4$ is increasingly important in cooler brown dwarfs. Another important elementary process in the dense atmospheres of substellar objects is the collision-induced absorption of $H_2$ for which accurate cross-sections have been evaluated recently. Based on the resulting non-grey model atmospheres, major spectroscopic and photometric properties of substellar objects are predicted, and their implications to future observations of substellar objects are briefly discussed.

## I. INTRODUCTION

Stars that occupy the cooler part of the HR diagram span a large range of luminosity from more than $10^{+5} L_\odot$ in supergiant stars to near $10^{-5} L_\odot$ in the coolest M (sub)dwarf stars. Because of the difference in luminosity as large as $10^{+10}$, M (sub)dwarfs are almost negligible as compared with red (super)giants in the visible sky and detailed observational studies of M (sub)dwarfs have also been difficult by the same reason. However, M dwarfs are far more numerous than red (super)giants, and the masses of most galaxies are essentially determined by M dwarfs. Further, it is recognized recently that the luminous matter, largely due to M dwarfs, is only a minor part of the total mass of the Universe, and a larger part (more than 90%) of the matter in the Universe should reside in invisible dark matter. Although the nature of the dark matter is quite unclear, recent discovery of MACHOs (Massive Astronomical Compact Halo Objects) in the halo of our Galaxy by the gravitational microlensing effect may lend a support to an idea that the dark matter may be baryonic, at least partly [Alcock et al., 1993; Aubourg et al., 1993]. Also, a possible presence of substellar objects beyond the hydrogen burning main sequence has been conceived and extensive searches of such objects, named brown dwarfs, have been done during the recent decade [e.g., Burrows and Liebert, 1993]. While no positive identification of brown dwarf is yet done, brown dwarf is an important candidate of the dark matter as well as of the MACHO.

For successful searches of brown dwarfs, and also for clarifying the nature of the MACHOs, it should be indispensable to understand the physical properties of the atmospheres of substellar objects and to be able to predict major observable properties in some details. For this purpose, one useful approach may be to have model atmospheres that properly incorporates the basic physical assumptions appropriate for substellar objects. One noticeable change in the equation of state (EOS) at low temperatures (below 1,000K) and high pressures (as high as $10^{+9}$ dyn/cm$^2 \approx 10^{+3}$ atms = 1 kbar), which may be realized in the atmospheres of substellar objects, is the predominance of CH$_4$ over CO even in the oxygen-rich environment, as has been shown earlier by the classical thermochemistry [Tsuji, 1964]. For this reason, CH$_4$ should be an important opacity source in substellar objects. Another important effect in the atmospheres of substellar objects is that the density is high enough to induce the electric dipole moment in non-polar molecules such as H$_2$ by collision, although the density may still be low enough for the EOS to be that of free gaseous particles so far as the pressures barely touch the kbar regime. In fact, collision-induced absorption (CIA) of abundant H$_2$ molecule is known to be an important opacity source already in M dwarf atmospheres [Tsuji, 1969; Linsky, 1969] and should play more important role in substellar atmospheres. However, it is only recently that more accurate quantum mechanical computations of the cross-sections for the CIA have been achieved [Borysow, 1994].

So far, several attempts on modelling the atmospheres of M (sub)dwarfs have been done [e.g., Mould, 1976; Allard, 1990], and attempts on realistic model atmospheres of substellar objects have just begun. Recently, an interesting case of non-gray model atmospheres of very low mass objects of zero metallicity has been investigated [Saumon et al., 1994]. We have explored a more realistic case of non-grey model atmospheres for non-zero metallicities that may correspond to substellar objects in the disk and halo. We have also applied the same method and input data to generate some models of M (sub)dwarfs to see the continuity between stellar regime for which some observational tests of model atmospheres can be possible and substellar regime for which no observation is available at present. In fact, M (sub)dwarfs are only observable relatives to brown dwarfs and it is even possible that some observed M dwarfs are actually young brown dwarfs on their cooling tracks. Of course, M (sub)dwarfs are important in their own right by various reasons. Especially, as M dwarfs and subdwarfs are the major populations of the disk and halo, respectively, and as they sample all the past stellar generations because of their long lifetime, M (sub)dwarfs are the most important tracers of constitution and evolution of matter in the Galaxy and galaxies.

## II. PROPERTIES OF COOL AND DENSE GASEOUS MIXTURES

Basic properties of the atmospheres of cool stars are largely governed by molecules, which can be assumed to be in chemical equilibrium. Here, CO molecule plays a decisive role because of its great stability: as almost all the carbon atoms are consumed in forming CO if the atmosphere is oxygen-rich (O>C), little carbon is left for other molecules containing carbon and only oxides can be formed in general. By the same reason, no oxide is formed if the atmosphere is carbon-rich (O<C) and molecules containing carbon prevail in such stars (carbon stars). However, this general rule can no longer be applied to chemical equilibria at high densities and low temperatures of substellar atmospheres, since CO is no longer the major reservoir of carbon because of the greater stability of CH$_4$ even in the case of O>C. Some details of the chemical equilibria at high pressure are shown in Fig. 1.

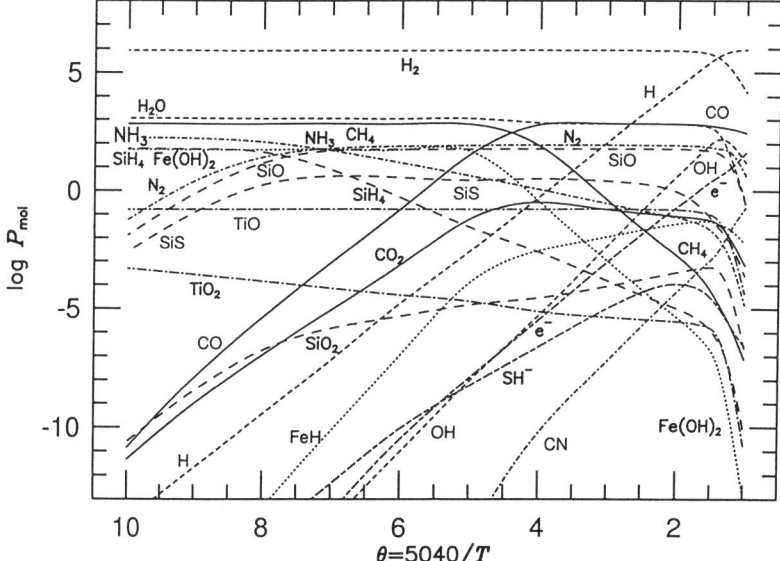

Fig. 1. Partial pressures of molecules in chemical equilibria are plotted against $\Theta = 5040/T$ in the gaseous mixture of the solar abundance under $P_g = 10^{+6}$ dyn/cm$^2$.

Atmospheric structure of cool stars is largely decided by the molecular opacities. We have considered the effect of molecular line absorption by a band model method [Tsuji, 1994] throughout. We have included methane (CH$_4$) as an important new opacity source at the very cool and dense atmospheres of substellar objects, in addition to those already important in M (sub)dwarf stars (infrared bands of H$_2$O, CO, OH, and SiO, together with the electronic bands of TiO, VO, FeH, CaH, MgH, and CN). Other polyatomics such as NH$_3$ and SiH$_4$ may be included in future (see Fig. 1). Another important opacity source is the CIA of H$_2$ and we have considered $\Delta v = 0$, 1, 2, and 3 transitions due to H$_2$-H$_2$ and H$_2$-He pairs for which accurate cross-sections are now available [e.g., Borysow and Frommhold, 1989,1990; Zheng and Borysow, 1994].

## III. MODEL ATMOSPHERES OF VERY LOW MASS OBJECTS

With the opacities outlined in the previous section, non-grey model atmospheres in radiative-convective equilibrium have been constructed. Some model atmospheres with $T_{\text{eff}}$ between 1,000 and 4,000K are shown in Fig. 2 for the cases of $Z/Z_\odot = 1.0$ and 0.01 ($Z_\odot$ represents the solar metallicity) by the solid and dotted lines, respectively. As is expected, atmospheric gas pressures are higher in the models of lower metallicity in general. The solid dot in each curve indicates the point where radiative and convective fluxes are equal. Clearly, convection is important in very low mass objects.

The spectral energy distributions for model atmosphere of $T_{\text{eff}} = 4,000$, 3,000, 2,000 and 1,000K are shown in Fig. 3 for the case of the solar metallicity. The heavy solid

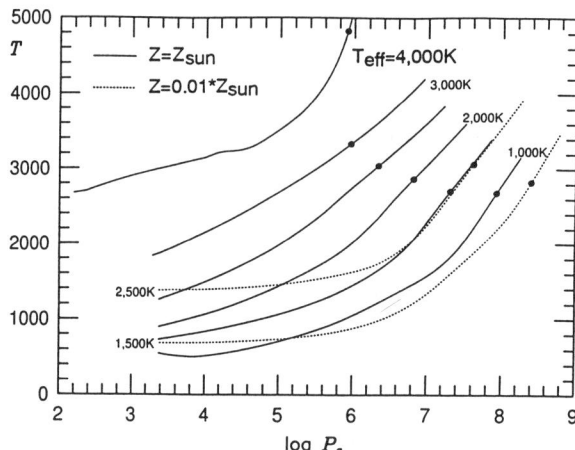

Fig. 2. Temperature-log $P_g$ structures of selected model atmospheres with $T_{eff}$ labelled. The case of $Z/Z_\odot = 1.0$ (solar metallicity) and 0.01 are shown by the solid and dotted lines, respectively. log $g$ =5.0 and $v_{micro}$ =1.0 km/sec throughout.

line is the line blanketed flux based on model atmosphere, and the dashed line is the blackbody radiation for $T = T_{eff}$. The dotted line is the continuum predicted by considering the continuous opacity alone but including the $H_2$ CIA. In the continuum, a peak at 1.6$\mu$m due to the minimum of $H^-$ absorption appears clearly in models of $T_{eff}$=4,000 and 3,000K. This peak is not so clear in the model of $T_{eff}$=2,000K, but a broad depression centered at about 2.4$\mu$m appears instead and tends to be stronger in cooler models. This depression is caused by the CIA of $H_2$ fundamental which is included as a continuous opacity source. A weak depression at 1.1 $\mu$m in the continuum of the model of $T_{eff}$=1,000K is due to the first overtone of the $H_2$ CIA. These results clearly demonstrate the importance of the CIA in the atmospheres of brown dwarfs.

On the line blanketed fluxes, the major molecular absorption changes from diatomic molecules (TiO, OH, CO etc.) at the high end of $T_{eff}$ to $H_2O$ at cooler M dwarfs, and finally to $CH_4$ in the coolest model of the substellar regime. Also, predicted line blanketed fluxes reveal large deviations from the blackbody radiations shown by the dashed lines, and the deviations are larger in models with lower $T_{eff}$. Especially, there appears a large excess of flux at around 1.2$\mu$m ($J$ band), where the opacity is minimum because the region longward of 1.4$\mu$m is highly opaque not only due to the combined effect of the $H_2O$ and $CH_4$ opacities but also due to the quasi-continuous $H_2$ CIA while the spectral region shortward of 1$\mu$m is also opaque due to the strong TiO bands.

To see the effect of metallicity, the spectral energy distributions for models of $T_{eff}$= 2,500 and 1,500K are shown in Fig. 4 and 5, respectively, for $Z/Z_\odot = 1.0$ and 0.01. Inspection of Fig. 4 reveals that the major molecular bands in the optical region change from oxides (TiO, VO etc.) in the metal rich M dwarfs to hydrides (CaH, MgH etc.)

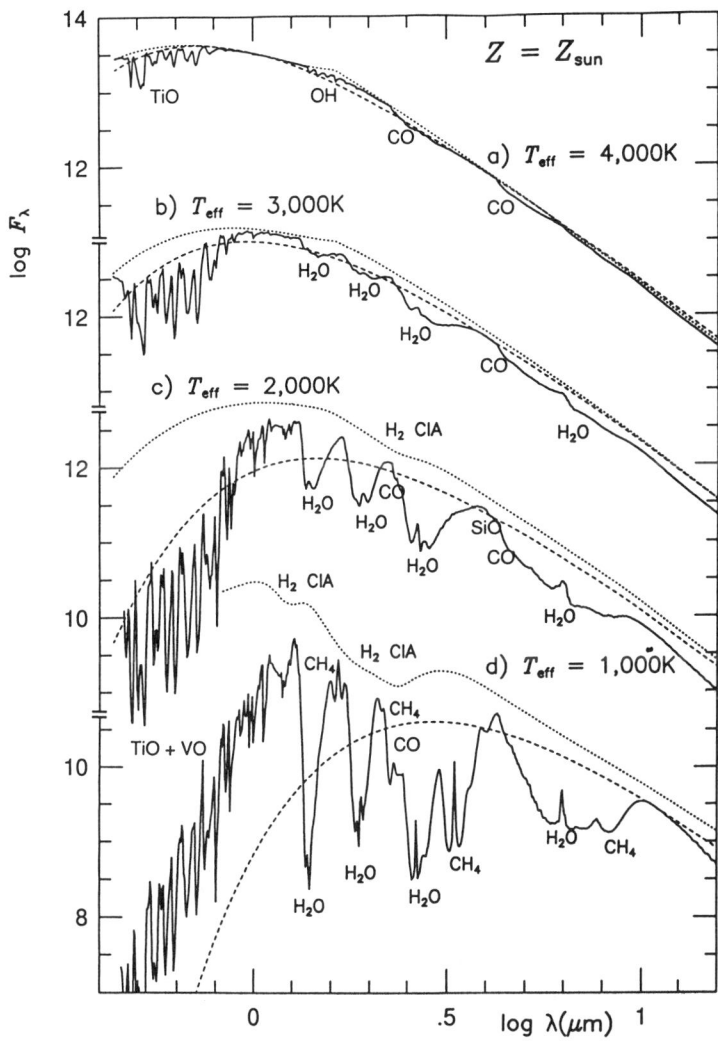

Fig. 3. Predicted spectral flux distributions, log $F_\lambda$ (erg/cm$^2$/sec/cm) vs. log $\lambda(\mu m)$, for models of the solar metallicity with $T_{eff}$ of a) 4,000K, b) 3,000K, c) 2,000K, and d) 1,000K. The solid, dotted, and dashed lines represent the line-blanketed flux, continuum, and black-body radiation, respectively. Major opacity sources are indicated.

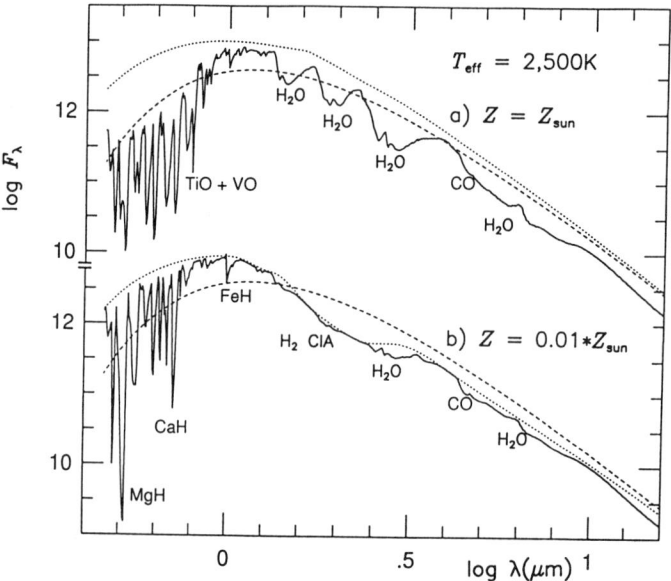

Fig. 4. The same as Fig. 3, but for models of $T_{\text{eff}} = 2,500$K (M dwarfs/subdwarfs) with a) $Z/Z_\odot = 1.0$, and b) $Z/Z_\odot = 0.01$.

in the metal poor subdwarfs. In Fig. 5, $CH_4$ bands remain conspicuous even in the case of $Z/Z_\odot = 0.01$ where other molecular bands suffer a large weakening. The effect of the $H_2$ CIA is relatively more important in more metal poor objects, not only in the substellar regime (Fig. 5) but also already in the stellar regime (Fig. 4). For this reason, infrared fluxes of the metal poor objects are highly depressed. All these results can readily be explained by a change of atmospheric structure, namely the increase of the gas pressures in the models of lower metallicity (Fig. 2). Our models of the low metallicity predict peaks at 1 and 4 $\mu$m, but such a large excess flux at the optical region near 0.7 $\mu$m as predicted by the zero metal models [Saumon et al.,1994; also see Fig. 2 of Hubbard et al. in this volume] never appears in our models, since the flux here is blocked by molecular bands. Thus, presence of a small amount of metal greatly modifies the spectral energy distributions of very low mass objects.

## IV. OBSERVATIONAL APPROACHES TO BROWN DWARFS

Most of the brown dwarf candidates so far known are relatively bright and may be on the cooling tracks of the evolving low mass stars. This fact makes it difficult to distinguish them from M dwarfs. The best way to avoid this difficult may be to

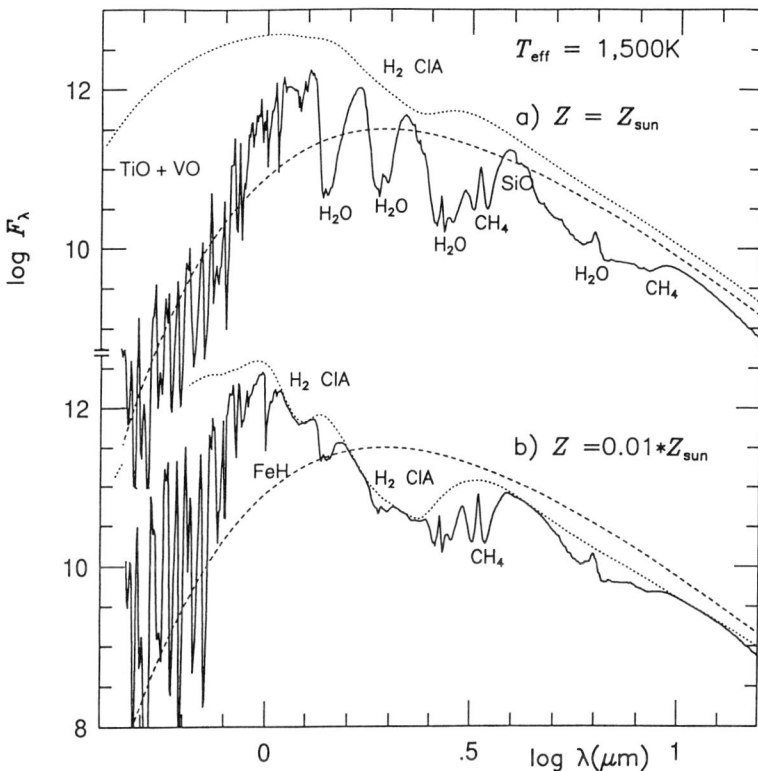

Fig. 5. The same as Fig. 3, but for models of $T_{\rm eff} = 1,500$K (substellar brown dwarfs) with a) $Z/Z_\odot = 1.0$, and b) $Z/Z_\odot = 0.01$.

search for brown dwarfs in the substellar regime (luminosity may be less than $10^{-5} L_\odot$) exclusive of stellar objects, and such observations can be possible by the progress in optical and infrared observations now under way. Also, the best way to confirm a candidate to be a true substellar object is to apply spectroscopic tests, and one possibility may be to observe $CH_4$ bands, which just appear to be visible in the model of $T_{\rm eff}$=1,800K, roughly the boundary between stellar and substellar objects for the solar metallicity [Burrow et al., 1993].

One interesting characteristic of the emergent flux distributions is that the brown dwarfs in the substellar regime are not necessarily very red but rather show a large flux excess around $J$ band (1.2$\mu$m). Further, such brown dwarfs may be not so bright at $K$ band (2.3$\mu$m) by the effect of the $H_2$ CIA in cooler and more metal poor models, and only $L'$ band (3.8$\mu$m) remains relatively bright in the infrared. Such characteristics that the brown dwarfs may be relatively bright in $J$ and $L'$ bands but not necessarily in $K$ band will have an obvious importance in searches of brown dwarfs, either by photometry or by imagings such as of multiple systems, clusters, galactic halos etc.

## V. CONCLUDING REMARKS

Very low mass stellar and substellar objects may represent the major constituents of the luminous and dark matter in the Universe, respectively. Then, these objects may be the most basic elements of the Universe and thus decide the constitution and ultimate fate of the Universe. Despite such basic importance, the nature of very low mass objects remains obscure, partly because of the difficulties in detailed modelling of cool and dense atmospheres of such objects. We tried to extend the model-atmosphere method, a standard approach to stellar atmospheric structure that has been proved to be useful in most stellar objects, to substellar brown dwarfs. We hope that our results will provide some clues for interpreting future observations of substellar objects such as brown dwarfs and MACHOs. For this purpose, however, continued efforts of improving the basis of the model atmosphere approach should be needed. For example, we need accurate cross-sections for the overtone transitions of the CIA by $H_2$-$H_2$ pair which, however, involve some difficult problems [Borysow, 1994], more data on molecular structures of such high-pressure species as $CH_4$, $NH_3$, and $SiH_4$ at elevated temperatures, proper treatment of dust formation in the dense substellar atmospheres, better theory of stellar convection, and so on. Thus, for better understanding the very tiny but key objects in the Universe, more detailed understanding on the elementary processes in dense gaseous mixtures should be quite essential.

## ACKNOWLEDGEMENTS

We thank Dr. A. Borysow for making available her computer codes to compute the cross-sections of the $H_2$ CIA and for helpful correspondences throughout this work.

## REFERENCES

Alcock, C., et al., 1993, Nature, **365**, 621.
Allard, F., 1990, Ph.D. thesis, Ruprecht-Karls-Universitat, Heidelberg
Aubourg, E., et al., 1993, Nature, **365**, 623.
Borysow, A., 1994, *Molecules in Stellar Environment*, ed. U.G.Jørgensen, (Berlin, Springer-Verlag), p.209.
Borysow, A., and Frommhold, L., 1989, Astrophys. J., **341**, 549.
Borysow, A., and Frommhold, L., 1990, Astrophys. J. Letter, **348**, L41.
Burrows, A., Hubbard, W.B., Saumon, D., and Lunine, J.I., 1993, Astrophys. J., **406**, 158.
Burrow, A., and Liebert, J., 1993, Rev. Mod. Phys., **65**, 301.
Linsky, J.L., 1969, Astrophys. J., **156**, 989
Mould, J., 1976, Astron. Astrophys., **48**, 443.
Saumon, D., Bergeron, P., Lunine, J.I., Hubbard, W.B., and Burrows, A., 1994, Astrophys. J., **424**, 333
Tsuji, T., 1964, Ann. Tokyo Astron. Obs., Ser. II, **9**, 1
Tsuji, T., 1969, *Low-Luminosity Stars*, ed. S. S. Kumar, (New York, Gordon & Breach), p.457.
Tsuji, T., 1994, *Molecules in Stellar Environment*, ed. U.G.Jørgensen, (Berlin, Springer-Verlag), p.79.
Zheng, C., and Borysow, A., 1994, Astrophys. J., in press.

# V. Helioseismology

# HELIOSEISMOLOGY AND SOLAR MODELS

H. Shibahashi

Department of Astronomy, University of Tokyo, Bunkyo, Tokyo 113, Japan

Fundamental properties and the present status of helioseismology are reviewed. The sun is regarded as an acoustic cavity, and its eigenfrequencies are recognized as the eigenlevels of the "acoustic potential." From the accurately measured frequencies of many eigenmodes of the sun, we can determine the shape of the potential by using inversion techniques. It is the sound speed distribution that is most accurately determined from the shape of the potential deduced from the helioseismic data. We show that we can, without significant uncertainties, deduce the density and the pressure distribution from the sound speed distribution thus determined. A procedure for constructing a solar model from helioseismic data is outlined.

## I. INTRODUCTION

The sun is oscillating by itself with a typical period of five minutes, though the amplitude is so tiny that we cannot recognize by our naked eyes the sun as a pulsating star. The oscillation itself was discovered in 1960, and we now know that the oscillation is a manifestation of a superposition of many eigenmodes of the sun. Helioseismology is a remote sensing, by which we probe from these oscillation data the invisible interior of sun, ranging from the core up to the subphotospheric region. Available observational information is the eigenfrequencies, the amplitudes and the phases of oscillations, and the linewidth of the power spectrum, and the eigenfrequencies have so far been most efficiently utilized as helioseismic information. From the information listed above, we can get information about the solar static structure such as density distribution, the dynamical structure such as how the solar interior is rotating, and the thermal structure such as properties of convection. It is obvious that the name of helioseismology comes from the analogy to seismology the earth, by which we investigate the interior of the earth from the earthquake data. However, it should be noted that helioseismology and the solar oscillation are different in some senses from the geo-seismology: First, the solar oscillations are almost always stochastically excited while the earthquakes occur sporadically. Secondly, in the case of the solar oscillations, there seem to be the excitation sources almost everywhere in a certain layer just beneath the photosphere, while the earthquake sources are mainly distributed only along certain localized ridges. Thirdly, we can see the solar surface at once and hence we can easily recognize the 2-D oscillation pattern of oscillations in the case of the sun, while the observational spatial resolution of the oscillation of the earth is limited by the distribution of the seismometers. From these facts, we are able to detect much more eigenmodes of the sun than the eigenmodes of the earth, and hence potentially we will be able to probe the solar interior more in detail than the interior of the earth. Therefore helioseismology is promising in the sense that we can diagnose the states of the dense plasma inside the sun.

## II. FUNDAMENTALS OF SOLAR OSCILLATIONS

The equations governing the solar oscillations are those of hydrodynamics. Once an equilibrium model of the sun is given, we treat the oscillation as linear perturbations and solve the equations as an eigenvalue problem. Since the thermal timescale in the sun is much longer than the oscillation timescale, which is of the order of five-minutes in the present case, the oscillation is almost adiabatic. There are two kinds of restoring forces to lead a given perturbed motions to be oscillatory: the one is the gaseous pressure, and the other is the buoyancy force. All the other forces such as electromagnetic forces or Coriolis forces are much weaker than these two forces in the present case of oscillations. The wave caused by repulsion due to the gaseous pressure is an acoustic wave, and the eigenmodes are called p-modes. It is the acoustic wave that is responsible for the observed five-minute oscillation of the sun. The buoyancy force causes a gravity wave, and the periods of these gravity modes (called g-modes) are longer than 40 minutes in the case of the sun. The detection of those long-period modes is still controversial. Hence, in this review, we concentrate on seismology using the solar acoustic modes, which have been definitely observed. For more details, see texts such as Unno et al. (1989), Christensen-Dalsgaard and Berthomieu (1991), and Gough (1993).

Since the equilibrium structure of the sun is almost spherically symmetric, the angular part of the eigenfunction is described in terms of the spherical harmonic functions. Therefore, the modal pattern of individual eigenmodes looks like those shown in Figure 1, and the eigenfunction of any scalar quantity $\Xi(r,\theta,\phi,t)$ has the form of

$$\Xi = \Psi(r) Y_\ell^m(\theta,\phi) \exp(i\omega t), \quad (1)$$

where $Y_\ell^m(\theta,\phi)$ denotes the spherical harmonics, $\theta$ the colatitude and $\phi$ the azimuth angle in the polar coordinate, $\omega$ the angular frequency, and $t$ the time.

Under some approximations, the equation governing the radial part of the eigenfunctions, $\Psi(r)$, turns to be a wave equation similar to Schrödinger equation in classical quantum mechanics:

$$\frac{d^2 \Psi}{dr^2} + \frac{1}{c^2}[\omega^2 - \Phi_\ell(r)]\Psi = 0, \quad (2)$$

where $c(r)$ denotes the sound speed. The squared eigenfrequencies correspond to the eigenlevels of a certain potential, $\Psi_\ell(r)$, which we call the acoustic potential here (cf. the left panel of Fig. 2). The shape of acoustic potential is essentially determined by the combination of the sound speed distribution and the degree $\ell$ of the spherical harmonics in the deep interior, and by the steep density gradient in the outer region. The eigenlevels are sequentially labelled by the number of modes. We call this number the radial order $n$. Hence we have totally three quantum numbers to identify a single eigenmode: the radial order $n$, the spherical degree $\ell$ and the azimuthal order $m$. It may be instructive to image the wave propagation as sketched in the right panel of

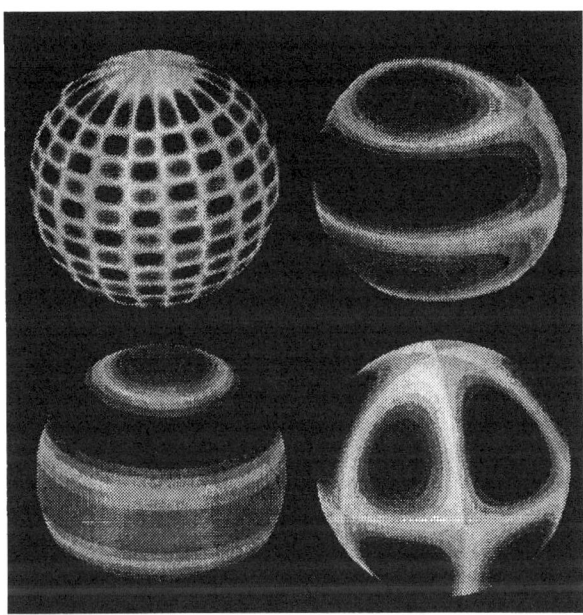

Figure 1: A schematic picture of the surface patterns of the eigenmodes of the sun,— $(\ell, m) = (10, 10)$ (top, left), $(\ell, m) = (4, 1)$ (top, right), $(\ell, m) = (4, 0)$ (bottom, left), and $(\ell, m) = (4, 3)$ (bottom, right). From Unno et al. (1989).

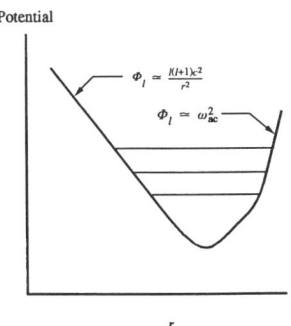
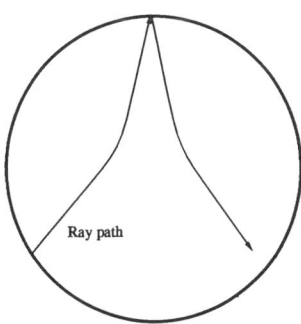

Figure 2: Left: A schematic picture of the acoustic potential and some eigenlevels. The acoustic potential is dominated by $\ell(\ell + 1)c^2/r^2$ in the deep interior, and by the acoustic cut-off frequency $\omega_{ac}$ that is related to the density gradient. Right: A schematic picture of a ray path emitted from the surface.

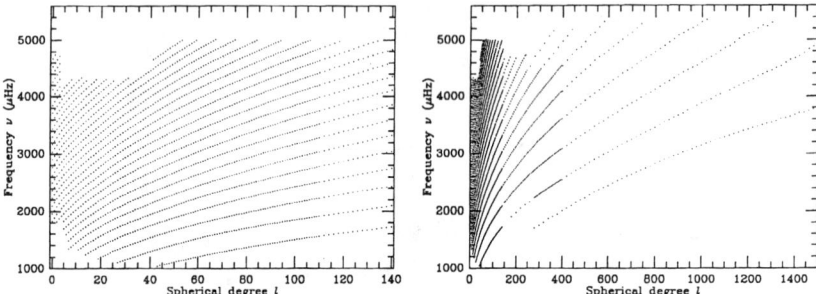

Figure 3: The observed p-mode frequencies in the range of $0 \leq \ell \leq 140$ (left) and in the range of $0 \leq \ell \leq 1500$ (right). The data were obtained by Jiménez et al. (1988) for $0 \leq \ell \leq 3$ and by Libbrecht et al. (1990) for $4 \leq \ell \leq 1500$. The ordinate is the frequency $\nu \equiv \omega/2\pi$ in units of $\mu$Hz. The lowest frequency ridge in the left panel corresponds to $n = 1$, while that in the right panel corresponds to f-modes. Errors in frequency are of the order of a few tens of nHz.

Fig. 2. This figure illustrates the ray path of the acoustic wave emitted from the solar surface. Since the sound speed becomes faster with the increase of depth, the ray is refracted, and propagates horizontally at a certain layer to turn back to the surface, at which the wave is reflected again downward. As seen in the left panel of Fig. 2, the layer to which the p-mode can penetrate is deeper with decreasing the degree $\ell$ and with increasing the order $n$. Only low-degree high-order modes can penetrate as deep as $r/R = 0.05$, in which nuclear reactions occur and most of the $^8$B neutrinos are generated.

## III. OBSERVATIONS OF SOLAR OSCILLATIONS

Observations of the solar oscillation are made by either measuring the Doppler shift of spectroscopic lines of the sun or measuring the brightness variation over the visible solar hemisphere. Whichever the observational technique is, the procedure to resolve the eigenmodes from the data is as follows. First, the spherical harmonic analysis of the spatial pattern is carried out over the visible hemisphere, and then the Fourier analysis of each spherical harmonic component is carried out with respect to time. As we noted previously, since we can get 2-D images of the sun, we can detect a quite large number of eigenmodes of the sun, $\ell \times m \times n$. However, the amplitudes of individual eigenmodes are quite small. The velocity amplitude is only of the order of 10 cm/sec, and the brightness variation amplitude is as small as $10^{-6}$.

Acoustic modes with large horizontal wavelengths (low degree $\ell$) are able to travel around the sun many times within their lifetimes, and they are truly global modes. The left panel of Fig. 3 shows the observed eigenfrequencies as a function of the spherical degree $\ell$. Since the eigenmodes are discrete, the eigenmodes are shown as discrete dots in this diagram. Each of the sequential ridges corresponds to a set of the radial order $n$. At higher $\ell$ the mode lifetime is shorter than the wave travel time around the sun. However, there still exists a distinct dispersion relation between $\ell$ and $\omega$ that depends on $n$ as a consequence of trapping in a wave guide beneath the photosphere (see the right panel of Fig. 3).

It should be emphasized that the observational accuracy is extremely high. The relative error of frequency of some mode is as small as $10^{-5}$. This means that we know more precisely about the eigenfrequencies of the sun than we know about the mass, radius or the luminosity of the sun. Furthermore, the numbers of detected modes are large. Hence, helioseismology should be regarded in some sense as an extraordinary precise science, which is quite seldom in the field of astrophysics. The frequency resolution is essentially determined by the length of the observing time. To get higher time resolution by continuing observations without night gaps, observations are sometimes carried out at the south pole, or by the networked observation sites which are distributed over the world. Observations from satellites are also performed. For more details, see Hill et al. (1991).

## IV. FORWARD PROBLEM

Once we get a good observational set of the eigenfrequencies, we can use these data to diagnose the solar interior. In principle, we have two approaches to this purpose. The one is called the forward problem. In this approach, we first make a series of solar models, then calculate theoretically eigenfrequencies of each model, and compare the calculated frequencies with the observed frequencies to look for the best-fit model. Input physics to make a variety of solar models are; the equation of state, the nuclear reacting rate, the treatment of convection, diffusion of elements, and the chemical composition distribution. A method of constructing a "standard solar model" is well established (Ulrich and Cox 1991). The main assumptions adopted in this method are as follows: the sun is hydrostatic and in thermal balance; the sun is assumed to be chemically homogeneous at the zero age; the composition in the envelope is unchanged during the evolution—that is, the chemical composition is the same as that in the envelope of the present sun, which is spectroscopically determined; the mass of the sun is constant during the evolution; we should use the standard theory for the nuclear processes, the opacity, the equation of state, and so on, for input microphysics. We make such a $1M_\odot$ star evolve by following the standard evolution theory till the age of $t = 4.5 \times 10^9$ yrs, which is estimated as the present solar age from the age of the oldest meteorites. The luminosity and the radius of the model depend mainly on the initial chemical composition and the treatment of convection in the sun, respectively. We adjust the initial composition and the mixing length of convection so that the model's luminosity and radius at the present solar age coincide with the present solar luminosity and radius, respectively.

The eigenfrequencies of the standard solar model thus constructed agree fairly well with the observed ones in the usual sense of astrophysics. The frequency differences are of the order of at most 10 $\mu$Hz, which corresponds to as small as 1 part of $10^3$ of the frequency themselves. However, these discrepancies are larger than the observational errors, and hence we should say that there remain discrepancies between the theoretical model and the true sun.

There are uncertainties in input physics in making a standard solar model along the recipe outlined in the above. Despite the models constructed in this way is called the "standard" solar models, there is a series of models which are slightly different each other due to the uncertainties in input physics. The sources of uncertainties follow; In the convection zone, which occupies about the outer 30% in the fractional radius, the equation of state of the plasma, the chemical composition, and the treatment of convection, or in other words, the constant value of entropy. In the deeper radiative interior, the uncertainties come mainly from the equation of state, the chemical composition, and the opacity. In the nuclear reacting are, the models are dependent on the nuclear reaction rates $\varepsilon$. Effects of these uncertainties on the global features of the model have been intensively studied by various researchers (e.g. Bahcall and Pinsonneault 1992, Bahcall and Bethe 1993, Turck-Cièze and Lopes 1993, Christensen-Dalsgaard 1990, Christensen-Dalsgaard and Däppen 1992). The differences in the computer codes also affect the output (see, e.g., Gabriel 1990, Reiter et al. 1994). Among these uncertainties, the equation of state will be discussed in detail by Werner Däppen in these proceedings. In spite of much effort of these investigations, we do not yet have definite answers about how and which parts of the solar models are different from the true sun and what are the physical causes of the differences.

Some other nonstandard solar models have been proposed to solve the neutrino problem. Among them, the presence of WIMPs in the sun has once been suggested as a way of improving the both of the discrepancies in the solar neutrino flux and in the p-mode spectra. However, recent careful investigations showed that the presence of WIMPs is not helpful to improve the discrepancy of the p-mode spectra.

## V. INVERSE PROBLEM

The inverse problem is the other approach to investigate the solar structure by using the oscillation data. In this approach, information of eigenfrequencies is utilized more directly than in the forward problem (Gough and Thompson 1991). As we discussed previously, the eigenoscillations of the sun can be considered as an analogy to the eigenvalue potential well problem. If we know all the eigenlevels in the potential problem, we can determine the shape of the potential. Since the acoustic potential is governed by the sound distribution in the deep interior, the possibility of determination of the shape of the potential is equivalent to the determination of the sound speed distribution in the sun.

Figure 4 shows the sound speed distribution thus determined. We can see the change in the gradient of the sound speed around $r/R \simeq 0.70$. This corresponds to the base of the convection zone. The right panel of Fig. 4 shows the relative difference between the sound speed distribution determined in this way and that of the Shibahashi et al's (1983) solar model.

The next step is to construct a solar model based on the seismic data by determining not only the sound speed distribution but also the distribution of all the other physical quantities. We should note that the experimentally measured quantities of the sun used

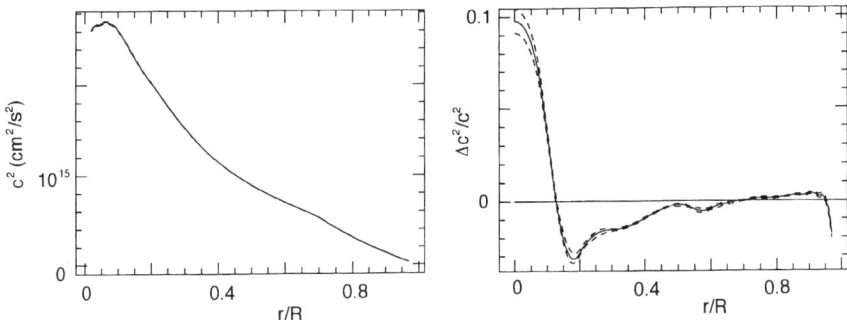

Figure 4: Left: The squared sound speed inverted by an asymptotic method from the observational frequencies shown in Fig. 3. Right: The relative difference between the squared sound speed inverted from the observational frequencies shown and that inverted from the eigenfrequencies of the model 1 of Shibahashi et al. (1983). The dashed lines correspond to the observational errors. After Vorontsov and Shibahashi (1990).

in the procedure of making the standard solar model are only the mass, the radius, and the luminosity. Most of the requirements are not based on the experimental data related directly to the present sun, but based on the theoretical consideration. Let us see another possibility to make a solar model based on more experimental data. The following is a recipe for making a seismic solar model. In this procedure, we use the solar mass, the solar radius, and the solar luminosity, and the sound speed distribution which is determined from helioseismology, and we deal with them as well-observed quantities. We assume that the sun is in hydrostatic equilibrium. Even if not, the sun settles down the equilibrium within its dynamical time scale which is of the order of one hour, and hence the assumption is highly reliable. This condition leads an equation combining the pressure gradient and the mass distribution. The mass distribution is governed by the density profile in the sun. Thus, the hydrostatic condition provides us with an equation relating the pressure profile and the density profile. Note that the sound speed is described in terms of the pressure, the density, and the adiabatic exponent. Hence, if we assume the adiabatic exponent's profile in the solar interior, we can get the pressure and the density profiles. The equation of state in the sun approximates that of the ideal gas but for the very outer layers. Hence the adiabatic exponent is almost constant except for near the surface and $\Gamma = 5/3$. Combining the hydrostatic condition and the sound speed data and assuming $\Gamma = 5/3$, we get the density and the pressure distribution in the sun. Figure 5 shows the preliminary results of this procedure.

Once we get the pressure and the density distribution in the sun, we can substitute those to the equations governing the thermal structure of the sun. These equations are the energy equation and the energy transfer equation, and they determine the

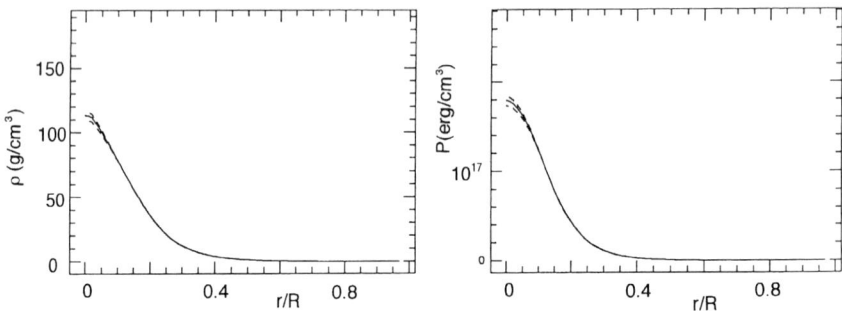

Figure 5: The density (top left) and the pressure (top right) in the Sun determined from the sound speed distribution (bottom) which is inferred from the helioseismic data shown in Fig. 3. The dashed lines correspond to the observational errors. After Shibahashi (1992).

temperature and the luminosity profile in the sun. We need physics of the opacity and the nuclear reaction rates and information about the initial chemical abundance. With the help of the above information, the equation for the thermal structure of the sun turns to be a differential equation for the temperature and the mean molecular weight profile. Note that these two quantities' distribution should be subject to the sound speed distribution inferred from helioseismology. We solve the differential equation subject to the sound speed distribution, and eventually we get the temperature and the mean molecular weight distribution. In the nuclear-reacting core, the chemical composition changes during evolution as a result of the nuclear reaction. Hence from the amount of the remaining hydrogen in the core, we can estimate the age of the sun. In this way, we can determine the distribution of various physical quantities in the sun, and diagnose the state of the plasma in the sun.

Once the solar model is constructed in this way, neutrino flux can be estimated. Comparing the neutrino flux thus estimated with observation, we can judge whether the solar neutrino problem arises from a deficiency of solar models. In the radiative zone in the outside of the nuclear-reacting core, the temperature can be determined from the pressure and the density, both of which have been determined by the sound speed distribution, if we assume the chemical composition. The luminosity there is equal to the surface luminosity of the sun. Then by using the transfer equation, we can determine the opacity there. We can calibrate the theoretically calculated OPAL opacity by using the opacity thus determined.

We have outlined how to reconstruct a solar model using helioseismic data. The above method is based on the asymptotic formulation based on Eq. (2). It should be noted that there is another approach based on the variational principle of the wave equation. This method has also been utilized to construct helioseismic solar models (Gough and Kosovichev 1990, Dziembowski et al. 1994, Antia and Chitre 1994).

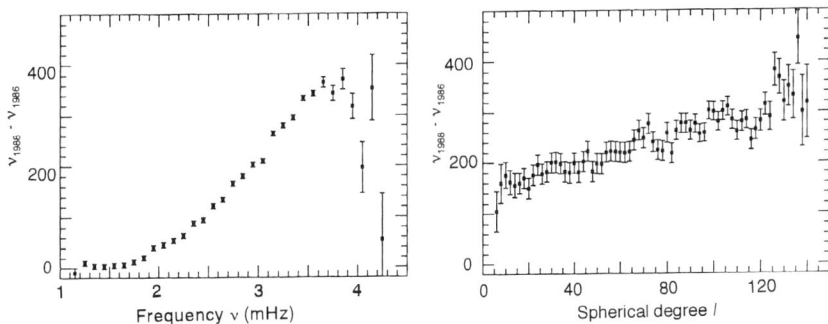

Figure 6: Slice-cut of the sun, showing the rotational profile inferred by helioseismolgy. After Gough, D. O., Kosovichev, A. G., Sekii, T., Libbrecht, K. G., & Woodard, M. F. (1993).

## VI. INTERNAL ROTATION OF THE SUN

The solar activity is thought to be caused by the dynamo generated by the interaction of the differential rotation and the convection. Hence the investigation of the rotation in the convection zone is important, and indeed it has been theoretically investigated by various researchers. The study of the solar internal rotation is also important for the understanding of the time evolution of the rotation in the stars. Helioseismology can be used to 'see' how the inside of the sun is rotating. If the sun were perfectly spherically symmetric, the eigenfrequency would be independent of the azimuthal order $m$. This degeneracy in $m$ is resolved by the rotation of the sun much as in the Zeeman effect of the atomic spectral lines. Hence, by using the rotationally induced $m$-splitting of the frequencies, we can determine the internal rotation of the sun. As discussed in section II, the extent of the wave propagating zone is strongly dependent on the mode. Furthermore, even for the modes belonging to the same degree $\ell$ and the same radial order $n$, the sectoral modes ($|m| = l$) are trapped in the polar regions, while the zonal modes ($m = 0$) are trapped in the equatorial zone. Thus, which part of the sun affects on the $m$-splitting is dependent on the mode, and then, by carefully investigating the $m$-splitting of various modes, we can determine the two-dimensional rotation profile in the sun. Figure 6 shows an example of the recent results of helioseismic investigation of the solar internal rotation. Roughly speaking, this result shows that the differential rotation with respect to the colatitude seen in the solar surface is conserved in the convection zone. This result is not consistent with the theoretically predicted rotation profile. More careful analysis and finer observations are being carried out to see if this apparent inconsistency is real.

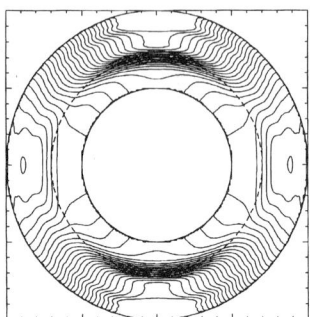

Figure 7: Right: The frequency change $\delta\nu \equiv \delta\omega/2\pi$ as a function of frequency $\nu(\equiv \omega/2\pi)$ for the averaged degree $5 \leq \ell \leq 60$ during 1988 and 1986. Left: The frequency change in the same duration as a function of degree $\ell$. After Libbrecht and Woodard (1990).

## VII. LONG-TERM VARIATION

Accurate measurements of the oscillations of the sun have been carried out by various researchers/groups over a time interval as long as a solar activity cycle (Pallé et al. 1989; Elsworth et al. 1990; Jefferies et al. 1990; Libbrecht and Woodard 1990). It is now clarified by these observations that the frequencies, the amplitudes, and the linewidths of the oscillation modes of the sun show a long term variation, which may be related to the solar activity. Figure 7 shows the frequency variation during 1986, which was at the solar minimum phase, and 1988, by which the solar cycle had reached considerable amplitude. This long-term frequency shift seems to be caused by the structural change of the layers near the photosphere and/or the chromosphere (Libbrecht and Woodard 1990; Gough 1990; Kuhn 1990, 1993; Shibahashi 1990; Goldreich et al. 1991; Evans and Roberts 1992). What the long-term frequency variation implies should be considered with other variation. The ACRIM data indicates that total irradiance increased during this interval by $10^{-2}$ percent (Willson and Hudson 1988), and the latitudinal temperature variation is similarly at variance. What does cause this long-term variation is still at issue.

## REFERENCES

Antia, H. M., and Chitre, S. M. 1994, Astrophys. J., in press.
Bahcall, J. N., and Bethe, H. A. 1993, Phys. Rev. D., **47**, 1298.
Bahcall, J. N., and Pinsonneault, M. H. 1992, Rev. Mod. Phys., **64**, 885.
Christensen-Dalsgaard, J. 1990, *Challenges to Theories of the Structure of Moderate-Mass Stars*, eds. D. Gough and J. Toomre (Springer, Berlin), p. 11.
Christensen-Dalsgaard, J., and Bertomieu, G. 1991, *Solar Interior and Atmosphere*, eds. A. N. Cox, W. C. Livingston, and M. S. Matthews (Univ. of Arizona Press, Tucson), p. 401.

Christensen-Dalsgaard, J., and Däppen, W. 1992, Astron. Astrophys. Rev., **4**, 267.
Dziembowski, W. A., Goode, P. R., Pamyatnykh, A. A., and Sienkiewicz, R. 1994, Astrophys. J., in press.
Elsworth, Y., Howe, R., Isaak, G. R., McLeod, C. P., and New, R. 1990, Nature, **345**, 322.
Evans, D. J., and Roberts, B. 1992, Nature, **355**, 230.
Gabriel, M. 1990, *Progress of Seismology of the Sun and Stars*, eds. Y. Osaki and H. Shibahashi (Springer, Berlin), p. 23.
Goldreich, P., Murray, N., Willette, G., and Kumar, P. 1991, Astrophys. J., **370**, 752.
Gough, D. O. 1990, *Progress of Seismology of the Sun and Stars*, eds. Y. Osaki and H. Shibahashi (Springer, Berlin), p. 283.
Gough, D. O. 1993, *Astrophysical Fluid Dynamics*, eds. J.-P. Zahn and J. Zinn-Justin (Elsevier, Amsterdam), p. 399.
Gough, D. O., and Kosovichev, A. G. 1990, *Proc. of IAU Colloq. No. 121, Inside the Sun*, eds. G. Berthomieu and M. Cribier (Kluwer, Dordrecht), p. 327.
Gough, D. O., Kosovichev, A. G., Sekii, T., Libbrecht, K. G., and Woodard, M. F. 1993, *GONG 92: Seismic Investigation of the Sun and Stars*, ed. T. M. Brown (Astron. Soc. Pacific, San Francisco), p. 213.
Gough, D. O., and Thompson, M. J. 1991, *Solar Interior and Atmosphere*, eds. A. N. Cox, W. C. Livingston, and M. S. Matthews (Univ. of Arizona Press, Tucson), p. 519.
Hill, F., Fröhlich, C., Gabriel, M., and Kotov, V. A. 1991, *Solar Interior and Atmosphere*, eds. A. N. Cox, W. C. Livingston, and M. S. Matthews (Univ. of Arizona Press, Tucson), p. 562.
Jefferies, S. M., Duvall, T. L., Jr., Harvey, J. W., and Pomerantz, M. A. 1990, *Progress of Seismology of the Sun and Stars*, eds. Y. Osaki and H. Shibahashi (Springer, Berlin), p. 135.
Jiménez, A., Pallé, P. L., Perez, J. C., Régulo, C., Roca Cortés, T., Isaak, G. R., McLeod, C. P., and van der Raay, H. B. 1988, *Proc. of IAU Symp. No. 123, Advances in Helio- and Asteroseismology*, eds. J. Christensen-Dalsgaard and S. Frandsen (Reidel, Dordercht), p. 205.
Kuhn, J. R. 1990, *Progress of Seismology of the Sun and Stars*, eds. Y. Osaki and H. Shibahashi (Springer, Berlin), p. 157.
Kuhn, J. R. 1993, *GONG 92: Seismic Investigation of the Sun and Stars*, ed. T. M. Brown (Astron. Soc. Pacific, San Francisco), p. 27.
Libbrecht, K. G., and Woodard, M. F. 1990, Nature, **345**, 779.
Libbrecht, K. G., Woodard, M. F., and Kaufman, J. M. 1990, Astrophy. J. Suppl. **74**, 1129.
Pallé, P. L., Régulo, C., and Roca Cortés, T. 1989, Astron. Astrophys., **224**, 253.
Reiter, J., Bulirsch, R., and Pfleiderer, J. 1994, Astron. Nach., **315**, 205.
Shibahashi, H. 1991, *Challenges to Theories of the Structure of Moderate-Mass Stars*, eds. D. Gough and J. Toomre (Springer, Berlin), p. 101.
Shibahashi, H., 1992, in *Frontiers of Neutrino Astrophysics*, ed. Y. Suzuki and K. Nakamura (Universal Academy Press, Tokyo), p. 93.

Shibahashi, H., Noels, A., and Gabriel, M. 1983, Astron. Astrophys., **123**, 283.
Turck-Chièze, S., and Lopes, I. 1993, Astrophys. J., **408**, 347.
Ulrich, R. K., and Cox, A. N. 1991, *Solar Interior and Atmosphere*, eds. A. N. Cox, W. C. Livingston, and M. S. Matthews (Univ. of Arizona Press, Tucson), p. 162.
Unno, W., Osaki, A., Ando, H., Saio, H., and Shibahashi, H. 1989, *Nonradial Oscillations of Stars* (2nd Edition) (Univ. of Tokyo Press, Tokyo).
Vorontsov, S. V., and Shibahashi, H., 1991, Publ. Astron. Soc. Japan, **43**, 739.
Willson, R. C., and Hudson, H. S. 1988, Nature, **332**, 810.

# V.2

# EQUATION-OF-STATE ISSUES IN HELIOSEISMOLOGY

W. Däppen

Department of Physics and Astronomy, University of Southern California
Los Angeles, CA 90089-1342, U. S. A.

Accurate measurements of observed frequencies of solar oscillations are providing a wealth of data on the solar interior. The frequencies depend on the solar structure, and on the properties of the plasma in the Sun. Except in the very outer layers, the stratification of the convection zone is almost adiabatic. There, the sound-speed profile is governed principally by the specific entropy, the (homogeneous) chemical composition and the equation of state. It is therefore essentially independent of the uncertainties in the radiative opacities. The sensitivity of the observed frequencies is such that it enables to distinguish rather subtle features of the equation of state. This opens the possibility to use the Sun as a laboratory for thermodynamic properties.

## I. INTRODUCTION

Solar acoustic oscillations have opened a new window into the Sun. By their nature they link the local sound speed in the interior with the observed oscillation frequencies. The spatial resolution of the solar disk allows the identification of a large number of individual oscillation modes, which are classified in terms of spherical harmonics. Modes in a large range of angular degrees, between $l = 0$ and a few thousand, are observed. The frequencies of these modes are centered around 3 mHz, which corresponds to periods around 5 minutes. They have been determined with high precision: typical relative errors are of the order of $10^{-4}$. The modes are confined to a cavity, which extends, broadly speaking, from the surface of the Sun, where the waves lose their material support, to the inner turning point which lies deeper the lower the angular degree $l$ is. Radial modes, with $l = 0$, have no inner turning point and their cavity is the entire Sun.

The observed solar oscillation modes are standing acoustic waves; hence the quantity most obviously probed is sound speed. Since the oscillations are largely adiabatic (except very near the surface), the frequencies are determined predominantly by the local adiabatic sound speed, which is a thermodynamic quantity. In addition, the frequencies depend on the density distribution in the Sun. Therefore, these *helioseismic* frequencies can be used as a diagnosis of the plasma of the solar interior. A high-quality thermodynamic potential is needed for the pressure-density relation (*i.e.* the equation of state, which is essential for determining the hydrostatic equilibrium between pressure gradient and gravity) and for thermodynamic quantities (mainly adiabatic sound speed).

Broadly speaking, theoretical inferences from the observed helioseismic frequencies can be made in two ways. In the *forward* approach, we build a solar model and compute its normal modes. Then the "best" model is the one that satisfies all observational constraints. In the *inverse* approach, we try to make as few theoretical assumptions as possible to infer the physical state of the solar interior directly from the oscillation frequencies.

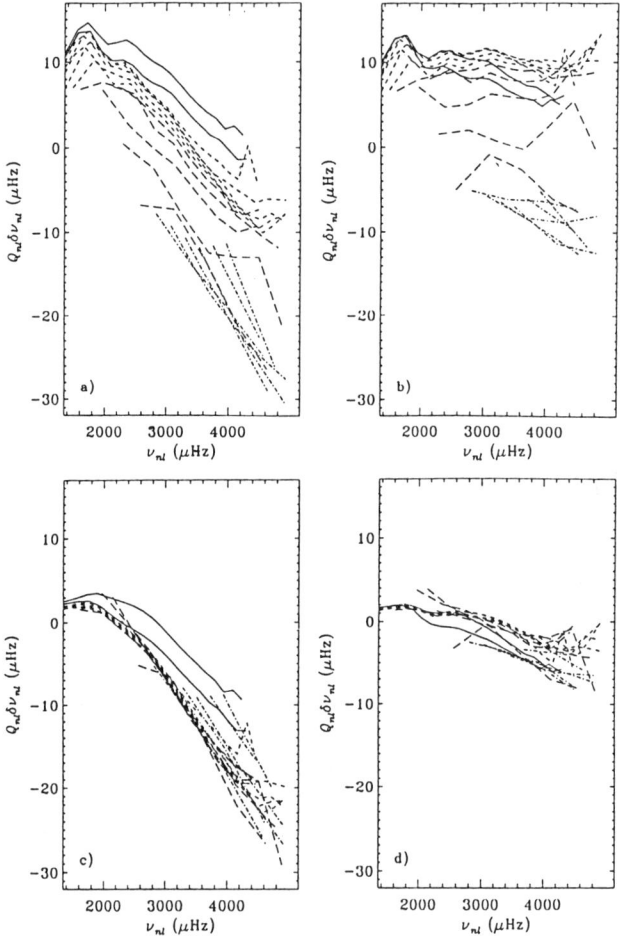

Fig. 1. Frequency differences between observed frequencies (in the compilation by Libbrecht et al., 1990) and four sets of computed frequencies, in the sense (observation) − (theory). The frequency differences are scaled by the factor $Q_{nl}$ (see text). The abscissa is cyclic frequency $\nu_{nl}$. The points have been connected with lines according to the value of the degree $l$: $l = 20, 30$: ─────; $l = 40, 50, 60, 80, 100$: ------------; $l = 120, 150, 200, 300, 400$: ‐ ‐ ‐ ‐ ‐ ‐ ; and $l = 500, 600, 700, 800, 900, 1000$: —·—·—·—. The models are distinguished by their equation of state and opacity (see text) (a) EFF equation of state, CT opacity; (b) EFF equation of state, LAOL opacity; (c) CEFF equation of state, CT opacity; (d) CEFF equation of state, LAOL opacity (from Christensen-Dalsgaard and Däppen, 1992).

A thorough introduction into theoretical helioseismology is the superb book by Unno *et al.* (1989). Reviews that deal specifically with the forward problem are, *e.g.*, those by Bahcall and Ulrich (1988), Christensen-Dalsgaard (1988), Christensen-Dalsgaard and Berthomieu (1991) and Turck-Chièze *et al.* (1992). The inverse approach is extensively discussed, *e.g.*, by Deubner and Gough (1984), Vorontsov and Zharkov (1989) and Gough and Vorontsov (1994). The reviews by Christensen-Dalsgaard (1991) and Christensen-Dalsgaard and Däppen (1992) specifically address the helioseismic determination of the equation of state.

## II. THE SUN AS AN EQUATION-OF-STATE EXPERIMENT

A. Comparison of theory with observations

One way to compare theory and observation is to take the difference between observed and computed solar oscillation frequencies. Figure 1 shows four such diagrams of frequency differences, each for a different theoretical model. Two equations of state and two different opacity tables were used in the models. The two equations of state were (i) the Eggleton, Faulkner and Flannery (1973) (EFF) equation of state (see IV.A.1), and (ii) the CEFF equation of state (IV.A.2), which is EFF plus a Coulomb term (Christensen-Dalsgaard, 1991; Christensen-Dalsgaard and Däppen, 1992). The opacities used were the tables by Cox and Tabor (CT) (1976), and of the Los Alamos Opacity Library (LAOL). Since I merely want to illustrate the sensitivity of the helioseismic method, it doesn't matter that these are not the most current opacities. A calculation based on the most recent Livermore (OPAL) opacities can be found in Berthomieu *et al.* (1993a).

I remark in passing that in such comparisons of observed with computed data ("O–C diagrams"), it is useful if an appropriate *scale factor* $Q_{nl}$ is taken out (see, *e.g.*, Christensen-Dalsgaard, 1988; Christensen-Dalsgaard and Berthomieu, 1991). This scale factor, which is essentially the inertia—or kinetic energy—of the mode with quantum numbers $nl$ (normalized to the same surface amplitude), contains the principal $l$ and frequency dependence of the individual mode frequencies $\nu_{nl}$.

The purpose of the illustration in Fig. 1 is to show the sensitivity of the helioseismic analysis with respect to changes in the physics of the model. A perfect model would yield a horizontal line, corresponding to all $\delta\nu_{nl} = 0$. Note that the discrepancies between theory and observation are huge compared to the observational errors which are nowadays significantly below 1 $\mu$Hz. Such a combination of quantity and quality of astrophysical data is truly exceptional.

B. Inversion of solar oscillation data

As mentioned above, the observed solar oscillation data can be directly inverted to yield the profile of internal sound speed. In fact, the resulting sound speed is so smooth that it allows differentiation with respect to radius. It is known that the resulting sound-speed gradient is closely related to the adiabatic gradient $\Gamma_1$ and its derivatives (Gough, 1984; Däppen and Gough, 1984). Thus, recently Vorontsov *et al.* (1994) and Baturin *et al.* (1994) obtained significant thermodynamic information with a nonlinear inversion. This inversion (Brodsky and Vorontsov, 1993; Gough and

Vorontsov, 1994) is based on an accurate asymptotic description of intermediate- and high-degree solar p modes, using the observational data currently available (Libbrecht et al., 1990).

The technique not only permitted to detect the influence of non-ideal Coulomb interactions between particles but also to establish a clear diagnostic potential to probe the mixture of heavy elements in the convection zone. Very recently, an attractive alternative approach by Hernández and Christensen-Dalsgaard (1994), based on a decomposition of the difference between observed and model frequencies, promises to be the finest diagnostic of the equation of state so far.

## III. MODELING A HELIOSEISMIC EQUATION OF STATE

The three basic material properties required in stellar models are the equation of state, opacity, and the nuclear-energy generation rate. Here, the focus is on the equation of state. I shall use the term equation of state in a slightly broader sense than usual, so that it encompasses not only pressure as a function of temperature and density, but also all thermodynamic quantities. These quantities must be consistent with each other, that is, their appropriate Maxwell relations have to be satisfied. Such *formal* consistency is always achieved if the equation of state and the thermodynamic quantities stem from a single thermodynamic potential. In trivial models (*e.g.* in a plasma assumed to be fully ionized everywhere) it is possible to write down a consistent equation of state and thermodynamic quantities independently. However, in more realistic cases, modeling a *thermodynamic potential* is the only practical way to obtain the equation of state and thermodynamic quantities.

A quick glance at Fig. 1 reveals that solar observations are indeed very sensitive to details of the equation of state. One might go further and conclude that the Sun prefers the CEFF to the EFF equation of state. However, such conclusions are fraught with danger, although probably not in this clear-cut case. The reason why one has to be prudent is that there are too many uncertainties in the solar model, coming, *e.g.*, from convection or opacity, so that one has to be alert to the possibility that by changing the equation of state one could trigger changes in the other physical parameters. If, say, the opacity is bad, one can not rule out that a *worse* equation of state could cause an overall better agreement with observations. Only when simultaneous progress with the other physical quantities is made, we will learn how to disentangle the different effects. However, for a sensitivity analysis, Figure 1 is already sufficient. The transitions from panels $a$ to $c$ and $b$ to $d$, respectively, are obtained by putting some additional nonideal effects (the Coulomb pressure) into the equation of state *with everything else unchanged*. The response of the Sun, as seen through the "eyes" of helioseismology, is huge.

A. Requirements on the solar equation of state

A stellar equation of state has to satisfy four conditions: (i) a large domain of applicability (in $\rho, T$), (ii) a high precision of its numerical realization, (iii) consistency between the thermodynamic quantities, and (iv) the possibility to take into account relatively complex mixtures with at least several of the more abundant chemical elements. More specifically, the first condition demands that the formalism can be used from the stellar surface (the photosphere), where $T$ is typically a few $10^3$ K and $\rho$ some $10^{-7}$ g/cm$^3$, to the center of a star where $T$ is, again typically, about $10^7$ K and $\rho$ some $10^2$ g/cm$^3$. The second condition demands that a given formalism can be cast in an algorithm that converges without ambiguity and with sufficient precision, so that all required thermodynamic derivatives (such as adiabatic gradients) can be computed. Note, that for this only *formal* precision is required: reality of the physical description is a different issue. The third condition, consistency, states that all thermodynamic quantities stem from a single thermodynamic potential. This condition is often violated in two- or more-zone formalisms, which contains a different physical theory in different parts of a star. An example is the *ad hoc* imposition of full ionization in the central region to mimic pressure-ionization, in combination with a conventional Saha equation in the envelope of the star. Such a formalism leads to a discontinuous thermodynamic potential and a violation of thermodynamic identities.

Violations of thermodynamic identities are inadmissible in calculations of stellar structure and oscillations. Calculations of stellar oscillation frequencies often exploit thermodynamic quantities to transform one variable into another (see, *e.g.*, Unno *et al.*, 1989). As an example, the adiabatic gradient $\Gamma_1$ is used to establish a connection between density and pressure changes, and it is an absolute necessity that the $\Gamma_1$ is consistent with the equation of state and other thermodynamic variables of the model. This example illustrates the necessity of formal consistency. Finally, the third and last condition, *i.e.* the possibility to describe rather realistic chemical compositions, is a bit less important for the equation of state itself. However, for opacity, heavy elements are very important, and a good equation of state plays an important role in any opacity calculation.

B. The role of the solar convection zone

Energy transport by radiation is treated adequately in the solar interior in the diffusion approximation; on the other hand, energy transport by convection is usually treated in a rather crude way, with an *a priori* unknown parameter, the so-called mixing length (see, *e.g.*, Cox and Giuli, 1968). Near the surface, convection is probably sufficiently vigorous to cause dynamic effects on the average hydrostatic equilibrium, yet such effects are often ignored. At the lower boundary of the convection zone, motion is normally supposed to stop at the point where convective instability ceases; there is no doubt, however, that motion extends into the convectively stable region through convective overshoot (see, *e.g.*, Berthomieu *et al.*, 1993b).

Despite the complications it introduces, in a certain sense convection simplifies the structure of the outer parts of the Sun. Regardless of the uncertain details of convective energy transport, there is no doubt that except in a thin boundary layer near its top the convection zone is very nearly adiabatically stratified (*e.g.* Gough and Weiss, 1976). One can show (Christensen-Dalsgaard, 1986) that the structure of the almost adiabatically stratified convection zone only depends on the equation of state, the composition and the constant value of the specific entropy, which in turn is essentially fixed by the value of the mixing-length parameter; particular, the convection zone structure is insensitive to the opacity. Another simplification of convection is that it makes the chemical composition homogeneous in the convection zone, although there is of course the possibility of gravitational settling (for a recent calculation, see Christensen-Dalsgaard *et al.*, 1993).

Beneath the convection zone, the stratification becomes highly dependent on radiative opacity. It is difficult to disentangle the helioseismic effects of equation of state and opacity, but if opacity can be nailed down relatively accurately, an equation of state diagnosis can also become possible. An example of an equation of state issue is the possibility of partial recombination of $He^+$ ions in the solar center (see Christensen-Dalsgaard and Däppen, 1992).

## IV. EQUATION OF STATE COMPARISONS

In the absence of a rigorous computation of the equation of state (to the needed accuracy), one can make comparisons between different models of the equation of state. Such comparisons will give us information about the overall uncertainty in the equation of state. But they also allow solar physicists to determine how uncertainties in the equation of state propagate into theoretically predicted oscillation frequencies. In this way, a "map" of the $T - \rho$ plane can be drawn, showing localized "interesting" regions, where nonideal effects of one or another kind are important. I will very briefly present the equations of state used in the comparisons. More details about them (and further references) can be found in the article by Christensen-Dalsgaard and Däppen (1992).

A. The equations of state used in the comparison

1. EFF

Eggleton, Faulkner and Flannery (1973) developed a simple equation of state in the chemical picture (EFF) that is formally consistent and includes an *ad hoc* pressure ionization device that works at least qualitatively correctly. The device is not based on a physical model (*e.g.* a description of an atom and its surrounding particles), but is imposed by forcing the anticipated result, *i.e.*, full ionization at high densities. In addition, the EFF equation of state incorporates a correct treatment of the partially degenerate electrons according to Fermi-Dirac statistics. Bound systems (atoms and ions) are always assumed to be in their ground state; the ground-state energy is constant and equal to the free-particle value.

2. CEFF

To overcome the lack of a Coulomb term in the EFF equation of state, Jørgen Christensen-Dalsgaard and I have added a Coulomb configurational term in the Debye-Hückel approximation (taken from the MHD equation of state, see below). Such an upgrade of the EFF equation of state was motivated by the fact that adding a Coulomb term to the EFF equation of state makes a significant contribution towards a more realistic equation of state (see below and the papers by Christensen-Dalsgaard, 1991; Christensen-Dalsgaard and Däppen, 1992). Of course the remaining disadvantages of the EFF equation of state still point to the need of more complete formalisms. However, the successful application of the CEFF equation of state to solar physics makes it very well suited as a reference equation of state.

3. MHD

The Mihalas-Hummer-Däppen (MHD) equation of state (Hummer and Mihalas, 1988; Mihalas et al., 1988; Däppen et al., 1990) is realized in the chemical picture with the free-energy minimization method. Occupation probabilities are introduced on the one hand to avoid the famous (or rather notorious) discontinuities that come along with simple cut-off recipes for internal partition functions. On the other hand they represent a result that should come from quantum mechanics, namely the fraction of atoms or ions for which a given state can exist (given the constraints of the surrounding particles). Only then, these "available" states are populated according to statistical mechanics. It is clear that such an approach is largely intuitive. However, its advantage is that complicated plasmas can be modeled, with *detailed* internal partition functions for a large number of atomic, ionic, and molecular species. All particles are allowed to interact with each other. Also, full thermodynamic consistency is assured by analytical expressions of the free energy and its first- and second-order derivatives. This not only allows an efficient Newton-Raphson minimization, but, in addition, the ensuing thermodynamic quantities are of analytical precision and can therefore be differentiated once more, this time numerically. Reliable third-order thermodynamic quantities are thus calculated. The MHD equation of state was realized for the international "Opacity Project" (see, *e.g.*, Seaton, 1987).

4. OPAL

The OPAL equation of state is realized in the physical picture. A detailed presentation is given by Rogers and Iglesias (*these proceedings*). In the physical picture, the concept of a perturbed atom in a plasma is not needed at all. Therefore, no assumptions about energy-level shifts or the convergence of internal partition functions have to be made. On the contrary, properties of energy levels and the partition functions come out from the formalism. The OPAL equation of state was developed by a group at Livermore as part of their opacity project (Rogers, 1986; Iglesias, Rogers and Wilson, 1987; Rogers and Iglesias, *these proceedings*). This equation of state does satisfy the requirements from stellar modeling that I mentioned above. Let me mention in passing that there are of course many other realizations of the physical picture (see, *e.g.*, Kraeft et al., 1986; Ebeling et al., 1991; Alastuey et al., 1994), but their solar model applications are still awaiting.

B. Results of the comparisons

Early comparisons showed a striking agreement between the MHD and OPAL equation of state for conditions as found in the hydrogen-helium ionization zones of the Sun (Däppen, Lebreton and Rogers, 1990; Däppen, 1990). For convenience, a representative result from this early comparison is shown in Figure 2, which compares the MHD and OPAL results with that of the simple EFF formalism (which is essentially a consistent ground-state-only Saha equation of state under these conditions). The absolute curves of part $a$ of Figure 2 are merely able to show the difference between MHD (or OPAL) and the simple EFF results. To see the difference between the MHD and OPAL results, one needs the magnified part $b$, which shows the *relative* differences between MHD and EFF, and between OPAL and EFF values, respectively. This relative plot now not only allows one to see the difference between MHD and OPAL results, but also their striking similarity.

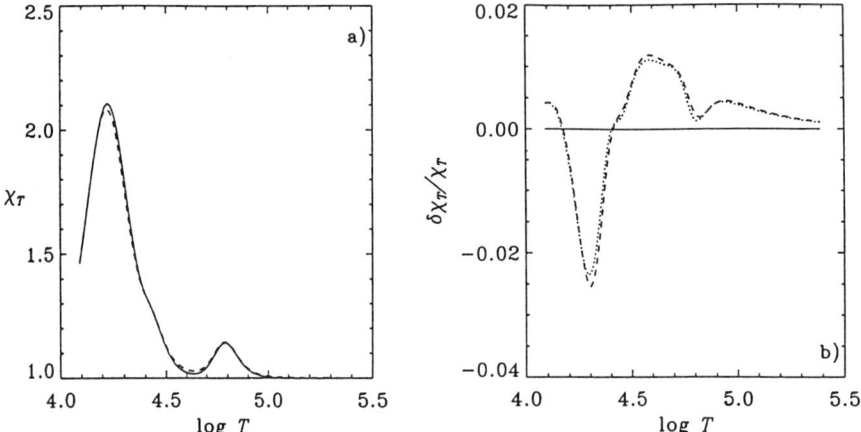

Fig. 2. Comparison of $\chi_T = (\partial \ln p/\partial \ln T)_\rho$ for $\rho = 10^{-5.5}$ g cm$^{-3}$. Absolute quantities (a) and relative differences (with respect to EFF) (b) are shown. See text for more details.

Later, it turned out that this agreement was nearly accidental. The reason for this was found by varying the parameters of the MHD equation of state. It followed that on the chosen isochore, all thermodynamic quantities are dominated by the Coulomb pressure correction (Däppen, 1990; Christensen–Dalsgaard, 1991; Christensen–Dalsgaard and Däppen, 1992). The Coulomb correction overshadows the effect of the excited states (which are of course treated differently in the MHD and OPAL approach). Note that the Coulomb term acts directly and indirectly, at least in the language of the chemical picture, because it is not mainly the free-energy of the Debye-Hückel term itself, but rather also the Coulomb-induced shift in the ionization equilibrium, which is responsible for the deviation from the unperturbed EFF result.

Of course, solar physicists were happy that two completely different formalisms delivered the same equation of state, but, by the same token, a first attempt to use the Sun as an equation-of-state test was also thwarted. This discovery suggested to upgrade the simple EFF equation of state with the help of the Coulomb interaction term. The resulting equation of state (called CEFF) has become a useful tool for solar physics (Christensen–Dalsgaard, 1991; Christensen–Dalsgaard and Däppen, 1992); at the same time, however, it became also clear that a helioseismic test of the important issue of chemical versus physical picture would be more difficult than first thought.

A more recent comparison of MHD and OPAL values (Däppen, 1992) has examined selected cases of higher densities (where sizeable discrepancies appear) and a first case of a mixture involving a representative solar heavy element (oxygen). It appears that for the heavier elements, the internal partition functions finally lead to the intuitively expected consequences for the thermodynamic quantities. This comparison for the first time establishes a clear case of disagreement between the MHD and OPAL results. Of course, only some 2 percent of the matter in the Sun consist of elements heavier that H and He, and therefore the signature of the MHD-OPAL discrepancy in $\Gamma_1$ is small (of the order of $10^{-3}$). Nevertheless, as has been demonstrated by Christensen-Dalsgaard and Däppen (1992) and Hernández and Christensen-Dalsgaard (1994), even the resulting tiny sound-speed differences are within reach of a helioseismic diagnosis.

## V. CONCLUSIONS

Even weakly-coupled plasmas can pose tough problems if high accuracy is demanded. Solar oscillations are an example of a case where the present observational material is much better than the theoretical models. The solar convection zone is especially well suited for a study of the equation of state. Both forward and inverse method provide powerful tools to examine fine details of the solar equation of state. Many results have already been obtained. The principal conclusion is that helioseismology not only requires a non-ideal Coulomb correction term in one form or another, but that it carries the potential to decide which is the best model for solar conditions.

## ACKNOWLEDGEMENT

I thank Vladimir Baturin, Jørgen Christensen-Dalsgaard, Douglas Gough and Forrest Rogers for stimulating discussions. This work was supported in part by the U.S. National Science Foundation grant AST-9315112.

## REFERENCES

Alastuey, A., F. Cornu, and A. Perez, 1994, Phys. Rev. E, **49**, 1077-1093.
Bahcall, J.N. and R.K. Ulrich, 1988, Rev. Mod. Phys., **60**, 297-372
Baturin, V.A., W. Däppen, D.O. Gough and S.V. Vorontsov, 1994, Mon. Not. R. Astr. Soc., submitted
Berthomieu, G., J. Provost, P. Morel, Y. Lebreton, 1993a, Astron. Astrophys., **268**, 775-791
Berthomieu, G., P. Morel, J. Provost, J.-P. Zahn, 1993b in *Proc. IAU Symposium No 137: Inside the Stars*, eds. Werner W. Weiss and Annie Baglin, ASP Conference Series, **40**, 60-62
Brodsky, M. and S.V. Vorontsov, 1993, Astrophys. J., **409**, 455
Christensen-Dalsgaard, J., 1986, in *Seismology of the Sun and the distant stars*, ed D.O. Gough, (Reidel, Dordrecht), 23-53
Christensen-Dalsgaard, J., 1988, in *Seismology of the Sun and Sun-like Stars*, eds Domingo, V. and Rolfe, E.J., ESA SP-286, Noordwijk, The Netherlands, 431-450
Christensen-Dalsgaard, J., 1991, in *Lecture Notes in Physics*, Vol. **388**: Challenges to Theories of the Structure of Moderate-mass Stars, eds Gough, D.O. and Toomre, J., (Springer, Heidelberg), 11-36
Christensen-Dalsgaard, J. and G. Berthomieu, 1991, in *Solar Interior and Atmosphere*, eds Cox, A.N., Livingston, W.C. and Matthews, M., Space Science Series, (University of Arizona Press, Tucson), 401-478
Christensen-Dalsgaard, J. and W. Däppen, 1992, Astron. Astrophys. Review, **4**, 267-361
Christensen-Dalsgaard, J., C.R. Profitt, and M.J. Thompson, 1993, Astrophys. J., **403**, L75
Cox, A.N. and J.E. Tabor, 1976, Astrophys. J. Suppl., **31**, 271-312
Cox, J.P. and R.T. Giuli, 1968, *Principles of Stellar Structure*, (Gordon and Breach, New York)

Däppen, W., 1990, in *Lecture Notes in Physics*, Vol. **367**: *Progress of Seismology of the Sun and Stars*, eds. Osaki, Y. and Shibahashi, H., 'Springer, Berlin), 33-40
Däppen, W., 1992, in *Astrophysical Opacities*, eds Mendoza, C. and Zeippen, C. (*Revista Mexicana de Astronomía y Astrofísica*), 141 – 149
Däppen, W. and D.O. Gough, 1984, in *Theoretical Problems in Stellar Stability and Oscillations*, Institut d'Astrophysique, Liège, p. 264-269
Däppen, W., Y. Lebreton and F. Rogers, 1990, Solar Physics, **128**, 35-47
Däppen, W., D.M. Mihalas, D.G. Hummer and B.W. Mihalas, 1988, Astrophys. J., **332**, 261-270
Deubner, F.-L. and D.O. Gough, 1984, Ann. Rev. Astron. Astrophys., **22**, 593-619
Ebeling, W., A. Förster, V.E. Fortov, V.K. Gryaznov and A.Ya. Polishchuk, 1991, *Thermodynamic Properties of Hot Dense Plasmas*, (Teubner, Stuttgart)
Eggleton, P.P., J. Faulkner and B.P. Flannery, 1973, Astron. Astrophys., **23**, 325-330
Gough, D.O., 1984, Mem. Soc. Astron. Ital., **55**, 13-35
Gough, D.O. and S.V. Vorontsov, 1994, Mon. Not. R. Astr. Soc., submitted
Gough, D.O. and N.O. Weiss, 1976, Mon. Not. R. Astr. Soc., **176**, 589-607
Hernández, P. and J. Christensen-Dalsgaard, 1994, Mon. Not. R. Astr. Soc., in press
Hummer, D.G. and D.M. Mihalas, 1988, Astrophys. J., **331**, 794-814
Iglesias, C.A., F.J. Rogers and B.G. Wilson, 1987, Astrophys. J., **322**, L45
Kraeft W.D., D. Kremp, W. Ebeling and Röpke G., 1986, *Quantum Statistics of Charged Particle Systems*, (Plenum, New York)
Libbrecht, K.G., M.F. Woodard and J.M. Kaufman, 1990, Astrophys. J. Suppl., **74**, 1129-1149
Mihalas, D.M., W. Däppen and D.G. Hummer, 1988, Astrophys. J., **331**, 815-825
Rogers, F.J., 1986, Astrophys. J., **310**, 723-728
Seaton, M., 1987, J. Phys. B: Atom. Molec. Phys., **20**, 6363-6378
Turck-Chièze, S., W. Däppen, E. Fossat, J. Provost, E. Schatzman, D. Vignaud, 1992, Physics Report, **230**, 57-235
Unno, W., Y. Osaki, H. Ando, H. Saio and H. Shibahashi, 1989, *Nonradial Oscillations of Stars, 2nd Edition*, (University of Tokyo Press, Tokyo)
Vorontsov, S.V., V.A. Baturin, D.O. Gough, W. Däppen, 1994, in *The equation of state in astrophysics*, eds. G. Chabrier and E. Schatzman, (Cambridge University Press, Cambridge), in press
Vorontsov, S.V. and V.N. Zharkov, 1989, Sov. Sci. Rev. E. Astrophys. Space Phys., **7**, 1-103

# VI. Planetary Sciences

# VI.1

# CURRENT UNCERTAINTIES IN THE INTERIOR PHYSICS OF BROWN DWARFS AND GIANT PLANETS

W.B. Hubbard,[†] A. Burrows,[‡] J.I. Lunine,[†] and D. Saumon[†]

[†] Lunar and Planetary Laboratory, University of Arizona
Tucson, AZ 85721, U. S. A.
[‡] Departments of Physics and Astronomy, University of Arizona
Tucson, AZ 85721, U. S. A.

The thermodynamics of brown dwarfs (BDs: substellar objects with masses less than about 80 Jupiter masses but greater than about 10 Jupiter masses) and giant planets (GPs) are dominated by the equations of state of liquid metallic hydrogen. However, the distribution and high-pressure properties of elements with $Z > 1$ are a dominant source of uncertainty in the structure and evolution of these objects.

Because of its high cosmic abundance, helium plays an important role in the energetics of GPs where hydrogen-helium immiscibility can occur; temperatures are generally too high in BDs for H-He phase separation. We have recently incorporated an accurate helium equation of state in BD/GP models, but, except for the phenomenon of immiscibility with hydrogen in the GP mass range, its gross effect on the models is minor. A much greater source of uncertainty comes from the "metals" (elements with $Z > 2$), which have a dominant influence on atmospheric and interior photon opacity. As a limiting case, we have investigated the effect of a total absence of metals in a suite of BD models (Saumon, D., P. Bergeron, J.I. Lunine, W.B. Hubbard, and A. Burrows, 1994, Astrophys. J., **424**, 333). Not only do such models simulate a putative primordial low-mass, low-metal, low-luminosity galactic halo population (some of which may have been recently detected via microlensing surveys), but they could also stand for a population of normal-composition BDs in which the atmospheric metals abundance has been greatly suppressed by sequestration of metals in the deep interior, via a hypothesized immiscibility of metals with metallic hydrogen.

Very little work has been done on immiscibility of metals in metallic hydrogen, and it is not known whether cooler BD atmospheres can be made extremely metal-poor by this process. However, if it occurs, it can have a dramatic effect in changing the evolutionary timescale and spectral properties of such BDs.

Our group has also investigated the effect of various initial deuterium abundances on the evolution of BDs and GPs. The marginal mass for deuterium fusion, using standard reaction rate theory, is $13 \times$ Jupiter, depending little on metals abundance. If the initial deuterium abundance is as high as $2 \times 10^{-4}$, a deuterium main sequence develops, and lasts about $6 \times 10^7$ years.

## I. INTRODUCTION

The equation of state of most of the mass of BDs and GPs lies in the range of parameter space $Z \sim 1$, $\Gamma \sim 1$ to 30, and $r_e \sim 0.1$ to 1. Here $Z$ is the atomic number, $r_e$ is the usual electron spacing parameter defined in terms of the electron number density $n_e$ by $4\pi a_o^3 r_e^3/3 = n_e^{-1}$, and $\Gamma$ is the usual electron plasma coupling parameter defined by $\Gamma = e^2/r_e a_o k_B T = (3.16 \times 10^5 \text{K})/r_e T$, where $e$ is the electron charge, $a_o$ the Bohr radius, $k_B$ Boltzmann's constant, and $T$ the temperature. The composition of the material is for the most part hydrogen, with an admixture of $\sim 25\%$ helium by mass, and (for solar-composition material), $\sim 1$ to 2% by mass of material with $Z > 2$. In accordance with astrophysical usage, we denote the mass fraction of hydrogen by $X$, the mass fraction of helium by $Y$, and the mass fraction of all elements heavier than helium (the so-called "metals" in astrophysical parlance) by $Z$. The latter $Z$ will always be expressed as a percentage in this paper so that it will not be confused with the atomic number.

The equation of state of pure hydrogen is well understood for this range of parameters (Ichimaru, 1994), and there is little likelihood that any of the thermodynamic variables important for the structure and evolution of BDs and GPs are in error by more than a few percent within most of their mass range. With the exception of the details of the so-called plasma phase transition (PPT; Saumon and Chabrier, 1992), the pressure range under which metallization of hydrogen occurs is noncontroversial and understood to be in the vicinity of $\sim 1$ to 3 Mbar (depending on $T$); this phase transition may have important observable consequences for GPs but not for BDs.

However, the equation of state of a dense mixture of hydrogen with other species is less well understood, and uncertainties in the behavior of the mixture strongly limit our understanding of the evolution of BDs and GPs. There are two types of mixtures of concern in this context: (1) a mixture of hydrogen with helium with $Y \approx 25\%$; (2) a mixture of hydrogen, helium, and metals, with $Y \approx 25\%$, and $Z \leq 2\%$. The H-He mixture has been studied extensively (Stevenson, 1975; Hubbard and DeWitt, 1985; Klepeis et al., 1991), and it is commonly believed that limited miscibility of helium in metallic hydrogen in the giant planets Jupiter and Saturn has led to formation of a helium-enriched inner region in Saturn (and possibly in Jupiter), with a corresponding depletion in the observable outer layers. Because helium is a major component of the mass of both Jupiter and Saturn, it could provide a significant gravitational energy source should it commence to unmix from the hydrogen via immiscibility in liquid metallic hydrogen (Stevenson and Salpeter, 1977). The temperature range under which helium develops limited solubility in metallic hydrogen does not appear to exceed $10^4$ K according to most calculations (Stevenson, 1975; Hubbard and DeWitt, 1985), and thus unmixing of helium from hydrogen would be in an intermediate stage in Saturn and only incipient in Jupiter. In contrast, the calculations of Klepeis et al. (1991) predict a maximum phase-separation temperature of about 40000 K at a pressure of 10.5 Mbar, and for solar-composition hydrogen and helium, insolubility would commence at $T \leq 15000$ K. This phase-separation temperature is about a factor of two higher than that predicted by the other calculations, and would imply very strong separation in both Jupiter and Saturn. This is not supported by observations.

In contrast to the outer metallic-hydrogen layers of GPs, the metallic-hydrogen layers of BDs do not cool below $10^4$ K within the age of the Galaxy, and so helium-hydrogen immiscibility is of potential relevance only to GPs (even if Klepeis et al. are correct).

In BDs of solar composition, the abundance of metals is so low that only an insignificant amount of gravitational energy is released if some of these elements have limited solubility. However, even trace amounts of metals dominate the photon opacity in the ideal-gas envelope of a BD. It is this outermost skin which regulates the rate of escape of the BD's heat into space, and therefore a significant change in its photon opacity will have major repercussions on the evolution history of a BD (as well as, obviously, on the emergent spectrum). Thus the commencement of limited miscibility of metals in a BD's interior regions is likely to lead to observable consequences.

Stevenson (1986) pointed out that Fe may become partially insoluble in BDs, and that this effect might have an influence on grain opacity in such BDs. However, he did not quantify the effect. At present, the synthesis of wavelength-dependent atmospheric opacities for BDs is an ongoing project of several groups (Saumon, et al., 1994; Allard, 1990), and is not yet solved in general. Our group has computed BD evolution trajectories for two limiting cases: solar-composition metallicity ($Y \approx 25\%$ and $Z \approx 2\%$), and zero metallicity (pure hydrogen and helium, with $Y \approx 25\%$ and $Z = 0\%$). We argue that these limiting cases bound the possibilities for BDs which gradually lose atmospheric metallicity due to development of a metal-rich core as a consequence of limited solubility of the metals in metallic hydrogen, and we discuss these possibilities in the following.

Finally, although deuterium is of course a fully soluble trace species in BDs and GPs, it serves as a brief but important nuclear energy source in BDs owing to its vulnerability to thermonuclear processing at much lower temperatures ($T \sim 5 \times 10^5$ K) than those for which the p-p reaction chain can proceed. In this paper we also report on some of our recent calculations of the marginal mass for deuterium burning in low-mass BDs.

## II. EVOLUTION OF BROWN DWARFS

Figure 1 shows luminosity $L$ in units of the solar luminosity $L_\odot$, as a function of time $t$ for two ensembles of BDs. The first ensemble (dashed lines) is computed for zero metallicity, while the second ensemble (solid lines) is computed for solar metallicity ($\odot$); these ensembles correspond to the sequences of models presented by Saumon et al. (1994). It is important to note that in both of the ensembles, the models are fully convective, and that the interior trajectories of pressure $P$, mass density $\rho$, and temperature $T$ follow an isentrope. The entropy corresponding to the isentrope is established by the atmospheric boundary condition, which depends on the opacity of the atmosphere. Thus the differences between the two ensembles of evolutionary tracks are entirely caused by the composition of the ideal-gas atmosphere. Equation of state differences in the deep interior are entirely due to the presence or absence of metals and are negligible.

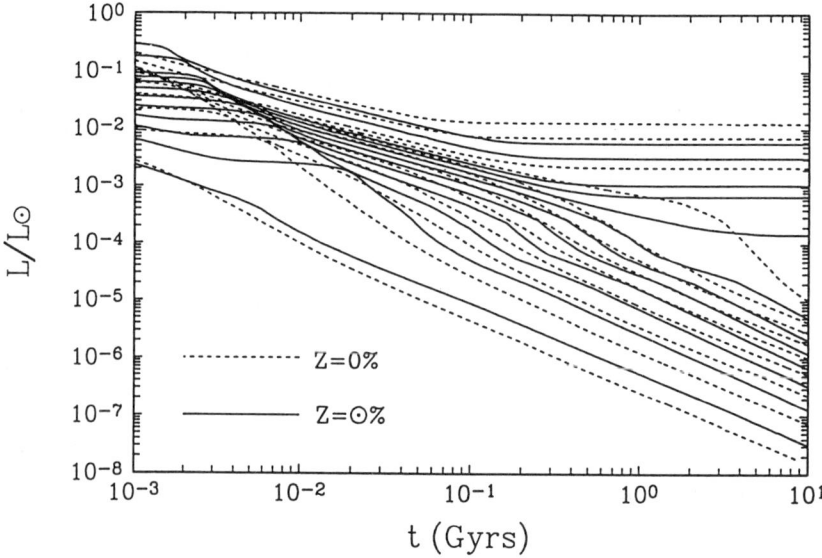

Fig. 1. Evolutionary tracks for zero-metallicity and solar-metallicity brown dwarfs and marginal hydrogen burners. The tracks are for (top to bottom) 0.20 $M_\odot$, 0.15 $M_\odot$, and 0.10 to 0.01 $M_\odot$ in steps of 0.01 $M_\odot$.

In Fig. 1, we see that the absence of metals causes the zero-metallicity BDs to cool more quickly than the solar-metallicity BDs, while the objects that are able to reach the main sequence and burn hydrogen via the p-p reaction are more luminous in the absence of metals which increase atmospheric opacity. The "ripple" which appears in the tracks at early phases ($t < 3 \times 10^{-2}$ Gyrs) marks the end of deuterium burning, which is computed for a presumed initial number abundance of deuterium to hydrogen of $D/H = 2 \times 10^{-5}$.

The absence of metals has a dramatic effect on emergent spectra of BDs. Figure 2, taken from Saumon et al. (1994), shows a comparison of black body flux curves and computed nongrey flux curves for BDs with atmospheric effective temperatures of 1500 K (upper curve) and 1000 K (lower curve), and a surface gravity of $10^5$ cm sec$^{-2}$. Note that the nongrey emission profile peaks in red and near-IR, while the corresponding black body curves peak well in the infrared, at wavelengths $\sim 5$ μm and beyond. The point of this figure is that very strong depletion of metals in the atmosphere of a BD can markedly disguise the true effective temperature of the object, perhaps foiling a search predicated on expected black body emission characteristics.

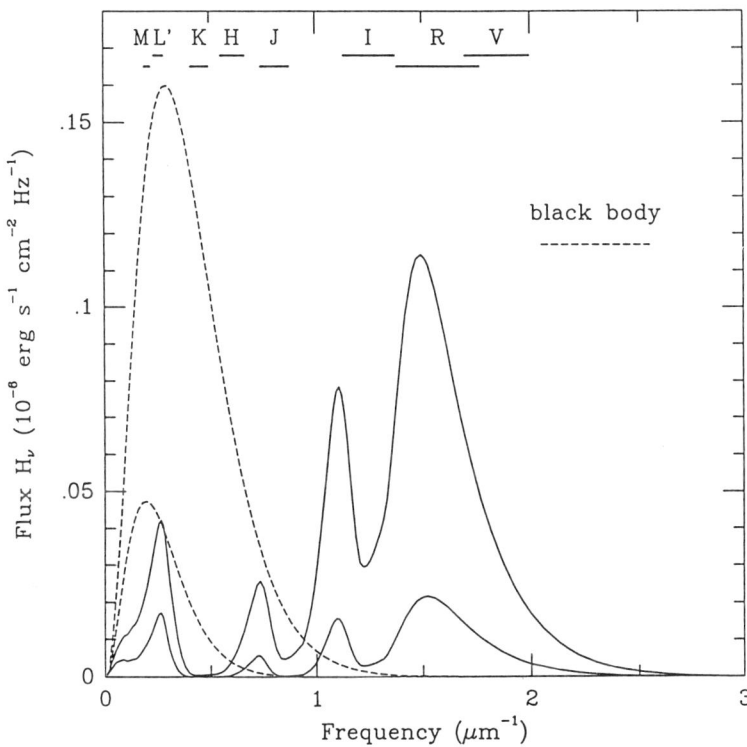

Fig. 2. Emitted spectra of zero-metallicity brown dwarfs (solid curves), and black bodies at the same effective temperature (dashed curves). Bandpasses of some standard astronomical filters used to search for brown dwarfs are shown for comparison.

## III. SOLUBILITY OF METALS IN BROWN DWARF INTERIORS

Figure 3 shows, for the same ensembles as in Fig. 1, the trajectories of the central temperatures $T_c$ as a function of $t$. Objects which have masses above the hydrogen-burning limit at $0.0767 M_\odot$ (solar metallicity) or $0.092 M_\odot$ (zero metallicity) reach constant central temperatures and become lower main-sequence stars, while objects below the limiting mass (BDs) cool steadily with time. Figure 4 shows the evolution of the same objects, but on the $T_c$ vs. $\rho_c$ plane, where $\rho_c$ is the central mass density. At the right-hand side of both Fig. 3 and Fig. 4 are shown predicted temperature ranges affecting the solubility of the labeled elements in metallic hydrogen, which we discuss in detail below. Note on Fig. 4 that the finite-metallicity trajectories fall on top of the zero-metallicity trajectories, but do not extend as far. As explained in section II, this is because the metals do not affect the equation of state to any extent, but do affect how rapidly the object evolves via atmospheric opacity.

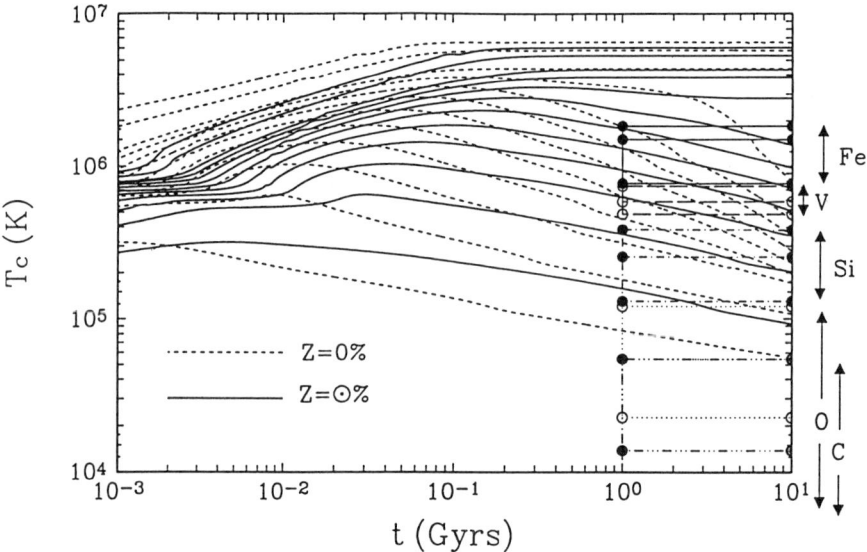

Fig. 3. Central temperature vs. time for brown-dwarf and lower main-sequence models. At the right side of the diagram are shown three characteristic temperatures for each element indicated in the right-hand margin.

To explore the possibility of limited solubility of metals in BD interiors, we have used the crude ion-sphere model of Stevenson (1976). In this model, we consider only binary mixtures of hydrogen and elements with $Z > 2$, ignoring the possible influence of helium. As shown by Stevenson, if the pressure is sufficiently high for electron screening to be negligible, then the Gibbs free energy of mixing approaches an asymptotic high-pressure limit given by

$$\Delta G = -\frac{27}{100}(4/9\pi)^{2/3} f(x) + k_B T [x \ln x + (1-x) \ln(1-x)], \tag{1}$$

where $x$ is the number concentration of the metal in hydrogen, and

$$f(x) = \frac{\langle Z^{5/3} \rangle}{\langle Z \rangle} - x Z_2^{7/3} - (1-x) Z_1^{7/3}. \tag{2}$$

Here $Z_1 = 1$ is the atomic number of the hydrogen, which has concentration $(1-x)$, and $Z_2$ is the atomic number of the metallic ions at concentration $x$. The averages are carried out with respect to the concentration.

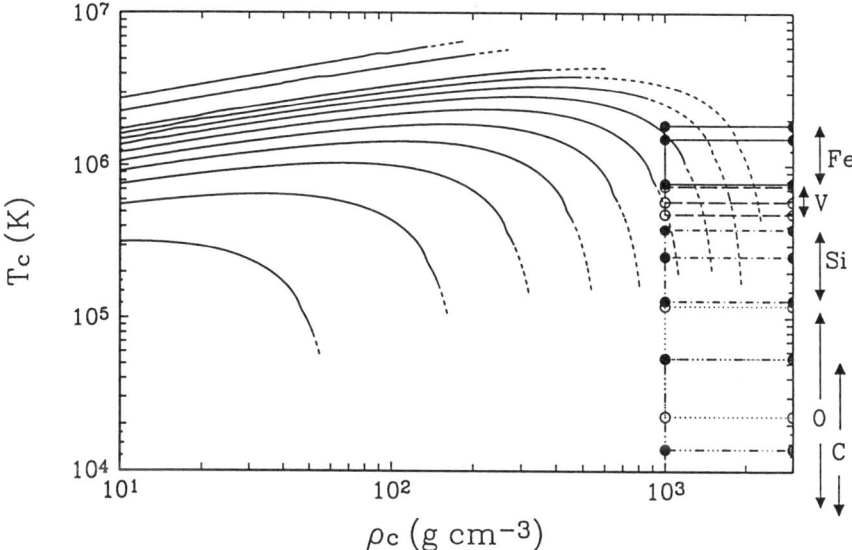

Fig. 4. Central temperature vs. central density curves for constant masses. Curves are terminated at an age of 10 Gyrs.

Despite the crudity of Stevenson's model, Hansen (1980) was able to confirm its essential details, and also to explore the pressure-dependence of the results. In this paper, we use Stevenson's simple asymptotic, pressure-independent expression for $\Delta G$ for convenience, while recognizing that Hansen found that metals tended to become *less* soluble in hydrogen with decreasing pressure. Accordingly, we have computed three characteristic temperatures for each of several impurity elements in metallic hydrogen in the asymptotic limit, as follows.

First, we assume that the element is fully pressure-ionized with no screening, so that $Z_2$ represents its full atomic number. We then use the double-tangent construction to $\Delta G$ to compute the temperature at which the concentration $x$ of the element becomes equal to its solar abundance with respect to hydrogen. This temperature then represents the characteristic temperature below which the element would begin to form an enriched core in the BD, depleting the element's atmospheric abundance. This first characteristic temperature is the uppermost of the three temperatures shown for each element in Figs. 3 and 4.

Second, we repeat the calculation by assuming that the element acts like a point ion with charge $Z_2 - 2$, as would be the case if pressure ionization removes all of the bound electron shells except the innermost one, but otherwise does not change the free energy. In this model, $Z_2$ in Eq. (1) is simply replaced with $Z_2 - 2$. The middle characteristic temperature shown in Figs. 3 and 4 is then the temperature at which the element of atomic number $Z_2$ has a solubility in metallic hydrogen equal to its concentration in solar-composition material, as evaluated using the double-tangent construction to $\Delta G$ with $Z_2 - 2$. Note that the uppermost temperature for carbon is thus identical to the middle temperature for oxygen.

Finally, we repeat the calculation by assuming that the ions have charge $Z_2 - 2$ and a concentration of $10^{-4}$ of solar. This defines the lowest characteristic temperature shown in Figs. 3 and 4. Thus far no low-mass stars or prospective BDs have been identified with such a low metallicity, but we estimate that such objects would begin to have spectra resembling the ones shown in Fig. 2.

According to the results shown in Figs. 3 and 4, some of the more important metals such as C and O will probably *not* be depleted in the atmospheres of BDs, except possibly for those at the lowest masses, $\sim 0.01 M_\odot$. However other important elements involved in grain formation may become depleted in more massive BDs. And, as Fig. 3 makes clear, as opacity-causing elements are depleted in the atmosphere, the BD's evolution accelerates, resulting in even more depletion.

As Fig. 4 shows, it is possible that even C and O will be depleted in the outer layers of Jupiter. However, this involves an extrapolation of Stevenson's theory from $r_e \sim 0.1$, where it may possibly have some qualitative validity, to $r_e \sim 1$, where it probably does not. This question constitutes a major uncertainty in current research on the interior of Jupiter and Saturn.

## IV. IGNITION OF DEUTERIUM IN LOW-MASS OBJECTS

Deuterium is converted to $^3$He in BDs at significantly lower temperatures than those required to make the thermonuclear p-p reaction proceed. It is interesting that this lowest thermonuclear threshold in astrophysics is reached by objects only about one order of magnitude more massive than Jupiter. The reactions that we consider are

$$p + p \to d + e^+ + \nu_e, \tag{3}$$

where $Q_p = 1.442$ MeV, and

$$p + d \to {}^3\text{He} + \gamma, \tag{4}$$

where $Q_d = 5.494$ MeV. Reaction rates are calculated using the theory of Fowler, Caughlan, and Zimmerman (1975), with the intermediate screening algorithm of Graboske et al. (1973). Primordial deuterium is consumed via reaction (4) at temperatures $\sim 5 \times 10^5$ K, and if sufficiently abundant, can even cause the BD's luminosity and interior temperatures to stabilize at constant values for a time. Reaction (3) does not proceed in a steady state in BDs by definition.

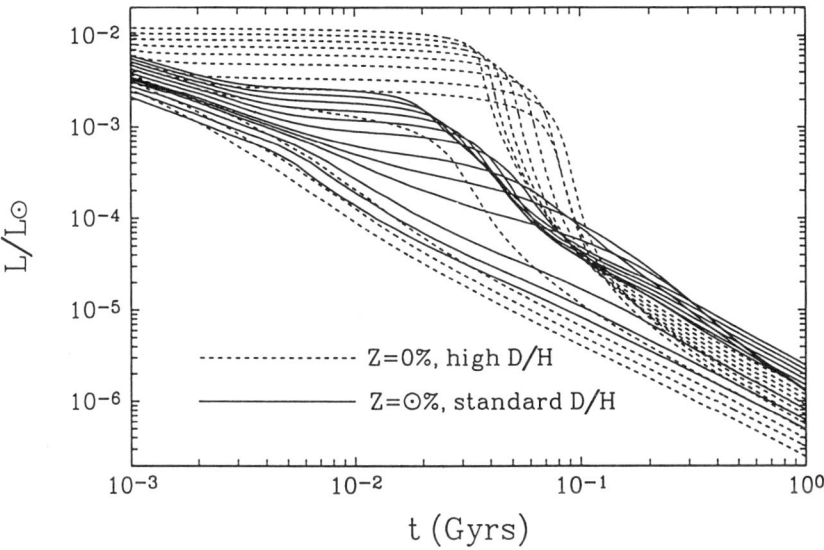

Fig. 5.   Luminosity vs. time curves for brown dwarfs near the deuterium-burning limit.

Figure 5 shows trajectories of constant mass for ensembles of BDs chosen to clearly illustrate this threshold. The solid curves show the evolution of luminosity with time for standard models with solar metallicity and D/H $= 2 \times 10^{-5}$, for masses ranging from 0.010 $M_\odot$ to 0.020 $M_\odot$ in steps of 0.001 $M_\odot$ (only one Jupiter mass!), plus an extra model with $M = 0.0135\ M_\odot$. As these curves show, there is a phase lasting $\sim 4 \times 10^7$ years when BDs in this mass range stay at luminosities about an order of magnitude larger than normal while they consume their initial stores of deuterium. Also shown in Fig. 5 (dashed curves) are an ensemble of zero-metallicity BDs with the same masses and elevated deuterium fractions (D/H $= 2 \times 10^{-4}$). The increased deuterium abundance is meant to correspond to an elevated primordial value reported by Songaila et al. (1994), and these zero-metallicity BDs thus correspond to hypothetical Population III objects, which might have formed out of low-metallicity, high-deuterium primordial Big Bang material not previously processed in stars. As Fig. 5 shows, the effect of the extra deuterium and reduced opacity is to increase luminosities in the deuterium-burning phase and to prolong lifetimes in this phase to values approaching $10^8$ years.

We have computed sequences of models with extremely small mass steps in order to accurately determine the deuterium-burning limit for low-mass BDs. Figure 6 shows the results for models with solar metallicity and standard deuterium abundance.

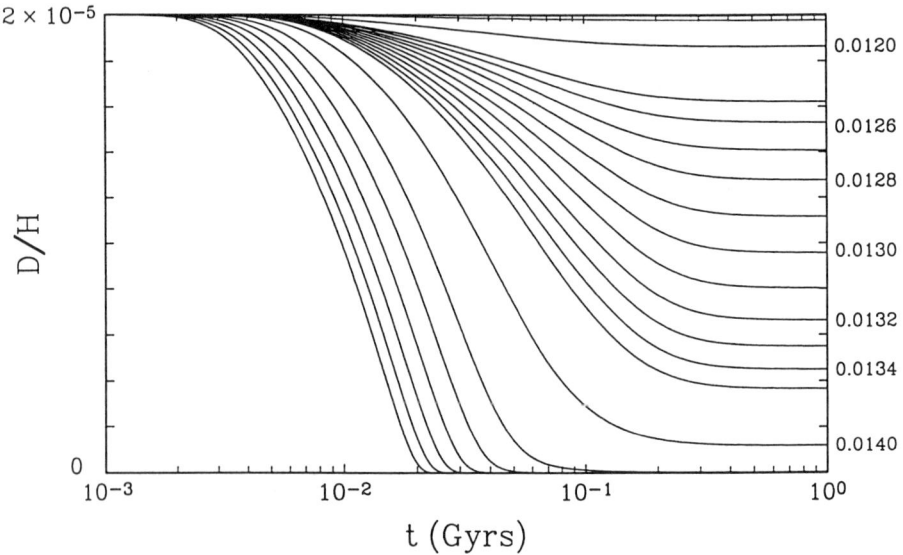

Fig. 6. Curves of deuterium abundance as a function of time, for a range of masses. Solar metallicity is assumed.

In addition to the masses shown in Fig. 5, we added a sequence of models with masses ranging from 0.0125 $M_\odot$ to 0.0135 $M_\odot$ in steps of $10^{-4} M_\odot$ (0.1 Jupiter mass!). BDs less massive than 0.0120 $M_\odot$ preserve their primordial deuterium, while by 0.0135 $M_\odot$ most of the deuterium is consumed. Thus the deuterium-burning limit is at 13 Jupiter masses.

Finally, we have repeated the calculation of deuterium consumption, using zero-metallicity models and an initial deuterium abundance one order of magnitude higher. Results are shown in Fig. 7, where we have used a mass interval of 0.005 $M_\odot$. The deuterium-burning limit shifts to a slightly higher mass, but still lies very close to 13 Jupiter masses.

## V. CONCLUSIONS

For the most part, questions involving the interior physics of BDs and GPs, as reflected in the gross properties of the equation of state and thermonuclear reaction rates, are well understood at present and do not present appreciable uncertainties. Thus it is interesting that despite this seemingly satisfactory situation on the theory side, observational surveys have in the main failed to produce large numbers of candidate objects. Although candidate objects have been identified in some relatively young clusters (e.g., in $\rho$ Oph; age = $3 \times 10^6$ years; Comeron et al., 1993; Burrows et al., 1993), they presumably have been detected only because of their youth and correspondingly high luminosity (see Fig. 1). The presumably large population of BDs in our galaxy with ages comparable to the age of the sun (5 Gyrs) or the age of the Galaxy ($\sim$ 10 Gyrs) has remained largely undetected (Burrows and Liebert, 1993). However, it is interesting that most attempts to detect BDs have been based on their predicted emitted radiation characteristics, and as yet no reliable synthetic spectra exist for such old BDs with effective temperatures $\sim$ 1000 K, with the exception of the $Z = 0\%$ synthetic spectra of Saumon et al. (1994).

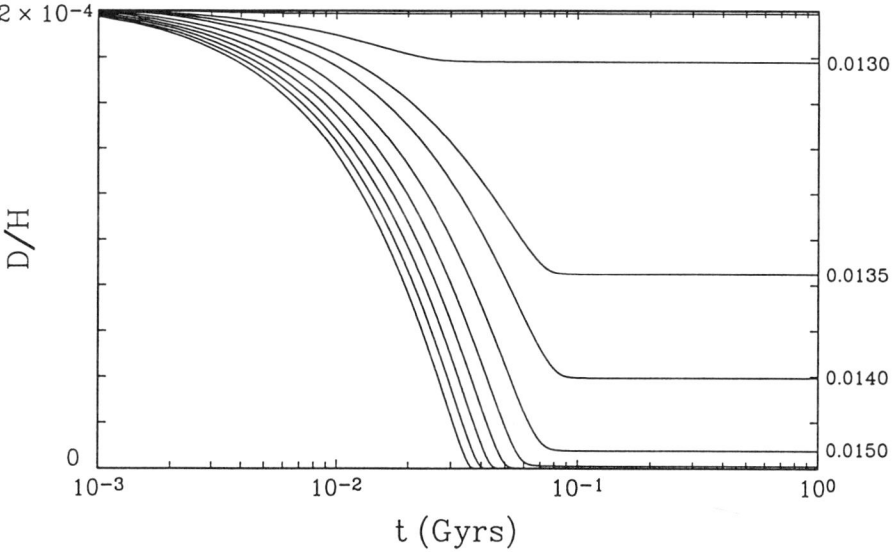

Fig. 7. Curves of deuterium abundance as a function of time, for a range of masses, assuming enhanced primordial deuterium abundance (Songaila et al., 1994) and zero metallicity.

Recently, innovative attempts to detect objects in the BD mass range located in the galactic halo have reported some initial successes (Alcock et al., 1993; Aubourg et al., 1993). These experiments are based on the gravitational lensing phenomenon, and rely upon continuous monitoring of a large number ($\sim 10^7$) of background stars in search of a rare event where the gravitational potential of a dark foreground object focuses the light from a background source. The amplitude and temporal scale of the focusing event depend upon the propagation distances, the mass of the focusing object, the miss distance, and the transverse velocity of the focusing object. Clearly not all of these parameters are known, and so a determination of the mass of the lensing object is only possible in a statistical sense. Nevertheless, this method does not depend upon the emitted radiation from BDs, and so is a prime approach for detecting old, cold BDs. In fact, the preliminary results reported by Alcock et al. and Aubourg et al. involve events in which the lensing objects were probably in the BD mass range, and may be consistent with a significant number of such objects in the galactic halo. More events will need to be observed before definite conclusions can be reached.

We have shown that theoretical calculation of the evolution of old, faint BDs and calculation of their emergent spectra as affected by metals presents great uncertainty in the confrontation between data and theory. Questions involving the limited solubility of metals in dense hydrogen thus assume critical importance for further progress. We also stress that related questions involving the solubility of helium and other elements in metallic hydrogen still loom large in the study of the interior structure and evolution of the giant planets Jupiter and Saturn.

Investigations of the phase diagrams of binary mixtures of liquid metallic hydrogen and traces of cosmically abundant elements, for hydrogen mass densities ranging from $\sim 1$ to $\sim 1000$ g cm$^{-3}$, are badly needed.

## REFERENCES

Alcock, C., et al., 1993, Nature, **365**, 621.

Allard, F., 1990, Ph.D. thesis, Ruprecht Karls Univ. Heidelberg

Aubourg, E., et al., 1993, Nature, **365**, 623.

Burrows, A., W.B. Hubbard, D. Saumon, and J.I. Lunine, 1993, Astrophys. J., **406**, 158.

Burrows, A., and J. Liebert, 1993, Rev. Mod. Phys., **65**, 301.

Comeron, F., G. Rieke, A. Burrows, and M. Rieke, 1993, Astrophys. J., **416**, 185.

Fowler, W.A., G.R. Caughlan, and B.A. Zimmerman, 1975, Ann. Rev. Astron. Astrophys., **13**, 69.

Graboske, H.C., H.E. DeWitt, A.S. Grossman, and M.S. Cooper, 1973, Astrophys. J., **181**, 457.

Hansen, J.P., 1980, J. de Physique, **41**, C2.

Hubbard, W.B., and H.E. DeWitt, 1985, Astrophys. J., **290**, 388.

Ichimaru, S., 1994, *Statistical Plasma Physics. Vol. II: Condensed Plasmas*, (Reading, Addison-Wesley), pp. 163-205.

Klepeis, J.E., K.J. Schafer, T.W. Barbee III, and M. Ross, 1991, Science, **254**, 986.

Saumon, D., and G. Chabrier, 1992, Phys. Rev. A, **46**, 2084.

Saumon, D., P. Bergeron, J.I. Lunine, W.B. Hubbard, and A. Burrows, 1994, Astrophys. J., **424**, 333.

Songaila, A., L.L. Cowie, C.J. Hogan, and M. Rugers, 1994, Science, **368**, 599.

Stevenson, D.J., 1975, Phys. Rev. B, **12**, 3999.

Stevenson, D.J., 1976, Phys. Letters, **58**A, 282.

Stevenson, D.J., 1986, in *Astrophysics of Brown Dwarfs*, M.C. Kafatos, R.S. Harrington, and S.P. Maran, eds., (Cambridge Univ. Press), pp. 218-232.

Stevenson, D.J., and E.E. Salpeter, 1977, Astrophys. J. Suppl., **35**, 239.

# VI.2

# CONSTITUTION AND EVOLUTION OF THE EARTH

Eiji Ito

Institute for Study of the Earth's Interior, Okayama University
Misasa, Tottori 682-01, Japan

The Earth (radius, 6370 km) is divided into the mantle of rock and the core of molten iron alloy at a depth of 2890 km. Mineralogical constitution of the mantle based on the high-pressure phase equilibrium studies is outlined. The sharp seismic discontinuity at 660 km depth, which separates the upper and lower mantle, is characterized by the dissociation of spinel. The negative $dP/dT$ slope for the dissociation boundary would play an important role to the stagnation of subducted slabs and considerable spreading of giant up-welling plumes in the upper mantle. Interpretation of the remarkable difference in Mg/Si between upper mantle peridotite and the cosmic abundance by fractional crystallization of perovskite and majorite in the magma ocean has been in debate till now. Alternatively, reaction between molten iron and silicate melt suggests that the discrepancy can be reconciled by partitioning $SiO_2$ into the proto core if the magma ocean would have extended to a depth greater than 1000 km. A possible chemical evolution of the core is discussed.

## I. INTRODUCTION

Direct information on the Earth's interior has been revealed through the analyses of seismic waves that travel deep in the Earth. Analyses of travel-time curves and of periods of free oscillation yield seismic velocity-depth and density-depth profiles. The seismic model PREM [Dziewonski and Anderson, 1981], which has been accepted to be the global model at the present, is reproduced in Fig. 1(A), together with a schematic drawing of a cross section of the Earth (Fig. 1(B)).

The "solid" earth (radius, 6370 km) is divided into several concentric layers. The primary division is between the mantle of rock and the metallic core at a depth of 2890 km. The mantle is separated into the upper and lower mantle by the sharp discontinuity around 660 km depth. The upper mantle below the discontinuity around 400 km depth is often designated as the transition zone. The region of 200 to 300 km right above the core-mantle boundary is marked as the D" layer, where the velocity gradient is generally accepted to be smaller than that of the lower mantle. As reflected waves are often observed from its tope [e.g., Nataf and Houard, 1993], however, the D" layer is considered to be the thermal and compositional boundary layer of the interaction between the mantle and core.

The core is virtually divided into the molten outer core and the solid inner core at 1220 km radius, both being composed dominantly of iron. However, density of the outer is about 10 % lower than that of iron under the corresponding conditions. Therefore substantial amounts of lighter elements are alloyed in the outer core [e.g.,

Jacobs, 1987]. The density and velocity of the inner core generally agree with those of iron indicating that the inner core is more pure.

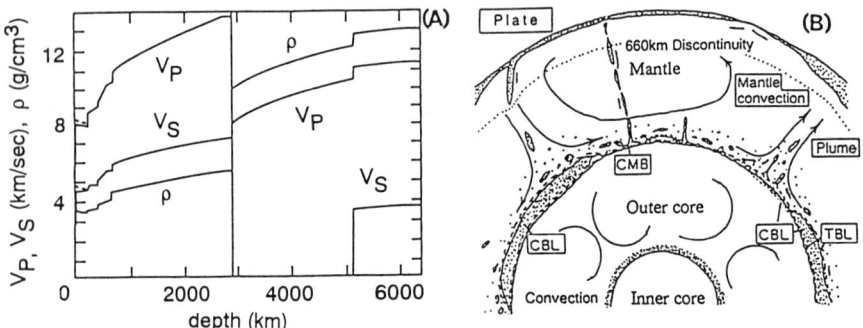

Fig. 1. (A) Seismic Earth model, PREM [Dziewonski and Anderson, 1981]. (B) Schematic drawing of a cross section of the Earth. CMB: core-mantle boundary, CBL: chemical boundary layer, TBL: thermal boundary layer.

In the Earth's interior, various dynamic motions are active as known in Fig. 1(B). These are subducting plates, convection, and up-welling plumes in the mantle, which cause volcanism, earthquake and various tectonic movements near the surface. There should be also active convection in the outer core to induce the Earth's magnetic field.

In this article, the constitution of the mantle is outlined with implication to a mantle dynamics inferred from seismic tomography and then the early evolution of the Earth is discussed focusing on the geochemical importance of the core separation.

## II. CONSTITUTION OF THE MANTLE

A. High-pressure phase transformations of mantle mineral

Xenoliths of the volcanic rocks with deep origin, such as diamond-bearing kimberite, indicate that the uppermost part of the mantle down to about 200 km depth is peridotite. The most typical peridotitic mantle model is pyrolite proposed by Ringwood [1975], whose chemical composition is shown in Table 1 along with cosmic abundance of refractory elements and chondritic mantle model (see later). He persisted that the whole mantle is basically homogeneous and has pyrolite composition. Pyrolite in the upper mantle crystallizes to an assemblage of $(Mg_{0.89}Fe_{0.11})_2SiO_4$ olivine (57 wt%), $(Mg,Fe)SiO_3$ Ca-poor orthopyroxene (17 wt%), $(Ca,Mg,Fe)_2Si_2O_6$ Ca-rich clinopyroxene (12 wt%), and $(Mg,Fe,Ca)_3(Al,Cr)_2Si_3O_{12}$ pyrope-rich garnet (14 wt%).

Pressure-induced polymorphic transformations and reactions of mineral phases are of basic importance to construct a mineralogical model of the mantle, because the abrupt increases in seismic velocities and density associated with these phase changes could be responsible to the discontinuities in the seismic Earth's model (cf. Fig. 1(A)). The phase changes in the upper mantle pyrolite assemblages up to the lower mantle pressure are actually categorized into two almost independent series; e.g., those in the olivine and the remainders.

Table 1. Cosmic abundances of major refractory elements and the chemical mantle models.

| Cosmic abundance(Si=$10^6$)* | | | Mantle models | |
|---|---|---|---|---|
| Element | Abundance | | Chondritic** | Pyrolite*** |
| Mg | 1,074,000 | $SiO_2$ | 50.5 | 45.1 |
| Si | 1,000,000 | $TiO_2$ | 0.2 | 0.2 |
| Fe | 900,000 | $Al_2O_3$ | 3.7 | 3.3 |
| Na | 57,400 | FeO* | 6.8 | 8.0 |
| Al | 84,900 | MnO | - | 0.15 |
| Ca | 70,400 | MgO | 35.3 | 38.1 |
| Ni | 49,300 | CaO | 3.0 | 3.1 |
| P | 10,400 | $Na_2O$ | 0.4 | 0.4 |
| K | 3,770 | $Cr_2O_3$ | 0.8 | 0.4 |
| Ti | 2,400 | Total | 99.9 | 98.8 |
| Cr | 13,500 | Mg/Si | 1.04 | 1.26 |
| Mn | 9,550 | | | |

* Anders and Grevesse [1989]   **Sun [1982]   *** Ringwood [1975]

Following Bernal's [1936] old prediction that the 400 km discontinuity would be caused by the transformation of olivine into the spinel structure, a large number studies had been carried out to construct a high-pressure phase diagram in the system $Mg_2SiO_4$–$Fe_2SiO_4$ since the 1960s [e.g., Ringwood and Major, 1970; Akimoto et al., 1976]. Through these works, it was found that a new phase is stabilized between the stability fields of olivine (a) and spinel (g) in compositions rich in $Mg_2SiO_4$ and its stability field extends in both pressure and composition with increasing temperature. The new phase was named modified spinel (b), because its crystal structure is closely related to that of spinel. The recently constructed phase boundaries for the olivine-modified spinel-spinel transformation [Katsura and Ito, 1989] is in the lower portion of Fig. 2. The mantle olivine marked by the dotted line in Fig. 2 transforms into the modified spinel and the spinel successively.

The $(Mg,Fe)_2SiO_4$ spinel dissociates at higher pressures into assemblages consisting of $(Mg,Fe)SiO_3$ perovskite, $(Mg,Fe)O$ magnesiowüstite, or $SiO_2$ stishovite depending on composition, as shown in the upper portion of Fig. 2 [Ito and Takahashi, 1989]. The magnesian spinels including the marked mantle composition dissociates into perovskite and magnesiowüstite within an extremely narrow pressure interval (< 0.14 GPa). It should be also noted that the dissociation boundary, which is little affected by iron content, has a definitely negative $dP/dT$ slope of $-3$ MPa/deg. The resultant assemblage is characterized by the magnesian perovskite and relatively iron rich magnesiowüstite; e.g., $(Mg_{0.97}Fe_{0.03})SiO_3$ and $(Mg_{0.83}Fe_{0.17})O$ at 1600 °C, respectively.

Fig. 3 is the phase diagram for the system enstatite (Mg-end member pyroxene)-pyrope (Mg-end member garnet)[Akaogi et al., 1987; Irifune and Ringwood, 1987]. The composition marked by the dotted line represents that of the olivine-subtracted pyrolite mantle assuming $Fe^{2+}$ and $Ca^{2+}$ cations to be equivalent to $Mg^{2+}$. The phase diagram indicates the following phase changes would occur in the mantle. Pyroxene coexisting with garnet dissolves rapidly in garnet phase at pressures higher than 10 GPa and the assemblage is completely converted into single phase of garnet at about 15 GPa. The garnet enriched in pyroxene component thus formed was named majorite. Under increasing pressure the majorite is converted into perovskite after coexistence with ilmenite and perovskite phases. It should be noted, however, that the perovskite single phase is stabilized at pressure fairly higher than that required to stabilize $MgSiO_3$ perovskite.

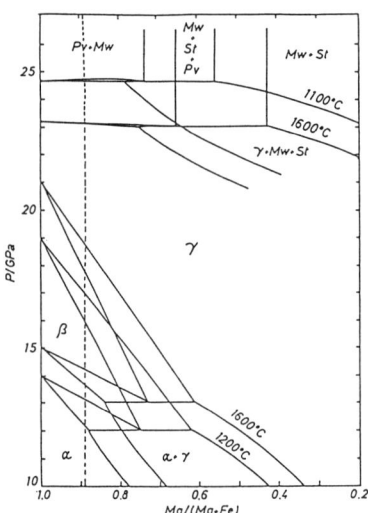 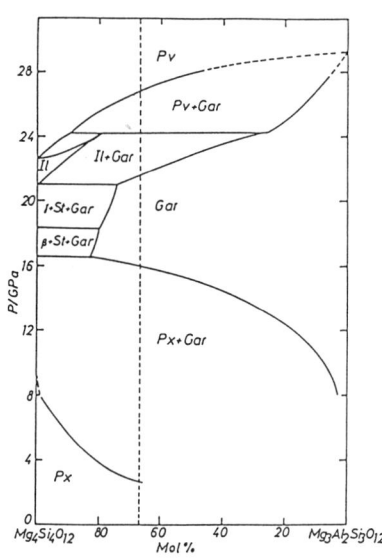

Fig. 2. Phase diagrams of the system $Mg_2SiO_4$–$Fe_2SiO_4$. $\alpha$: olivine, $\beta$: modified, $\gamma$: spinel, Pv: (Mg,Fe)$SiO_3$ perovskite, Mw: (Mg,Fe)O magnesiowüstite, St: $SiO_2$ stishovite. Data sources are Katsura and Ito [1989] and Ito and Takahashi [1989]. Dotted line denotes the composition of mantle olivine (see text).

Fig. 3. Phase diagram of the system $Mg_4Si_4O_{12}$–$Mg_3Al_2SiO_3O_{12}$ at $1500°C$. Px: pyroxene, Gar: garnet, $\beta$: $Mg_2SiO_4$ modified spinel, $\gamma$: $Mg_2SiO_4$ spinel, St: $SiO_2$ stishovite, Il: ilmenite, Pv: perovskite. Data sources are Akaogi et al. [1987] and Irifune and Ringwood [1987]. Dotted line represents the composition of olivine-subtracted pyrolite (see text).

The phase relations just outlined make it possible to construct a mineralogical model for a given mantle composition. The pyrolitic model is shown in Fig. 4. The assemblage of olivine, two pyroxenes and garnet in the upper mantle changes to modified spinel or spinel plus majorite in the transition zone. The 400 km discontinuity is chiefly caused by the olivine-modified spinel transformation. The mantle majorite actually dissociates into magnesian perovskite and accessory phases such as $CaSiO_3$ perovskite, $Al_2O_3$-rich phase, and $SiO_2$ stishovite, because the magnesian perovskite and ilmenite phases cannot incorporate large $Ca^{2+}$ ion. The 660 km discontinuity is responsible to the dissociations of both spinel and majorite. Consequently, the magnesian perovskite is by far the most dominant constituent of the lower mantle, followed by relatively iron-rich magnesiowüstite. As the perovskite is known to be stable even at pressure of the bottom of the mantle [Knittle and Jeanloz, 1987], this phase is believed to be the most dominant material of our planet.

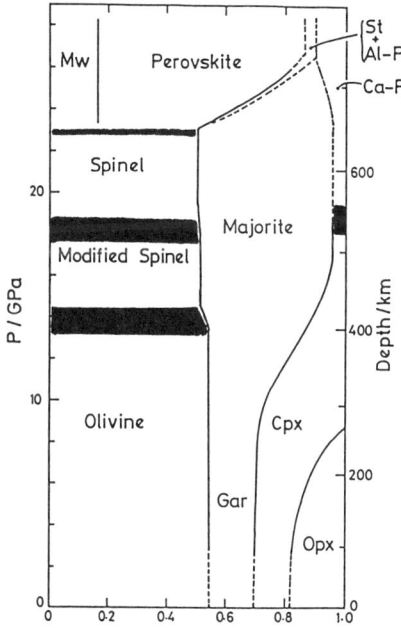

Fig. 4. Volumetric mineral constitution of pyrolite mantle down to the lower mantle. Gar: garnet, Opx: orthopyroxene, Cpx: clinopyroxene, Mw: magnesiowüstite, St: stishovite, Ca-P: $CaSiO_3$ perovksite, Al-P: $Al_2O_3$-rich phase. After Ito and Takahashi [1987].

The spinel dissociation completes within a very small depth range (less than 4 km), which is consistent with the seismic evidence that the 660 km discontinuity is so sharp that reflected waves from it are frequently observed [e.g., Nakanishi, 1988]. Therefore, temperature at the depth is estimated to be about 1600 °C from the dissociation boundary [Ito and Katsura, 1989].

B. Implication to mantle dynamics

The phase boundary with a negative slope in the mantle acts as a barrier to material transportation across it. Therefore, a large negative slope would cause a double-layered convection above and beneath the phase boundary. Numerical calculations [e.g., Christensen and Yuen, 1985; Nakakuki et al., 1994] indicate that the critical slope at the 660 km discontinuity for the double-layer convection is in a range −4 to −8 MPa/deg and, for the slope −3 MPa/deg of the spinel dissociation, leaking or intermittent penetrative convection can occur between the upper and lower mantle.

Recent seismic tomography reveals three-dimensional perturbations in velocities over most of the mantle and visualizes the following features of subducting slabs and large-scale rising plumes [e.g., Fukao et al., 1992; Fukao, 1992]. Although some slabs subduct into the lower mantle, many of them bend to subhorizontal and are stagnant right above the 660 km discontinuity over approximately 1000 km length. Moreover, beneath the stagnated slabs is studded with high velocity anomalies down to the bottom of the mantle, suggesting that the accumulated slab would collapse and fall on to the core-mantle boundary episodically. On the other hand, the large up-welling plume which is originated from a wide region above the core-mantle boundary, ascends in the lower mantle forming a column-like shape. However, the plume spreads in the upper portion of the lower mantle and splits into smaller ones in the upper mantle. These tomographic images are consistently interpreted that the subducted slab and the up-welling plume are subjected to positive and negative buoyant forces caused by the genative slope of the spinel dissociation boundary.

## III. EARLY EVOLUTION OF THE EARTH

A. Stratification of the primitive mantle

As the Earth was formed from the dust component of the solar nebula, it is reasonable to assume that the primitive bulk Earth had a composition of develatilized solar nebula [Hart and Zindler, 1986]. In other words, the cosmic abundance of refractory elements [Anders and Grevesse, 1989] put on a general constraint on the chemistry of the Earth. The mantle compositions along this line are designated as chondritic models, because most of them were derived from the composition of C1 chondrite which have not been fractionated [e.g., Sun, 1982].

As compared in Table 1, the most significant difference between the chondritic and pyrolitic models is in Mg/Si ratio, the chondritic model having definitely lower Mg/Si ratio. Therefore, on condition that the upper mantle is peridotitic, the chondritic model requires that substantial amount of silicon is stored in the lower mantle. In this case, the lower mantle is largely enriched in pyroxene component and can be composed almost entirely of $(Mg,Fe)SiO_3$ perovskite [e.g., Anderson and Bass, 1986].

Such a large-scale chemical layering of the mantle would be caused by fractional crystallization in the deep magma ocean which would have covered the early Earth to a depth greater than 1000 km [Ohtani, 1985]; the lower mantle would be formed by sedimentation of the perovskite and/or majorite prior to the solidification of the residual silica-deficient liquid as the upper mantle. Melting studies on mantle peridotite [Ito and Takahashi, 1987] and chondritic mantle composition [Ohtani and Sawamoto, 1987] have revealed that majorite and perovskite appear as the liquidus phases at pressures higher than 15 GPa, supporting the fractionation of these phase in the magma ocean.

Nevertheless, the partitioning data of major and trace elements between liquidus perovskite, majorite, and liquid indicate that the upper mantle shows no geochemical signatures which should have caused by separation of large amounts of perovskite and majorite [Kato et al., 1988]. However, the possibility of stratification of the primitive mantle should be investigated in more detail especially with regard to the following. The liquidus phase might be strongly composition-dependent and magnesiowüstite as the liquidus phase could be possible [Agee, 1990]. The possible density crossover between liquid and crystal at high pressure [e.g., Agee and Walker, 1993] is crucial for the differentiation. Moreover, metal-solid separation in the magma ocean is seriously affected by vigorous convection and resultant element distribution does not always reflect difference in partition coefficients [Abe, 1993].

B. Chemistry and evolution of the core

Although the chemical reaction at the core-mantle interface would have continued for a long term of the Earth's history, the effect of the reaction on the core composition may not be crucial so as far as the rate-determining process is solid state diffusion in the mantle. According to the magma ocean hypothesis, core segregation had rapidly proceeded by sinking of molten iron through a partially or completely molten silicate [e.g., Sasaki and Nakazawa, 1986]. It is highly likely that most of light elements of the core would have incorporated in the process of core formation. Therefore, the chemical reactions between molten iron and silicate melt at high pressure should have put an important constraint on the core composition and might give a clue to solve the evolution of the core.

We [Ito and Katsura, 1991; Ito et al., 1994] have systematically investigated partitioning of Si, Mg, and O between molten iron and silicate melt up to 26 GPa, using fine powdered mixtures of pure iron and $MgSiO_3$ enstatite or $(Mg_{0.9}Fe_{0.1})_2SiO_4$ olivine as the starting materials. Important observations on the quenched products are as follows: (1) Certain amounts of silicon ($> 1$ wt%) dissolved in molten iron from silicate melt at pressures higher than 20 GPa, whereas no evidence

These results indicate that Si and O are important light elements of the terrestrial core. The entry of Si in the core could have substantial influence on the mantle chemistry. In Fig. 5, the distribution coefficients of Si between silicate melt and molten iron, $D_{Si}$, are plotted against pressure. Natural log of $D_{Si}$ decreases linearly with increasing pressure. The horizontal broad line in Fig. 5 denotes the value of $D_{Si}$ which produce the pyrolite mantle from the chondritic one by partitioning $SiO_2$ into the proto-core. Therefore, the marked difference in Mg/Si ratio between the upper mantle peridotite and the chondritic mantle model can be reconciled, if the magma ocean had extended to a depth greater than 1000 km.

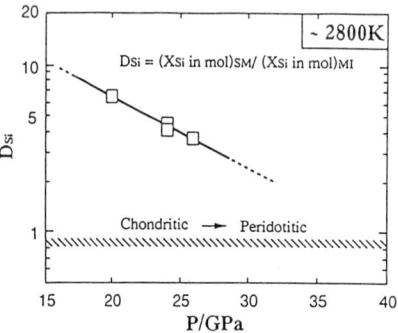

Fig. 5. Change in distribution coefficient of Si between silicate melt and molten iron, $D_{Si}$, at about 2800 K with pressure. The horizontal hatched line corresponds to the value which produces the peridotitic mantle from the condritic one by partitioning $SiO_2$ into the proto core. After Ito et al. [1994].

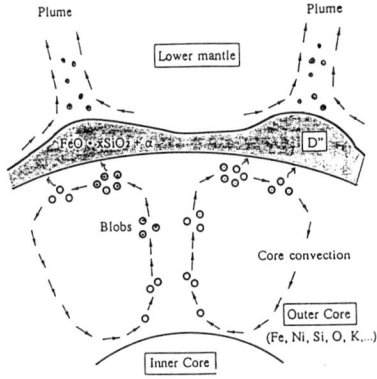

Fig. 6. An idealized cartoon of the chemical evolution of the core (see text).

Following the successive cooling of the Earth, formation and growth of the solid inner core of relatively pure iron would have proceeded gradually, which might made the outer core be somewhat supersaturated with respect to Si and O. Therefore, along the ascending stream of the outer core convection, the blobs of immiscible liquid with composition $FeO-\chi SiO_2$ ($\chi > 1$), which were observed as small spherules in our experiment, would emerge and become dominant. The blobs should accelerate the convection in the outer core because of its lower density than that of the mother iron liquid. This compositional convection may be of essential importance for driving the dynamo in the Earth's core. After the blobs rise to the top of the outer core, they accumulate at the bottom of the mantle and gradually solidify to form the D" layer. Therefor the D" layer is composed predominantly $SiO_2$ and FeO and is growing downward, a part of which, in turn, might be continuously scratched off by the up-welling plumes and be assimilated to the lower mantle. The fate of the D" layer would be an important factor to assess the lower mantle chemistry.

## IV. CONCLUDING REMARKS

The determinations of equation of state and rheological property for the deep mantle phases are active fields of the high-pressure mineral physics, because the results are indispensable to quantitative interpretation of both the static and dynamic nature of the mantle.

The Earth has been evolved by heat release associated with material transportation in various styles and scales since it was born ca. 4.5 billion years ago. It's present status reflects all the time-integrated results of many events and complicated processes, which, in turn, provides many clues to solve the evolution. However, the following issues may be of special significance. (1) Characterization of the promodidial material of the Earth should be put forward. Investigation of extraterrestrial materials including other planets is quite important as well as the constrains from cosmochemical and astrophysical view points. (2) Clarification of the accretion and the state of proto-Earth is undoubtedly set the initial condition of the evolution. The successive core formation may have special importance, because it separated the Earth into almost closed systems, the mantle and the core.

## REFERENCES

Abe, Y., 1993, in Evolution of the Earth and Planets, E. Takahashi, R. Jeanloz, and D. Rubie, Eds. (Geophysical Monograph 74/IUGG, 14, Washington, D.C.), p. 41.
Agee, C., 1990, Nature, **346**, 834.
Agee, C., and D. Walker, 1993, Earth Planet. Sci. Lett., **114**, 315.
Akaogi, M., A. Navrotsky, T. Yagi, S. Akimoto, 1987, in High-Pressure Research in Mineral Physics, M. H. Manghnani and Y. Syono, Eds. (TERRAPUB, Tokyo/American Geophysical Union, Washington, D. C.).
Akimoto, S., Y. Matsui, and Y. Syono, 1976, in Physics and Chemistry of Minerals and Rocks, R. G. J. Strens, Ed., (Wiley-Interscience, London), p. 327.
Anders, E., and N. Grevesse, 1989, Geochim. Cosmochim. Acta, **53**, 197.
Anderson, D. L., and J. D. Bass, 1986, Nature, **320**, 321.
Bernal, J. D., 1936, Observatory, **59**, 268.
Dziewonski, A. M., and D. L. Anderson, 1981, Phys. Earth. Planet. Interior, **25**, 297.
Fukao, Y., 1992, Science, **258**, 625.
Fukao, Y., M. Obayashi, H. Inoue, and M. Nenbai, 1992, J. Geophys. Res., **97**, 4809.
Hart, S. R., and A. Zindler, 1986, Chem. Geol., **57**, 247.
Irifune, T., and A. E. Ringwood, 1987, Earth Planet. Sci. Lett., **86**, 365.
Ito, E., and T. Katsura, 1989, Geophys. Res. Lett., **16**, 425.
Ito, E., and T. Katsura, 1991, Proc. Jpn. Acad., **67**, 153.
Ito, E., K. Morooka, O. Ujike, and T. Katsura, 1994, J. Geophys. Res., submitted.
Ito, E., and E. Takahashi, 1987, Nature, **328**, 514.

Ito, E., and E. Takahashi, 1989, J. Geophys. Res., **94**, 10,637.
Jacobs, J. A., 1987, The Earth's Core, (Academic Press Inc., London), p. 413.
Kato, T., A. E. Ringwood, and T. Irifune, 1988, Earth Planet. Sci. Lett., **89**, 123.
Katsura, T., and E. Ito, 1989, J. Geophys. Res., **94**, 15,663.
Knittle, E., and R. Jeanloz, 1987, Science, **235**, 668.
Nataf, H.-C., and S. Houard, 1993, Geophys. Res. Lett., **20**, 2371.
Nakakuki, T., H. Sato, and H. Fujimoto, 1994, Earth Planet. Sci. Lett., **121**, 369.
Nakanishi, I., 1988, Geophys. J., **93**, 335.
Ringwood, A. E., 1970, Phys. Earth Planet. Interior, **3**, 109.
Ringwood, A. E., and A. Major, 1970, Phys. Earth Planet. Interior, **3**, 89.
Ohtani, E., 1985, Earth Planet. Sci. Lett., **38**, 70.
Ohtani, E., and H. Sawamoto, 1987, Geophys. Res. Lett., **14**, 733.
Sasaki, S., and K. Nakazawa, 1986, J. Geophys. Res., **91**, 231.
Sun, S. S., 1982, Geochim. Cosmochim. Acta, **46**, 179.

# VII. High-Pressure Metal Physics

# VII.1

# THE DENSE HYDROGEN PLASMA: TRANSLATIONAL, ORIENTATIONAL, AND ELECTRONIC STRUCTURE

N.W. Ashcroft

Laboratory of Atomic and Solid State Physics, Clark Hall
Cornell University, Ithaca, NY 14853-2501, U. S. A.

At conditions typical of stellar interiors the states of hydrogen can normally be thought of as reasonably symmetric, but as density and temperature decline the electronic component becomes progressively more inhomogeneous eventually leading to pairing of protons in a metallic environment. On further reduction a ground-state metal-insulator transition takes place. Associated with proton pairing is orientational ordering which in a ground state is itself embedded within a framework of translational or crystalline order. Both aspects influence the electronic structure and near the metal-insulator transition can be subject to excitonic and polaronic distortions. The experimental evidence on the rotational characteristics suggest an electron problem in which the correlation length for associated potential correlations is exceedingly short.

## I. INTRODUCTION

The purpose of the following is to examine the possible phase-structure of dense hydrogen as a function of *declining* density starting, however, with extremely high density degenerate conditions typical of some stellar interiors. Accordingly let $\{\vec{r}_{1i}\}$ be the coordinates of $N_1$ electrons (mass $m_1$) in a volume $V$. Similarly let $\{\vec{r}_{2i}\}$ be the coordinates of $N_2$ protons (masses $m_2$) in the same volume. We imagine a neutral ($N_1 = N_2$) canonical ensemble of electrons and protons; the temperature is $T$ ($\beta = 1/k_BT$). At a non-relativistic level the microscopic Hamiltonian for this dual Fermion case is the simplest rendition of a nucleus-electron system [Chihara, 1985] namely

$$\hat{H} = \sum_{\alpha=1,2}\left\{\sum_{i=1}^{N_\alpha}(-\hbar^2/2m_\alpha)\vec{\nabla}^2_{\alpha i} + \frac{1}{2}\sum_{\alpha'}(-1)^{\alpha+\alpha'}\int_V \vec{dr}\int_{V'}\vec{dr}'\phi_c(\vec{r}-\vec{r}')\hat{\rho}^{(2)}_{\alpha\alpha'}(\vec{r},\vec{r}')\right\}$$
(1)

where $\phi_c(\vec{r}-\vec{r}') = e^2/|\vec{r}-\vec{r}'|$ is the fundamental Coulomb interaction. All configurational information resides in the two-particle density operator $\hat{\rho}^{(2)}_{\alpha\alpha'}(\vec{r},\vec{r}') = \hat{\rho}^{(1)}_\alpha(\vec{r})\hat{\rho}^{(1)}_{\alpha'}(\vec{r}') - \delta_{\alpha\alpha'}\delta(\vec{r}-\vec{r}')\hat{\rho}^{(1)}_\alpha(\vec{r})$ where the one-particle density operators for electrons ($\hat{\rho}^{(1)}_1(\vec{r})$) and protons ($\hat{\rho}^{(1)}_2(\vec{r})$) are defined by

$$\hat{\rho}^{(1)}_\alpha(\vec{r}) = \sum_{i=1}^{N}\delta(\vec{r}-\vec{r}_{\alpha i}), \quad (N_1 = N_2 = N).$$
(2)

251

The *structural* character of the states of (1) can be described by specifying the appropriate statistical average

$$\rho_\alpha^{(1)}(\vec{r}) = \langle \hat{\rho}_\alpha^{(1)}(\vec{r}) \rangle = Tr \hat{\rho}_\alpha^{(1)}(\vec{r}) e^{-\beta H} / Tr \ e^{-\beta H} \qquad (3)$$

where a quantum or classical trace (as conditions dictate) is taken over the combined states of (1). Thus if $\rho_2^{(1)}(\vec{r})$ for protons has no microscopic variation in space the corresponding phase for protons would be considered a fluid. The Hamiltonian given in (1) lacks direct reference to the spins of the two classes of Fermions; however, the emerging phases can reveal spin-ordering. The evident symmetry of underlying Hamiltonian for dense hydrogen leads to formal scaling results for the ground state energy and correlation functions [Moulopoulos and Ashcroft, 1990].

To begin the treatment of (1) we need to establish an appropriate context of high densities, and to do this it is convenient to modify the form of $\hat{H}$. Let $\bar{\rho} = N_1/V = N_2/V$. Then as can be readily verified, in the thermodynamic limit (1) can be rewritten as

$$H = \sum_{i=1}^{N_1} (-\hbar^2/2m_1) \vec{\nabla}_{1i}^2 + \frac{1}{2} \int_V \vec{dr} \int_V \vec{dr}' \phi_c(\vec{r}-\vec{r}') \left\{ \hat{\rho}_{11}^{(2)}(\vec{r},\vec{r}') - 2\hat{\rho}_1^{(1)}(\vec{r})\bar{\rho} + \bar{\rho}^2 \right\} \quad (4a)$$

$$+ \sum_{i=1}^{N_2} (-\hbar^2/2m_2) \vec{\nabla}_{2i}^2 + \frac{1}{2} \int_V \vec{dr} \int_V \vec{dr}' \phi_c(\vec{r}-\vec{r}') \left\{ \hat{\rho}_{22}^{(2)}(\vec{r},\vec{r}') - 2\hat{\rho}_2^{(1)}(\vec{r})\bar{\rho} + \bar{\rho}^2 \right\} \quad (4b)$$

$$- \int_V \vec{dr} \int_V \vec{dr}' \phi_c(\vec{r}-\vec{r}') \left\{ \hat{\rho}_1^{(1)}(\vec{r}) - \bar{\rho} \right\} \left\{ \hat{\rho}_2^{(1)}(\vec{r}') - \bar{\rho} \right\} \quad (4c)$$

and in this form two *plasma* problems represented by (4a) and (4b) are separately well defined.

From $\bar{\rho}$, length scales can be established starting with either (4a) or (4b) and from these a definition of density scales. Taking electrons first, we write $(4\pi/3) r_{s1}^3 a_1^3 = 1/\bar{\rho}$ where $a_1 = \hbar^2/m_1 e^2$ is the standard Bohr radius. For convenience we will adopt the convention $r_s = r_{s1}$. It is well known that for average electron densities chosen in a range corresponding to $r_s \sim 1$, possible ground states of (4a) include the homogeneous ($\rho_1^{(1)}(\vec{r}) = const$), paramagnetic Fermi-liquid. In a single-particle description the Fermi energy is $\varepsilon_F = (9\pi/4)^{2/3}/r_s^2 (\text{Ry})$ or $50.1/r_s^2$ eV; the associated pressure $p = \{3(9\pi/4)^{2/3}/10\pi r_s^5\} p_a = 51.8/r_s^5$ Mbar can be used as an illustrative guide to the overall pressure. Here $p_a = e^2/2a_o^4$ is the corresponding atomic unit with the practical value 14,720 GPa (or 147.2 Mbar). At very much lower densities ($r_s \gtrsim 100$) the symmetry of (4a) is broken and the system crystallizes ($\rho_1^{(1)}(\vec{r})$ is periodic and the system possesses diagonal long-range order); the spin arrangement of what has become a Wigner crystal is antiferromagnetic. Kohn and Luttinger [1965] were the first to observe that uniform pairing states of (4a) with off-diagonal long-range order may also arise in principle.

With due recognition of the change in length scale implied by the proton Bohr radius ($\hbar^2/m_2 e^2 = (m_1/m_2) a_1$) the corresponding states of (4b) are immediate. Thus at $r_s \sim 1/1836 = 5.45 \times 10^{-4}$ (some $2 \times 10^{11}$ times the density of ordinary solid hydrogen) the protons in (4b) treated non-relativistically will possess the same properties as an electron Fermi liquid at $r_s \sim 1$. Correspondingly at $r_s \gtrsim 0.0545$ the proton problem described by (4b) will be expected to form a Wigner crystal (again a broken symmetry) with the nuclear (proton) spins antiferromagnetically aligned. This will occur at $\sim 2 \times 10^5$ times the density of ordinary solid hydrogen, and at pressures in the range $3 \times 10^8$ Mbar. These are conditions approaching those found in certain degenerate stars. Since the Einstein frequency for a proton Wigner crystal is simply $\omega_{p2}/\sqrt{3}$ (where $\omega_{p2}$ is the plasma frequency for objects with the mass and charge of protons) it follows that an estimate of the zero-point energy in the crystalline phase is, per proton, $3(m_1/m_2)^{1/2}/r_s^{3/2}$ in Rydbergs. This is 75 eV (or $0.85 \times 10^6$ K) at $r_s = 0.0545$ and 5,500 eV (or $0.65 \times 10^8$ K) at $r_s = 5.45 \times 10^{-4}$.

If (4a) or (4b) are treated classically, then for temperatures in excess of $(e^2/r_{s\alpha}a_\alpha)/k_B\Gamma_m$, (where $\Gamma_m$ is approximately 172) the systems take up the fluid state of the classical one component plasma (for a comprehensive review see Ichimaru [1994]). Otherwise at comparable temperatures and higher densities the phases are classical crystals. In fluid states the pair distribution functions possess the characteristics of highly correlated classical fluids. The quantum fluid states of (4a) and (4b) satisfy

$$(\partial g_{\alpha\alpha}(r)/\partial r)_0 = g_{\alpha\alpha}(0)$$

where $g_{\alpha\alpha'}$ are the pair distribution functions. Thus for protons ($g_{22}$) a cusp condition controls the contact value (see below) which in turn is important in the physics of fusion. Since a cusp possesses spatial variation down to vanishing distances, classical and quantum aspects (and in particular the enhancement factor) are not straightforwardly separated in this application.

We turn now to the role of the term (4c) which couples the two plasmas and which assumes increasing importance as density declines. This is easily assessed by noting that in the presence of any potential external to the electron system the characteristic electron response length is gauged by $k_{TF}^{-1}$ where the Thomas-Fermi wave-vector $k_{TF}$ is given by $k_{TF} = (4k_F/\pi a_1)^{1/2}$. Hydrogen will be truly dense in the sense of electron response to proton fields if the mean interproton spacing (here $2r_s a_1$) is less than $k_{TF}^{-1}$. We therefore require $r_s$ values less than $(4\pi^2/9)^{1/3}/16 = 0.12$. For corresponding densities lower than this significant inhomogeneity in the electron system is anticipated (and confirmed by Figure 1 as we shall see).

## II. INHOMOGENEOUS ELECTRON STATES IN DENSE HYDROGEN

Within the adiabatic separation of time scales, the fields of the protons appear, as noted, as the sole source of potential external to the electrons. For a given configuration of protons (say a set of coordinates we denote by $\{\vec{R}\}$) the ground state energy $E_1$ of the electron system is then a unique functional of the corresponding inhomogeneous electron density $\rho_1^{(1)}(\vec{r})$, i.e.

$$E_1 = E_1[\rho_1^{(1)}(\vec{r})] = E_1[\rho_1^{(1)}(\vec{r}; \{R\})]$$

which is the content of the Hohenberg-Kohn theorem [1964]. An entirely equivalent independent-particle description is obtained if we start with a set of $N_1$ such Fermions placed in an external potential $V(\vec{r}) = V_{\{\vec{R}\}}(\vec{r})$, each satisfying

$$\{-\hbar^2/2m_1 \vec{\nabla}^2 + V(\vec{r})\}\psi_i = \varepsilon_i \psi_i \tag{5}$$

with $V(\vec{r})$ determined self-consistently and including an exchange-correlation contribution originating with the corresponding energy $E_{exc}[\rho_1^{(1)}(\vec{r})]$. This is the procedure of Kohn and Sham [1965]; for periodic structures (5) leads to a Bloch-electron description. Though the development is in principle exact, the exchange-correlation term $\delta E_{exc}[\rho_1^{(1)}(r)]/\delta\rho^{(1)}(\vec{r})$ is unknown in detail. In applications to dense hydrogen, it has frequently been evaluated using the local-density approximation (LDA), i.e. for the curves in Figure 1, particularly those close to $r_s \sim 1 \to 2$, an approximation stating that

$$E_{exc} = \int d\vec{r} \rho_1^{(1)}(\vec{r}) E_{exc}\left(\rho_1^{(1)}(\vec{r})\right) \tag{6}$$

where $E_{exc}(\rho_1^{(1)})$ would be the exchange-correlation energy per electron for a uniform interacting system at average density $\rho_1^{(1)}$. In principle the systematic solution of equations (5) will lead to the density $\rho_1^{(1)}(\vec{r})$ (for a given structure) and then through the functional statements to the energy $E_1$ uniquely associated with this density corresponding to that structure.

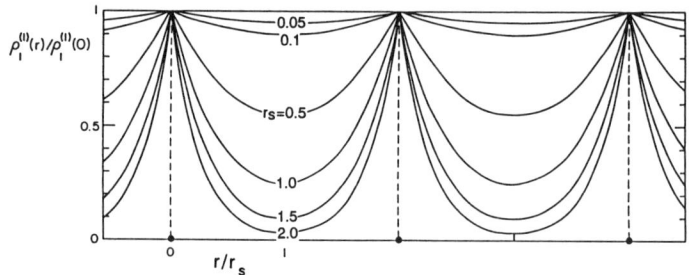

Figure 1. Approximate electron density profiles $\rho_1^{(1)}(r)$ within a spherical-cell model and determined for various densities $\bar{\rho}$ (as parametrized by the electron spacing parameter $r_s$). The densities reflect the cusp theorem, at the protons, and the periodicity requirement (i.e. pairing is not permitted).

If, for the moment an adiabatic principle is invoked, and protons are assigned to the sites $\{\vec{R}\}$ of a perfect simple high-symmetry lattice, then this density $\rho_1^{(1)}(\vec{r})$ is certainly periodic, i.e.

$$\rho_1^{(1)}(\vec{r}) = \langle \hat{\rho}_1^{(1)}(\vec{r}) \rangle = \rho_1^{(1)}(\vec{r} + \vec{R}), \qquad \forall \vec{R}. \tag{7}$$

The choice of a body-centered cubic (BCC) Bravais lattice and densities typified by $r_s \sim 1 - 2$ together characterize the state of *metallic hydrogen* as originally envisaged by Wigner and Huntington in their important paper of 1935. In the intervening years, most theoretical calculations for the static model seem to indicate that at these densities the structure in the metallic state is neither simple nor particularly symmetric (see, for example, Brovman et al., 1972; Ebina and Miyagi, 1989). But, for all lattices in this problem (including BCC) it is a quite general property of the quantum mechanics of the many-particle Coulomb problem that the electronic density $\rho_1^{(1)}(\vec{r})$ satisfies a cusp condition [Kato, 1957; Kimball, 1973; Carlsson and Ashcroft, 1982] at each site $\vec{R}$. Thus, at each proton $\partial \rho_1^{(1)}(r)/\partial r|_0 = (-2/a_1)\rho_1^{(1)}(r)|_0$ where $r$ denotes a radial coordinate measured from such a site. Since protons are, on average, separated by $2r_s a_o$, a form can be proposed for the density $\rho_1^{(1)}$ which *approximately* combines both the cusp condition (which is quantum mechanical in origin) and the periodicity condition (11) (which is structural), and which makes but one additional physical assumption linked to the absence of core states namely that the density is otherwise monotonic. For $r$ taken along a line of centers, the function $\rho^{(1)}(r) \propto \cosh a(1 - r/r_s a_1)$ satisfies both physical constraints, the cusp condition leading to the requirement $\tanh a = 2r_s/a$ where $a$ may be written as $a = 2\gamma r_s$. The dimensionless quantity $\gamma$ is clearly close to unity for large $r_s$ (low densities).

Though obviously approximate, Figure 1 is a useful guide to the expected role of electron inhomogeneity in the discussion of lower density conditions that follows. For the model above it depicts corresponding plots of relative ground state densities $\rho_1^{(1)}(r)/\rho_1^{(1)}(0)$ for various values of $r_s$, including $r_s = 0.05$ which as noted above is close to the expected melting density of the protons in the problem represented by (4b). A quantitative determination of $\rho_1^{(1)}(\vec{r})$ within the same spherical-cell model requires, of course, a complete solution for the electron states. This has been given for band-minimum states within the Wigner-Seitz approximation by Styer and Ashcroft [1984] resulting (for $r_s \sim 1$) in densities very similar in form to those in Figure 1. By using a method originating with Bardeen [1938] it is also possible to assess the role of a periodic array of singular Coulomb attractions (without screening) on the effective masses that might otherwise be assigned to nearly-free-electron band states associated with an assumed monatomic metallic phase. Thus

$$\frac{m}{m^*} = \frac{\rho_1^{(1)}(r_s)}{\bar{\rho}}\left\{\chi_s/2 - e_o - 1 + (2 + e_o)\frac{M(1 - e_o; 4; \chi_s)}{M(2 - e_o; 4; \chi_s)}\right\} \qquad (8)$$

where $e_o = (-\varepsilon_o(r_s)/(e^2/2a_o))^{-1/2}$ with $\varepsilon_o(r_s)$ the energy of the band minimum (and obtained from the solution of the Schrödinger equation). In (8), $\chi_s = \chi(r_s) = 2r_s/e_o$ and $M(a, b, \chi)$ is Kummer's hypergeometric function. The key point here is that for very small $r_s$, $(m^*/m)$ quickly approaches unity; but for $r_s \sim 1.5$, $m^*/m$ is already $\sim 1.2$, and thereafter rises very rapidly with declining density, as expected physically. The role played by the cell boundary density is clearly important; free electron estimates can significantly overestimate the kinetic energy, even at quite high densities. For $r_s \sim 0.05$ and particularly at higher densities it is evident (and expected from the screening argument above) that the electron density is largely homogeneous and apart from overall constant energies, term (4b) in the complete Hamiltonian begins to represent an accurate description of the system. But as densities decline from $r_s = 0.05$ into the region where the corresponding length scale matches the atomic scale ($r_s \sim 1$) it is also evident that electron inhomogeneity rises substantially, and with it the possibility of structural instability. (An *electronic* instability is certainly expected for a monatomic arrangement around $r_s \sim 2.5$, the density anticipated for a Mott transition [Svane and Gunnarsson, 1990]). Even before this a structural instability is anticipated when the cusp structure is dominant and is most easily seen by way of example.

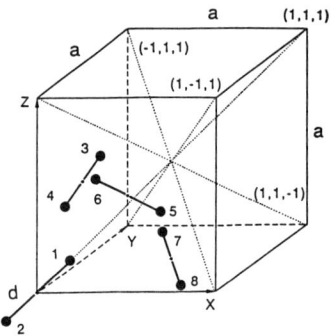

Figure 2. An example of a cubic phase of dense hydrogen (the $\alpha - N_2$ or Pa3 structure). If proton pairs are adiabatically separated along directions parallel to body diagonals, the eventual result is a simple cubic (where the intra-proton separation is given by $2d = \sqrt{3}a/2$). A further high symmetry states is found at $2d = \sqrt{3}a/4$.

Suppose the lattice $\{\vec{R}\}$ is actually a simple cubic: then a *continuous* adiabatic pathway exists along body diagonals for protons in which a second underlying simple cubic is reached, but one which possesses an eight-point basis comprised of four associated pairs. (The energy per proton is *not* necessarily monotonic along such a path.) If $a$ is the lattice constant of the new lattice, then the pairs are at basis points $\vec{b}_1, \ldots, \vec{b}_8$ given by (see Figure 2) $\pm \alpha a(1,1,1), \frac{a}{2}(0,1,1) \pm \alpha a(1,-1,1), \frac{a}{2}(1,0,1) \pm \alpha a(1,1,-1)$, and $\frac{a}{2}(1,1,0) \pm \alpha a(-1,1,1)$, where $\alpha = d/a\sqrt{3}$ (with $2d = 2d(r_s)$ the intra-pair spacing). It is easily verified that the original starting simple cubic is recovered when the intra-pair spacing is taken as $d = \sqrt{3}a/4$, at which point the lattice constant is $a/2$. In this configuration the independent particle picture of the electronic structure corresponds to the case of a single-electron per primitive cell, and providing the density is still above the Mott transition limit will reflect a metallic state. But it is equally clear that an energetic advantage is possible in principle by permitting $d$ to depart from the high symmetry value of $\sqrt{3}a/4$, the overall density otherwise being held constant. From Figure 1 it is apparent that because of the unique curvature of the density $\rho_1^{(1)}(r)$ then for low enough $\bar{\rho}$ electrons are expected to gain exchange and electrostatic energy from overlap, much as they do in the example of the free hydrogen molecule (where, it may be noted, the relative density $\rho_1^{(1)}(r)/\rho^{(1)}(0)$ at mid-bond is already 0.61). The eventual penalty to be balanced when protons are permitted to move in order to exploit this benefit is the ultimate rise in electrostatic energy. Kinetic energies also change but this argument nevertheless suggests perhaps the onset of the next broken symmetry associated with (4) (others will become evident). From the band viewpoint the argument bears certain similarities to the one leading to Peierls transition in one-dimensional systems. The evident difference is that an overall band-gap is no longer mandatory, though as we shall now see, at a critical density it must occur.

## III. PAIRING STRUCTURES

Though (4) also admits of generalized pairing states [Moulopoulos and Ashcroft, 1991] we shall consider for the moment states where protons only are paired; the time average locations of the mass centers of these pairs will continue to form a lattice. If the average density $\bar{\rho}$ is allowed to decrease in an unlimited fashion, a ground-state of (4) is reached which is self-stabilized. The two plasmas are now completely interdependent and the ensuing one-atmosphere ground state of (4) is ordinary solid hydrogen; the low temperature and other properties of this system including its spin arrangements have been extensively reviewed by Silvera [1980]. Though the *average* electron density remains quite high ($r_s = 3.13$) solid hydrogen is an insulator with a large energy gap. The proton pairs that are formed are far from static; indeed they exhibit translational dynamics, and also rotational characteristics the latter persisting to remarkably high densities and having a significant bearing on the overall electronic structure. The pairs with their accompanying electron charge have important *internal physical* characteristics that bear on the expected transitions in hydrogen. For example, if isolated, a hydrogen molecule has a static electronic polarizability $\alpha$ of $5.42 a_1^3$ (in $D_2$ it is $5.39 a_1^3$) reflecting substantial internal electronic fluctuation. For a unit charge held static at a distance $r$ from the center of such a pair, polarization attraction ($\sim -2e^2\alpha/r^4$) can be substantial. For a position midway between near neighbor molecules in a $Pa3$ structure it amounts to $-2.7$ Ry.

A ground state *electronic* trace of (1) leads to an effective Hamiltonian for the subsequent development of the dynamics of the protons, including those of paired states. In many discussions of the energetics and dynamics of dense hydrogen a common assumption is that the resulting many-proton potential can be represented as a sum of pairwise inter- *and* intra-molecular terms. The former are normally linked to the familiar dispersion-force analysis of the problem, and hence to the polarizability and related physics of molecular pairs. This exercise is usually carried out at low densities, and the assumption is routinely made that inter-molecular potentials determined this way are then *transferable* to high densities (where 'high' now implies compressions upwards of an entire order of magnitude).

It is easy to see by example that such an assumption must fail even within currently accessible densities. Let $\overrightarrow{D}_{jj'}$ be the interaction between instantaneous dipoles at, say sites $\vec{R}_j$ and $\vec{R}_{j'}$, and let $\overrightarrow{D}(\vec{k})$ be its site Fourier transform. Then [Maggs and Ashcroft, 1987] the ground-state energy of the fluctuating dipole system is proportional to

$$\int \vec{dk} \int d\omega/2\pi Tr \ \ln \{\overrightarrow{1} - \alpha(i\omega)\overrightarrow{D}(\vec{k})\} \qquad (9)$$

where $\alpha(\omega)$ is the dynamic polarizability of the molecules. The standard (e.g. van der Waals) result emerges from (9) by a development in powers of $\alpha\overrightarrow{D}$. It is clearly only valid if the largest eigenvalue of $\alpha\overrightarrow{D}$ is less than unity. For close-packed structures and $k \to 0$ this condition fails to be met when $\bar{\rho}/\bar{\rho}_o \geq 5.7$. It follows that for $r_s \leq 1.76$, or $p \gtrsim 40$ GPa (see below), the assumption justifying the form of some commonly used potentials cannot be sustained an observation that will remain true even for a range of densities lower than $\rho/\rho_o = 5.7$. As a consequence attempts to uniquely separate intra- and inter-molecular effects if based on these assumptions are probably also not sustainable although they have been attempted [Moshary et al., 1993]. Electrons do not divide up in this manner, of course, and the difficulty is further compounded by the fact that many-body interactions are frequently omitted. Finally, a potential expansion is attempting to mimic state (or density) dependent terms originating with bands which, by $r_s = 1.76$ are already quite wide ($\sim 16$ eV).

In the passage from the high density conditions discussed above, to conditions near the self-stabilizing density, it is apparent that a metal to insulator transition has occurred, and in principle has occurred in a paired phase. The experimental determination of the corresponding critical density, or equivalently the critical pressure, has been a major goal for several years, and has been given considerable impetus by the innovative use of ultrahigh pressure manifestations of the diamond anvil cell [for a recent review see Mao and Hemley, 1994]. To date definitive signatures of an insulator-metal transition remain elusive, though there is evidence of a systematic closure of the overall band-gap as the pressure increases. As discussed by Mao and Hemley, the diamond cell admits electromagnetic probes; x-rays have given evidence that up to moderate pressures ($r_s \gtrsim 1.73$) the structure is in the hexagonal class, i.e. stacked hexagonal layers where the structural units located on each site are proton pairs. Their detailed structural (e.g. orientational) disposition is actually not so far determined by the x-ray studies. However considerable information has been gained on their dynamics and other characteristics from spectroscopy, both infra-red (IR) and Raman. From the latter it appears that for densities as high as $r_s \sim 1.46$ the proton pairs persist in rotational or at least highly librational states [Hemley et al., 1990; Mao and Hemley, 1994] suffering at most only small changes with pressure. Recognizing that the electronic structure reflects the instantaneous location of the protons this observation already makes an interesting statement if the rotational motion is assumed isotropic (notice that the limiting case of the *opposite* assumption, namely that the rotor is planar, requires one to consider the excited states of the operator $\hat{L}_z^2$ rather than of the operator $\hat{L}^2$). For, the rotational energy scale is set by the rotational constant $\hbar^2/m_2 d^2 = 2(m_1/m_2)(a_1/d)^2$Ry$= 2.2$ mRy, from which we conclude from experiment that in dense hydrogen any energy barriers to rotation cannot be much larger than this. Thus the actual choice of real-space structure must be such that the total electronic energy reflecting a particular but periodic orientation of pairs, also cannot change significantly on the mRy scale of energies when subsequent changes in orientation are permitted. This requirement is obviously satisfied for spherically symmetric environments around any given molecule. For *crystalline* environments it is likely to be best satisfied by symmetric structures with high coordination. Accordingly although the real space structure is suggested to be hexagonal for $r_s > 1.73$, the persistence of rotational states above this suggests the possibility of adjustment or rearrangement to preserve high coordination and reasonably symmetric structures.

The rotational constant also sets the scale of temperature above which the system may be treated (for these motions) classically. The paired units possess no average electric dipole, but they do carry a permanent quadrupole $Q$ of magnitude $\sim 0.4 \; ea_1^2$. The problem of determining the energetics of *static* permanent axially symmetric quadrupoles has been discussed by Miyagi and Nakamura [1967]. Orientational order in a system of quadrupoles will not *necessarily* be commensurate with a given space lattice whose configuration is fixed by much larger translational energies; the ground state(s) might then be likened to a frustrated spin glass. Eventually a phase transition to a system of freely rotating quadrupoles is expected as temperature increases and then the picture of unhindered quadrupoles leads to a useful insight on the notion of effective interactions. To see this let the mutual potential energy of a pair of axisymmetric quadrupoles be $\phi(r,\Omega_1,\Omega_2)$ where $\Omega_1$ and $\Omega_2$ are polar angles of their primary axes. Then it follows that the classical partition function for the pair, with separation held fixed, is

$$Z(r) = \int d\Omega_1 \int d\Omega_2 \, \exp\left[-\beta\phi(r,\Omega_1,\Omega_2)\right]. \tag{10}$$

If temperatures $T$ are chosen so that the exponent is reasonably small, then the partition function can be expanded, and the integration over solid angles $\Omega_1$ and $\Omega_2$ can be carried out for the first few terms. The first surviving term leads to an effective state dependent attraction $-aQ^4/r^{10}k_BT$ where the constant $a$ (of order 0.15) has been given by Keesom [1921]. Though obviously small (which follows from the size of $Q$ compared to expected values of pair potentials) it grows with decreasing temperature and it has an interesting magnitude when viewed in terms of structural energy *differences*; and a short range form favoring high coordination. The point is that depending on time scales, this general approach to the construction of effective potentials can be used for other degrees of freedom.

One of the most striking discoveries in dense hydrogen is a significantly temperature dependent anomaly (of magnitude about $2\frac{1}{2}\%$ or less) in the intra-pair breathing mode, or vibron. It is seen with both Raman and IR probes and in both hydrogen and deuterium [Mao and Hemley, 1994; Cui *et al.*, 1994]. To set the scale in terms of those relevant to total energy calculations in dense hydrogen which we discuss next, it might be mentioned that this particular proton mode is associated with an energy $\hbar\omega_{vib}$ which is typically 40 mRy. And again a general inference can be made namely that if hydrogen were viewed from the normal perspective of a phonon description, then the change in vibron frequency would be seen as the consequence of the altering of the elements in a dynamic matrix. If coupling elements between rotational and translational degrees of freedom were clearly present in such a matrix then associated with the vibron anomaly we should also expect an additional (temperature dependent) anomaly in certain phonon modes, and also in the rotational modes themselves.

## IV. THE BAND OVERLAP TRANSITION

Though it is not expected that a band-gap can close in a continuous fashion (see below) the general principle of metallization of dense hydrogen by systematic band overlap was advanced in 1975 by Ramaker *et al.* in terms of zone occupancy, and for energy bands in a specific structure by Friedli and Ashcroft [1977]. In recent treatments of this problem, the assumption is made that the protons are actually resident on the sites of a perfect, *static*, crystal structure (a Bravais lattice augmented by a physical basis of varying degrees of complexity). As noted above experiment strongly suggests that dense hydrogen is actually in a *rotational* state (or a librational state of large amplitude) and the characteristic energies of these motions are typically 0.002 Ry per molecule. The detailed role this motion plays with respect to what in a system with perfect translational symmetry we would regard as *gap* has yet to be clarified; we will return to this important point below. Within the perfect crystal assumption, the gap itself also depends quite sensitively on the crystal structure chosen [Ashcroft, 1990], with structures based on the hexagonal classes permitting particularly small structure factors that can lead to premature overlap as density increases. These classes nevertheless appear to be favored by total energy calculations based on implementations of the density function approach discussed above for periodic structures [Barbee et al., 1989; Nagara and Nakamura, 1992; Kaxiras et al., 1991; Kaxiras and Gao, 1994]. The corresponding calculations are largely carried out within the local-density approximation and the variations in energy for different structures, including structures which tilt the axes of proton pairs to varying degrees are typically at the level of the rotational energy scale (a few mRy). For the most part the motion associated with the vibron is omitted. The accuracy of the local density approximation has been partly examined by Chacham and Louie [1991] who confirm for hydrogen the general tendency of the LDA to quite seriously underestimate band-gaps in any chosen structure. It is a significant impediment in a problem where at the one-electron level the major focus is precisely on the closing of a gap.

In seeking the lowest energy structure the dynamics of the protons must play a significant role [Straus and Ashcroft, 1977; Surh et al., 1993]. The vibron zero-point energy amounts to about 10.0 mRy per proton; the zero-point energy from phonons is rising rapidly with density [Hemley and Mao, 1994] and at $r_s \sim 1.5$ raises this figure to about 11.2 mRy per proton. The associated zero-point amplitudes are quite substantial and tend to average over details that might be considered significant in static lattice estimates. If electronic and lattice dynamical energies are actually treated on the same footing, there appears to be a tendency at high density to once again favor more symmetric structures [Straus and Ashcroft, 1977; Natoli et al., 1993]. Whatever structure emerges from minimization procedures, the results are now becoming subject to a constraint emerging from a low temperature equation of state which is essentially fixed (for $\rho/\rho_0 \leq 6$) by experiment [Mao and Hemley, 1994]. Beyond these densities, and making the assumption that there are no further significant transitions, it can be linked to density functional calculations [Ashcroft, 1990]. Figure 3 shows an example. To arrive at a useful impression of the sheer scale of compressional energies associated with the hydrogen problem, the shaded area represents an equivalent $p - V$ work equal to the binding energy per proton of the hydrogen molecule.

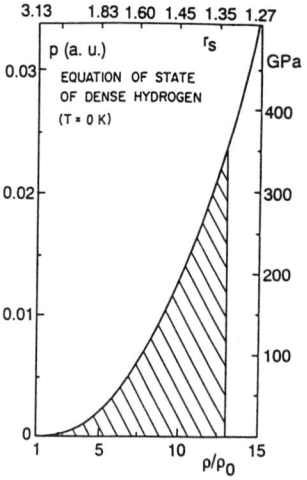

Figure 3. Equation of state ($T=0$ K) in a paired state of hydrogen. For $\rho/\rho_0 \leq 6$ the curve reflects an amalgamation of experimental data [Hemley and Mao, 1994]. For densities higher than this, where kinetic energy dominates the electron energy, the curve matches onto total energy results utilizing density functional theory [Ashcroft, 1990]. Per proton, the pressure-volume work is given by $\rho_0^{-1} \int_0^{\rho/\rho_0} p(x)dx/x^2$. The shaded area gives, as a guide, the compressional equivalent of the binding energy per proton in an isolated molecule.

For the most part the determinations of the band-gap closure condition utilize highly developed numerical procedures assuming static crystalline structures. However, it is also possible to obtain a plausible bound analytically, at least for symmetric structures. This can be demonstrated by suppressing for the moment the rotational physics and taking the $Pa3$ structure described above as an example; as noted it is a simple cubic with an 8-point basis $\{\vec{b}_i\}$. But it may also be viewed as a hexagonal structure (sectioning the cubic structure along (111) planes) and when this is done the $Pa3$ structure consists of hexagonal arrays with proton pairs both tilted *and* rotated. Then in a plane-wave basis equation (9) reduces to the secular equation for single-particle energies $\varepsilon_{\vec{k}}$

$$\det\left\{(\varepsilon_o(\vec{k}+\vec{K}')-\varepsilon_{\vec{k}})\delta_{\vec{K}'\vec{K}''}+V_{\vec{K}'\vec{K}''}\right\}=0 \qquad (11)$$

where $\varepsilon_o(\vec{k})=\hbar^2/k^2/2m_1=(a_1k)^2$ Ry, is a free electron energy, and $\vec{K}'$ is a reciprocal lattice vector ($\vec{K}'=(2\pi/a)(l,m,n)$). In (11) the off-diagonal matrix elements are

$$V_{\vec{K}'\vec{K}''}=(N_2/V)S(\vec{K}'-\vec{K}'')v(\vec{K}'-\vec{K}'')=(3/4\pi r_s^3 a_1^3)S(\vec{K}'-\vec{K}'')v(\vec{K}'-\vec{K}'') \qquad (12)$$

where $v(\vec{k})$ is the Fourier transform of the interaction of a an electron with (i) a proton $(-4\pi e^2/k^2)$, *and* (ii) the self consistently determined electronic charge distribution around the proton $((4\pi e^2/k^2)f(k))$; $f(k)$ will be discussed below. The quantity $S(\vec{K}'-\vec{K}'')$ is the geometric structure factor per basis point

$$S(l,m,n)=\frac{1}{4}\Big\{\cos 2\pi\alpha(l+m+n)+(-1)^{l+m}\cos 2\pi\alpha(-l+m+n)$$

$$+(-1)^{m+n}\cos 2\pi\alpha(l-m+n)+(-1)^{n+l}\cos 2\pi\alpha(l+m-n)\Big\} \qquad (13)$$

where, again, $\alpha=d(r_s)/\sqrt{3}a=\{(3/4\pi)^{1/3}/2\sqrt{3}\}(d/a_1r_s)$. The simple cubic with lattice constant $a/2$ (referred to above as a limiting case of $Pa3$) corresponds to the choice $\alpha=1/4$. A high symmetry structure is also obtained with $d=\sqrt{3}a/8$, or $\alpha=1/8$ ($2\pi\alpha=\pi/4$). It follows that for $\vec{K}=(2\pi/a)(1,0,0)$ (and star) the structure factor vanishes; the same is true for $\vec{K}(2\pi/a)(1,1,0)$ (and star). However for $\vec{K}=(2\pi/a)(2,0,0)$ (and star) we have $S(2,0,0)=2\cos^2 2\pi\alpha-1$ while for $\vec{K}=(2\pi/a)(1,1,1)$ and (star) $S(1,1,1)=\cos^3 2\pi\alpha$.

Accordingly, the *lower* bands of the band-structure can be represented by an equivalent face-centered-cubic description where $V_{111}=(3/4\pi r_s^3 a_1^3)S(111)v(111)$ and $V_{200}=(3/4\pi r_s^3 a_1^3)S(200)v(200)$. An immediate indication of the consequences of the especially symmetric disposition of the proton pairs in the $Pa3$ structure follows from an examination from the structure factor per atom obtained from a *rotational* average of a two-point basis $\vec{b}$ and $-\vec{b}$ ($|\vec{b}|=d$) situated on each site of a face-centered-cubic lattice. For a reciprocal lattice vector $\{\vec{K}\}$ this is $j_o(Kd)$ and we find the following close correspondence: (i) for $\vec{K}=(2\pi/a)(1,1,1,), S(\vec{K})=0.65$ and $j_o(Kd)=0.63$, and (ii) for $\vec{K}=(2\pi/a)(2,0,0), S(K)=0.5$ and $j_o(Kd)=0.53$. This proximity is significant in establishing a coherent potential type of approximation to the more general problem of determining an equivalent band-structure for a system exhibiting very obvious rotational dynamics (see below). If $d$ takes the value $\sqrt{3}a/8$ then $S(2,0,0)$ is also zero (but $|S(2,1,0)| = |S(1,1,1)|$). In this case the structure possesses a Jones zone with 32 faces.

To directly incorporate the effects of pressure (or density) it is convenient to introduce a unit of wave vector $k_X = 2\pi/a = (3\pi^2/4)^{1/3}/a_1 r_s$, and a corresponding unit of energy, $\varepsilon_X = \hbar^2 K_X^2/2m_1 = (3\pi^2/4)^{2/3}/r_s^2$ Ry. The form of the matrix element is then

$$\bar{V}_K(r_s) = -6r_s(4/3\pi^2)^{4/3} S(\bar{K})(1-f(\bar{K}))/\bar{K}^2 = -0.416 r_s S(\bar{K})(1-f(\bar{K}))/\bar{K}^2, \quad (14)$$

where $\bar{V} = V/\varepsilon_X$ and $\bar{K} = K/k_X$. In the same units the energy at the $X$-point of the first Brillouin zone for the FCC structure is then 1; for the $L$ point it is 3/4, etc. Viewed from this standpoint of the FCC zone, the band-overlap found by Friedli and Ashcroft [1977] involves the $X$ and $L$ points. To within second-order corrections the condition for the onset of band overlap as a function of density is now very easy to establish: the lower level at $X$ (i.e. $1 - |\bar{V}_{200}(r_s)|$) must rise to surpass the upper level at $L$ (i.e. $\frac{3}{4} + |\bar{V}_{111}(r_s)|$) i.e.

$$|\bar{V}_{111}(r_s)| + |\bar{V}_{200}(r_s)| = 1/4. \quad (15)$$

To determine the solution $(r_s)$ to (15) it is necessary to obtain the function $f(k)$ reflecting the response charge. Two approaches are possible, both yielding similar results. The first is to continue with the expectations suggested by Figure 1, namely that for the reciprocal lattice vectors of interest the response charge density is largely controlled by the cusp theorem. It is then easy to see that given $\alpha = 2\gamma/a_1$,

$$f(k) = \left\{\frac{\alpha^2}{k^2 + \alpha^2}\right\}^2 = \left\{\frac{r_s^2}{r_s^2 + \bar{k}^2(3\pi^2/4)^{2/3}/4}\right\}^2,$$

since $\gamma$ is very nearly 1 for $r_s$ values in the vicinity of 1. The second is to obtain $f(k)$ from linear response, namely $f(k) = k_{TF}^2/(k^2 + k_{TF}^2)$ in the Thomas-Fermi approximation. In the units introduced above $f(k)$ is then

$$f(k) = r_s/(r_s + \bar{k}^2 \pi^2/4^{4/3}).$$

For either the cusp-form or linear-response form of $f$, the solution to (15) for the onset of band overlap is $r_s \approx 1.8$. But condition (15) clearly neglects the additional degeneracies associated with the higher zones (which are not described by FCC, but by the argument given above, cannot depart too far from this). It also neglects the general effects of higher bands which lead to band-repulsion (i.e. further band gap widening), and to a rescaling of all free-electron energies as discussed above. Once again, a bound on this can be obtained by simply using the effective mass values expected for the monatomic case ($m^*/m \sim 1.2$ for $r_s \sim 1.5$). Then (15) becomes

$$|\bar{V}_{111}(r_s)| + |\bar{V}_{200}(r_s)| = m^*/4m$$

and the overlap now occurs at $r_s = 1.5$, quite close to the value found by Friedli and Ashcroft [1977]. But as we shall now see, before the actual closing occurs, there is the possibility of further instability, but one resulting from effects now going decidedly beyond the independent-particle approximation.

From the independent-particle viewpoint (with an even number of electrons per unit cell of assumed crystalline structures) the systematic decline in density leads to a progressive loss in the Fermi surface area of a metallic state through the eventual closing of a 'negative gap' $\varepsilon_G$ (see Figure 4a). For dense hydrogen most band-structure calculations to date locate the associated band extrema at quite distinct points in the Brillouin zone (separated, say, by $\vec{k}_o$). Depending on the average structural arrangement of protons, there can be several such extrema, with different degrees of isotropy. If residual Coulomb interactions are now restored to this picture, collective excitations can arise. But if conditions (here changes in density as impelled by pressure) should cause one of these collective modes to decline, and indeed eventually vanish, then an instability will be the result. Viewed from the insulating side (Figure 4b) the same transition is seen as the condensation of excitons which will proceed when the exciton binding energy and gap energy are coincident [Knox, 1966]. Thus, incorporation of Coulomb interactions into these one-electron descriptions results in a picture where electrons are seen to be pairing with holes, and then condensing to form a new many-particle state of lower overall energy. This description is very much akin to the BCS theory of the pairing of electrons with electrons under the action of *effective* attractive interactions, except that here in this excitonic case the resulting state lacks off-diagonal long-range order. The original idea that additional many-body states might interpose themselves in a gap-closing metal-insulator transition (and possibly delay it) was first advanced by Mott [1961]. Since then the subject has been extensively reviewed (see for example, Halperin and Rice, [1968]).

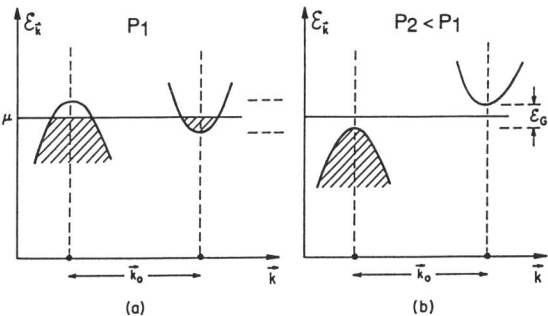

Figure 4. Band gap opening as a function of declining pressure in a one-electron picture. The left panel ($p=p_1$) depicts electron and hole bands with a 'negative gap' (and a Fermi surface). The right panel ($p=p_2 < p_1$) shows an emerging gap $\varepsilon_G$. Restoration of Coulomb interactions can lead to an instability and charge distortion.

According to band-theory, exactly this type of gap opening is predicted in dense hydrogen, as pressure is lowered. Where (in density or pressure) it should occur is not yet known, but for symmetric structures it appears that compressions in the vicinity of $\rho/\rho_0 \gtrsim \sim 8$ are required. An important point to recognize is that because of the indirect nature of the gap, the onset of any gap-closing instability is necessarily accompanied by a charge-distortion with a wave-vector characteristic of $\vec{k}_o$. It may not be obviously commensurate with the symmetry of the lattice, per se, but an interesting feature associated directly with the rotational character of hydrogen is that with very little energy penalty, small adjustments in the orientational disposition of proton pairs can clearly occur to take maximum benefit of this onset. And the reason they may choose to do so is not difficult to see. For, accompanying any such charge distortion (or wave) in the electron system must be a local imbalance in the electronic response charge *around* proton pairs and subsequent orientation can take place to maximize electrostatic gain. Put another way, prior to gap closure a state of mildly ionic character in hydrogen must be the consequence of any gap-closing instability. In a Hubbard model of the hydrogen molecular solid, partial occupancy of one of the single twice-occupied orbitals is beginning to take place. Though there is plainly a level of conjecture in projecting the consequences of this instability it is interesting to observe that the end result of such a process fully exploiting the emerging Madelung energy is yet another broken-symmetry state where the hydrogen ions $H^-$ and $H^+$ now play an essential role. En route the emergence of partially ionized *paired* units is suggested by the above argument, and this charge separation will be subject to dynamics on a temporal scale set by the vibron frequency. These matters are taken a little further in the Appendix.

## V. FLUCTUATION AND STRUCTURE

Infra-red (IR) activity is expected to be associated with the broken symmetry electronic states discussed above and in the Appendix, and indeed the recent experimental observation of strong IR absorption at the vibron anomaly might be seen as a point in its favor. From the band-structure results, it is in the neighborhood of this anomaly that a single-particle gap *should* be closing. If so then the possibility obviously arises that the gap energy will actually cross the vibron energy as noted also by Mao and Hemley [1994]. With a strong linear coupling between the two systems this confluence can lead to a a resonant mixing. In this situation the possibilities of new many-body states becomes somewhat richer. Perhaps the most interesting is the direct passage from an insulator (or semiconductor) to a superconductor, the first example of such a transition evidently being proposed by Little [1964]. It is difficult to give any credible estimate of a transition temperature, but we note that $\hbar\omega_{vib}$ is 4000 K, and ordering temperatures are typically $10^{-2}$ of the exchange boson energy scale. Here the physics is controlled by electron-optical phonon coupling. However, it is unlikely that any resulting *gap* in the pairing-structure would lead to infra-red activity in the range seen.

But the electronic structure and the vibron come together in a somewhat more direct way. As observed above, a ground-state electronic trace of (1) leads to an effective Hamiltonian governing the subsequent protonic motion. It has traditionally been separated into intra- and inter-molecular parts, though as noted the division is more intuitive than rigorous. The first point to note is that if electronic excitation energies are becoming comparable to the emerging scale of proton excitation energies, then the standard adiabatic approach for the determination of electron response (and then effective intra- and inter-molecular potentials) will certainly stand correction. It would be physically more appropriate to consider elementary excitations that possess a polaronic character.

The second concerns the likely progression of the intra-molecular potential which is used to determine the magnitude of the vibron, and its excitations, as probed by Raman and IR. The shaded area on Figure 3, which gives a measure of the compressional work required to match the binding energy per proton of a free molecule, gives strong impetus to the notion that the effective pairing interaction controlling proton motion within a pair will steadily diminish as density increases. Exactly this trend can be seen with the aid of a Heitler-London model and the use of the cusp condition. Figure 5a [Ashcroft, 1990] shows a progression of such curves; Figure 5c shows the progression of the corresponding depths with $r_s$, suggesting immediately a vibrational instability at $r_s \gtrsim 1.3$ (or $p \sim 320$ GPa). But before this there arises yet another new possibility, and it is deeply linked to the requirements of the symmetry of the structure. With the ideal $Pa3$ structure specifically in mind (we will return to the rotational issue shortly) it is clear that as a function of increasing adiabatic separation of protons a symmetric arrangement can eventually be reached (this matter is discussed further by Ashcroft [1991]) with, as stated above, the possibility of subsidiary extrema en route particularly if they coincide with other high symmetry structures. Thus along paths connecting equivalent degenerate arrangements, the situation must eventually be as represented in Figure 5b. But for small enough barriers this leads at once to the development of correlated tunnel-splitting; in Raman effect this would be associated with an overlapping of signals and in infrared absorption to an increasingly large signal associated with extensive proton displacements (fluctuations into what might well be local minima associated with states of putative metastable metallic character). If the rotational physics is now appended to this picture (through additional but smaller effects on the critical barrier height) the result is straightforward to predict. Some pairs will be in conditions favoring tunneling; some will be delayed. Thus at any instant we would expect a representative configuration to display a fraction of reasonably well defined pairs and the balance as disproportionated protons. Configurations of this character are evidently seen in the simulation studies of Hohl *et al.* [1993]. In dense deuterium it would require a somewhat smaller value of the rapidly diminishing barrier to achieve the same effect, that is to say, a somewhat higher pressure. Since the structure at these densities is not known it certainly remains possible that for symmetry reasons *induced* IR activity could still account for the observations [Zallen, 1968], but as Hemley *et al.* [1994] point out, the magnitude of the absorption has more the character that is expected of charge transfer. Indeed they also propose an interesting charge-transfer process involving hydrogen molecules rather than the limiting ionic case suggested by Baranowski (see Appendix A).

It is by now clear that except for the last almost all of the arguments given above are largely framed within the common assumption of perfect crystalline order. But the experiments contradict this; it is an evidently well established experimental fact that even at 10-fold compression hydrogen continues to support excitations of a fully rotational character. For deuterium beyond about 5-fold compression, the excitations probably correspond to large amplitude librations about time average directions that, however, are not necessarily guaranteed to form an ordered array (the system might well conform to a dynamic orientational glass). Low frequency Raman measurements have led to an interpretation that in hydrogen at least angular momentum continues to remain a plausibly good quantum number (in the ground state $J = 0$ is assumed dominant). The electronic consequences of this are significant for the assignment $J = 0$

means that the directions of pair axes are *uncorrelated* in their motions which in turn means that the disordering potential for the electron problem is one with an exceedingly short correlation length.

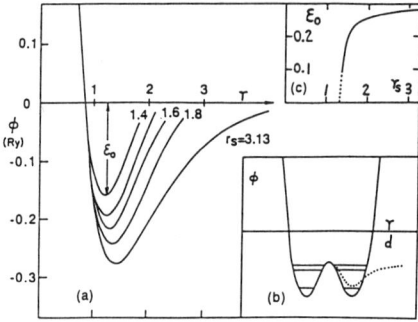

Figure 5. Progressive reduction of an effective intra-molecular potential $\phi$ with increasing density, and obtained within a Heitler-London approximation [Ashcroft, 1990]. (b) Progression of the corresponding depths $\varepsilon_o$ as a function of density and displaying an expected limit to the existence of pairing dynamics ($p \gtrsim 320$ GPa). (c) Developement of tunneling states when the effective potential $\phi$ (now displaying the double well structure reflecting crystal symmetry) binds at most two vibrational modes. Non-monotonic or multiple well structure is also a possibility and is indicated by the dotted line.

In establishing an independent electron viewpoint of this problem (i.e. prior to restoring the many-body effects leading to additional charge distortion) two major electronic manifestations of the rotational physics can be envisioned. The first is an obvious site-diagonal disorder; the second is site off-diagonal disorder most easily revealed by viewing the problem in a tight-binding framework where overlaps of electronic wavefunctions assigned to molecular pairs are now clearly dependent on relative orientation. Together their presence would normally be expected to lead to two primary classes of electron states, as has been argued earlier [Ashcroft, 1993]. These are (i) an itinerant class, forming the bulk of the wide-band states, and (ii) a localized class forming states mainly in the mid-gap region it is exists. These will be separated in general by two mobility edges, and it is quite reasonable to expect that as pressure increases an eventual confluence of the two will ensue. This leads to the question of key importance for the one-electron problem: do the effects of rotational disorder act to delay or advance a band overlap transition expected from predictions based on the assumption of a true crystalline state? If this issue can be settled in the independent particle context it is then necessary to revisit the question when residual Coulomb interactions are reintroduced (as discussed above).

The first issue can be treated [Edwards and Ashcroft, 1994] within the coherent potential approximation (CPA) which is not at all restricted to the effects of weak disorder (for a review, see Elliot *et al.* 1974). The intent is to find the best crystalline electronic mean-field problem that is physically equivalent to the rotationally disordered problem. In view of the experimental assignment of $J = 0$ to each site, it is plausible to take this to be a virtual crystal with a spherically symmetric basis. (We have seen earlier in the case of $Pa3$ the proximity of its structure factors to that of a rotationally average basis on FCC, suggesting that a 4-pair cluster in the molecular CPA should do very well.) The optimum choice of the potential to be associated with each site in the virtual crystal is determined by demanding that on average (that is to say, when a complete rotational average is carried out) there is no scattering from the aggregate perturbation. The requirement is equivalent to demanding that the average t-matrix associated with the difference in the actual potential, and the proposed virtual crystal, shall vanish. In general it is known from the application of the CPA to other problems involving disorder in an otherwise crystalline environment that band widths tend to narrow and band gaps tend to increase. It is also known that CPA tends to neglect interference between scattering from different sites. However, in this problem, it should be reiterated that the disorder possesses an exceedingly short correlation length so that the standard prediction of CPA (namely a *sharp* band edge, and hence in fact only one class of states) may be close to the actual physics.

The route to the Wigner-Huntington monatomic state of hydrogen as a function of *increasing* pressure has therefore proven to be far from direct, and also surprisingly complex given the simplicity of (1). Near the projected metal insulator transition we certainly expect fluctuations to play an important role in a system which is, after all, highly quantum. Martensitic transformations occur in the light alkali metals to which the transformed state of hydrogen may eventually be likened. Solid-solid transitions in a monatomic state *can* therefore be expected, and at relatively modest temperatures. To which, of these states the molecular solid might eventually transform at a given temperature is, at present, a matter of some uncertainty. Electronically the rotational aspects of this problem appear to manifest themselves in quite subtle ways. In the many-body viewpoint the charge distortion argument advanced above will have to be applied on a configurational basis; on proton time scales it leads to a migration of ionicity. Since the associated charge resides on pairs undergoing concomitant vibrational motion it also means the possibility of infrared activity on a completely corresponding time-scale. Yet to be resolved is whether these conclusions are consistent with the apparent growth in infrared activity with increasing pressure beyond the onset. This appears to imply a corresponding increase in the degree of static charge transfer, a process that surely must face an ultimate limit.

## APPENDIX: Ionic and Partially Ionic States

The idea that a fully charged separated ionic phase might *eventually* form in dense hydrogen has already been put forward by Baranowski [1992]. It is interesting to examine the basic energetics of Baranowski's proposal. The point is that between the molecular (covalent) phase and fully dissociated metallic phase there may be a phase which is ionic, and which we might liken to LiH. Now the energy of the *metallic* monatomic phase can be written (in Ry/electron)

$$a_k/r_s^2 - (a_e + a_m)/r_s + \epsilon_c(r_s) + \epsilon_{bs}(r_s) + \epsilon_p(r_s) \qquad (16)$$

where the first pair of terms dominate. The first is kinetic energy where $a_k = (3/5)(9\pi/4)^{2/3}$. The second consists of exchange ($a_e = (3/2\pi)(9\pi/4)^{1/3}$) and electrostatic Madelung energy ($a_m$). For simple symmetric structures $a_m \sim 1.79$. In (16) $e_c$ is a correction for correlation and $\varepsilon_{bs}$ for electron inhomogeneity (the band structure energy) which is also relatively small. The much smaller term $e_p(r_s)$ is the energy bound up in proton dynamics. As is well known, (16) has a metastable minimum at $r_s \approx 1.63$; at this minimum the first two terms lead to an energy $-0.84$ (separated neutral atoms would lead to -1).

Now consider Baranowski's ionic phase $H^+H^-$ (a state of radically broken symmetry which we might refer to as "protium hydride") which possesses an overall gap yet (as is the case in LiH) also possesses relatively free-electron-like bands originating with overlap of neighboring $H^-$ wave functions. We imagine building such a phase by removing electrons from *half* of the atoms (at a cost of 1 Ry per atom) and placing them on the remaining half (at a benefit $e_a$, the electron affinity, augmenting the -1 Ry bound state energy). In this case the energy per electron is again

$$a_k/r_s^2 - (a_e + \bar{a}_m)/r_s - \frac{1}{2}(1 + e_a) + e_c(r_s) + \varepsilon_{bs}(r_s) + e_p(r_s). \qquad (17)$$

The key difference resides in the Madelung energy, and the internal energies $(1 + e_a)$ of the $H^-$ ions. If we have a rocksalt structure with lattice constant $a$ then the electrostatic energy is usually given per ion pair as $-\alpha_m e^2/r_{nn}$ where $r_{nn}$ is the nearest neighbor distance $(a/2)$ and $\alpha_m$ is the conventional Madelung constant. Using this we have $\bar{a}_m = \alpha_m(3/4\pi)^{1/3}$ ($\approx 1.084$), and we find that (17) is minimized at $r_s = 2.21$. In this case the first three terms lead to an energy $-0.98$ Ry, i.e. at a metastable minimum evidently *lower* than the fully dissociated metal, assuming reasonably comparable values for the correlation, band-structure and proton-dynamics corrections. As noted, this is an 'endpoint' state, and as such is not consistent with the observed rotational characteristics of dense hydrogen (which might, however, be more representative of intermediate points on a path to full ionicity).

If these estimates can be justified by more careful calculations then Baranowski's suggestion, that an ionic phase may interpose itself between molecular and metallic may merit further investigation. Notice that the idea of stabilizing such a phase through the basic terms in (17) can be applied at once to LiH, the difference being that Li$^+$ is not a point ion but has a core ($r_c \approx 0.60 a_1$) inside of which valence electrons are largely excluded. In setting up the Madelung energy problem, this exclusion leads (in the grouping of all $q \to 0$ terms in a reciprocal space formulation for an overlapping system) to an additional term $a_3/2r_s^3$ in (17) [Ashcroft and Langreth, 1967] where $a_3 = 3(r_c/a_1)^2$. Minimizing this energy with respect to $r_s$ leads to a density $r_s = 2.4$ where the experimental value is 2.38. This average electron density is *higher* than that found in metallic Li itself, and in Mg, Cd, and Hg. That LiH possesses reasonably free-electron-like bands has already been established by Baroni et al. [1985]. It is interesting to observe for the hydride the crucial role being played by electrostatics in further stabilizing in this dense environment the $H^-$ ion which is large in extent, which is bound at the seemingly fragile level of 0.7 eV, and which possesses *not a single bound excited state*.

As noted we are referring here to a fully ionic state, one whose spatial conformation is not obviously in agreement with the reported persistence of rotational order. In this context it is interesting to observe that for the $Pa3$ structure, the wave vector $k_o$

characteristic of the onset of a charge-density instability, has a magnitude $(2\pi/a)\sqrt{3}/2$ and therefore equivalent to a wavelength $(2/3)\sqrt{3}a$. This is just $(2/3)$ of the cell diagonal (which accommodates 3 hexagonal layers) and it follows that successive minima and maxima of the charge distortions fall on alternating hexagonal layers, and hence, in a hexagonal-cell viewpoint, on neighboring pairs. Provided an overall band-gap is maintained it is anticipated that if distortions associated with the additional symmetry related band maxima and minima are included, then an arrangement of partially ionic character will emerge where pairing is maintained. (The possibility of an eventual metallic state based on $H_2^+$ pairs was already considered in 1968 [Ashcroft, 1968].) In effect the unit cell is doubled; the argument just given can clearly be repeated for the subsequent larger cell, and so on. It leads to a picture of *static* ionic distortions, and is complementary to the dynamic model introduced recently by Hemley *et al.* (1994).

## ACKNOWLEDGMENTS

This work has been supported by the National Science Foundation which is gratefully acknowledged. I wish to thank Professor M. Teter, Professor A. Ruoff, Dr. R. J. Hemley, and Mr. Byard Edwards for many helpful discussions.

## REFERENCES

Ashcroft, N. W. and D. C. Langreth, 1967, Phys. Rev., **155**, 682.
Ashcroft, N. W., 1968, Phys. Rev. Lett., **21**, 1748.
Ashcroft, N. W., 1990, Phys. Rev. B, **41**, 10963.
Ashcroft, N. W., 1991, *Frontiers of High Pressure Research*, (New York, Plenum Press), p. 115.
Ashcroft, N. W., 1993, J. Non-Cryst. Solids, **156-158**, 621.
Baranowski, B., 1993, Polish Journal of Chemistry, **66**, 1737.
Barbee, T. W., A. Garcia, M. L. Cohen, and J. L. Martins, 1989, Phys. Rev. Lett., **62**, 1150 (1989).
Bardeen, J., 1938, J. Chem. Phys., **6**, 367.
Baroni, S., G. Pastori Parravicini, and G. Pezzica, 1985, Phys. Rev. B, **32**, 4077.
Brovman, E. G., Yu. Kagan, and E. Xholas, 1972, JETP, **34**, 1300.
Carlsson, A. E. and N. W. Ashcroft, 1982, Phys. Rev. B, **25**, 3474.
Chacham, H. and S. G. Louie, 1991, Phys. Rev. Lett., **66**, 64.
Chihara, J., 1985, J. Phys. C, **18**, 3103.
Cui, L., N. H. Chen, S. J. Jeon, and I. F. Silvera, 1994, Phys. Rev. Lett., **72**, 3048.
Ebina, K. and H. Miyagi, 1989, Phys. Lett. A, **142**, 237.
Edwards, B. and N. W. Ashcroft, 1994, to be published.
Elliot, R. J., J. A. Krumhansl, and P. L. Leath, 1974, Reviews of Modern Physics, **30**, 75.
Friedli, C. and N. W. Ashcroft, 1977, Phys. Rev. B, **16**, 662.
Halperin, B. I. and T. M. Rice, 1968, Solid State Physics, **21**, 115.
Hemley, R. J., H. K. Mao, and J. F. Shu, 1990, Phys. Rev. Lett., **65**, 2670.
Hemley, R. J., Z. G. Soos, M. Hanfland, and H. K. Mao, 1994, Nature, **369**, 384.
Hemley, R. J. and H. K. Mao, 1994, Rev. Mod. Phys., **66**, 671.
Hohenberg, P. and W. Kohn, 1964, Phys. Rev., **136**, B864.

Hohl, D., V. Natoli, D. M. Ceperley, and R. M. Martin, 1993, Phys. Rev. Lett., **71**, 541.
Ichimaru, S., 1994, *Statistical Plasma Physics, Vol. II: Condensed Plasmas*, (New York, Addison-Wesley).
Kato, T., 1957, Commun. Pure Appl. Math, **10**, 151.
Kaxiras, E., J. Broughton, and R. J. Hemley, 1991, Phys. Rev. Lett., **67**, 1138.
Kaxiras, E. and Z. Guo, 1994, Phys. Rev. B, **49**, 11822.
Keesom, W. H., 1921, Phys. Zeits, **XXII**, 129.
Kimball, J. C., 1973, Phys. Rev. A, **7**, 1648.
Knox, R. S., 1966, Solid State Physics, **Suppl. 5**, 100.
Kohn, W. and J. M. Luttinger, 1965, Phys. Rev. Lett., **15**, 524.
Kohn, W. and L. J. Sham, 1965, Phys. Rev., **140**, A1133.
Little, W. A., 1964, Phys. Rev. A, **134**, 1416.
Maggs, A. C. and N. W. Ashcroft, 1987, Phys. Rev. B, **36**, 7586.
Mao, H. K. and R. J. Hemley, 1992, American Scientist, **80**, 234.
Mao, H. K. and R. J. Hemley, 1994, Rev. Mod. Phys., **66**, 671.
Miyagi, H. and T. Nakamura, 1967, Prog. Theor. Phys., **37**, 641.
Moshary, F., N. H. Chen, and I. F. Silvera, 1993, Phys. Rev. B, **48**, 12613.
Mott, N. F., 1961, Phil. Mag., **6**, 287.
Moulopoulos, K. and N. W. Ashcroft, 1990, Phys. Rev. B, **41**, 6500.
Moulopoulos, K. and N. W. Ashcroft, 1991, Phys. Rev. Lett., **66**, 2915.
Nagara, H. and T. Nakamura, 1992, Phys. Rev. Lett., **68**, 2468.
Natoli, V., R. M. Martin, and D. Ceperley, 1993, Phys. Rev. Lett., **70**, 1952.
Ramaker, D. E., L. Kumar, and F. E. Harris, 1975, **34**, 812.
Silvera, I., 1980, Rev. Mod. Phys., **52**, 393.
Straus, D. and N. W. Ashcroft, 1977, Phys. Rev. Lett., **38**, 415.
Styer, D. E. and N. W. Ashcroft, 1984, Phys. Rev. B, **29**, 5562.
Surh, M. P., T. W. Barbee, and C. Mailhiot, 1993, Phys. Rev. Lett., **70**, 4090.
Svane, A. and O. Gunnarsson, 1990, Solid State Commun., **76**, 851.
Thouless, D. J., 1974, Phys. Rep., **13**, 93.
Wigner, E. and H. B. Huntington, 1935, J. Chem. Phys., **3**, 764.
Zallen, R., 1968, Phys. Rev., **173**, 824.

## VII.2

# PROGRESS ON HYDROGEN AT ULTRAHIGH PRESSURES

Russell J. Hemley and Ho-kwang Mao

Geophysical Laboratory and Center for High-Pressure Research
Carnegie Institution of Washington, 5251 Broad Branch Rd., N.W.
Washington, D.C. 20015, U. S. A.

We examine recent progress and current challenges in experimental studies of hydrogen to static pressures of 250 GPa (2.5 Mbar). Three phases persist to megabar pressures: phase I, the high-temperature, low-pressure phase; phase II, a low-temperature, high-pressure phase; and phase III, the high-pressure phase stable above 150 GPa. Recent data on the infrared response in phases II and III, specifically measurements on $H_2$, $D_2$, and HD, are examined. Infrared spectra reveal orientational and structural differences between $H_2$ and $D_2$ in phase II. All isotopic systems exhibit a large increase in vibron absorption in phase III, which is indicative of major changes in structure and bonding properties, including charge transfer. The I-II-III triple point is found in both $H_2$ and $D_2$; in addition, detailed Raman studies of $D_2$ show evidence for two additional invariant points along the phase III line. The vibrons of phase III show characteristic order-parameter behavior as a function of temperature.

## I. INTRODUCTION

Delineating the range of stability of molecular bonding, in addition to the orbital shell structure of atoms, in highly compressed matter represents an important challenge in modern condensed-matter science. Since the first calculations of dense plasmas [Fowler, 1926], the behavior of the hydrogen under extreme conditions has played a key role in understanding this phenomenon. At low density, hydrogen is a quantum molecular solid with the molecules in free rotational states. With increasing pressure, the molecules are expected to lose their orientational freedom, and the molecular solid may become metallic [Mao and Hemley, 1994]. At the highest pressures, this molecular solid is predicted to transform to a non-molecular, or atomic, metal [Wigner and Huntington, 1935]. The behavior of the material at these high pressures is also of great importance for understanding stellar and planetary interiors.

The extraordinary range of compression over which hydrogen can be studied with new static high-pressure techniques affords an unprecedented, if not unique, view on the variation of chemical bonding in condensed matter. The observation of spectroscopic signatures of the molecular bond to multimegabar pressures (> 250 GPa [Hemley and Mao, 1988]) demonstrates the persistence of the H-H covalent interaction to very high densities ($\rho/\rho_0 > 10$, where $\rho_0$ is the zero-pressure density of the solid as $T \to 0$). Despite the tenacity of the molecular bond at the highest pressures, recent experimental work demonstrates that several transitions in the molecular solid occur with increasing pressure. The results are evidence for a sequence of symmetry breaking transitions associated with the evolution of the system ultimately toward a dense plasma state. Moreover, suggestions of new phenomena are evident from recent measurements [Mao and Hemley, 1994].

Here we examine recent progress in the experimental study of hydrogen at ultrahigh pressures. A general review of studies over a range of pressures is given by Mao and Hemley [1994]; here we focus on very recent results and those obtained exclusively at megabar pressures ($> 100$ GPa). Three phases of hydrogen are important under these conditions: phase I, the high-temperature, low-pressure phase; phase II, a low-temperature high-pressure phase; and phase III, the high-pressure phase stable above 150 GPa. The latter is characterized by a remarkable enhancement in absorption of the intramolecular stretching mode (vibron). In the following section, we briefly discuss experimental techniques, specifically recent developments in high-pressure devices and spectroscopic methods. We then review the recent infrared measurements on hydrogen, new data on $D_2$ and HD, and detailed Raman and infrared studies of the phase diagram. This is followed by a brief summary of observed spectroscopic anomalies in such studies. Finally, relationships among these observations and current issues are discussed in the concluding section.

## II. EXPERIMENTAL TECHNIQUES

The experiments were performed using diamond-anvil cell methods. With the diamond cell, numerous types of studies can be carried out to determine physical and chemical properties over an increasingly wide range of pressures (i.e., to $> 300$ GPa) [Jayaraman, 1983; Mao, 1989; Mao and Hemley, 1994]. In recent years, we have developed new classes of diamond cells which permit routine experiments on hydrogen and other materials in excess of the 200 GPa and at variable temperature. An example of one such device, a modified multimegabar Mao-Bell cell is shown in Fig. 1. Its salient features include beveled diamonds, disc seats for the anvils, and a double lever arm. The new anvil mount is an example of the incremental, but significant improvements implicit in the new designs; such a change is particularly well-suited for use with ultrahigh-pressure studies because of the increased ease and stability of alignment of beveled anvils with ultrasmall tips (to $< 20$ µm).

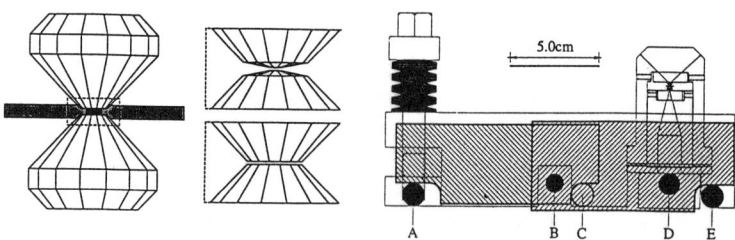

Fig. 1. New diamond anvil cell design [Mao *et al.*, 1994]. Left: (top) flat anvils used for lower pressure studies; (bottom) beveled anvils. Center: opposed anvil configuration. Right: two-stage lever arm device.

Studies of hydrogen at megabar static pressures have been carried out using various spectroscopic techniques. Measurements of the vibrational spectra provide information on the state of bonding in the material under pressure, and are sensitive to the crystal structure and degree of orientational order. The vibrational excitations for hydrogen are shown in Fig. 2. Spectroscopic techniques document the existence of phase transitions and can be used to determine the phase diagram over a wide range of pressures and temperatures. An example is the discovery of phase III above 150 GPa, which is characterized by a discontinuous shift in the Raman-active vibron [Hemley and Mao, 1988; Hemley and Mao, 1989; Lorenzana et al., 1989].

The development of analytical techniques utilizing high-intensity synchrotron radiation sources at short and long wavelengths has constituted a breakthrough of singular importance for high-pressure research. For example, synchrotron x-ray diffraction experiments performed to 42 GPa ($\rho/\rho_0 = 6$) have established that the lattice structure of hydrogen and deuterium is hexagonal close-packed [Mao et al., 1988; Hemley et al., 1990; Hu et al., 1994]. Synchrotron radiation provides up to four orders of magnitude higher brightness in the infrared than does conventional infrared sources [Williams, 1990], and thus it is ideally suited for study of microscopic samples in the diamond cell (which are typically less than 10 picoliters in volume above 100 GPa). Because of the broad spectral distribution, the source is also especially useful for studying bands over a wide wavelength range as well as broad excitations in such samples. The recent development of synchrotron infrared techniques for ultrahigh-pressure spectroscopy, and in particular of hydrogen, has been highly significant [Hanfland et al., 1992, 1993]. As a result of coupling of internal vibrations of the hydrogen molecules in condensed phase, the vibrational excitation is dispersed into a range of energy forming a band. The IR vibron probes the out-of-phase vibration close to the origin of the band; the difference between IR and Raman vibrons gives a measure of the intermolecular coupling [van Kranendonk, 1983]. Moreover, infrared measurements have revealed a number of new phenomena, as discussed in the following sections.

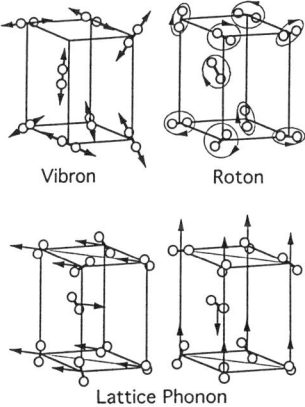

Fig. 2. Principal vibrational excitations ($k = 0$) in hydrogen in the rotationally disordered hexagonal close-packed structure [Mao and Hemley, 1994]. The pure molecular vibration excitation (vibron) is designated $Q_{\Delta v}(J)$, where $v$ is the vibrational quantum number, $\Delta$ the difference between the final and the initial levels, and $J$ the rotational quantum number. The rotational spectrum for the free molecule is given by $BJ(J+1)$, where $B$ is the rotational constant. The lattice phonons include the Raman-active $E_{2g}$ and the $B_{2g}$ modes.

## III. INFRARED SPECTRA OF HYDROGEN

Figure 3 shows representative infrared spectra for solid $H_2$ over a wide energy range with increasing pressure. Note that absorption corresponding to the intramolecular stretching mode vanishes in the free molecule. The IR activity of the vibron in phase I arises from induced interactions. Below 150 GPa, there are broadened rotational excitations with a pressure dependence identical to that of the pure $S_0(0)$ and $S_0(1)$ transitions observed in the Raman spectrum [Hemley et al., 1990]. At 110 GPa and 85 K, a weak, sharp peak appears, which is indicative of a phase transition (Figs. 3 and 4) [Hanfland et al., 1993]. The additional vibron indicates that the symmetry of the phase is lower than hcp, but the vibron-roton combination bands change little, suggesting only minor structural changes and partial orientational ordering. The most significant feature of the spectra shown in Fig. 3 is the dramatic increase in intensity of the pure vibron fundamental near 4500 cm$^{-1}$ at 150 GPa. In addition, the character of the spectrum in the region of the vibron-roton bands changes completely at this pressure. Finally, we note that the intensity of a band at $\sim$ 3250 cm$^{-1}$ increases with pressure; this is discussed in a later section.

The frequency shifts of the bands with pressure are shown in Fig. 4. The band origin of the vibron, which is measured by IR, rises continuously without turnover up to 140 GPa [Hanfland et al., 1992]. However, above 140 GPa the frequency of the IR band begins to decrease, consistent with bond weakening above this critical pressure. The intermolecular coupling shows a dramatic increase from 3 cm$^{-1}$ to 500 cm$^{-1}$ in this pressure range, which drives the turnover of the Raman vibron [Hanfland et al., 1992]. The IR vibron exhibits a low-temperature discontinuity [Hanfland et al., 1993], the magnitude of which is close to that measured in the Raman spectrum [Hemley and Mao, 1988]. Moreover, the IR vibron frequency decreases with increasing pressure, indicating that the discontinuous frequency drop in the transition reflects softening of the intramolecular vibration. Relatively subtle changes in the frequency shifts associated with I-II phase transition contrast with the large changes at the II-III transition. In addition to the intensity enhancement, the frequency shifts of the lower frequency vibron-lattice combination bands in phase III have a completely different pressure dependence (Fig. 4). On the other hand, the IR lattice phonon frequency seems to shift continuously from phase II to III, suggesting similar lattice structures in both phases; analogous results were obtained in Raman measurements [Hemley et al., 1990].

## IV. ISOTOPE EFFECTS

The determination of isotope effects on physical and chemical properties in this system provide a powerful means for testing hypotheses and assignments because of the large relative mass differences between hydrogen and deuterium. There is a pronounced isotope effect on predicted and observed ordering transitions for hydrogen and deuterium, with the heavier isotope ordering at significantly lower densities as a result of its smaller rotational constant [see, Mao and Hemley, 1994]. Infrared spectra measured for phase II of $D_2$ are shown in Fig. 5 [Hemley et al., 1994]. The spectrum of the heavier isotope has a much larger number of peaks than does hydrogen.

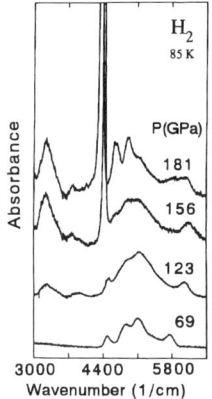
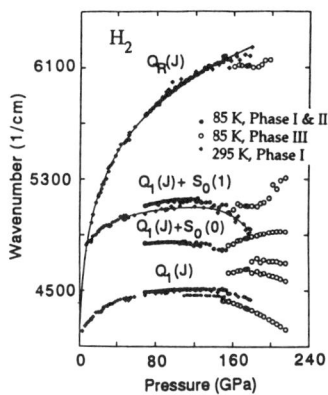

Fig. 3. Infrared absorption spectra of hydrogen from 69 to 181 GPa.

Fig. 4. Pressure dependence of the infrared vivrational frequencies for hydrogen.

The measurements indicate that the structure of deuterium is different, and indeed has lower symmetry than that of hydrogen in phase II. In addition, there is evidence for a sequence of transitions with increasing pressure within phase II, as reflected by the increasing number of satellite peaks that appear with pressure. Thus there are important and surprising differences between the two isotopes in phase II.

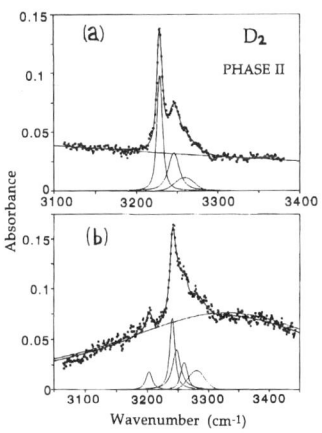
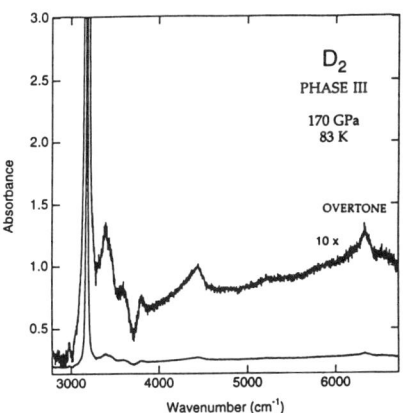

Fig. 5. Infrared spectra of phase II of $D_2$ at 83 K. (a) 92 GPa. (b) 149 GPa.

Fig. 6. Infrared spectrum in phase III of $D_2$ at 170 GPa and 83 K.

Originally, the infrared measurements on $D_2$ were carried out at higher pressures in part to test whether the intense feature observed in $H_2$ is a vibrational excitation or a pure electronic transition (e.g., an interband excitation, in which case the similarity between the Raman and infrared vibrons in accidental). A representative spectrum of $D_2$ at 170 GPa is shown in Fig. 6 [Hemley et al., 1994]. Indeed, we observe the strong enhancement of the vibron when the phase transition is crossed as determined by Raman measurements on the same sample; that is, both a discontinuity in frequency and an intensity increase occur concomitant with the discontinuity observed in the Raman vibron [Hemley and Mao, 1989]. No strong new features are observed at 4000–5000 cm$^{-1}$ (i.e., in the region of the $H_2$ vibron), which confirms that the intense feature in phase III is a vibrational excitation. In addition, new combination-band excitations involving the vibron and lattice modes appear in concert with the phase II-III transition, as in $H_2$. In contrast to hydrogen, we are able to observe the overtone of the vibron ($v = 0 \rightarrow 2$) both below and above the phase transition. A single band is observed at close to twice the vibron fundamental ($v = 0 \rightarrow 1$) transition. Like the vibron fundamental, the frequency of the overtone also shifts discontinuously at the phase transition. The pressure dependence of the frequencies for $D_2$ are shown in Fig. 7.

The vibron absorption for HD is of interest because, unlike $H_2$ and $D_2$ which do not have dipole moments and therefore no infrared-active vibron in the isolated molecule, HD possesses a weak dipole moment and therefore an infrared-active vibron [Karl and Poll, 1967]. Under equilibrium conditions, a equal mixture of hydrogen and deuterium will combine to form a mixture of 50% HD, and 25% each of $H_2$ and $D_2$. Measurements on a such a mixture are shown in Fig. 8. Indeed, a discontinuity and intensity increase are observed in all three vibrons. The intensity enhancement is noticeable first in the $H_2$ species, followed by HD and then $D_2$. Notably, the structure of the bands differs from one another and from that observed for the pure samples (Figs. 3 and 6) as a result of the disorder. Indeed, measurements on the isotopic alloy reveal spectroscopic features throughout the vibron band as a result of the breakdown in selection rules arising from the lack of translational invariance. The $H_2$ peak can be resolved into a doublet and there is a broad component on the high frequency side of the sharper $D_2$ vibron (see below). The absorption of the HD component at lower pressures is quite weak and is negligible in comparison to the strong intensity observed in phase III (for all three molecular species).

## V. PHASE DIAGRAM

The phase diagram of $H_2$ and $D_2$ has been extensively studied to 200 GPa using both infrared and Raman spectroscopy (Fig. 9) [Mao and Hemley, 1994; Goncharov et al., 1994; Li et al., 1994]. The phase lines were determined by changes in the vibron spectra (frequency discontinuities, changes in temperature dependence of frequencies, and linewidths) from isobaric temperature scans. For example, the II-III transition line exhibits a large discontinuous red shift in the Raman vibron with increasing temperature. The measurements confirm the existence of the triple point where the I-II, II-III and I-III phase lines intersect [Mao and Hemley, 1994; Goncharov et al., 1994; Cui et al., 1994]. Recently, Goncharov et al. [1994] carried out a detailed study of the behavior above the triple point in $D_2$. Frequency shifts from temperature scans from 166 to 188 GPa are shown in Fig. 10. The vibron discontinuity is clearly evident above the triple point near 130 K; this contradicts the conclusion drawn by Cui et al. [1994]

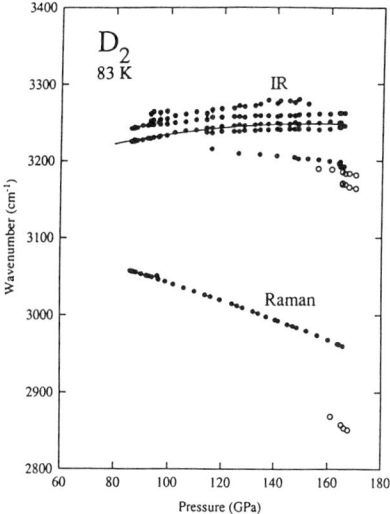

Fig. 7. Pressure dependence of the infrared and Raman vibron frequencies of $D_2$. The halved frequency of the overtone is shown by the line (phase II), and the lower open circles (phase III) in the infrared branch.

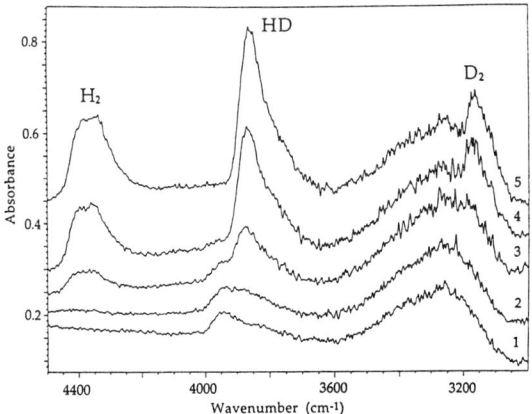

Fig. 8. Infrared absorption spectra of a HD, $H_2$, and $D_2$ mixture at a series of pressures through the II-III phase transition near 150 GPa and 85 K.

that the critical point reported previously [Hemley and Mao, 1990; Lorenzana et al., 1990] is actually a triple point.

Further information has been obtained by examining the temperature dependence of the vibron discontinuity along the I(II)-III phase line. The discontinuity decreases along the phase line, and no anomaly is observed at the triple point near 130 K. For pressures above 185–190 GPa and temperatures of 225–250 K, two distinct peaks cannot be resolved. Detailed analysis shows that at the phase transition, phases I and III coexist with the relative intensities of the two vibrons dependent upon the nature of the pressure gradient and the phase line; the observed spectra are significantly affected in the regions where the frequency shifts rapidly with pressure [Goncharov et al., 1994]. The results strongly suggest that the vibron discontinuity vanishes at these pressures. We propose that this corresponds to a second invariant point in the system, above which temperature the character of the I-III phase transition changes. The invariant point is a tricritical point if the I-III transition goes from first to second order, or another triple point, which implies the existence of an as-yet undetected phase line emanating from this point. In addition to the changes observed above 185 GPa and 225 K, the slope of the phase boundary changes near 170 GPa and 180 K (Fig. 9). The result suggests a third possible invariant point in hydrogen along the I(II)-III phase line. Both the discontinuity above the I-II-III triple point and the behavior near the higher-temperature invariant points were not observed in recent work by Cui et al. [1994] because temperature scans in the relevant $P$-$T$ range were not performed.

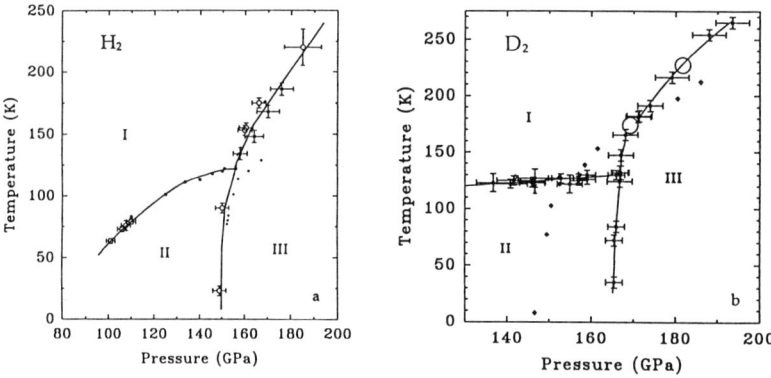

Fig. 9. Megabars phase diagrams. (a) $H_2$ [Li et al., 1994]. Filled circles (with error bars), infrared; open circles, Raman; smaller circles from Lorenzana et al. [1989]. (b) $D_2$ [Goncharov et al., 1994; Mao and Hemley, 1994]. Circles (with error bars), Raman; large open circles show the two possible higher temperature invariant points; diamonds from Cui et al. [1994].

Previous work has emphasized the first-order (discontinuous) change of the high-pressure transitions at low temperature (particularly to phase III [Hemley and Mao, 1988, 1989; Lorenzana et al., 1989]). The results described above show that continuous changes in the I-III transition occur above the high-temperature invariant point. Moreover, a strong temperature dependence of the vibron frequencies is found within *both* phases II and III at lower temperatures [Goncharov et al., 1994]. Measurements carried out with increasing temperature at constant pressures reveal strong temperature dependencies of the vibron frequencies in phase III prior to the transition to phase I. Below the higher-temperature invariant point on the phase I-III line, for example, an additional discontinuous change occurs, whereas at temperatures above this point, the discontinuous change is not observed. These results are also consistent with previous Raman observations that the rotational (librational) spectra change little at the onset of the transition as a function of pressure [Hemley et al., 1990].

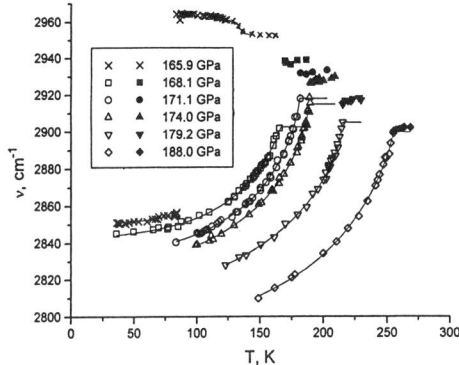

Fig. 10. Temperature dependence of the Raman vibrons at the I(II)-III phase transition in $D_2$ [Goncharov et al., 1994].

## VI. ADDITIONAL TRANSITIONS

Recent infrared and Raman measurements reveal new phenomena in diamond-cell samples of hydrogen and deuterium at megabar pressures (> 150 GPa). As mentioned above, experiments on $H_2$, for example, show a new intense band system at ~3250 cm$^{-1}$ beginning at pressures close to 100 GPa (Fig. 3). The intensity of the bands increases markedly with increasing pressure at low temperature (85 K). Experiments performed on $D_2$ and HD reveal a band system near the same frequency, also beginning near 100 GPa. In the $D_2$ experiments, the temperature dependence of the band was studied closely, and a very unusual temperature response was observed. The absorption strength increased with increasing temperature to 190 K. With further increase in temperature the intensity dropped and, at 210 K the peak disappeared. These results are particularly intriguing because they may be compared with previously observed anomalous low-frequency Raman band (near 250 cm$^{-1}$); both sets of features appear in the same pressure range, and exhibit similar critical temperature behavior ($T_c \sim 200$ K).

The new data may provide insights on the previously observed low-frequency Raman features [Hemley and Mao, 1992]. The lack of an isotope effect on the frequency demonstrates that the excitation does not involve hydrogen or deuterium vibrations (in the bulk sample, at the sample-diamond interface, or for material dissolved within the anvil). Some natural diamonds exhibit features in this general spectral range; but the new features we document at high pressures were not observed in the anvils at low pressures. The new bands could signal a transition in diamond occurring at the very tip of the anvil where it is in contact with the sample. Previously, we found evidence for structural transitions in diamond at the highly stressed tips of the anvils; this behavior is associated with intense luminescence and new low-frequency Raman features

[Mao and Hemley, 1991]. The temperature dependence and reversibility differ from the previously documented changes in diamond. This leads to the hypothesis that the new infrared features, as well as the previously observed Raman bands, are associated with unusual behavior of the diamond-sample interface or the bulk hydrogen sample, perhaps a non-vibrational (i.e., electronic) excitation. In addition, increase in the near-IR reflectivity measured above 150 GPa at room temperature [e.g., Mao et al., 1990; Hanfland et al., 1991] needs to be examined in future experiments. In particular, it will be important to examine the possible connection between the appearance of the 250 cm$^{-1}$ mode, the near-infrared reflectivity, and the new mid-infrared absorption features.

## VII. DISCUSSION

Recent work, in particular isotopic studies of $D_2$ and HD, confirm that the pressure evolution of hydrogen may be understood as a sequence of symmetry breaking transitions in the system. Specifically, there is symmetry breaking associated with orientational order (and possibly crystallographic change), as well as effects associated with charge transfer. More generally, the latter is evidence for changes in electronic structure and bonding under pressure. In addition, detailed study of the phase diagram of the materials reveal the topology of phases I, II, and III; the persistence of up to three invariant points on the phase III line is a particularly surprising result. The stability of the H-H bond in the dense solid to pressures well above 200 GPa continues to be a robust conclusion. Whether or not band overlap occurs still remains an open question [Hanfland et al., 1991]. It is important to note that anomalies in the optical, infrared, and low-frequency Raman spectra are not yet fully understood and may be associated with metallization or other changes in electronic properties. Indeed, it is possible that these features (regardless of their origin) reflect the same changes in physical state of the samples.

The increase in the IR vibron absorption in phase III is one of the most important observations and has now been reproduced in a large number of samples, including $H_2$, $D_2$, and HD. The symmetry of phase I (hcp) precludes an allowed vibron absorption (i.e., by factor group analysis); the observation of weak vibron absorption in this phase arises from induced interactions and disorder, as described above. Hence, the absorption increase on passage into phase III strongly suggests a transition to a crystal structure with a symmetry allowed vibron band. Indeed, the observation of a single strong vibron band (though broad) constrains the symmetry (space group) of the high-pressure phase. For example, this rules out *Pa3* and *c*-axis ordered hcp structures [cf., Mao and Hemley, 1994], which have no allowed IR vibron bands. Such an analysis, however, provides no information on the strength of the absorption. The absorption intensity can be modeled phenomenologically in terms of dynamical charges, as has been done for other elemental solids such as graphite and Se [Hemley et al., 1994].

The onset of intense IR vibron absorption at 150 GPa parallels the spectral properties of organic charge-transfer salts, including pressure-induced neutral-to-ionic transitions. The dramatic rise in absorption can be examined in terms of an increase of vibronic coupling between a charge-transfer electronic transition and the vibron [Hemley et al., 1994; Soos et al., 1994]. The system may be viewed within a localized picture involving pressure-induced ionization of a hydrogen molecule embedded in a matrix. At pressures below those expected for the formation of a high-carrier metal, there is partial ionization. Intramolecular charge transfer would obtain if, for example, the atoms within a molecule become crystallographically distinct. Baranowski [1992] has proposed that phase III is an ionic state consisting of $H^+H^-$ species. Such a change is expected to generate strong dipole absorption and vibron softening. Since the observed vibron discontinuity represents only a $\sim 5\%$ change in frequency, the formation of a fully ionic state is unlikely in phase III. On the other hand, the transfer of a small amount of charge remains a hypothesis to be tested. Indeed, it is important to examine both intramolecular and intermolecular charge transfer (for phases II and III).

Another possibility is strong electron-phonon coupling due to mixing of the vibron with a degenerate electronic excitation arising from closing of the gap (i.e., when the gap and vibron energies are equal). In this hypothesis, the crossing of these states drives the transition, or is coincident with it. The similar onset pressure of the absorbance increase and vibron discontinuity in $H_2$ and $D_2$ (with their large difference in vibron frequencies) argues against this proposal. There have been several theoretical predictions that molecular hydrogen at megabar pressures could exhibit novel physical properties (i.e., even in the absence of depairing) [see, Ashcroft, 1993; Moulopoulos and Ashcroft, 1991]. One possibility is exciton condensation and the formation of a charge-density wave, which may be expected on general grounds as the band gap closes [see, Hemley and Mao, 1990]. With strong electron-phonon coupling, charge fluctuations may produce local distortions of the lattice (polarons) and provide a means for the system to maintain an insulating state to the highest available pressures. The extent to which the symmetry breaking represents a static or frozen-in structural change as opposed to a dynamic or fluctuation phenomenon is a crucial distinction to be made in each of these proposals. The additional transitions described above may also be relevant to the new phenomena predicted by theory, although spurious contributions to the spectra (e.g., from diamond) need to be examined further [Hemley and Mao, 1992].

## ACKNOWLEDGMENTS

We are grateful to M. Hanfland, A. F. Goncharov, M. Li, J. H. Eggert, and I. I. Mazin for their excellent contributions to the work described above, and for many useful discussions. This research was supported by the N.S.F. and N.A.S.A.

## REFERENCES
Ashcroft, N. W., 1993, J. Non-Cryst. Solids, **156-158**, 621.
Baranowski, B., 1992, Polish J. Chem., **66**, 1637.
Cui, L., N. H. Chen, S. J. Jeon and I. F. Silvera, 1994, Phys. Rev. Lett., **72**, 3048.
Fowler, R. H., 1926, Mon. Not. R. Astron. Soc., **87**, 114.
Goncharov, A. F., I. I. Mazin, J. H. Eggert, R. J. Hemley and H. K. Mao, 1994, to be published.

Hanfland, M., R. J. Hemley and H. K. Mao, 1991, Phys. Rev. B, **43**, 8767.
Hanfland, M., R. J. Hemley and H. K. Mao, 1993, Phys. Rev. Lett., **70**, 3760.
Hanfland, M., R. J. Hemley, H. K. Mao and G. P. Williams, 1992, Phys. Rev. Lett., **69**, 1129.
Hemley, R. J., M. Hanfland and H. K. Mao, 1994, to be published.
Hemley, R. J. and H. K. Mao, 1988, Phys. Rev. Lett., **61**, 857.
Hemley, R. J. and H. K. Mao, 1989, Phys. Rev. Lett., **63**, 1393.
Hemley, R. J. and H. K. Mao, 1990, Science, **249**, 391.
Hemley, R. J. and H. K. Mao, 1992, Phys. Lett. A, **163**, 429.
Hemley, R. J., H. K. Mao, L. W. Finger, A. P. Jephcoat, R. M. Hazen and C. S. Zha, 1990, Phys. Rev. B, **42**, 6458.
Hemley, R. J., H. K. Mao and J. F. Shu, 1990, Phys. Rev. Lett., **65**, 2670.
Hemley, R. J., Z. G. Soos, M. Hanfland and H. K. Mao, 1994, Nature, **369**, 384.
Hu, J., H. K. Mao, J. F. Shu and R. J. Hemley, 1994, in *High-Pressure Science and Technology—1993*, edited by S. C. Schmidt *et al.* (American Institute of Physics, New York), p. 441.
Jayaraman, A., 1983, Rev. Mod. Phys., **55**, 65.
Karl, G. and J. D. Poll, 1967, J. Chem. Phys., **46**, 2944.
Li, M. M., A. F. Goncharov, R. J. Hemley, and H. K. Mao, 1994, in *National Synchrotron Light Source—Annual Report, 1994*, Brookhaven National Laboratory, in press
Lorenzana, H. E., I. F. Silvera and K. A. Goettel, 1989, Phys. Rev. Lett., **63**, 2080.
Lorenzana, H. E., I. F. Silvera and K. A. Goettel, 1990, Phys. Rev. Lett., **65**, 1901.
Mao, H. K., 1989, in *Simple Molecular Systems at Very High Density*, edited by A. Polian, P. Loubeyre and N. Boccara (Plenum, New York), p. 221.
Mao, H. K. and R. J. Hemley, 1991, Nature, **351**, 721.
Mao, H. K. and R. J. Hemley, 1994, Rev. Mod. Phys., **66**, 671.
Mao, H. K., R. J. Hemley and M. Hanfland, 1990, Phys. Rev. Lett., **65**, 484.
Mao, H. K., R. J. Hemley and A. L. Mao, 1994, in *High-Pressure Science and Technology—1993*, edited by S. C. Schmidt *et al.* (American Institute of Physics, New York), p. 1613.
Mao, H. K., A. P. Jephcoat, R. J. Hemley, L. W. Finger, C. S. Zha, R. M. Hazen and D. E. Cox, 1988, Science, **239**, 1131.
Moulopoulos, K. and N. W. Ashcroft, 1991, Phys. Rev. Lett., **66**, 2915.
Soos, Z. G., J. H. Eggert, R. J. Hemley, M. Hanfland and H. K. Mao, 1994, J. Chem. Phys., submitted for publication.
van Kranendonk, J., 1983, *Solid Hydrogen* (Plenum, New York).
Wigner, E. and H. B. Huntington, 1935, J. Chem. Phys., **3**, 764.
Williams, G. P., 1990, Nucl. Instrum. Methods Phys. Res. A, **291**, 8.

# VII.3

# THE POTENTIAL OF HIGH-POWER BEAMS FOR STUDYING MEGABAR MATTER, INCLUDING LOW-ENTROPY HYDROGEN COMPRESSION

J. Meyer-ter-Vehn and A. Oparin

Max-Planck-Institut für Quantenoptik, D-85748 Garching, Germany

Recent progress in the generation of uniform shock waves driven by laser-generated X-rays and producing pressures in the range of 10 to 1000 Mbar is reviewed; high-power ion beam drivers are also considered. Stimulated by these results, multiple-shock compression of solid hydrogen is simulated numerically, aiming for pressures of 1 – 10 Mbar at temperatures below 1 eV and exploring the transition from the molecular to the atomic phase and plasma phase transitions.

## I. DYNAMIC GENERATION OF MEGABAR MATTER

High-pressure laboratory experiments to determine equations of state and other properties of pressurized matter were limited to the region below 10 Mbar, so far. Static pressures of 2 – 3 Mbar have been generated in diamond-anvil cells [Hemley and Mao, 1992]. Higher pressures up to 10 Mbar have been achieved with dynamic methods making use of high explosives and two-stage gas guns for shock compression. Altshuler [1992] recently reviewed the shock method, and Fig. 1, taken from this review, summarizes laboratory absolute measurements (shaded region) up to 1992; also shown are a few (mostly relative) measurements in the range of 10 – 1000 Mbar which were obtained from underground nuclear explosions. The method to obtain planar shocks with highest pressures has been to accelerate flyer plates which strike the sample [see e.g. Ahrens, 1987].

Presently, high-power beams offer new possibilities to perform dynamic experiments in the laboratory up to pressures of 1000 Mbar, as indicated by the hatched region in Fig. 1. The dashed horizontal lines mark the pressure results of two recently published shock experiments, using laser-generated X-rays to drive the shocks. Experiments at Livermore [Cauble et al., 1993] aimed for maximum pressure; 25 kJ pulses from the Nova laser heated small hohlraum configurations to produce X-rays which drive flyer plates, and shocks due to impact on solid gold were observed, corresponding to pressures of 750 Mbar. Other experiments with 250 J pulses at MPQ/Garching and 2 kJ pulses at ILE/Osaka [Löwer et al., 1994] were designed to achieve optimal shock uniformity and achieved shock pressures between 5–20 Mbar by exposing samples directly to the X-rays; they will be discussed in more detail below.

The development of high-power beams is strongly pushed by research on inertial confinement fusion (ICF). The goal is to compress hydrogen fuel to more than 1000 times solid density and pressures above 100 Gbar. Igniting fusion requires megajoule pulses and corresponding machines have still to be built, whereas kilojoule pulses now

# VII.3 HIGH-POWER BEAMS FOR STUDYING MEGABAR MATTER

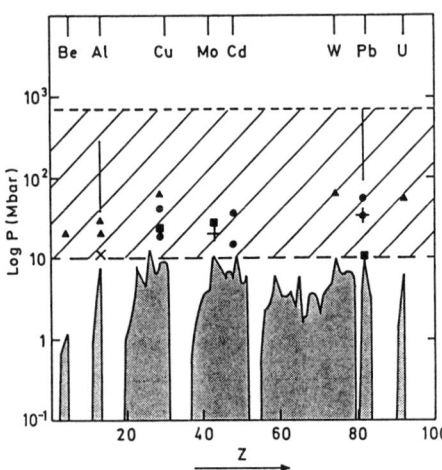

Fig. 1: Pressure ranges of dynamic experiments (taken from Altshuler [1992]). Shaded regions refer to laboratory absolute measurements, dots and bars to (mostly relative) results from underground explosions. The hatched area indicates the pressure region now opening for shock experiments driven by laser-generated x-rays; boundaries correspond to results of Löwer et al. [1994] (lower) and by Cauble et al. [1993] (upper).

available from a number of existing machines are sufficient for basic high pressure experiments. Key problems such as uniform energy deposition and suppression of target preheat are similar for both applications, and solutions first developed for ICF may help high pressure research.

## II. UNIFORM SHOCKS DRIVEN BY LASER-GENERATED X-RAYS

Experimental results obtained by the German-Japanese collaboration [Löwer et al., 1994] may demonstrate the progress achieved recently. Target configurations used for single-beam ASTERIX shots at MPQ and for GEKKO shots with four beams at ILE are depicted in Fig. 2. The pulsed laser light is focused into a gold cavity (*hohlraum*) a few mm in size, where it is partially converted into thermal X-rays when interacting with the inner surface of the cavity. In the experiments described, radiation temperatures between 90 – 150 eV were obtained. Shock waves are driven by radiative ablation. The flash of light emitted by the shock-heated material during shock breakout at the outer surface was registered by an optical streak camera. Step targets were used to measure shock velocities. So far, no attempt was made to measure a second parameter; pressures and temperatures were determined by comparison with Hugoniot data deduced from the SESAME equation of state tables [SESAME, 1983].

Streak records, corresponding to 7.5 Mbar shocks in gold, are shown in Figs. 3a and 3b. They were obtained with GEKKO using the D-shaped cavity sketched in Fig. 2b with (a) a step gold foil of 8 and 15 $\mu$m thickness and (b) a gold wedge the thickness of which varies linearly from 9 to 21 $\mu$m over a distance of 300 $\mu$m. One should notice the remarkable uniformity of the shock breakout signals both spatially and in time. Löwer et al. [1994] deduced that the shock velocity is spatially uniform to better than ±0.6% corresponding to an irradiation uniformity of better than ±1.6%. This uniformity is a major achievement demonstrating the advantage of *indirect* drive

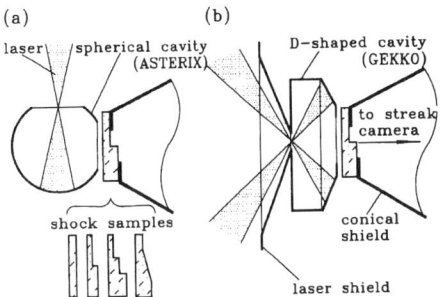

Fig. 2: Configuration of laser-heated gold cavity and shock sample. (a) Spherical cavity irradiated by the single-beam ASTERIX laser, bottom: various types of shock samples used in the experiments. (b) D-shaped, rotationally symmetric cavity irradiated by four beams from the GEKKO laser. From Löwer et al. [1994].

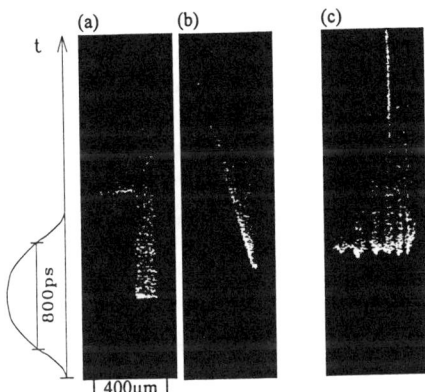

Fig. 3: Streak camera records of visible light emitted by the shock sample. On the left-hand side the temporal shape of the GEKKO laser pulse. (a) Record from a single-step gold sample corresponding to a 7.5 Mbar shock driven by a GEKKO cavity. (b) Same as (a), but for a wedge sample. (c) Record obtained from direct laser irradiation of a flat sample by the ASTERIX laser beam showing a non-uniform signal caused by irradiation non-uniformity. From Löwer et al. [1994].

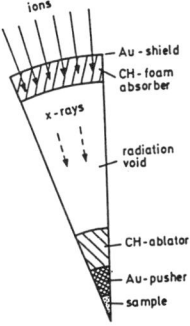

Fig. 4: Sector of a cylindrical hohlraum configuration proposed for radial irradiation with light ions (schematic plot). X-rays emitted from the beam-heated foam absorber drive an inner plastic ablator which implodes pusher and sample. For use at PBFAII (Sandia), the target size would be about 1 – 2 cm in diameter and height. Sample observation is possible along the cylinder axis.

via X-rays. For comparison, a record from a flat sample *directly* irradiated by the ASTERIX laser beam is shown in Fig. 3c; even though the ASTERIX laser has a good beam quality with intensity modulations of less than 10 %, strong rippling of the shock front is seen which degrades the signal for quantitative evaluation.

A major difficulty of these experiments consists in detecting the weak shock signals in the close neighbourhood of high-intensity laser light. Apparently, the authors succeeded to screen the shock detection system from scattered laser light by heavy shielding. However, non-thermal keV X-rays emitted from the laser interaction spots inside the gold cavity may also interfere with the shock measurement by preheating the sample material before the shock arrives. Such preheat effects were indeed observed. They depend on sample material and thickness as well as on the angle under which the sample area facing the hohlraum sees the laser interaction spot and receives primary hard X-rays. Shots made with the spherical target of Fig. 2a and thin 10 $\mu$m aluminum samples showed preheat light emission from the rear surface instantaneously with the laser pulse, considerably before shock arrival; the preheat signal could be suppressed by increasing the sample thickness to 25 $\mu$m and did not occur at all when driving the same samples with the D-shaped cavities. These results suggest that preheat can be controlled by adequate hohlraum design and sufficiently thick samples.

It appears that experiments of this type have a high potential for future high-pressure research. Pressures can be boosted by using the method of flyer plate impact, as demonstrated by Cauble *et al.* [1993]. However, higher pressures tend to imply thinner targets, shorter measuring times, and therefore reduced accuracy. This may be compensated by beam pulses with higher energy. For absolute EOS measurements with single planar shocks along the principal Hugoniot, also the material velocity behind the front or another second parameter has to be determined. For exploring wider regions of the phase diagram, multiple-shock techniques [Nellis *et al.*, 1992] have to be applied and will be discussed for the case of solid hydrogen in Section IV.

## III. HIGH-POWER ION BEAMS

At present, the development of laser drivers is most advanced. However, light and heavy ion beam drivers will also play an important role in the future, and target parameters achieved so far are of interest for high-pressure research. In particular, light ion beams generated in pulsed-power diodes have made considerable progress, and pulses with tens of kilojoules have been focused on spots of less than 1 cm$^2$. Beam parameters on target achieved with KALIF at KFK [Bluhm *et al*, 1992] and with PBFAII at Sandia [Mehlhorn *et al.*, 1992, 1994] are listed in Table I. It is seen that impressive deposition values in units of TW/g have been achieved, well suited for driving radiation cavities and achieving pressures in the multi-megabar regime. In particular the long pulse times (relative to lasers) are favorable for low-entropy compression, as discussed further below for hydrogen.

Shock wave experiments using direct drive have been reported recently by Bluhm *et al.* [1992]. However, the uniformity problem is even more severe for ion beams than for laser beams. For quantitative experiments, indirect drive will be the direction to go. Cylindrical targets leaving an open axis for observation along the cylinder axis have

### Table I Light Ion Beam Parameter

|  | KALIF[1] | PBFAII[2] | PBFAII[3] |
|---|---|---|---|
| Ion | p | p | Li |
| Energy | 1.7 MeV | 5 Mev |  |
| Focused power | 1 TW/cm$^2$ | 5 TW/cm$^2$ | 2.5 TW/cm$^2$ |
| Focus size | 6 mm | 6mm |  |
| FWHM | 50 ns | 15 ns | 24 ns |
| Specific power |  |  |  |
| Foam | 200 TW/g | 160 TW/g | 700 TW/g |
| Aluminium | 110 TW/g |  |  |
| Gold | 40 TW/g |  |  |

1) Bluhm et al. [1992], 2) Mehlhorn et al. [1992], 3) Mehlhorn et al. [1994]

been used at Sandia. A generic target for future experiments is sketched in Fig. 4. The beam is radially coming in and is stopped in a thin foam layer bounded outwards by a gold shield. Beam non-uniformities are smoothed by radiation transport in the hohlraum, the x-rays drive the inner target cylinder consisting of a plastic ablator, a gold pusher (representing the "flyer"), and the sample material in the center; the cylinder radius is 5–10 mm.

Heavy ion beams generated by accelerators are also considered for producing high energy density in matter [Arnold and Meyer-ter-Vehn, 1988], and target experiments have been performed at GSI/Darmstadt [Meyer-ter-Vehn et al., 1990]. High-intensity beams have still to be developed. So far, deposition powers of GW/g were achieved.

## IV. LOW-ENTROPY HYDROGEN COMPRESSION

It is proposed to use the new techniques for multiple-shock experiments, and the present section is devoted to an exploration of possibilities by means of numerical simulation. Compression by single shocks is limited to compression ratios of 4-5; increasing the strength leads to excessive heating. Keeping the strength of the first shock relatively low and making use of sequential reflections between hard boundaries is a way to compress materials at much lower entropy and to higher densities. This method of reverberating shocks was successfully applied by Weir et al. [1993] to compress solid hydrogen. Pressures of 1.2 Mbar at temperatures of 3000 K were reached by flyer plate impact driven by a two-stage gas gun. Conductivity measurements indicated that the transition from non-conducting molecular hydrogen to the metallic atomic phase was approached, but not yet reached.

A. Phase transitions in hydrogen

The simulations presented here are based on the SESAME [1983] equation of state. It was constructed by Kerley [1972], using separate models for the three phases: molecular solid, atomic solid, and fluid. It is in reasonable agreement with EOS data obtained from static compression as well as from single- and double-shock experiments. These

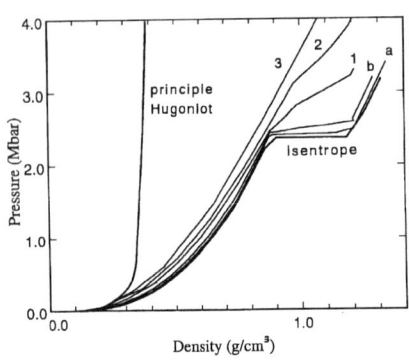

Fig. 5: Phase plane trajectories for different compression schemes of 10 K solid hydrogen. Isentropic and single-shock compression (principle Hugoniot) set the boundaries; other curves: (a) cylindrical 1 km/s liner impact (see Fig. 11), (b) double-sided 2.5 km/s planar impact, single-sided impact by (1) 5 km/s, (2) 7 km/s, (3) 10 km/s flyer on tamped planar hydrogen. Calculation based on SESAME EOS [Kerley, 1983] having a first-order transition between the molecular and the atomic phase at about 10 times solid density.

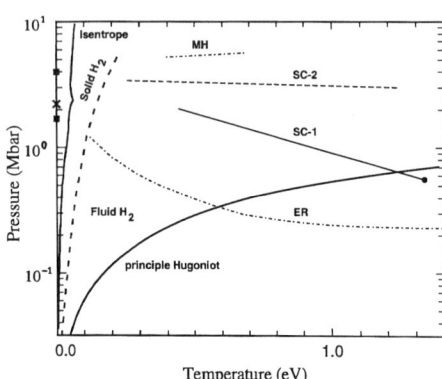

Fig. 6: Temperature-pressure phase diagram of hydrogen (taken from Saumon and Chabrier [1992]) with isentrope and principle Hugoniot as in Fig. 5. Predicted phase transition at $T = 0$ is indicated by solid squares [Min et al., 1986] and cross [Friedli and Ashcroft, 1977]. Different predictions for plasma phase transitions are shown: ER Ebeling and Richert [1985], MH Marley and Hubbard [1988], SH Saumon and Chabrier [1992]. The dashed line indicates the melting curve.

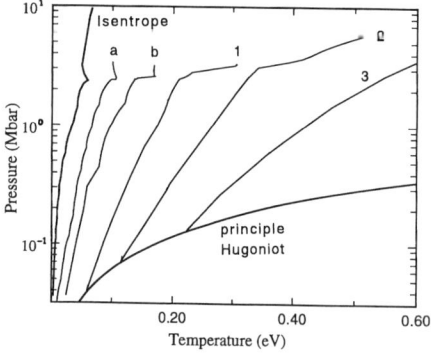

Fig. 7: Lower temperature part of the phase diagram of Fig. 6 showing different hydrogen compression trajectories as explained in Fig. 5.

data probe the EOS in the molecular solid and the fluid phase. Whether the first-order phase transition predicted to occur between the molecular and the atomic phase at low temperature is also real is an open question. This transition is clearly seen in Fig. 5, where different compression trajectories are plotted in the density-pressure phase plane. They represent the compression of a 10 K hydrogen solid either isentropically or by a single shock (principle Hugoniot) or by different compression schemes. The trajectories 1, 2, and 3, refer to single-sided flyer impact with different flyer velocities, as described in more detail further below.

Related transitions, so-called plasma phase transitions, have been predicted to occur in the fluid phase of hydrogen. Results by Saumon and Chabrier [1992] and others are plotted in Fig. 6. They are based on fluid calculations which apply to temperatures well above the melting curve and also predict first-order transitions between two phases having distinctly different conductivities. No experimental evidence for these transitions in hydrogen exists so far, and multiple-shock experiments driven by laser or ion beams as described above may be a way to explore this region. The principle Hugoniot and the isentrope starting from the 10 K solid are also plotted in Fig. 6, and Fig. 7 displays the trajectories of different compression schemes which cover the whole space between isentropic and single-shock compression.

B. Simulation of low-entropy hydrogen compression

The simplest way to achieve multiple-shock compression is to place the sample layer between flyer and some hard tamper, assuming planar geometry. A corresponding simulation is shown in Fig. 8, where a 75 $\mu$m thick gold flyer, having a uniform velocity of 5 km/s, is incident at time $t = 0$ on a 150 $\mu$m thick layer of solid hydrogen. The first shock, running into the hydrogen at about 11 km/s, is reflected by the gold tamper and is then running back and forth between tamper and flyer several times, before maximum compression is obtained with a density of 1.2 g/cm$^2$, a pressure of 3.3 Mbar, and a temperature of about 3500 K. At this stage, the sample moves together with flyer and tamper at half the impact velocity. One should notice that the highly compressed state lives for almost 20 ns. Trajectories of the hydrogen state are shown as curve 1 in the $(p, \rho)$ phase plane in Fig. 5 and for $(p, T)$ in Fig. 7. It is seen that the hydrogen state reaches the phase transition region, but at a temperature which is already too high to observe a sharp transitional behaviour. This is even more so for impact velocities of 7 and 10 km/s; corresponding trajectories are given by curve 2 and 3, respectively.

In order to reach the transitional region in hydrogen at lower temperature, the compression scheme has to be modified. First, let us consider two flyer plates incident symmetrically from opposite sides on a plane layer of solid hydrogen. Taking 100 $\mu$m thick layers for hydrogen and gold flyers, state trajectories are given as curves (b) in Figs. 5 and 7. For double-sided impact, a flyer velocity of 2.5 km/s is sufficient to reach the required 2-3 Mbar, and the temperature is 1500 K when the transition from molecular to atomic phase sets in. According to the SESAME EOS, it implies a density jump of almost 40%. The density evolution of the latter case is displayed in Fig. 9, plotting density versus time for each Lagrangian cell perspectively. The hydrogen cell numbers run from 1 at the central plane of symmetry to 50 at the outer boundary,

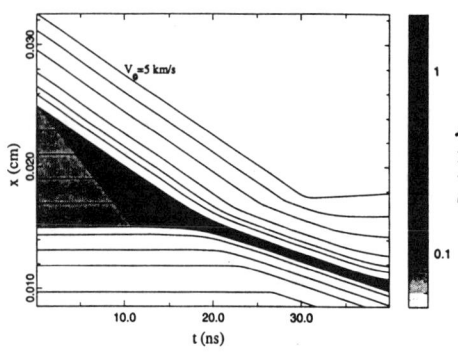

Fig. 8: Single-sided planar impact of a 75 μm thick gold flyer with a velocity of 5 km/s on 150 μm thick hydrogen tamped by a gold layer on the opposite side. The solid lines are flow trajectories. Different shading refers to different density. At maximum compression, the hydrogen is in the metallic phase with pressure 3.3 Mbar and temperature 0.3 eV.

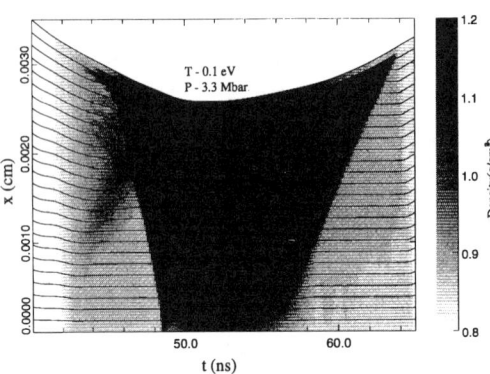

Fig. 9: Hydrogen density evolution for symmetric impact of two 2.5 km/s flyers striking from opposite sides. The density is plotted versus time for each Lagrangian cell (50 hydrogen cells); cell 1 lies at the plane of symmetry and cell 50 at one of the outer boundaries.

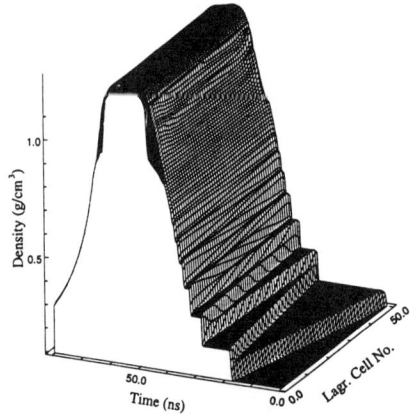

Fig. 10: Flow diagram for cylindrical compression at highest density (Aoki, Meyer-ter-Vehn [1994]). The flyer layer (not shown), incident with 1 km/s, strikes the solid hydrogen cylinder with 100 μm radius. The metallic phase (strongly shaded) sets in through a compression shock and decays through an outgoing rarefaction shock. Maximum pressure and temperature are indicated (also compare curves (a) in Figs. 5 and 7).

where the flyer strikes. One sees how the initial sequence of reverberating shocks is finally merging into an adiabatic compression regime, producing a very uniform state of hydrogen with a density of almost 1.3 g/cm$^2$ which exists for about 10 ns. One should notice, however, that the present 1D calculation does not account for rarefaction from the sides. From the sound velocity of 27 km/s at highest compression, we estimate that the rarefaction front runs about 300 – 500 μm into the sample and that the lateral extension of the sample has to be 1-2 mm initially to insure the formation of a highly compressed core. Tamping of lateral outflow may be considered; however, the tamping material needs to remain transparent to allow for optical observation of the compressed sample.

Another option is cylindrical implosion. A hollow cylindrical liner (flyer) is accelerated radially inwards, possibly in a hohlraum configuration as the one shown in Fig. 4, and compresses the sample in the center, which can be observed by light transmission along the axis. A detailed discussion of the corresponding hydrodynamics was given by Aoki and Meyer-ter-Vehn [1994]. Due to cylindrical convergence, energy is concentrated more efficiently than in planar geometry, and a liner velocity of 1 km/s is sufficient to achieve 3-4 Mbar in hydrogen. Though the strength of the converging shock increases when approaching the axis, the temperature of most of the sample stays at about 1000 K. Corresponding phase trajectories are seen in Figs. 5 and 7 as curves (a). The flow diagram of the hydrogen cells during the time of maximum compression is given in Fig. 10. The strongly shaded region refers to the metallic phase, and we observe interesting differences in the dynamics when comparing with planar compression. Whereas the metallic phase sets in and decays almost uniformly over the sample volume in the planar case (see Fig. 9), it occurs through dynamic fronts in Fig. 10. In particular, the decay occurs through a rare faction shock front which moves from the center to the outer boundary between 56ns and 64ns. The mass flow in this front is from high density (metallic regime) to low density (molecular regime). Since the density jump amounts to 40 %, a density hole is developing from the center outwards and should be observable, provided that the SESAME EOS is correct and the strong first-order phase transition is real.

## V. OUTLOOK

The recent progress in laboratory shock experiments, demonstrating very uniform compression of matter by thermal X-rays, offers new possibilities to measure Hugoniot data in the region of 10–1000 Mbar. Both high power lasers and ion beams are of interest to generate the driving x-rays. The results also suggest that more complex experiments using multiple-shock compression may become feasible to investigate phase diagrams in a wider region which was not accessible so far. The phase diagram of hydrogen at pressures between 1–10 Mbar and temperatures below 1 eV is of particular interest, since phase transitions from molecular to atomic structure have been predicted in the solid and in the fluid regime, but have not been observed experimentally, so far. In particular, it is not clear whether the structural change occurs in form of a first-order phase transition or not. Certainly, metallization with a more or less rapid change in conductivity is expected to occur, and the gas gun experiments by Nellis *et al.* clearly indicate that the transition is approached at pressures above 1 Mbar. It is hoped that the simulations presented here will stimulate laser and ion beam experiments to clarify the phase diagram of hydrogen and other materials.

## ACKNOWLEDGEMENTS

The authors acknowledge stimulating discussions with S. Anisimov, T. Löwer, and R. Sigel on the topic of this paper as well as the contribution of T. Aoki during the earlier stages of this work. The work was supported in part by the Bundesministerium fr Forschung und Technologie and by EURATOM.

## REFERENCES

Aoki, T., and J. Meyer-ter-Vehn, 1994, Phys. Plasmas **1**, 1962.
Ahrens, Th.J., 1987, *Methods of Experimental Physics*, Vol. 24, Part A (Academic Press).
Altshuler, L.V., 1992, in *Shock Compression of Condensed Matter - 1991*, edited by Schmidt, S.C., Dick, R.D., Forbes, T.W., Tasker, D.G. (North Holland, Amsterdam).
Arnold, R., and Meyer-ter-Vehn, J., 1988, Z. Physik D
Bluhm, H., et al., 1992, Proc.of 9th International Conference on High Power Particle Beams, Washington DC, May 25-29, p.51.
Cauble, R., Phillion,D. W., Hoover, T.J., Holmes, N.C., Kilkenny, J.D., and Lee, R.W., 1993, Phys. Rev. Letters, **70**, 2102.
Ebeling, W., and Richert, W., 1985, Phys. Lett. **108A**, 80, and Phys. Stat. Sol. **128**, 467 (1985).
Hemley, R.J., and Mao, H.K., 1992, in *Shock Compression of Condensed Matter - 1991*, edited by S.C. Schmidt, R.D. Dick, T.W. Forbes, D.G. Tasker (North Holland, Amsterdam), p.27.
Kerley, G.I., 1972, A Theoretical Equation of State for Deuterium, Los Alamos Scientific Laboratory, Report LA-4776, UC-34, available from NTIS.
Löwer, Th., Sigel, R., Eidmann, K., Földes, I.B., Hüller, S., Massen, J., Tsakiris, G.D., Witkowski, S., Preuß, W., Nishimura, H. Shiraga, H., Kato, Y., Nakai, S., and Endo, T., 1994, Phys. Rev. Letters, **72**, 3186.
Marley, M.S., and Hubbard, W.B., 1988, Icarus **73**, 53.
Mehlhorn, T., et al., 1992, Proc. Beams'92, p.31.
Mehlhorn, T., et al., 1994, Proc. Beams'94, to be published.
Meyer-ter-Vehn, J., Witkowski, S., Bock, R., Hoffmann, D.H.H., Hofmann, I., Müller, R.W., Arnold, R., Mulser, P., 1990, Phys. Fluids B**2**, 1313.
Nellis, W.J., Mitchell, A.C., McCandless, P.C., D.J. Erskine, D.J., Weir, S.T., 1992, Phys. Rev. Lett. **68**, 2937.
Saumon, D., and Chabrier, G., 1992, Phys. Rev. A**46**, 2084.
SESAME Report on the Los Alamos Equation-of-State Library, 1983, Report No. LALP-83-4 (T4 Group LANL, Los Alamos).
Weir, S.T., Nellis, W.J., and Mitchell, A.C., 1993, *High Pressure Science and Technology - 1993*, edited by S.C. Schmidt, J.W. Shaner, G.A. Samara, and M. Ross (American Institue of Physics, New York), in press.

# VIII. Condensed-Matter Plasmas

# VIII.1

# INTERATOMIC INTERACTIONS IN DENSE MERCURY VAPORS

F. Hensel[†] and M. Yao[‡]

[†] Institute of Physical Chemistry and Materials Science Center
Philipps-University of Marburg, 35032 Marburg, Germany
[‡] Department of Physics, Kyoto University, Kyoto 606, Japan

The vapor phase of fluid mercury is distinct from that of normal insulating fluids such as argon in that its electronic structure and interparticle interaction depend strongly on density. The most striking manifestation of this density dependence is the nonmetal to metal transition which shows up when the vapor is compressed to the region of the liquid-vapor critical point. The paper discusses recent experimental results in the liquid-vapor critical region of metals which show that the existence of the metal-nonmetal transition noticeably influences the electronic, thermodynamic, structural and interfacial features of the fluid. The main emphasis is on surface induced phenomena. Reflectivity experiments on mercury against an optically transparent sapphire window close to the vapor-liquid coexistence curve reveal clearly the existence of a prewetting transition of mercury on the sapphire substrate. The transition line, which terminates at high temperature at a prewetting critical temperature $T_{pwc}$ lying below the bulk critical temperature $T_c$ and at low temperature at the wetting transition temperature $T_w$, lies close to the bulk vapor liquid coexistence curve.

## I. INTRODUCTION

The problem of the interrelation of the metal-nonmetal transition and the liquid-vapor phase transition in fluid mercury has received considerable theoretical attention in the past. The pioneering study of Landau and Zeldovitch [1943] suggested the possibility of separate first-order electronic and liquid-vapor transitions in mercury. Subsequent theoretical attempts to model the statistical mechanics of the metal-nonmetal transition in fluids reach similar conclusions but are still insufficient to provide a clear-cut answer to the question from theory [*e.g.* Mott, 1974; Nara *et al.*, 1977; Yonezava and Ogawa, 1982].

Measurements such as those of the electrical conductivity (fig. 2) and the equation of state (fig. 1) [Hensel and Franck, 1966; Kikoin and Sechenkov, 1967; Yao and Endo, 1982; Götzlaff, 1988] clearly show that there is no sharp (first-order) electronic transition except across the liquid-vapor phase change for fluid mercury, *i.e.* the liquid-vapor phase separation tends to separate the nonmetallic and metallic fluids. Near the critical point the conductivity drops sharply, thus showing a strong effect of the phase transition on the electronic structure. The close correlation between the behaviour of the density and that of the conductivity convincingly shows that the variation of density is the dominant factor governing the metal-nonmetal transition.

In practice, however, very careful measurements are required to separate the effects of density and temperature in the critical region. Part of the difficulty arises because not only the compressibility but also the pressure derivative of the electrical conductivity become very large in that region. This implies that small pressure errors cause large density and conductivity errors. Consequently, for a reliable correlation of the conductivity and density, precise temperature and pressure control is essential. This is not easily achievable, because fluid mercury is difficult to experiment with. A combination of high temperature and pressure is required to bring the fluid anywhere near its critical point, which is located at the temperature $T_c=1751$ K, the pressure $p_c=1673$ bar and the density $\rho_c=5.8$ g cm$^{-3}$.

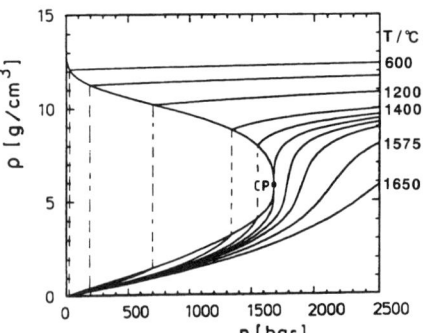

Fig. 1. Equation-of-state data of fluid mercury at subcritical and supercritical temperatures as a function of pressure.

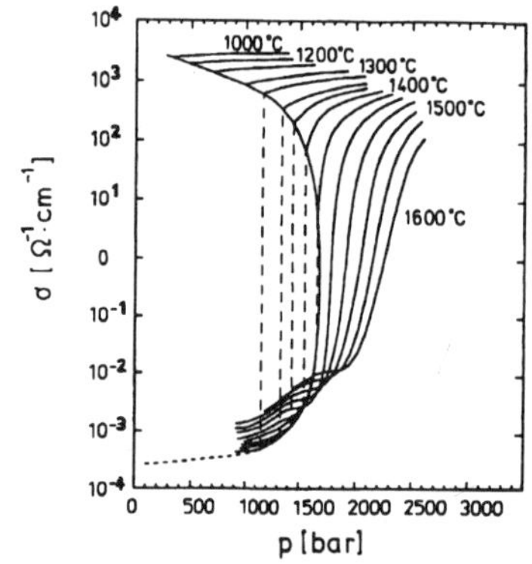

Fig. 2. Electrical conductivity isotherms of fluid mercury at subcritical and supercritical conditions as a function of pressure.

It is evident from the foregoing that there are substantial changes in the nature of bonding of mercury upon evaporation. For example, far below the liquid-vapor critical point (e.g. near the triple point), when the density difference between the coexisting phases is large, liquid mercury is reasonably well described by the nearly-free-electron (NFE) approximation, whereas in the vapor phase at sufficiently low densities, the valence electrons occupy spatially localized atomic orbitals. In such a situation, near the triple point, the liquid-vapor phase transition coincides precisely with a metal-nonmetal transition. Both the density and the electronic structure change on passing from one phase to the other. The vapor phase is nonmetallic and well characterized by highly polarizable atoms which interact through weak van der Waals forces, whereas valence electrons in the coexisting metallic liquid phase are dissociated. The structure of the liquid is regarded as built up of single screened ions, each diffusively uncoupled from every other with interactions thought to arise from screened Coulomb potentials.

However, as noted above, this description applies only to conditions far below the critical point. The challenge to our understanding, however, arises in the region closer to the critical point and in the supercritical region. Nearer to the critical point, the liquid density is much less and the vapor density much greater. At what density does the vapor become metallic, or the liquid nonmetallic?. It is this variation of the interaction in the neighbourhood of the metal-nonmetal transition that is difficult to deal with theoretically and which has important consequences for the liquid-vapor critical point phase transition.

## II. ELECTRICAL PROPERTIES

The most significant experiments relevant to these questions are measurements of electrical properties which signal the transformation from a metallic to a nonmetallic state. For that purpose we consider in fig. 3 the separate effects of temperature and density. At the highest density $\rho=13.6$ g cm$^{-3}$, corresponding to the liquid range near room temperature, the conductivity $\sigma$ is about $10^4$ $\Omega^{-1}$ cm$^{-1}$ and free-electron theory gives an electron mean free path $\lambda$ of about 7 Å which exceeds only slightly the mean interatomic spacing. Application of the NFE model leads to the conclusion that $\sigma$ can be satisfactorily explained within the context of the Ziman theory if the effect of the atomic $d$ states is included in the pseudopotential [Evans, 1970]. Mercury is thus essentially a NFE metal, despite the comparatively small mean free path. The NFE character is further confirmed by the observation that in this density range the low-frequency optical conductivity $\sigma(\omega)$ shows Drude-like behaviour [Hefner et al., 1980] and the Hall coefficient $R_H$ [Even and Jortner, 1972] retains the free-electron value. A comparison of the density dependence of $\sigma$, $\sigma(\omega)$, $R_H$ and the Knight shift $K$ [Warren and Hensel, 1982] shows that for densities down to about 11 g cm$^{-3}$ the properties of mercury can be described by the NFE theory of metals but, with further decreasing density, a rather gradual diminution of metallic properties occurs in the density range between 11 and 9 g cm$^{-3}$. For still smaller densities, the behaviour of $\sigma$, $\sigma(\omega)$ and $K$ is characteristic of a substance with semiconducting properties.

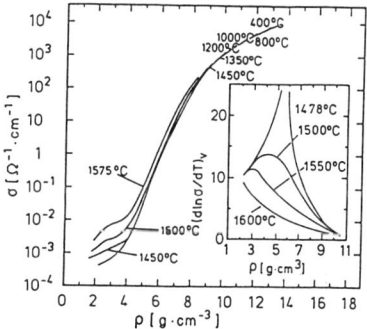

Fig. 3. Electrical conductivity of fluid mercury at constant subcritical and supercritical temperatures as a function of density. The temperatures are in degrees Celsius. The inset shows the constant-volume temperature coefficient at constant temperatures as a function of density.

As is well known, this type of metal-semiconductor transition is predicted by the Bloch-Wilson band model to occur for an expanded divalent metal such as mercury when the 6$s$ valence and 6$p$ conduction bands no longer overlap. In a crystal, a real energy gap appears and widens as the density decreases. Mott [1966] has proposed that the general features of the crystalline model survive in the liquid state with band edges smeared out by disorder. Thus, the density $N(E)$ of states is expected to tail into the gap owing to the loss of long-range order. The tails overlap in the region of the Fermi energy $E_F$ replacing the real energy gap of the crystal by a pseudogap or a minimum in $N(E)$ at $E_F$. The pseudogap depends strongly on density. When the magnitude of $N(E)$ in the pseudogap decreases with sufficient expansion to a negligibly small value, the optical properties of mercury must become compatible with the opening up of an energy gap. The latter is observed for expanded mercury for densities lower than 9 g cm$^{-3}$. For these densities the shape of the $\sigma(\omega)$ curves is characteristic of a substance either with a real energy gap or with a range of energies, which is so thinly populated with states that their contribution to the optical properties is negligibly small. This view is completely consistent with the observation of a sharp drop in the Knight shift [Warren and Hensel, 1982] for densities smaller than 9 g cm$^{-3}$.

In principle, one might hope to determine the energy gap experimentally by identifying it with the activation energy of the conductivity which successfully describes the temperature dependence of the conductivity in crystalline and amorphous solid semiconductors. However, in high-temperature liquids such as mercury not too far from its critical point, strong fluctuations in local density become important. Hence, the application of solid state concepts suffers from serious limitations. This is evident immediately on consideration of the constant-volume temperature coefficient $[\partial \ln \sigma / \partial T]_V$ plotted versus the density in the inset in figure 3 for various temperatures. The strong temperature dependence of $[\partial \ln \sigma / \partial T]_V$ around the critical density of 5.8 g cm$^{-3}$ is an obvious indication of the strong interplay between the liquid-vapor critical density fluctuations and the electrical characteristics.

## III. LIQUID-VAPOR ASYMMETRY OF MERCURY

As we have seen in the foregoing sections, liquid and gaseous mercury is very distinct from normal insulating fluids such as argon in that its electronic structure and interatomic interactions are strongly depending on the thermodynamic state of the system. It is obvious that there are substantial changes in the nature of bonding of mercury on evaporation, at which point there is a transition from a metallic bonding state to an insulating state characterized by highly polarizable atoms which interact via van der Waals forces. This strong state dependence of the interatomic interaction has been a primary motivation for energy-dispersive x-ray diffraction measurements of the static structure factors $S(Q)$ of fluid mercury over the whole liquid-vapor density range [Tamura and Hosokawa, 1994]. Typical results are presented in fig. 4 in form of the Fourier-transforms of $S(Q)$, i.e. the pair correlation functions $g(R)$, at different temperatures and densities. The liquid structure just above the melting point (e.g. at $20^o$C and 13.55 g cm$^{-3}$) is regarded as built up of single screened ions with interactions described by effective density dependent, spherically symmetric pairwise potentials. This view is prompted by the experimental observation that relatively little change in the local atomic arrangement occurs during melting. For mercury, the molar volume increases by only about 3.6%, and the average near-neighbor distance, given by the position of the first peak of the radial distribution function $g(R)$, is almost identical to the average near neighbor distance $R_{sol}$ in the crystalline structure close to melting. Consequently, the liquid metallic phase is treated as a monatomic state which typifies the solid structure.

The most significant aspect of the structural data for mercury (fig. 6) is that within the range of NFE behaviour between 13.55 g cm$^{-3}$ and 11 g cm$^{-3}$ thermal expansion proceeds by a structural evolution in which the average coordination number decreases roughly in proportion to the density while the average near-neighbor distance remains nearly constant. Such changes in shape and position of the pair correlation function are usually observed for fluids for which the nature of the effective pair interaction does not markedly change with density [Winter et al., 1987]. However, it is immediately evident from a glance at fig. 4 that this structural trend changes in a noteworthy way for densities smaller than 11 g cm$^{-3}$, i.e. when the diminution of metallic properties sets in. The decrease in the average coordination number becomes distinctly smaller whereas the average near-neighbor distance gradually increases approaching in the low density insulating vapor phase the value of the equilibrium distance $R_{vdW}$ of the weakly attractive potential of the van der Waals $Hg_2$ dimer. Thus the changes in the overall trends of the radial distribution function of mercury reflect the strong thermodynamic state-dependence of the effective interaction. However, the most significant experiments relevant to the effect of the variation of the interaction in course of the metal-nonmetal transition are those on the liquid-vapor coexistence curve of mercury [Götzlaff, 1988].

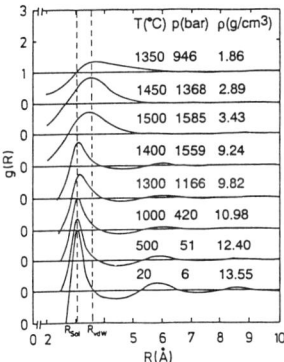

Fig. 4. Pair correlation function of mercury at different temperatures and densities.

Despite extreme conditions at the critical points of fluid mercury the liquid-vapor coexistence curve has been measured to a resolution $\Delta T/T_c = |T - T_c| / T_c \simeq 5 \times 10^{-4}$, which is close enough to demonstrate the important differences between mercury and insulating molecular fluids. Figure 5 shows a reduced plot ($\rho/\rho_c$ versus $T/T_c$) of the coexisting liquid densities $\rho_L$ and vapor densities $\rho_V$ together with the curve of the mean densities $\rho_d = (\rho_L + \rho_V)/2$ versus the reduced temperature for the metal mercury and for the inert gas xenon. The curve for mercury has been established indirectly from the intercepts of measured isochores with the vapor pressure curve. The density values have been assigned to the respective liquid and vapor branches of the coexistence curve by comparing the slopes of the isochores with those of the vapor pressure curve at the point of intersection.

The comparison of the coexistence curves in fig. 5 shows a lack of correspondence between mercury and xenon which supports the view that the interaction potentials operating in these fluids differ fundamentally. The fact that the cohesion in metallic systems consists of a sum of long-range Coulomb interactions has led to the speculation that critical points of metals could be in a different universality class than that of insulating fluids with potentials which decay with interatomic separation as $R^{-6}$. This is clearly not the case for mercury. The shape of its coexistence curve in the critical region can be described with high accuracy by the scaling law

$$\rho_L - \rho_V = B \left(\frac{\Delta T}{T_c}\right)^\beta$$

with the same exponent $\beta$ as molecular fluids (fig. 6a), where $B$ is a constant. The experimentally determined exponent $\beta$ for mercury is 0.36, a value only slightly higher than that found for the three dimensional Ising-model.

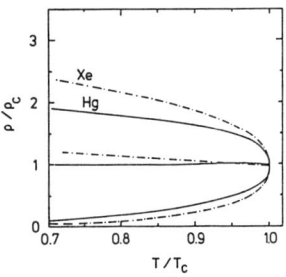

Fig. 5. Reduced densities of the coexisting vapor and liquid of the inert gas Xe compared with that of the metal Hg. The plot also shows the mean value of the liquid and vapor densities $\rho_d = (\rho_L + \rho_V)/2$.

The difference between the coexistence curves of xenon and mercury becomes visible mainly in the behaviour of the diameter which is strongly affected by the strong thermodynamic state-dependence of the effective particle interaction in mercury.

The plot in figure 6b demonstrates that fluid mercury violates the empirical law of rectilinear diameter. By contrast, the deviations from this law are extremely (mostly immeasurably) small for the coexisting curves of essentially all nonmetallic one-component fluids [Goldstein et al., 1987]. The law states that the locus of the tie-line midpoints $\rho_d = (\rho_L + \rho_V)/2$ is a linear function of $T$. Since both $\rho_L$ and $\rho_V$ approach the limiting density $\rho_c$ at the liquid-vapor critical point, the law can be written

$$\rho_d - \rho_c = D_1 \frac{\Delta T}{T_c}.$$

Modern theory of liquid vapor critical phenomena based on renormalization group studies [Nicoll, 1981] predicts that the temperature derivative of the diameter $(\partial \rho_d/\partial T)$ diverges at least as fast as the constant-volume specific heat $C_V$. That is, as the reduced temperature $T/T_c$ goes to zero, the diameter varies as

$$\rho_d - \rho_c = D_0 \left(\frac{\Delta T}{T_c}\right)^{1-\alpha} + D_1 \left(\frac{\Delta T}{T_c}\right) + \cdots$$

where $\alpha=0.11$ is the same exponent that describes the behaviour of the constant-volume specific heat $C_V$.

It is evident from a glance at figure 6b that close to the critical point the diameter behaviour in mercury is characterized by the $(1 - \alpha)$ term. However, the competing variations of the electronic structure of the liquid and vapor phases with density and temperature cause a strong wiggle in the diameter for $T < T_c$. Far below $T_c$ the diameter has a positive slope: in this region of temperature, when the density difference between the coexisting phases is large, the coexisting liquid is metallic ($\rho \geq 11$ g cm$^{-3}$), while the insulating vapor consists of atoms interacting through weak van der Waal forces. At higher temperatures, where the liquid is in the electronic transition range, the diameter actually slopes towards higher densities, opposite to the behaviour of molecular fluids. It seems reasonable to assume that the skewing of the diameter towards higher densities in the metal-nonmetal density transition region between 11 and 8 g cm$^{-3}$ is connected with the strong volume dependence of the energy gap between the 6s and 6p bands which leads to a contribution to the configurational or thermal pressure of the system. The broadening of the 6s band with decreasing volume is a result of the repulsive forces between mercury atoms at small interatomic separations. Consequently, excitation of electrons into the 6p state, whose energy is decreasing with increasing volume, has the effect of removing some of this repulsion, thereby lowering the pressure. The decrease in repulsion may be viewed as a decrease in the effective hard-core diameter of the atoms or as a softening in the effective interparticle forces.

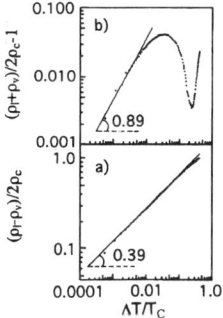

Fig. 6. a) Power-law analysis of the order parameter of mercury. b) Power-law analysis of the diameter of mercury.

## IV. SURFACE INDUCED PHENOMENA

Whilst the experimental results presented in the foregoing sections demonstrate that considerable progress has been made in understanding the bulk properties of mercury, and especially the number, the location and the order of the electronic and thermodynamic phase transitions, the properties of nonuniform fluid mercury, *i.e.* mercury for which the number density exhibits spatial variation, are less well understood. On the other hand many important phenomena are associated with strongly nonuniform situations. For example, in small clusters or in the vicinity of a substrate the properties of mercury may be completely different. An interface has indeed an important influence on the mercury atoms: it suppresses, over a half-space, the interaction with neighboring atoms, and it exerts on them an attraction or a repulsion. Both effects influence strongly the electronic and the thermodynamic properties of mercury.

Particularly interesting in this respect are small clusters which are characterized by huge surface to volume ratios. Mercury has a closed-shell $s^2$ atomic configuration and has a fairly large $s$-$p$ promotion energy $\Delta_{sp}(\text{Hg})=6.7$ eV. Therefore one expects that small clusters of mercury are insulating and bounded through weak van der Waals forces. This feature contrasts clearly with the metallic properties associated with the corresponding bulk material where metallic bonding results from the overlap between $s$- and $p$- bands.

The relation of cluster size to the valence electronic structure of noninteracting, isolated clusters has been studied directly by employing modern supersonic beam techniques. For example, the ionization thresholds for argon-, krypton-, xenon- and mercury-clusters [Ganteför et al., 1989; Rademann et al., 1987] produced in supersonic beams are shown in figure 7. Here, the ionization energies determined by energy-resolved mass spectrometry and by photoelectron-photoion-coincidence spectroscopy are plotted as a function of the cube root of the reciprocal number $n$ of atoms in the cluster. This kind of plot permits a comparison of the measured size dependence of the ionization potentials $I_P$ with the predictions of Born's theory of solvation for the difference in solvation energies of a charge $e$ in an infinite dielectric and in a sphere of the same dielectric [Mackor et al., 1988]

$$I_P = W + \frac{0.5e^2}{R} \times \frac{\epsilon - 1}{\epsilon}$$

which for an ideally conducting sphere ($\epsilon \to \infty$) leads to

$$I_P = W + \frac{0.5e^2}{R}$$

where $R$ represents the radius of an idealized sphere that contains $n$ atoms, $W$ represents the work function of the corresponding bulk material, $e$ the elementary charge, and $\epsilon$ the dielectric constant of the cluster material.

It is immediately evident from a glance at fig. 7 that in the small cluster region mercury- and the rare gas clusters show qualitatively the same behaviour (dashed curves), but the $I_P$ values for mercury exhibit a strong decrease by more than 2 eV in the size region between about 13 and more than 100 atoms approaching the curve (solid line) calculated for ideally conducting ($\epsilon \to \infty$) spheres. The experimental results indicate that for Hg clusters the electronic properties of the bulk metal evolve gradually with increasing particle size in the size range $13 < n < \infty$. The strong variation in the electronic structure in course of this transformation from non-metallic to metallic properties is also evidenced by studies of inner shell $5d$ autoionization spectra [Brechignac, 1985], of optical absorption spectra [Rademann et al., 1992] and of the atomic cohesion energy [Haberland et al., 1990].

The strong variation in the electronic structure with cluster size influences also the kinetic features of the vapor-liquid phase transition of mercury. Measurements of the critical supersaturation for the homogeneous nucleation of mercury in the temperature range 260 to 400 K [Martens et al., 1987] show that none of the current theories for homogeneous nucleation satisfactorily predicts the observed critical supersaturation. The measured values are about three orders of magnitude lower than the values predicted by the conventional Becker-Döring-Zeldovitch (BDZ) theory. It is noteworthy that the change in the value of the bulk liquid surface tension necessary to bring the classical nucleation theory in agreement with the experimental observation is about 40%.

In contrast, most molecular liquids require only a very small adjustment of the bulk liquid surface tension to bring nucleation theory and experiment into agreement. An important difference between mercury and the molecular fluids is that in the former the size-dependent metal-nonmetal transition occurs in small clusters. Profound changes in the electronic structure of small clusters take place with increasing size, which manifest themselves in a correspondingly strong size-dependence of all properties most probable also including density and surface tension. Any rigorous theory of nucleation of metal vapor must take into account that the very existence of the size-dependent metal-nonmetal transition can noticeably influence the homogeneous nucleation process in supersaturated metal vapor. The kinetic formalism of the BDZ-theory can possibly be retained, but the formation Gibbs free energies of small clusters containing 2-100 atoms must be calculated *ab initio* employing direct statistical mechanical evaluation of the partition functions of these small clusters, a procedure which is even difficult to realize for molecular clusters as large as those important in nucleation processes. However, for metal clusters the situation is even more complicated because in contrast to most molecular clusters, for which to a first approximation the behaviour can be described by reference to a single simple dispersion interaction potential for all cluster sizes, in metal clusters the effective interaction becomes size-dependent. The occurence of the nonmetal to metal transition with increasing size implies that the nature of the inter-particle interaction must change dramatically from a van der Waals-type to metallic interaction. The nucleation process in supersaturated metal vapor is fundamentally distinct from that of molecular vapor in that the interparticle potentials in the critical condensation nucleus are not quantities related to intrinsically atomic properties, but rather depend strongly on the nature of the electronic structure of the cluster.

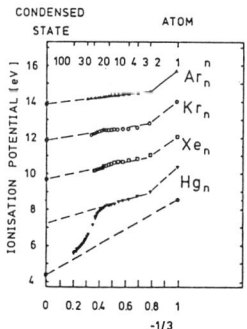

Fig. 7. Comparison of the size dependence of the effective ionization potentials of clusters of Ar, Kr, Xe, and He. The ionization potential is plotted versus the reciprocal cube root of the number of atoms $n$ in the cluster.

Another phenomena of current interest is the behaviour of fluid mercury at the interface of a rigid solid. It is well known that on substrates of low surface energy, onto which ordinary metals can be condensed easily, metallic mercury often has an extremely low sticking coefficient. Indeed, the closest example in common experience of a nonwetting substance is mercury on glass at room temperature. For such a situation it has been argued [Cahn, 1977; Ebner and Saam, 1977; Pandit *et al.*, 1982] that a wetting transition should occur at the interface of the rigid solid in contact with the two fluid phases (liquid and vapor) at a temperature near the critical temperature of the bulk fluid. The assumption that for mercury

$$\gamma_{sv} < \gamma_{sl} + \gamma_{lv}$$

where $\gamma_{sv}$, $\gamma_{sl}$ and $\gamma_{lv}$ are the substrate-vapor, substrate-liquid and liquid-vapor surface tensions, respectively, defined for liquid vapor coexistence conditions, implies, via Young's equation [Cahn, 1977], nonzero values for the contact angle $\theta$ (for mercury on glass, quartz, and sapphire $\theta$ is far beyond $90^o$). Then the liquid is said to partially wet (mostly called nonwetting) the substrate. Using critical scaling arguments for the surface tensions, Cahn (1977) showed that for some higher temperature $T$ below the critical temperature $T_c$ the inequality must become an equality; $\theta$ is then zero and the liquid completely wets, i.e. a transition to complete wetting occurs at the wetting temperature $T_w$. Depending on the relevant interaction potential (gas-solid surface, gas-gas interactions) the wetting transition is expected to be a continuous or a first order phase transition [Schick, 1990]. In the latter case the structure of the fluid interface changes discontinuously from that corresponding to an adsorbed liquid film of finite thickness below $T_w$ to one of infinite thickness above $T_w$. The temperature derivative $(\partial \gamma_{sv}/\partial T)$ of the substrate-vapor surface tension also changes discontinuously at $T = T_w$. Should first order wetting occur, then it is reasonable to expect that it is accompanied by a line of prewetting transitions [Ebner and Saam, 1977; Cahn, 1977], i.e. the discontinuous jump in film thickness which occurred at coexistence will persist off of coexistence. The only difference is that the film thickness cannot jump to an infinite thickness off of coexistence because the liquid is not thermodynamically stable off of coexistence. But it can jump from thin to thick. The prewetting line ends at the prewetting critical point.

The existence of such a prewetting transition in mercury has been detected recently [Hensel and Yao, 1994] by employing reflecting light from a vertical sapphire-mercury vapor interface. In order to locate the prewetting line, reflectivity measurements were performed at constant temperature $T$ by increasing $p$, the pressure of the vapor, towards its value at bulk saturation $p_{sat}$ or alternatively at constant $p$ by decreasing $T$ towards $T_{sat}$. The line of first-order prewetting transitions $p_{pw}(T)$ [$< p_{sat}(T)$] extends from the point of the (first order) wetting transition $(T_w, p_{sat})$ to a prewetting critical point $(T_{pwc}, p_{pwc})$. Typical results for two different experimental paths are shown in figure 8. The shape of curve 1 is characteristic for nonwetting off of coexistence, i.e. $T < T_w$. The measured reflectivity for $p < p_{sat}$ (solid line) coincides with that calculated employing Frenel's formula for normal incidence. There is no indication of a precursor of wetting. The sudden increase in reflectivity (dashed line) indicates bulk phase separation, i.e. condensation to the liquid state.

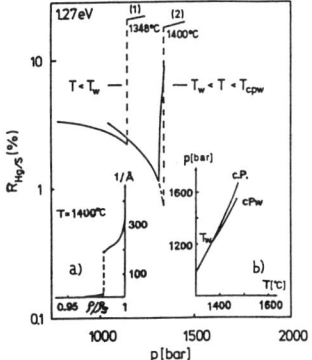

Fig. 8. Reflectivity of mercury vapor against an optically transparent sapphire substrate at constant temperatures $T$ as a function of pressure near vapor-liquid coexistence for $T < T_W$ (curve 1) and for $T_W < T < T_{pwc}$ (curve 2). See text for details. Inset a) shows the wetting-layer thickness calculated with a slab model employing the theory of the reflectivity of thin absorbing films. Inset b) shows the bulk vapor pressure curve terminated by the critical point c.p. together with the prewetting line which terminates at high temperature at the prewetting critical point $cp_W$ and at low temperature at the wetting temperature $T_W$.

In contrast, curve 2 shows the typical shape of the reflectivity curves observed for $T_w < T < T_{pwc}$. Distinctly before condensation sets in (dashed line), the reflectivity curve changes discontinuously the sign of its slope at the prewetting pressure $p_{pw}$. The values calculated for the reflectivity at the sapphire-mercury vapor interface deviate significantly in this range from the experimental curve. By all means, the features displayed by curve 2 are completely consistent with the abrupt formation of a wetting film with the density and optical properties of the coexisting liquid at $T_{sat}=1400^oC$. Since the optical data of liquid mercury are known at coexistence [Hefner et al., 1980; Ikezi et al., 1978] the wetting layer thickness can be calculated by employing the theory of optical reflectance and transmittance properties of thin absorbing multilayer films [Berning, 1963]. Wetting layer thickness calculated by this theory for path 2 are shown in the inset a) of figure 8. Complete wetting occurs as coexistence is approached at $p/p_{sat} = 1$. In addition, there is a jump in the film thickness when it crosses the prewetting line at a bulk pressure which is 2 percent lower than its value at saturation. This jump gets smaller as the prewetting line is crossed farther from coexistence and vanishes completely at the prewetting critical point. The prewetting line determined in this way is displayed in the inset b) of figure 8. It extends from the wetting transition at $T_w$ to the prewetting critical point at approximately $1470^oC$ and 1600 bar. A noteworthy observation is that $T_w$ lies in the range where the gradual metal-nonmetal transition occurs in liquid mercury at coexistence. This again is an indication for the intimate interplay between the vapor-liquid critical phenomena and the changes in the electronic structure associated with the metal-nonmetal transition.

## ACKNOWLEDGMENT

Financial support by the Deutsche Forschungsgemeinschaft and the Fonds der Chemischen Industrie is gratefully acknowledged. M. Y. is thankful to the Alexander von Humboldt-Stiftung for a grant for part of this work.

## REFERENCES

Berning, P.H., 1963, *Physics of Thin Films*, ed. by Georg Hass, Vol. 1 (Academic Press, New York).
Brechignac, C., M. Broyer, Ph. Cahuzac, G. Delacretaz, P. Labastie, J.P. Wolf, and L. Wöste, 1985, Chem. Phys. Lett., **120**, 559.
Cahn, J.W., 1977, J. Chem. Phys., **66**, 3667.
Ebner, C. and W.F. Saam, 1977, Phys. Rev. Lett., **38**, 1486.
Evans, R., 1970, J. Phys. C: Met. Phys. Suppl., **2**, 137.
Even U. and J. Jortner, 1972, Phys. Rev. Lett., **28**, 31.
Ganteför, G., G. Bröker, E. Holub-Kappe, and A. Ding, 1989, J. Chem. Phys., **91**, 7972.
Goldstein, R.E., A. Parola, N.A. Ashcroft, M.W. Pestak, M.H.W Chen, J.R. de Bruyn, and D.A. Balzarin, 1987, Phys. Rev. Lett., **58**, 41.
Götzlaff, W., 1988, Doctoral thesis, University of Marburg.
Haberland, H., H. Kornmeier, H. Langosch, M. Oschwald, and G. Tanner, 1990, J. Chem. Soc. Faraday Trans., **86**, 2473.
Hefner, W., R.W. Schmutzler, and F. Hensel, 1980, J. Physique Coll., **41**, C8 62.
Hensel, F. and E.U. Franck, 1966, Ber. Bunsenges. Phys. Chem., **70**, 1154.
Hensel, F. and M. Yao, 1994, to be published.
Ikezi, H., K. Schwarzenegger, A.L. Simons, A.L. Passner, and S.L. McCall, 1978, Phys. Rev. B, **18**, 2494.
Kikoin, I.K. and A.R. Sechenkov, 1967, Phys. Metals Metall., **24**, 74.
Landau, L. and G. Zeldovitch, 1943, Acta Phys. Chem., USSR, **18**, 1940.
Mackor, G., A. Nitzan, and L.E. Brus, 1988, J. Chem. Phys., **88**, 5076.
Martens, J., H. Uchtmann, and F. Hensel, 1987, J. Phys. Chem., **91**, 2489.
Mott, N.F., 1966, Phil. Mag., **13**, 989.
Mott, N.F., 1974, *Metal-Insulator-Transitions*, (Taylor and Francis, London).
Nara, S., T. Ogawa, and T. Matsubara, 1977, Prog. Theo. Phys., **57**, 1474-1489.
Nicoll, J.F., 1981, Phys. Rev. A, **24**, 2203.
Pandit, R., M. Schick, and M. Wortis, 1982, Phys. Rev. B, **26**, 5112.
Rademann, K., B. Kaiser, U. Even, and F. Hensel, 1987, Phys. Rev. Lett., **59**, 2319.
Rademann, K., O. Dimopoulou-Rademann, M. Schlauf, E. Even, and F. Hensel, 1992, Phys. Rev. Lett., **69**, 3208.
Schick, M., 1990, *Liquids At Interfaces*, ed. by J. Charvolin, J.F. Joanny, and J. Zinn-Justin, (North-Holland, Amsterdam).
Tamura, K. and Hosokowa, 1994, J. of Physics: Condensed Matter, **6**, A 241.
Warren, W.W. and F. Hensel, 1982, Phys. Rev., **B 26**, 5980.
Winter, R., T. Bodensteiner, W. Gläser and F. Hensel, 1987, Ber. Bunsenges. Phys. Chem., **91**, 1327-1330.
Yao, M. and H. Endo, 1982, J. Phys. Soc. Japan, **51**, 966.
Yonezawa, F. and T. Ogawa, 1982, Prog. Theo. Phys., Suppl., **72**, 1.

# VIII.2

# LIQUIDS NEAR THE CRITICAL POINT

H. Endo[†] and M. Yao[‡]

[†] Faculty of Engineering, Fukui Institute of Technology, 910 Fukui, Japan
[‡] Department of Physics, Faculty of Science, Kyoto University
606-01 Kyoto, Japan

The electronic and structural properties of expanded fluid Hg and Se are discussed utilizing the recent experimental results. Anomalous behaviors in the transport and optical properties due to large density fluctuations are observed near the critical points which are accompanied by structural modification.

## I. INTRODUCTION

Study on the expanded metals was begun in late 1960s. Since that time, considerable effort has been devoted to develop the experimental techniques under extreme conditions and the experimental accuracy has been much improved. Up to now many properties have been measured especially for Hg, alkali metals and chalcogen.

The main topics in this paper are concerning expanded Hg and chalcogen. The phase diagram on $P$-$T$ plane and $P$-$V$ plane is shown schematically in Fig. 1. As seen in Fig. 1, the density of liquid can be changed widely by a suitable combination of temperature and pressure. The density of Hg is 13.7g/cc at the triple point (TP) and is reduced to 5.6g/cc at the liquid-gas critical point (CP). In Table 1 the critical temperature, pressure and density are tabulated for some elements whose critical point can be attained by static heating and pressurizing methods. In general, the critical point of metallic element is located at high temperature and pressure compared with nonmetallic element, owing to the metallic cohesion.

Table 1. Critical temperature ($T_c$), critical pressure ($P_c$) and critical density ($d_c$) for various materials.

|    | $T_c$(K) | $P_c$(bar) | $d_c$(g/cm$^3$) |
|----|----------|------------|-----------------|
| K  | 2280     | 161        | 0.19            |
| Rb | 2017     | 124.5      | 0.292           |
| Cs | 1924     | 92.5       | 0.379           |
| Hg | 1751     | 1670       | 5.8             |
| S  | 1310     | 204        | 0.60            |
| Se | 1860     | 380        | 2.0             |
| Ar | 150.7    | 48.0       | 0.531           |

An attractive evidence in expanded fluid is an appearance of various clusters which are originated from large density fluctuations near the critical point. When one has an atom and add other atoms successively, as shown in Fig. 2 schematically, how the structure and properties are changed? Especially interesting question is "At what number of atoms the system begins to exhibit bulk-like properties?"

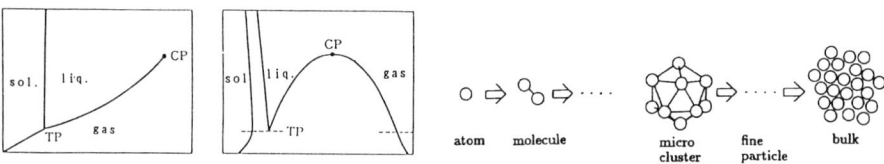

Fig. 1. Phase diagram on P–T and P–V plane.

Fig. 2. Atoms, molecules and clusters which are formed in expanded fluid.

## II. EXPANDED FLUID Hg

X-ray and neutron diffraction experiments of expanded fluid Hg suggest that the nearest neighbor distance, $R_1$ is almost independent of density, while the average co-ordination number, $N_1$ shows a nearly linear dependence on density [1]. This result indicates that the fluid Hg is not expanded uniformly.

Knight shift, $K$ and optical absorption data give useful information about the metal-nonmetal (M-NM) transition in expanded metal. As seen in Fig. 3 the Knight shift $K$ of Hg, which is proportional to the $s$-electron part of the density of states at the Fermi level $E_F$, decreases with density and vanishes near 9g/cc. In this figure the optical gap ($E_g$) deduced from the optical absorption data is also shown [3]. The absorption edge shifts rapidly toward lower energies with increasing density. The optical gap becomes zero around 8g/cc. These indicate that the gap of density of states opens around 9g/cc. Therefore, the M-NM transition in fluid Hg occurs around 9g/cc owing to the loss of overlapping between the valence and conduction bands.

The optical absorption spectra of fluid Hg at higher densities up to 7.5g/cc are shown in Fig. 4[3,4]. The absorption edge shifts rapidly towards lower energies with increasing density. It is noted that an additional absorption band is observed in the near-infrared (NIR) region at densities above 4g/cc. This NIR band grows remarkably with increasing density. The NIR band has a negative slope in contrast to the absorption edge with positive slope. It is interesting to investigate how the NIR band is influenced by the addition of impurity element. Figure 5 shows the optical absorption spectra of Hg-0.22%Bi amalgams at 1540 °C [4]. The numbers given on each curve are pressure in bar and the density in g/cc is indicated in the parenthesis. The shape of the NIR band is remarkably changed by the addition of multivalent element Bi, which may suggest that the NIR band is sensitive to the balance between the numbers of electrons and holes.

Figure 6 shows the temperature variation of the thermopower, $S$ at constant pressures [4]. At 1680 bar, which is slightly higher than the critical pressure of about 1670 bar, $S$ changes dramatically from $-200$ $\mu$V/K to $-3000\mu$V/K in a narrow temperature ranged around 1480°C and then becomes small negative with further increasing temperature. At higher pressures such an anomalous behavior in $S$ becomes smaller. Unlike some of the previous data, $S$ remains negative in the whole temperature and pressure range under the present investigation. In order to know at what temperature, pressure and densities $S$ exhibits anomalous behavior, the contours of constant $S$ are shown by the solid lines on the temperature and pressure plane in Fig. 7 [4]. The isochore curves of 4 and 6g/cc are also shown by the dashed lines in the figure. The contours of $-300\mu$V/K nearly coincide with the isochore of 6g/cc. It is evident that the absolute value of $S$ is large in a narrow region bounded by the isochores of 6g/cc and 4g/cc and becomes extremely large in the near vicinity of the critical point. The anomalous behavior in the thermopower may be associated with the dynamical critical phenomena in the systems where the electrons play an important role.

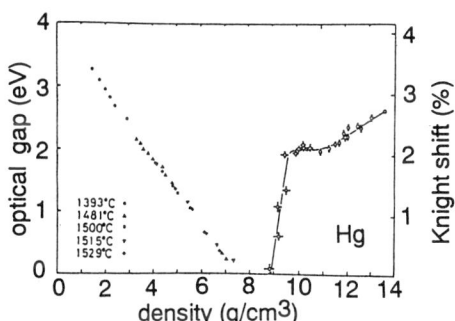

Fig. 3. Density variation of the optical gap and the Knight shift for fluid Hg.

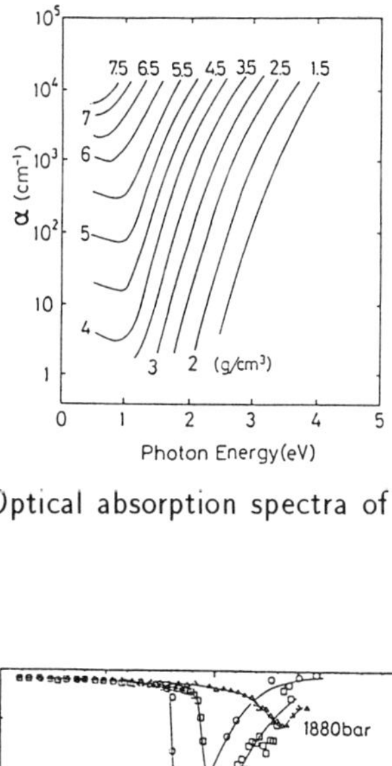

Fig. 4. Optical absorption spectra of fluid Hg.

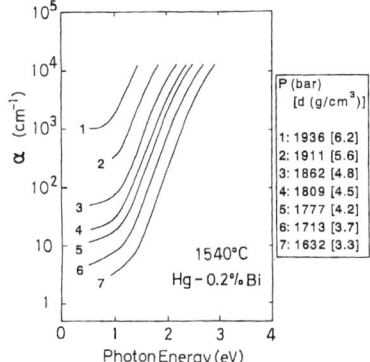

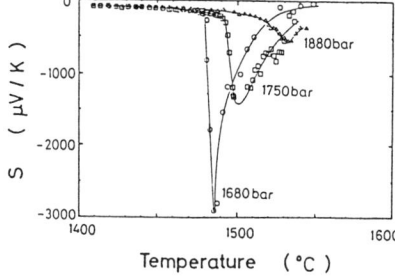

Fig. 5. Optical absorption spectra of Hg-0.2%Bi.

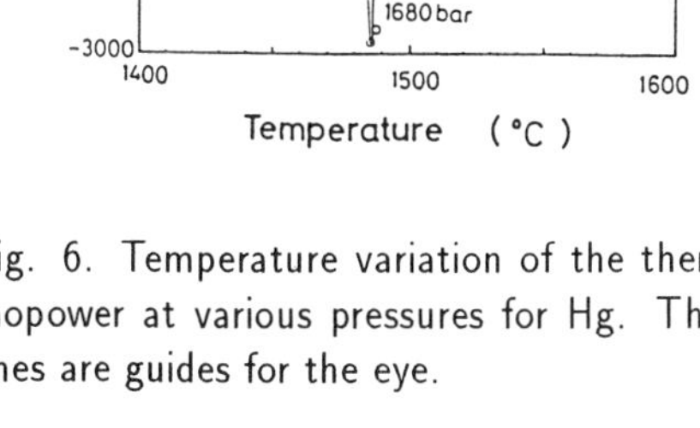

Fig. 6. Temperature variation of the thermopower at various pressures for Hg. The lines are guides for the eye.

## III. EXPANDED FLUID CHALCOGEN

Many experimental and theoretical investigations on liquid chalcogen such as sulfur, selenium and tellurium have been done. The interest in the liquid chalcogen systems stems from the presence of the close correlation between chain conformations and electronic states. Recent computer simulations for the chalcogen chain or ring molecule with various geometries have demonstrated that the band structure and the charge distribution strongly depend on the configuration [5-7]. Experimentally, photoelectron spectra of small Se ring molecules have been measured in supersonic molecular beam, and large differences are found in the structures of the spectra between Se molecules with even and odd numbers of atoms [8].

Chalcogens have usually chain structure with a variety of conformations. The trigonal structure is known as most stable phase (see Fig. 8). The repulsive exchange interaction between lone pair (LP) orbitals occupied by two electrons plays an important role in stabilizing the helical conformation. In liquid state fluctuations in bond length, bond angle and dihedral angle are large. A recent molecular-dynamics simulation for liquid Se has confirmed that the dihedral angle distortions are more prominent than both bond length and bond angle [9].

Recently some optical measurements have been done for liquid chalcogens at high temperature and high pressure [10,11]. Data of the optical absorption and optical conductivity for liquid Se indicate that the optical gap of liquid Se closes near the critical point (CP). It is interesting to investigate the structural and electronic properties of liquid chalcogen in the wide density range by changing temperature and pressure, since substantial changes in the electronic states are induced by modification of the intra- and inter-chain correlations.

Figure 9 shows the contours of constant density denoted by dashed line on $P$–$T$ plane [12]. When one approaches to near CP by raising temperature, isochore curve bends. Near this region thermodynamic data suggest that the volume contracts. The hatched area shows nearly metallic region in which dc conductivity $\sigma$ is higher than $100\Omega^{-1}cm^{-1}$ [13]. The dotted line in the figure represents the region where the optical gap $E_g$ closes, which is deduced from the absorption data [10]. It is interesting that $E_g \sim 0$ lies near the region which the anomalous behavior is observed in the isochore curve. Liquid Se near the melting point exhibits similar semiconducting behavior to that of crystalline Se. From the measurements of viscosity [14], magnetic susceptibility [15] and NMR [16], an average chain length may be estimated to be $10^4$–$10^5$ atoms near the melting point. The structure of liquid Se is strongly affected by changing temperature and the mean chain length decreases rapidly with increasing temperature, producing a number of dangling bond states. It is interesting that expanded liquid Se near CP exhibits the transition from semiconducting to metallic state which may be accompanied by a radical change of the polymeric structure.

Figure 10 shows the estimated chain length as a function of temperature which corresponds to the number of atoms included in a chain [15-19]. The chain length of Se near CP is extremely small, which is close to that of liquid Te. It should be pointed out that liquid Te shows metallic behavior near the melting point. However, the microscopic origin of the metallicity for liquid Te has not been solved yet.

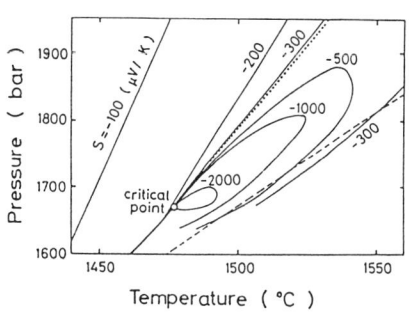

Fig. 7. Contours of constant thermopower are shown on P–T plane by the solid lines and the isochores of 4 and 6g/cc by the dashed and dotted lines, respectively.

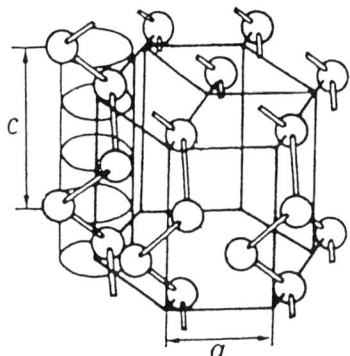

Fig. 8. Trigonal structure of chalcogen.

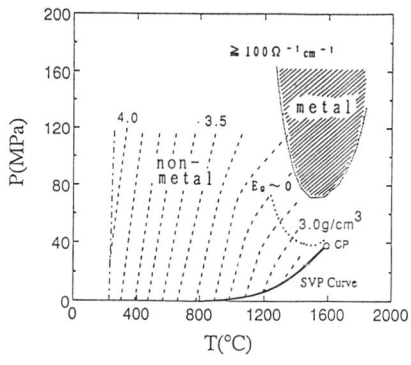

Fig. 9. Contours of constant density on P–T plane.

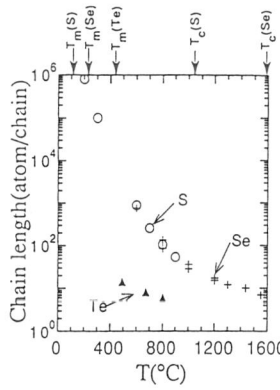

Fig. 10. Variations of chain length of S, Se and Te with temperature.

Figure 11 shows the variation of $\sigma(\omega)$ for liquid S deduced from the reflectivity data at 1100°C and at the pressure from 100 to 25MPa [11], where the density decreases from 1.24 to 0.1g/cc. The optical conductivity $\sigma(\omega)$ is closely related to the joint density of states. The $\sigma(\omega)$ curve at 100MPa has a peak around 3.9eV, which is associated with the optical transition from the lone pair state (LP) to the antibonding state ($\sigma^*$). The peak height of $\sigma(\omega)$ rapidly diminishes with decreasing pressure. It is noted that the peak position is nearly independent of pressure between 100 and 300MPa. Since application of pressure mainly changes the inter-chain distance, no shift in the peak position suggests that the optical transition in liquid S is not associated with the transition between neighboring chains, that is, the optical transition in liquid S arises from that in a single or isolated chain.

Figure 12 shows the temperature variation of the optical conductivity $\sigma(\omega)$ for liquid Se at the pressure of 170MPa [11]. The $\sigma(\omega)$ curve at 500°C rises near 1.7eV which is close to the optical gap, and has a broad peak around 3.5eV, corresponding to LP-$\sigma^*$ transition. The peak shifts to lower energy side and broadens with increasing temperature. It is noted that a remarkable shift of peak is observed above 1100°C. The increase of $\sigma(\omega)$ at low photon energies corresponds to the appearance of the Drude term which implies that liquid semiconducting Se transforms to metallic states. The difference in $\sigma(\omega)$ spectra between liquid Se and S may be related to larger extension of the outer $p$ orbitals in Se than that in S. It is reasonable assumption that the interchain coupling in liquid Se is rather stronger than in liquid S. At higher temperature the frequent excitation from the neighboring chain occurs which result in the bond weakening and bond breaking.

Figure 13 shows the sketch of characteristic aspects of the structural variation in expanded liquid Se. The change in bond length and bond angle of liquid Se by raising temperature may be small compared with that in dihedral angle. At higher temperature the average distance between neighboring chains increases and chain length becomes short. In the gas states under super-critical region there appears Se$_2$ dimer. Near CP at high temperatures and pressures short chains aggregate each other owing to large density fluctuations and metallic clusters are formed as mentioned above. It is considered that the atomic and electronic structures of this metallic cluster may be similar to those of Te droplet near the melting point.

Figure 14(a) shows the results of $\sigma(\omega)$ for liquid Se-Te mixtures at 700°C and 40MPa [11]. Each $\sigma(\omega)$ curve has one broad peak. The peak position lies around 3eV on Se rich side, and lies around 2eV with the Drude term on Te rich side. This is consistent with the result for the semiconductor to metal transition region on the temperature and concentration plane denoted by hatching in Fig. 14(b) which was decided from dc conductivity measurements [20]. The Drude term is observed in $\sigma(\omega)$ curve for the Te rich mixtures.

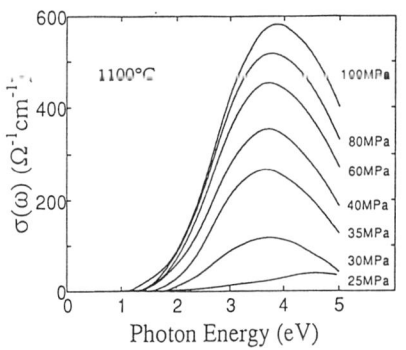

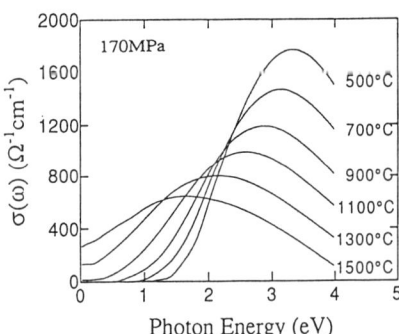

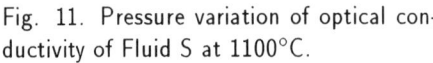

Fig. 11. Pressure variation of optical conductivity of Fluid S at 1100°C.

Fig. 12. Temperature variation of optical conductivity at 170MPa in liquid Se.

It should be emphasized that there exits a strong correlation between the electronic states and atomic arrangements in liquid Te. The droplet with small size was used as a sample in order to get deep supercooled state. Figure 15 shows the correlation function, $S(Q)$ for liquid Te at 430, 350 and 280°C together with $S(Q)$ for liquid Se [21]. The profile of $S(Q)$ for the droplets at 430°C is quite similar to that for the bulk liquid Te. When the temperature is lowered, the height of the first peak of $S(Q)$ decreases and it becomes lower than that of the second peak. The profile of $S(Q)$ for Te supercooled down to 280°C is similar to that for liquid semiconducting Se near the melting point. The detailed analysis reveals that the first peak of $S(Q)$ is mainly associated with the interchain correlation.

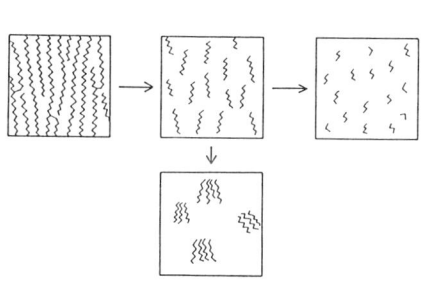

Fig. 13. Structural variations in expanded liquid Se

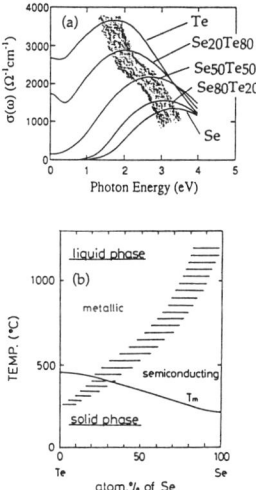

Fig. 14. (a) Concentration variation of optical conductivity of liquid Se-Te mixtures at 40MPa and 700°C. (b) Semiconductor to metal transition region for Se-Te mixtures (including supercooled states) on the consentration-temperature plane, shown by the hatching.

EXAFS is a good tool to investigate the local structure of disordered systems. Nevertheless, there have been only a few applications of the EXAFS technique to liquids. A serious problem is the fact that EXAFS signals from a liquid are in general strongly damped due to the inherent disorder of the atomic arrangement and thermal agitation. EXAFS can give useful information on the neighboring configuration around a particular atom as long as the atoms are covalently bonded. The analysis of EXAFS data for liquid Te reveals that a two-fold coordinated chain structure subsists and there exit shorter bonds of length 2.8Å and longer bonds of length 3.1Å [21]. As shown in Fig. 16 the number of shorter bonds increases as the temperature is lowered and the number of longer bonds is substantially diminished near the freezing temperature [21].

We have studied the dynamical structure of liquid Te by means of inelastic neutron scattering measurements. Figure 17 shows the vibrational density of states (V-DOS) as a function of energy at the scattering angle 35° for trigonal Te at 400°C, supercooled liquid Te at 400°C and liquid Te at 467°C[22]. For trigonal Te, V-DOS consists of three groups of bands centered at about 16, 10, 5meV. For supercooled liquid Te at 400°C, there remain the three bands in V-DOS. This indicates that Te has a chain structure even in the liquid state. The peaks are, however, broad compared with those for trigonal Te. The intensity of the bond stretching mode and bond bending mode are remarkably reduced. The peak position of the torsional and bond bending mode are nearly the same as those for trigonal Te. On the other hand, the peak of the bond stretching mode seems to move slightly towards higher energy. For liquid Te at 467°C, the V-DOS is even more broadened and yet still the three-bands structure is more or less preserved. At the scattering angle 35°, the peak at about 16meV is strongly reduced and a new distinct peak appears at 13meV. It is interesting that in Raman scattering study of liquid Te Yashio and Nishina [23], and Magana and Lannin [24] observed a peak near 14meV which is not seen for trigonal Te. The low energy peak in the bond stretching modes at 467°C is direct evidence for the weak bonds in the chain of liquid Te. The weak bond is due to the charge transfer from the lone pair orbitals in a chain to the $\sigma^*$ orbitals in the neighboring chain. It is concluded that semiconductor to metal transition observed for liquid Se near CP is arisen from the formation of the clusters which have similar nature with metallic Te droplet.

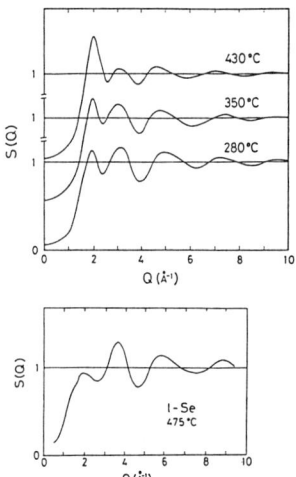

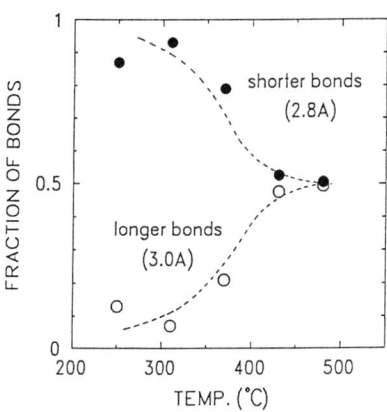

Fig. 15. Correlation function, $S(Q)$ for liquid Te at 430, 350 and 280°C together with $S(Q)$ for liquid Se. Pressure variation of optical conductivity of Fluid S at 1100°C.

Fig. 16. Fraction of the number of shorter bonds and longer bonds as function of temperature.

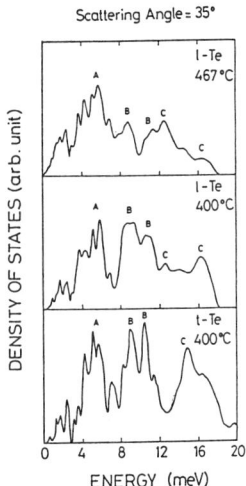

Fig. 17. Vibrational density of states of trigonal Te at 400°C, supercooled liquid Te at 400°C and liquid Te at 467°C at the scattering angle 35°.

## REFERENCES

[1] K. Tamura and S. Hosokawa, J. Non-Cryst. Solids, **156-158** (1993) 646.
[2] W.W. Warren,Jr. and F. Hensel, Phys. Rev. B, **26** (1982) 5980.
[3] M. Yao, H. Hayami, and H. Endo, J. Non-Cryst. Solids, **117/118** (1990) 473.
[4] M. Yao, K. Takehana, and H. Endo, J. Non-Cryst. Solids, **156-158** (1993) 807.
[5] M. Springborg and R.O. Jones, J. Chem. Phys., **88** (1988) 2652.
[6] D. Hohl, R.O. Jones, R.Car, and M. Parrinello, Chem. Phys. Lett., **139** (1987) 540.
[7] D. Hohl, R.O. Jones, R.Car, and M. Parrinello, J. Chem. Phys. **89** (1988) 6823.
[8] J. Becker, K. Rademan, and F. Hensel, Z. Naturforsch, **46a** (1991) 453.
[9] D. Hohl and R.O. Jones, Phys. Rev. B, **43** (1991) 3856.
[10] K. Tamura, J. Non-Cryst. Solids, **117/118** (1990) 450.
[11] H. Ikemoto, I. Yamamoto, M. Yao, and H. Endo, J. Phys. Soc. Jpn., **63** (1994) 1611.
[12] R. Fischer, R.W. Schmutzler, and F. Hensel, J. Non-Cryst. Solids, **35 &36** (1980) 1295.
[13] H. Hoshino, R.W. Schmutzler, and F. Hensel, Ber. Bunsenges. Phys. Chem., **90** (1986) 587.
[14] J.C. Perron, J. Rabit, and J.F. Rialland, Philos. Mag. B, **46** (1982) 321.
[15] W. Freyland and M. Cutler, J. Chem. Soc. Faraday Trans., **76** (1980)756.
[16] W.W. Warren, Jr. and R. Dupree, Phys. Rev. B, **22** (1980) 2257
[17] M. Misawa, J. Phys.: Condens. Matter, **4** (1992) 9491.
[18] H. Radscheit and J.A. Gardner, J. Non-Cryst. Solids, **35&36** (1980) 1263.

[19] D.M. Gardner and G.K. Fraenkel, J. Am. Chem. Soc., **78** (1956) 3279.
[20] M. Yao, M. Misonou, K. Tamura. K. Ishida, K. Tsuji, and H. Endo, J. Phys. Soc. Jpn., **48** (1980) 109.
[21] H. Endo, J. Non-Cryst. Solids, **156-158** (1993) 667.
[22] H. Endo, T. Tsuzuki, M. Yao, Y. Kawakita, K. Shibata, T. Kamiyama, M. Misawa and K. Suzuki, J. Phys. Soc. Jpn., submitted.
[23] M. Yashiro and Y. Nishina, in *The Physics of Selenium and Tellirium* edited by E.Gerlach and P.Grosse (Springer,1979), p.206.
[24] J.R. Magana and J.S. Lannin, Phys. Rev. Lett., **51** (1983) 2398.

# VIII.3

# VOLUME DEPENDENCE OF THE STRUCTURE OF LIQUID METALS

K. Tsuji

Department of Physics, Keio University, 3-14-1 Hiyoshi, Yokohama 223, Japan

Recent progress in the structural study of liquid metals under pressure is reported. X-ray diffraction for liquid metals has been measured under pressure by using synchrotron radiation. Pressure dependence of static structure factor $S(Q)$ and that of pair distribution function $g(R)$ have been obtained. The changes of the nearest neighbor distance $R_1$ with volume $v$ are compared for several liquid metals; liquid alkali metals with s-valence electrons and liquid polyvalent metals with p-valence electrons. These volume dependences of the structural data are also compared with those for expanded fluids.

## I. INTRODUCTION

In liquid metals, application of pressure causes remarkable increase in the electron density $n$ and changes the screening length and the electronic levels such as the Fermi level. These changes in the electronic structure reflect on the ion-ion pair potential and on the static structure factor of liquid metals. Although measurements of X-ray or neutron diffraction yield information on the atomic arrangements in liquid state, there are some difficulties in measuring diffraction of liquid metals under pressure [Brown and Barnett, 1972; Egelstaff et al., 1974; Tsuji et al., 1975; Endo and Tsuji, 1985]: (1) generation of high pressure and high temperature, (2) elimination of strong background diffraction from the pressure-transmitting medium and the sample container, (3) prevention of reactions of the liquid specimen with surrounding materials, etc. Recently these difficulties have been conquered by using a cubic-type high-pressure apparatus and synchrotron radiation with high brightness, high energy and small divergence [Tsuji et al., 1989]. This method has been applied to the investigations of the pressure-induced structural changes in several liquid metals [Tsuji et al., 1988; Tsuji, 1990; Tsuji et al., 1990; Yaoita et al., 1990; Yaoita et al., 1991; Yaoita et al., 1992; Yaoita et al., 1993].

## II. EXPERIMENTAL

MAX80, an apparatus for high-pressure and high-temperature experiments installed in National Institute for High Energy Physics (KEK), was used. The high pressure vessel was a cubic-type apparatus [Shimomura et al., 1985a, 1985b; Shimomura, 1986] which compressed a cubic pressure transmitting medium by six anvils. Each tungsten carbide anvil had a center flat of $4 \times 4$ or $6 \times 6$ mm$^2$. A sample was filled in a 2-mm-o.d., 1-mm-i.d. Teflon or BN capsule. The capsule was embedded in the cube, which was made of a mixture of amorphous boron and epoxy resin. A Chromel-Alumel thermocouple was placed at the center of the sample assembly and taken out through the gaps between the anvils. An internal pressure calibrant of NaCl was placed at lower part of the assembly.

High-energy and high-intensity X-rays of synchrotron radiation from a bending magnet in the TRISTAN accumulation ring (AR) or from a high-field wiggler in the Photon Factory (PF) storage ring were used. At both beamlines, X-rays with high energies up to 120 keV are available. Because the absorption coefficient is small for high energy X-rays, a thick specimen can be used. A sharp slit system eliminated the background diffraction from the pressure-transmitting medium and the sample container. Diffraction intensity from liquid metals was measured at ten diffraction angles of $2\theta$ from $2°$ to $15°$. After the correction for the energy distribution of incident X-rays, absorption by the specimen and pressure-transmitting medium, etc., $S(Q)$ for liquid metals was obtained. Details of experiment and data analysis were described elsewhere [Tsuji et al., 1989].

## III. RESULTS AND DISCUSSION

A. Alkali metals

Atoms of alkali metals have s-valence electrons. Their wave functions spread spherically around the ions. A hard-sphere model was successfully applied to describe the structure of liquid alkali metals [Ashcroft and Leckner, 1966]. The effective diameter of the hard sphere should vary with pressure due to the change in the screening effect. Some calculations were done for the volume dependence of the hard-sphere diameter in alkali metals [Dickey et al., 1967].

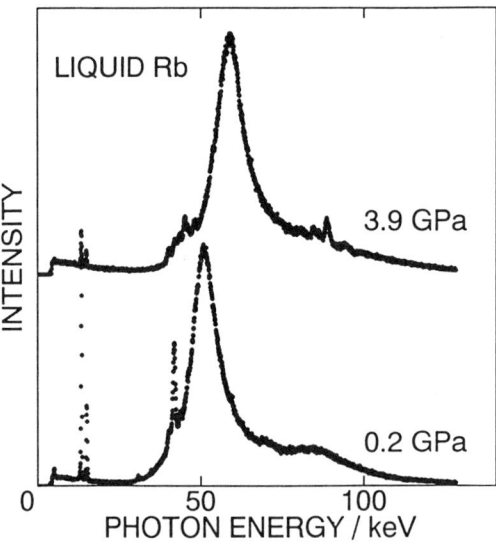

Fig. 1. X-ray diffraction pattern of liquid Rb at 0.2 GPa and 370 K (lower) and at 3.9 GPa and 540 K (upper). The data were taken by energy-dispersive method at $2\theta = 3.5°$.

In Fig. 1, examples of raw data are shown for liquid Rb. Diffraction intensity curve consists of broad peaks from the liquid Rb and small peaks from the Teflon container. Sharp peaks at the low energy region are fluorescence X-rays of RbK$\alpha$ and RbK$\beta$. With increasing pressure, the position of the first peak shifts towards higher energy.

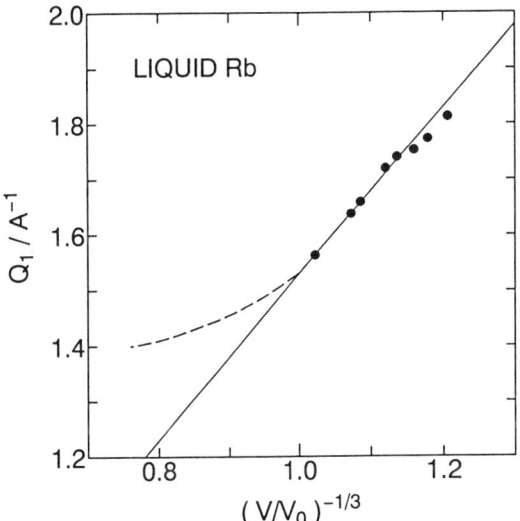

Fig. 2. The position of the first peak $Q_1$ for liquid Rb as a function of $(v/v_0)^{-1/3}$. The solid line shows the expected values from a uniform-contraction model. The broken line shows the results for expanded liquid Rb by Freyland et al. [1979] for reference.

In Fig. 2, the position of the first peak $Q_1$ is shown as a function of the $(v/v_0)^{-1/3}$, where $v_0$ is the volume at atmospheric pressure. If the liquid contracts uniformly, $g(R)$ varies proportionally along $R$ axis as $v^{1/3}$. Then $S(Q)$ varies proportionally along $Q$ axis as $v^{-1/3}$. The solid line in Fig. 2 shows the expected values from uniform-contraction model. The peak position for liquid Rb changes along this line at low pressure and deviates from the line at higher pressure.

In alkali metals, conduction electrons with s-character spreads far from the ion core. For example, at atmospheric pressure, the packing fraction of the ion core is only 15 % for Rb. At low-pressure region, the compression may arise mainly in the conduction electrons. At 5 GPa, $v/v_0$ is 0.6, and the packing fraction of the ion core volume increases to 25 %. As the effective hard-sphere diameter is the sum of the ion core diameter and the thickness of the screening electrons around the ion core, this value is comparable to the typical packing fraction of the effective hard-sphere in the hard sphere model, 45 %. Therefore, the repulsion between the ion cores should play an important role in the volume dependence of the structure.

In contrast to the compression by pressure, the volume expansion of the liquid can be realized by heating the liquids at high pressure preventing the vapourization. Studies on the structure of liquid Rb and Cs have been made by diffraction measurements at extremely high temperature changing $v/v_0$ by more than three times. In Fig. 2, the results for expanded Rb by Freyland *et al.* [1979] are also shown for reference. In the expanded liquid Rb, the shift of the peak position is small and deviates from the uniform-expansion model. When the liquid is heated to sufficiently high temperature for the thermal energy to be comparable to the height of the attractive part of the ion-ion potential, the distribution of the position of the neighboring atoms spread farther from the central atoms, causing an increases in entropy, $S$. Thus the free energy decreases as its $-TS$ term decreases. On the other hand, when the pressure is applied to the liquids at constant temperature, the volume contacts so that the $pv$ term in the free energy decreases. These difference in the volume changes is considered to cause the different volume dependence of $Q_1$ between at high pressures and at high temperatures.

In Fig. 3, the results for liquid Cs are shown. The volume dependence is almost the same as that for liquid Rb. The value of $Q_1$ deviates from the uniform-contraction model at $v/v_0$ above 1.15.

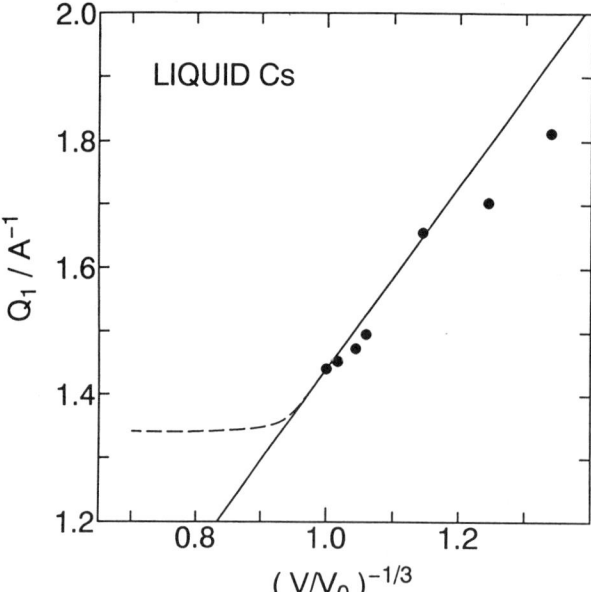

Fig. 3. The position of the first peak $Q_1$ for liquid Cs as a function of the $(v/v_0)^{-1/3}$. The solid line shows the expected values from a uniform-contraction model. The broken line shows the results for expanded liquid Cs by Winter *et al.* [1987] for reference.

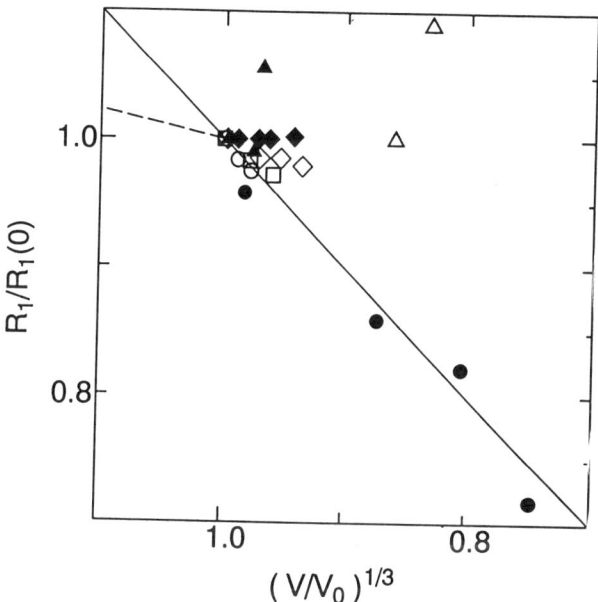

Fig. 4. The position of the first peak $R_1$ for several liquid metals as a function of $(v/v_0)^{1/3}$. Liquid Na (open circle), liquid K (open square), liquid Cs (solid circle), liquid Ga (open diamond), liquid Bi (solid diamond), liquid Se (open triangle), and liquid Te (solid triangle). The solid line shows the expected values from a uniform-contraction model. The broken line shows the results for expanded liquid Rb by Freyland et al. [1979] and liquid Cs by Winter et al. [1987] for reference.

In Fig. 4, the positions of the first peak $R_1$ for liquid alkali metals are shown as a function of $(v/v_0)^{1/3}$. The solid line shows the expected values from a uniform-contraction model. The broken line shows the results for expanded liquid Rb by Freyland et al. [1979]) and liquid Cs by Winter et al. [1987].

In liquid alkali metals with s-valence electron, $R_1$ varies as $v^{1/3}$. In fact, this is true in liquid Na, liquid K [Tsuji et al., 1975], liquid Rb and liquid Cs [Tsuji et al., 1990]. However, a slight deviation was observed in liquid Cs. On the transition in solid state from bcc to fcc at 2.3 GPa with a small volume change of 1%, the first coordination distance increases corresponding to the first coordination number change from eight to twelve. The relatively small change of $R_1$ near $(v/v_0)^{1/3} = 0.85$ suggests that a similar transition occurs in the liquid state. The remarkable structural change near $(v/v_0)^{-1/3} = 0.75$ is considered to be due to 6s-5d electronic transition which was predicted from a band calculation for crystalline Cs [McMahan, 1986]. This change is also consistent with the two species model, which predicted a remarkable structural change of liquid Cs at about 3 GPa [Rapoport, 1968].

B. Liquid polyvalent metals

At atmospheric pressure, gallium crystallizes to an orthorhombic structure consisting of a stacking of distorted hexagonal close-packed layers, with the bonds within the layers considerably weaker than the bonds between the layers. A gallium atom has one nearest neighbor at 2.47 Å and six second neighbors at 2.70 ~ 2.79 Å. Gallium has several high-pressure phases and metastable crystalline phases [Young, 1991].

At atmospheric pressure, bismuth crystallizes to a rhombohedral structure identical to arsenic, which can be assumed to be a distorted simple cubic structure. Four high pressure-phases of bismuth are known [Young, 1991].

Selenium and tellurium are semiconductors with crystal structures consisting of spiral chains. Each atom is tightly bonded to two neighbors within a chain with covalent bonds of two p-electrons. Another two p-electrons form lone-pair electrons and the bonds between adjacent chains are relatively weak. The crystal structure for Se and Te is also regarded as a distorted simple cubic structure.

In metals, the ion-ion potential $\phi(r)$ shows the Friedel oscillation in the asymptotic form of $\cos(2k_F r)/r^3$ at large distance. For polyvalent metals such as gallium and bismuth, the magnitude of Fermi wave vector $k_F$ is large and the period of the oscillation is short, which leads fine structures in $\phi(r)$. The application of pressure changes $k_F$ and $\phi(r)$ remarkably. This may be the major reason why such metals have many polymorphs under pressure. The change in $\phi(r)$ affects the structure in liquid state. Actually the shape of the shoulder or the subpeak in $S(Q)$ for liquid gallium and liquid bismuth changes remarkably with increasing pressure.

In Fig. 4, volume dependence of $R_1$ is also shown for several liquid polyvalent metals.

In liquid Ga with s- and p-valence electrons, $R_1$ decreases with volume contraction. However, the change is smaller than expected from the uniform-contraction model.

In liquid Bi with p-valence electrons, the change in $R_1$ with volume contraction is small. In liquid bismuth, the bonds are formed mainly with p-electrons. As the wave functions of p-electrons spread along x-, y- and z-direction, the interaction between atoms is anisotropic. This may be the reason of the small volume variation of $R_1$.

In liquid Te and liquid Se, $R_1$ increases with decreasing volume. The increase in $N_1$ with pressure suggests that a new covalent-like bonds between atoms in the adjacent chains are formed as the interchain distance approaches the intrachain distance [Tsuji et al., 1988, Yaoita et al., 1991].

Recently Brazhkin and coworkers reported first-order phase transitions in liquid sulfur [Brazhkin et al., 1991], liquid selenium [Brazhkin et al., 1989], liquid tellurium [Brázhkin et al., 1992a], liquid bismuth [Umnov et al., 1992], and liquid iodine [Brazhkin et al., 1992b]. According to the diffraction measurements [Yaoita et al., 1993], there is no drastic change of $S(Q)$ across the proposed phase boundary of liquid tellurium. However, the temperature variation of the first peak of $S(Q)$ shows the deviation from the usual thermal-expansion model.

Intense synchrotron radiation from the next generation ring will lead to considerable advances in the structural study of liquid metals under pressure in the next few years [Shimomura et al., 1992].

## ACKNOWLEDGEMENTS

For their significant contributions to the work described in this paper, the author is deeply indebted to his colleagues and collaborators, K. Katayama, K. Yaoita, O. Shimomura, T. Kikegawa, H. Kanda and H. Nosaka.

## REFERENCES

Ashcroft, N. W., and Lekner, J., 1966, Phys. Rev., **145**, 83.
Brazhkin, V. V., Voloshin, R. N., and Popova S. V., 1989, JETP Lett., **50**, 424.
Brazhkin, V. V., Voloshin, R. N., Popova S. V., and Umnov, A. G., 1991, Phys. Letters A, **154**, 413.
Brazhkin, V. V., R.N.Voloshin, R. N., Popova, S. V., and Umnov, A. G., 1992a, J. Phys. Condens. Matter., **4**, 1419.
Brazhkin, V. V., Popova, S. V., Voloshin, R. N., and Umnov, A. G., 1992b, High Pressure Res., **6**, 363.
Brown, K. H., and Barnett, J. D., 1972, J. Chem Phys., **57**, 2016.
Dickey, J. M., Meyer, A., and Young, W. H., 1967, Proc. Phys. Soc., **92**, 460.
Egelstaff, P. W., Page, D. I., and Heard, C. R. T., 1974, J. Phys. C, **4**, 1453.
Endo, H., and Tsuji, K., 1985, in *Handbook of Thermodynamic and Transport Properties of Alkali Metals*, (Oxford, Blackwell Scientific Publications), p. 321.
Freyland, W., Hensel, H., and Gläser, W., 1979, Ber. Bunsenges. Phys. Chem., **83**, 884.
McMahan, A. K., 1986, Physica, **139&140B**, 31.
Rapoport, E., 1968, J. Chem. Phys., **48**, 1433.
Shimomura, O., Yamaoka, S., Yagi, T., Wakatsuki, M., Tsuji, K., Fukunaga, O., Kawamura, H., Aoki, K., and Akimoto, S., 1985a, Mater. Res. Soc. Symp. Proc., **22**, 17.
Shimomura, O., Yamaoka, S., Yagi, T., Wakatsuki, M., Tsuji, K., Kawamura, H., Hamaya, N., Fukunaga, O., Aoki, K., and Akimoto, S., 1985b, in *Solid State Physics under Pressure*, (Reidel/KTK), p. 351.
Shimomura, O., 1986, Physica, **139&140B**, 292.
Shimomura, O., Tsuji, K., and Hamaya., N., 1992, High Pressure Res., **8**, 703.
Tsuji, K., Endo, H., and Minomura, S., 1975, Philos. Mag., **31**, 441.
Tsuji, K., Shimomura, O., Tamura K., and Endo, H., 1988, Z. Phys. Chem. N. F., **156**, 495.
Tsuji, K., Yaoita, K., Imai, M., Shimomura, O., and Kikegawa, T., 1989, Rev. Sci. Instrum., **60**, 2425.
Tsuji, K., 1990, J. Non-Cryst. Solids, **117&118**, 27.
Tsuji, K., Yaoita, K., Imai, M., Mitamura, T., Kikegawa, T., Shimomura, O., and Endo, H., 1990, J. Non-Cryst. Solids, **117&118**, 72.
Yaoita, K., Tsuji, K., Imai, M., Kikegawa, T., and Shimomura, O., 1990, High Pressure Res., **4**, 339.
Yaoita, K., Imai, M., Tsuji, K., Kikegawa, T., and Shimomura, O., 1991, High Pressure Res., **7**, 229.
Yaoita, K., Tsuji, K., Katayama, Y., Imai, M., Chen, J.-Q., Kikegawa, T., and Shimomura, O., 1992, J. Non-Cryst. Solids, **150**, 25.

Yaoita, K., Tsuji, K., Katayama, Y., Koyama, N., Kikegawa, T., and Shimomura, O., 1993, J. Non-Cryst. Solids, **156-158**, 157.

Young, D. A., 1991, *Phase Diagrams of the Elements*, (Berleley, Univ. California Press), and references therein.

Umnov, A. G., Brazhkin, V. V., Popova, S. V., and Voloshin, R. N., 1992, J. Phys. Condens. Matter, **4**, 1427.

Winter, R., Hensel, F., Bodensteiner, T., and Glaser, W., 1987, Ber. Bunsenges. Phys. Chem., **91**, 1327.

# VIII.4

# ELECTRON-HOLE PLASMAS IN ELEMENTAL SEMICONDUCTORS

J.P. Wolfe

Physics Department and Materials Research Laboratory
University of Illinois at Urbana-Champaign
Urbana, IL 61801-3080, U. S. A.

Photoexcitation of a semiconductor produces pairs of oppositely charged mobile carriers-electrons and holes. At sufficiently low temperatures, the electrons and holes bind pairwise to form short-lived excitons, analogous to positronium. The density of the gas of excitons may be increased by increasing the optical excitation level, and at sufficiently high excitation and low temperature in Si or Ge, the excitons condense into droplets of electron-hole liquid. The electron-hole liquid is a metallic two-component plasma obeying Fermi statistics. In this paper, I will review some recent experiments which examine the critical region for electron-hole liquid formation. The principal aim is to determine the phase diagram for this highly quantum fluid, which involves both liquid-gas and metal-insulator transitions.

## I. INTRODUCTION—EXCITONS AND PHOTOLUMINESCENCE

A semiconductor is so-named because, at normal temperatures, only a small fraction of the valence electrons are free to move throughout the crystal. Indeed, in a pure semiconductor crystal at zero temperature, there are *no* mobile charge carriers. This is because an energy gap, $E_g$, separates a completely filled valence band and a completely empty conduction band. If an electron is removed from the valence band, an otherwise neutral region of the crystal becomes positively charged, forming a "hole". Thus, the motion of electrons in a nearly filled valence band is conveniently described as the motion of holes, analogous to vacancy motion in a nearly filled parking lot.

Photoexcitation of a semiconductor promotes an electron from the filled valence band into the empty conduction band. Only a photon with energy $h\nu$ greater than $E_g$ can produce a free electron-hole pair, and the difference, $h\nu - E_g$, is imparted as kinetic energy to the two carriers in the initial generation process. This excess kinetic energy is lost quickly—generally within a nanosecond—by emission of phonons (lattice vibrations), and the "gas" of free carriers approaches a kinetic energy distribution given by the lattice temperature. When a "thermalized" electron falls back into the hole (an event called "recombination") a photon with energy roughly equal to the band-gap is emitted. In the elemental semiconductors Si and Ge, the *photoluminescence spectrum* directly reveals the *energy distribution* of photoexcited particles.

Figure 1 shows an image of the photoluminescence caused by recombination of electrons and holes in Si at room temperature [Steranka, 1985]. The boundaries of the crystal are made visible by scattered luminescence light. A narrow laser beam from a Nd:YAG laser (1=1.06 μm) is incident from the left and, having a photon energy only slightly above $E_g$, the laser photons have a rather long absorption length in Si. The electrons and holes produced by the laser light diffuse away from the narrow excitation region and eventually recombine, producing the luminescence light.

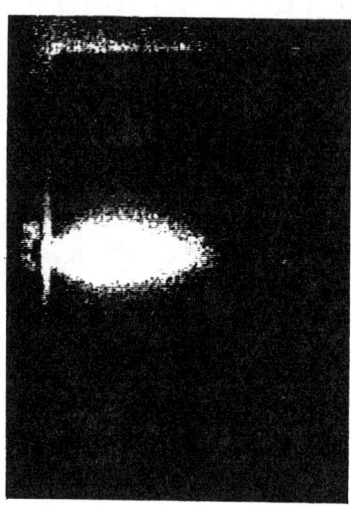

Fig. 1. Photoluminescence from recombining electron-hole pairs in Si at 300 K. The crystal dimensions are 1.7 × 2.9 × 3.0 mm$^3$.

The photoexcited electrons and holes are negative and positive particles, respectively, and therefore they attract each other. Before they recombine, an electron and a hole can form a bound state—an "exciton"—analogous to hydrogen or (more accurately) positronium. The binding energy of the exciton, $\varphi_\chi$, compared to that of a hydrogen atom, $\varphi_o$= 13.6 eV, is approximately given by,

$$\varphi_\chi = (\mu/m_o)(1/\varepsilon)^2 \varphi_o ,$$

where $\mu$ is the effective reduced mass of the carriers, $m_o$ is the free electron mass, and $\varepsilon$ is the dielectric constant of the crystal. For Si and Ge, the measured excitonic binding energies are 15 meV and 4 meV, respectively. The Bohr radii are about 50 Å, and 150 Å, respectively. The exciton is the fundamental (neutral) particle of a semiconductor.

## II. ELECTRON-HOLE LIQUID AND THE LIQUID-GAS PHASE DIAGRAM

At low temperatures and moderate excitation levels, a majority of the photoexcited electron-hole pairs in high-purity Si or Ge are bound into "free excitons" (FE), which are able to diffuse in the crystal before recombining. As the density of the excitonic gas is raised (by increasing the excitation power) interactions between excitons become important. Biexcitons, analogous to molecular hydrogen, or even more complex excitonic molecules can form. Moreover, at sufficiently high density, the excitonic gas is predicted to undergo an *insulator-metal transition* [Mott, 1974], whereby the excitons and molecules ionize, leaving an electron-hole plasma (EHP).

At low temperatures the excitonic gas in Si and Ge can also undergo a gas-liquid transition in which droplets of dense *electron-hole liquid* (EHL) are formed [Hensel et al., 1977]. The droplets are stabilized by the coulomb attraction between electrons and holes. Photoluminescence spectra of Si shown in Figure 2 illustrate this transition. At 25 K, the spectrum corresponds to a fairly high density gas of free excitons and electron-hole plasma. At 10 K, a clear bifurcation into two distinct phases is observed: the lower energy (high density) component is the electron-hole liquid and the higher energy (low density) component is free excitons. The critical temperature for formation of EHL in Si is about 24.5 K. In Ge, $T_c$ is about 6 K. A myriad of interesting experiments have been performed on electron-hole liquid, as reviewed, for example, in Hensel et al. [1977] and Jeffries and Keldysh [1983].

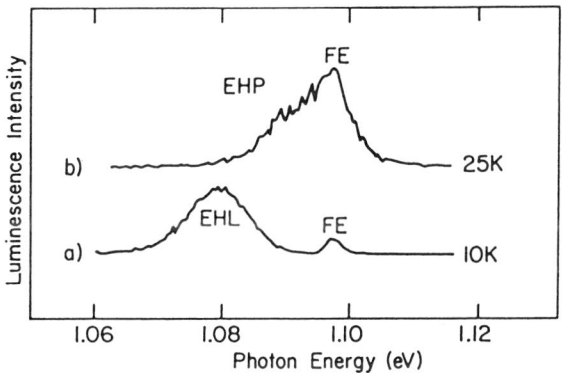

Fig. 2. Photoluminescence spectra of Si a) below, and b) above the critical temperature for electron-hole liquid (EHL) formation. FE = free excitons, EHP = electron-hole plasma [Steranka, 1985].

A summary of these ideas is shown in the schematic phase diagram of Figure 3. The solid and dashed lines represent a standard Guggenheim [1945] plot of liquid-gas coexistence densities, based on the spectroscopically measured density and critical temperature for EHL in Si [Smith, 1988]. The equilibrium density of electron-hole liquid at low temperature, shown at the right side of the diagram, is $n_l = 3 \times 10^{18}/\text{cm}^3$. The dotted line is a theoretical estimate of the insulator-metal-transition density for Si given by,

$$n_m = (1.19)^2 \varepsilon k_B T / 8\pi e^2 a_\chi^2,$$

which is based on the Deybe-Huckel screening approximation. $a_\chi$ is the excitonic Bohr radius. The EHL is a strongly ionized two-component plasma described by Fermi-Dirac statistics. Its ground state energy and density are determined by the interplay of kinetic, exchange and correlation energies, as in a metal.

Experimental determination of the actual excitonic phase diagrams in Si and Ge is not easy. Measurements of $n_l$ and $T_c$ are fairly straightforward (see below), and the

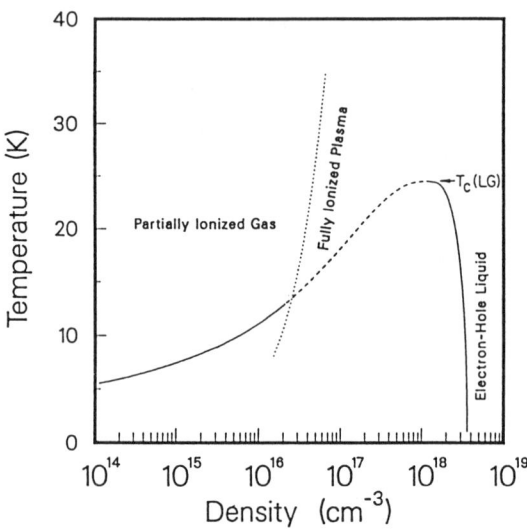

Fig. 3. Schematic phase diagram for excitonic matter in Si [Smith and Wolfe, preprint]. The metal-insulator transition (dotted line) is predicted to occur well below the liquid density, in large part because the multiplicity of the electron and valence bands creates a highly quantum liquid with $r_s < 1$. ($n_l = 3/4\pi(r_s a_x)^3$)

far left side of the phase diagram (the saturated gas density) at low temperature is predicted from a standard thermodynamic arguments to be,

$$n_{\text{sat}} = AT^{3/2} e^{-\varphi_l/k_B T},$$

where $A$ is a constant and $\varphi_l$ is the liquid binding energy with respect to free excitons. However, the shape of the phase diagram in the overlapping liquid-gas and ionized-plasma region is difficult to determine. The experimentalist is faced with deconvolving several components from a complex spectrum, which, due to high excitation levels and inherently non-uniform particle densities, may not represent a well defined equilibrium situation.

How the metal-insulator transition affects the liquid-gas boundary is the subject of much theoretical interest and speculation (see, for example, the discussion on pages 39–49 by Rice in Hensel et al., 1977). Unlike the cases of fluid metals—Hg, Rb, and Cs—whose metal-insulator transitions occur very close to their liquid-gas critical densities, the predicted metal-insulator transition in the excitonic system occurs at densities well below the EHL critical point. One might reasonably ask if the excitons undergo a continuous ionization transition, or whether a first-order transition distinct from the liquid-gas transition could arise, leading to a phase diagram with two critical points and two condensed phases of different densities. The latter idea has been postulated by

Rice [1974], based on a suggestion by Landau and Zeldovich [1943], and supported by several theories [Kraeft et al., 1975; Balslev, 1980]. Further progress in understanding the structure and thermodynamics of this novel quantum fluid rests on experimental guidance.

In the remainder of this paper I will review experiments by Smith and Wolfe [1986,1988] at Illinois which explore the critical region of electron-hole liquid in the elemental semiconductor Si. The results suggest that two critical points do indeed occur. Similar experiments on Ge have been reported by Simon et al. [1992], with the same conclusion. Earlier studies of the phase diagram in Si were performed by Shah, Combescot and Dayem [1977], and Dite, Kulakovsky and Timofeev [1977]. Early studies of the excitonic phase diagram in Ge were conducted by Thomas et al. [1978].

## III. TIME-RESOLVED STUDIES OF THE PHASE DIAGRAM IN Si

To determine the phase diagram, one must systematically identify and isolate the photoluminescence from each excitonic species. Figure 4a shows the energy levels associated with excitons, molecules, and electron-hole liquid. The exciton component has a spectral width of about $k_B T$, where $T$ is the temperature of the gas. The electron-hole liquid—a degenerate two-component plasma—has a spectral width given by its Fermi energy, which has the dependence

$$E_F \approx \left(\hbar^2/2m^*\right) n^{2/3}$$

for a pair density $n$. By properly considering the effective masses of the carriers and the degeneracies of the bands [Hensel et. al., 1977], one can determine the density of the liquid from the measured spectral line width. The renormalized gap $E_g'$ of the EHL is also related to the pair density through well established theoretical models. The molecular binding energy, $\varphi_m$, has been determined spectroscopically and agrees reasonably well with rather sophisticated theories of this 4-particle complex (2 electrons and 2 holes); its spectral characteristics are also well established [Thewalt, 1978; Gourley, 1979].

The relative numbers of electron-hole pairs in the excitonic species (excitons, molecules, EHL) depend on the excitation conditions and the effective volume occupied by the carriers. To achieve high densities, it is necessary to use short, intense laser pulses with $h\nu$ well above $E_g$. Typically one uses a cavity-dumped Ar$^+$ laser, which has a sub-micron absorption length in Si and Ge, and a pulse width of about 15 ns. For Si, spectra are collected by time-resolved photon counting.

Figure 4b shows a series of time-resolved spectra following a 15-ns laser pulse. [Smith, 1988; Smith and Wolfe, 1986] At early times a broad electron-hole-liquid line is dominant. As time progresses, the e-h pairs in the liquid recombine with a lifetime of 140 ns and also evaporate into excitons, leaving the FE component at late times to decay in a lifetime of a few microseconds. During this whole process, the volume

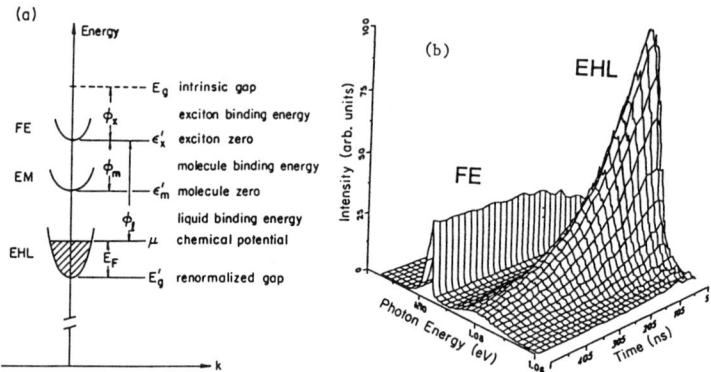

Fig. 4. a) Energy levels for excitons, biexcitons and electron-hole liquid [Simon et al.,1992]. b) Time resolved photoluminescence spectra in Si at 14.5 K following a 15 ns excitation pulse [Smith, 1988].

occupied by the "cloud" of electron-hole droplets and excitons is expanding away from the excitation surface, as revealed by time-resolved imaging [Steranka, 1985].

Such spectral studies show that a *gas temperature* close to the *lattice temperature* is established within 10–20 ns and that the EHL establishes its equilibrium density on an equally short timescale [Steranka, 1985]. Hence, it is reasonable to assume that at each instant of time a local equilibrium is established between the excitonic species. A sequence of time-resolved spectra, therefore, can be used to observe the constituents and thermodynamics of the gas over a broad range of densities. It must be kept in mind, however, that each spectrum corresponds to some average pair density over space.

We begin by examining the excitonic gas at low densities (late times) and then analyze the increasingly complex spectra at high densities (early times). In this way, it is possible to isolate the various excitonic species as they appear at increasing gas densities. The other adjustable experimental parameter is the crystal temperature, which is controlled by a cryostat with a helium exchange gas. The temperature is measured with a carbon resistor and agrees well with that obtained from the free exciton (FE) lineshape.

Figure 5b shows the luminescence spectrum of free excitons at 28 K. the solid line is the data at $t = 2000$ ns and the circles are the Maxwell-Boltzmann distribution,

$$I(E) = CE^{1/2} \exp\left(-E/k_B T\right),$$

at the measured lattice temperature. The spectrum in Fig. 5a is obtained at an earlier time of 600 ns (i.e., a higher gas density) and displays an added luminescence component on the low energy side of the exciton line. The light curve represents the theoretical lineshape for *biexcitons* in Si, which become observable at higher gas densities. The position and width of this biexciton component are the same as those obtained previously in extensive studies of excitonic molecules [Thewalt and Rostworowski, 1978; Gourley and Wolfe, 1979]. In fact, the relative intensities of the exciton spectrum and the new component suggests that there may be a considerable number of 3-particle "trions" present [Thomas et al., 1977], so we label the low energy luminescence EC, for excitonic complexes. These low density data establish how the relative intensities of the EC and FE scale, which is very useful in deconvolving high density spectra.

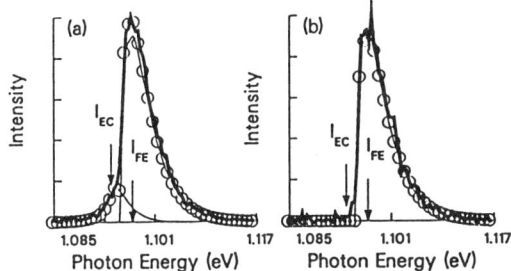

Fig. 5. Time-resolved luminescence spectra at $T = 28$K showing free excitons (FE) and excitonic complexes (EC)—molecules and trions. a) 600 ns after laser pulse. b) 2000 ns after the laser pulse [Smith and Wolfe, 1986].

Let us now look at earlier times, corresponding to higher gas densities. First we examine the changes in the spectrum at 150 ns as the temperature is raised through the liquid-gas critical temperature. Figure 6a shows a simple spectrum of EHL and FE. As discussed before, the broader EHL line is due to the large range of velocities in this degenerate Fermi liquid ($E_F = 23.5$ meV corresponding to $n_\Gamma = 3.0 \times 10^{18}$ cm$^{-3}$). As the temperature is raised to 21.7 K, Fig. 6b, the EHL becomes less dense and the FE/EC gas components dominate. At this temperature, however, a new luminescence component (shaded) arises between the EHL and FE. This same component (i.e., same lineshape) is present at 25.2 K (Fig. 6c), which is above the critical temperature for electron-hole liquid.

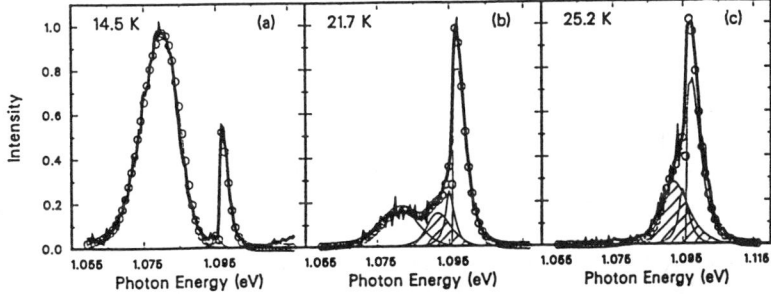

Fig. 6. Luminescence spectra at $t = 150$ ns. Solid lines are data; open circles are fits involving EHL, EC, FE, and (shaded) a "condensed plasma (CP)" component. a) 14.5 K. b) 21.7 K. c) 25.2 K [Smith and Wolfe, 1986].

To probe the origin of this new excitonic species, we examine the early-time spectra at 28 K, shown in Figure 7a. The luminescence below the FE/EC spectra are plotted. At the earliest measurement time of 25 ns, a broad peak is observed which agrees with the theoretical spectrum of an electron-hole plasma with $n = 8.5 \times 10^{17}$ cm$^{-3}$. Aside from the overall intensity, only one adjustable parameter (density) is required to fit this spectrum because the density dependences of the renormalized gap (arrow) and the Fermi energy (spectral width) are known from a study of the EHL. [Vashishta and Kalia, 1982]

Because the system in Fig. 7a is at $T > T_C$, one might expect that the electron-hole plasma created by the high optical excitation will decay in density until the metal-insulator transition is crossed, whereupon the e-h pairs in the plasma will bind into excitons. However, as shown in Fig. 7a, the low energy edge of the spectrum (arrow) approaches a constant energy, corresponding to a plasma density of about $2.5 \times 10^{17}$ cm$^{-3}$. Figure 7b shows the time evolution of the plasma density determined in this way for several crystal temperatures ranging from 28 K to 45 K. At 45 K and above, no stable density is reached.

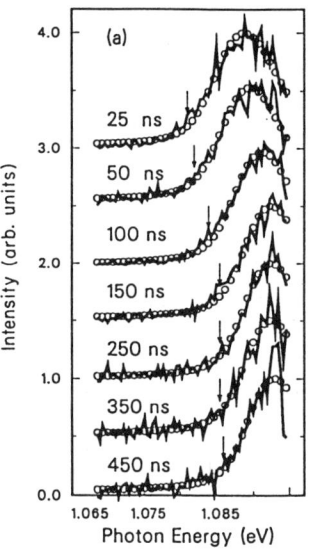
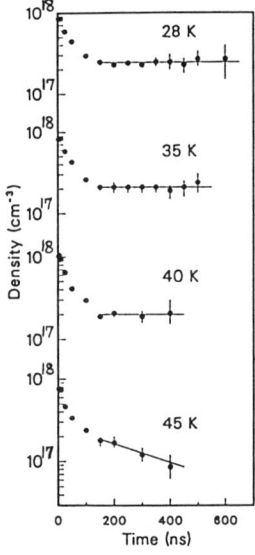

Fig. 7. a) Luminescence just below the FE/EC lines at $T = 28$K, with FE/EC contributions subtracted. Open circles are theoretical plasma lineshapes determined by a single parameter—pair density $n$. The arrows indicate the renormalized gap. b) plot of the plasma density derived from spectra such as those in (a). Below 45 K, the plasma densities level off in time, indicating a condensation into droplets of weakly ionized plasma [Smith and Wolfe, 1986].

Smith and Wolfe [1986] have interpreted this remarkable behavior as evidence for a second condensed phase (CP) in Si. Between 150 ns and 600 ns, the CP intensity decays by over an order of magnitude. Therefore, if the spectra at times later than 150 ns in Figure 7a indeed arises from an electron-hole plasma, then a spatial condensation of this plasma must be occurring because the average density of the gas is decaying. The corresponding phases (weakly and strongly ionized plasmas) implied by this unusual phase diagram has been theoretically proposed by Landau and Zeldovich [1943], Rice [1974], Kraeft et al [1975, 1978] and Balslev [1980].

## IV. DISCUSSION

The principal experimental finding of this work is the following: all of the spectra in the critical region are well fit by a combination of *known spectra* (excitons, excitonic molecules and electron-hole liquid) plus a *a single additional component of electron-hole plasma* whose spectral position and width are defined by one adjustable parameter—pair density. The existence of a plasma component in addition to EHL is expected because the metal-insulator transition of excitons is predicted to be well below the critical density of EHL. This straightforward and simple analysis of the data leads one to the inescapable conclusion that the weakly ionized plasma condenses into constant-density droplets over a wide temperature range, as shown in phase diagram of Figure 8.

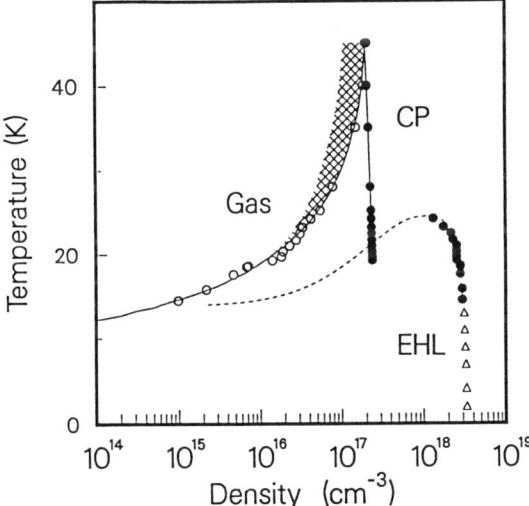

Fig. 8. Proposed phase diagram of excitonic matter in Si. Solid circles are densities determined from the spectral fits. Open circles and shaded region show the range of calculated pair densities in the excitonic gas. Triangles are low temperature EHL densities [Smith and Wolfe, 1986].

An alternative explanation of this data has been offered by Hernandez [1987] and by Steele, McMullan and Thewalt [1987]. They propose that the extra luminescence attributed to CP is actually due to large excitonic complexes, a possibility also considered by Smith and Wolfe [1986]. Steel et al. [1989] have demonstrated the existence of such complexes—termed polyexcitons—in the green photoluminescence spectrum of Si. In this spectrum, shown in Figure 9, two e-h pairs recombine simultaneously, giving off one photon with nearly twice the bandgap energy.

At the present time it is difficult to disprove either the existence of a polyexciton contribution to the infrared spectrum of Smith and Wolfe or the occurrence of electron-hole plasma in the green spectrum of Steel and Thewalt. There are some significant experimental differences, however, which must be kept in mind. The green photoluminescence is emitted from only a micron-sized layer near the crystal surface, whereas the infrared spectrum samples the entire excitonic cloud, which extends over 100 $\mu$m into the crystal. Also, the reported green spectra are integrated over time and obtained with continuous excitation. So, the relative numbers of polyexcitons and weakly ionized plasma should be different in the two experiments. In particular, the pulsed experiments achieve higher average densities—clearly sufficient to produce a dense plasma—whereas the CW-excitation experiments may not reach this density regime. Using the binding energies measured by Steele $et\ al.$, Smith and Wolfe have estimated the ratio of large to small complexes and conclude that the infrared luminescence intensity from large complexes should be about an order of magnitude smaller than the observed extra luminescence; however, there is considerably uncertainty in such estimates.

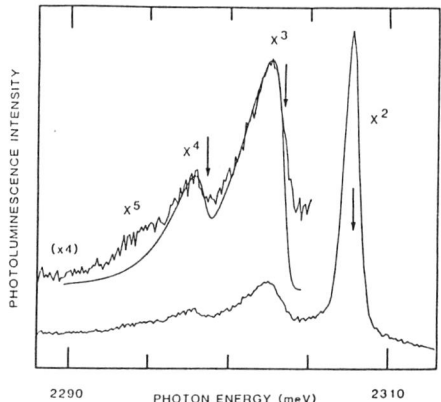

Fig. 9. Green photoluminescence spectrum showing emission from biexcitons ($x^2$), triexcitons ($x^3$), tetraexcitons ($x^4$), etc. The crystal temperature is 20 K and the excitation is a continuous 800 mW at 700-nm wavelength. The green light is strongly absorbed by Si, so this spectrum samples a thin region of excitonic matter near crystal surface [Steele, McMullan and Thewalt, 1987].

It is difficult to see how a combination of polyexciton spectra can reproduce the spectra in Figure 7a. The relative numbers of large and small polyexcitons should depend on temperature and density, implying that the spectral shape of their combined luminescence would be quite variable, apparently inconsistent with the data. Certainly the luminescence spectra at 25 ns in Figure 7a ($T = 28$ K) is due to electron-hole plasma because the pair density is quite high at these early times before the gas has expanded significantly into the sample [Shah and Dayem, 1976; Steranka, 1985]. The smooth transition into the similarly shaped CP component would have to be fortuitous, requiring an appropriate distribution of polyexcitons. The time-resolved experiments of Smith and Wolfe have the notable advantage of approaching the critical region from the high density side, which facilitates the observation of condensed plasma.

It would be helpful to devise an independent probe of the system—one which would distinguish between plasma droplets and large molecules. Light scattering is a candidate. Rayleigh scattering from droplets in Ge have provided valuable information on their size [Hensel, et al., 1977]; however, similar experiments on Si have generally failed, probably due to the much smaller droplet size. The present experiments contain no information about the size of the proposed CP droplets, so there is some chance of observing Rayleigh scattering.

Finally, it would be interesting to consider where the line of demarcation is between large polyexcitons and small metallic clusters. In other words, just how many pairs does it take to create an electronic structure approximating that of a weakly ionized plasma? Clearly there are a number of interesting issues remaining for this novel excitonic system in semiconductors.

The Illinois work has been supported by the National Science Foundation, presently Grant DMR92-07458.

## REFERENCES

Balslev, I., 1980, Phys. Stat. Sol. (b) **101**, 749.
Dite, A. F., V. D. Kulakovskii, and V. B. Timofeev, 1977, Zh. Eksp. Teor. Fiz. **72**, 1156 [Sov. Phys. JETP **45**, 604].
Gourley, P. L., and J. P. Wolfe, 1979, Phys. Rev. **B20**, 3319.
Guggenheim, E. A., 1945, J. Phys. Chem. **13**, 253; *Thermodynamics* (Wiley, New York, 1967).
Hensel, J. C., T. G. Phillips, G. A. Thomas, and T. M. Rice, 1977, *Solid State Physics*, edited by H. Ehrenreich, F. Seitz, and D. Turnbull (Academic, New York), Vol. 32.
Hernandez, J. P., 1987, Phys. Rev. Lett. **58**, 2822.
Jeffries, C. D., and L. V. Keldysh, 1983, Electron-hole Droplets in Semiconductors, Modern Problems in Condensed Matter Sciences, Vol. 6 (North Holland).
Kraeft, W. D., K., Kilimann, and D. Kremp, 1975, Phys. Stat. Sol. (b) **72**, 461; R. Zimmermann, K., Kilimann, W. D. Kraeft, D. Kremp, and G. Ropke, 1978, Phys. State Sol. (b) **90**, 175.
Landau, L. D., and G. Zeldovich, 1943, Acta. Phys. Chim., URSS **18**, 194.
Mott, N. F., 1974, *Metal-Insulator Transitions*, (Taylor and Francis, London).
Rice, T. M., 1974, in *Proceedings of the Twelfth International Conference on the Physics of Semiconductors, Stuttgart* edited by M. H. Pilkuhn (Teubner, Stuttgart), p. 23.
Shah, J., M. Combescot, and A. H. Dayem, 1977, Phys. Rev. Lett. **38**, 1497.

Shah, J., and A. H. Dayem, 1976, Phys. Rev. Lett. **37**, 861.
Simon, A. H., S. J. Kirch, and J. P Wolfe, 1992, Phys. Rev. **B46**, 10098.
Smith, L. M., 1988, Ph.D. Thesis, University of Illinois.
Smith, L. M., and J. P. Wolfe, 1986, Phys. Rev. Lett. **57**, 2314; preprint.
Steele, A. G., W. G. McMullan, and M. L. W. Thewalt, 1987, Phys. Rev. Lett. **59**, 2899.
Steranka, F. M., 1985, Ph.D. Thesis, University of Illinois; F. M. Steranka, and J. P. Wolfe, 1986, Phys. Rev. **B34**, 1014.
Thewalt, M. L. W., and J. A. Rostworowski, 1978, Solid State Commun. **25**, 991.
Thomas, G. A., J. B. Mock, and M. Capizzi, 1978, Phys. Rev. **B18**, 4250.
Thomas, G.A., and T. M. Rice, 1977, Solid State Commun. **23**, 359.
Vashista, P., and R. K. Kalia, 1982, Phys. Rev. B25, 6492.

# IX. Complex Fluids and Solids

# IX.1

# COMPLEX FLUIDS: ANOMALOUS RELAXATION, PERCOLATION, AND WETTING

F. Yonezawa, S. Fujiwara, S. Gomi, and K. Omata

Department of Physics, Keio University, 3-14-1 Hiyoshi, Kohoku-ku
Yokohama 223, Japan

Our recent research on complex fluids is reviewed. This article is divided into three parts. The first part is devoted to some theoretical models of the anomalous relaxation observed in various kinds of disordered systems. In the second part, percolation theory, by which disordered systems are also modeled, is focused and, in particular, we report on the determination of percolation thresholds with high accuracy. In the last section, we discuss a topic concerning wetting which is an old but developing research area of complex fluids.

## I. ANOMALOUS RELAXATION IN THE FRAMEWORK OF ONE-BODY PICTURES

A. Introduction

Recently, much attention has been given to the problem of anomalous relaxation which is observed in quite a wide variety of phenomena, such as the structural relaxation in supercooled liquids [Götze, 1991] and the time decay of the thermoremanent magnetization in spin glasses [Binder and Young, 1986].

In these phenomena, the relaxation functions are expressed by the non-exponential form. Two types of empirical formulae are widely used to fit experimental data.
(i) Cole-Cole form for the complex susceptibility $\chi(\omega)$ [Cole and Cole, 1941],

$$\chi(\omega) = \frac{1}{1 + (-i\omega\tau)^\alpha} \quad (0 < \alpha < 1), \tag{1}$$

(ii) Kohlrausch law for the relaxation function $F(t)$ [Kohlrausch, 1847; Williams and Watts, 1970],

$$F(t) = \exp\left[-\left(\frac{t}{\tau}\right)^\beta\right] \quad (0 < \beta < 1), \tag{2}$$

where $\omega$ is a frequency, $\tau$ a relaxation time and $\alpha$ and $\beta$ are parameters which characterize these relaxations. The susceptibility $\chi(\omega)$ and the relaxation function $F(t)$ are related to each other through Fourier-Laplace transformation. When $\chi''(\omega)$ is described by Cole-Cole form, $F(t)$ has the power law dependence on time in the range of $t \ll \tau$ and $t \gg \tau$. Note that the case $\alpha = \beta = 1$ corresponds to the normal Debye type relaxation [Debye, 1929].

Since a number of experimental data are described by these simple empirical laws, it is highly expected that there must be some common nature in anomalous relaxations observed in apparently quite different phenomena. This common nature, if it exists at all, can be found when the microscopic mechanisms of anomalous relaxations are clarified and this problem is the subject of the present section.

B. Modeling

A purpose of this subsection is to propose models in order to obtain a clue to a better understanding of anomalous relaxation. Although the many-body effects undoubtedly play important roles for this phenomenon, it is interesting and instructive to clarify to what extent we can describe the essential aspects of the anomalous relaxation in a framework of the one-body pictures.

Let us consider a system, such as supercooled liquids, in which the density of particles is so high that the structural relaxation becomes anomalous. The motion of a particle is considerably restricted by the existence of other particles. We take these many-body effects into account through the constraint for particles each of which independently executes random walk on lattices . Two possible constraints are considered.

(a) temporal restriction

The time interval between jump events is not constant but is distributed so widely that there is a possibility that a particle does not change its position for a long time.

(b) geometrical restriction

The lattice has a number of sites to which particles are not allowed to move.

In the following subsections, we make use of the concept of "*fractal* " in order to realize these constraints. As is well known, fractal is characterized by two properties; 1) self-similarity and 2) non-integer dimension. To clarify the effects of these properties on the structural relaxation, we take up several models and carry out analytical and/or simulational calculations of the several quantities relevant to the relaxation phenomena:
(1) the relaxation function (the Fourier component of the particle density)

$$F(\mathbf{k},t) = \langle \exp(i\mathbf{k} \cdot \{\mathbf{r}(t) - \mathbf{r}(0)\}) \rangle, \tag{3}$$

(2) the complex susceptibility

$$\chi(\mathbf{k},\omega) = 1 + i\omega \int e^{i\omega t} F(\mathbf{k},t) dt, \tag{4}$$

(3) the particle density

$$G(\mathbf{r},t) = \langle \delta(\mathbf{r} - (\mathbf{r}(t) - \mathbf{r}(0))) \rangle, \tag{5}$$

where $\mathbf{r}(t)$ is the position of the particle at time $t$, $\mathbf{k}$ is the wave number and $\langle \cdots \rangle$ stands for the sample average.

Note that in the case of normal random walk (RW), the relaxation function becomes exponential.

C. Fractal time random walk

We consider the model with temporal restriction [Gomi et al., 1993]. Our model is a random walk on a regular lattice (with the lattice constant $d$), but the time interval between jump events is taken as a random variable following the probability distribution $\psi(t)$. We choose the form of $\psi(t)$ as

$$\psi(t) = \begin{cases} At^{-1-a} & (t_0 < t) \\ 0 & (t_0 > t) \end{cases}, \qquad (6)$$

where $A = at_0^a$ is the normalization constant and the cutoff parameter $t_0$ is required in order to make $\psi(t)$ be well-defined. Because of the regularity of the lattice, the motion of the particle for each direction is considered to be equivalent and therefore we confine our discussion to random walks on a 1-dimensional lattice without loss of generality. At each jump event, the particle moves to a left or right neighbor site along the lattice with equal probability.

In the case of $a < 1$, this model is known as the fractal time random walk (FTRW) [e.g., Shlesinger, 1988]. and was used by Scher and Montroll to study non-Gaussian transport phenomena observed in some amorphous semiconductors [Scher and Montroll, 1975]. Although a large number of works have been done concerning with this model [e.g., Scher et al, 1991; Klafter and Zumofen, 1994], most of them are limited to the study of transport properties.

Let us first show the results of simulation for the case of $a < 1$. The relaxation function $F(k,t)$ for $a = 0.7$ are shown in Fig. 1(a) for various values of $k$. When $t$ is scaled by an appropriately chosen value $\tau$, these curves fall onto one master curve (Fig. 1(b)), showing that the scaling law is fulfilled for $F(k,t)$. In this figure, $t$ is scaled by $\tau$ which is defined later by Eq. (9).

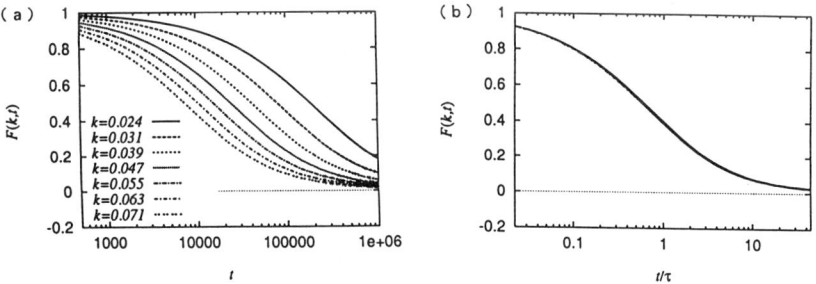

Figure 1: (a) Relaxation function $F(k,t)$ vs. $\log t$ for various values of $k$: $a =0.7$, $k = n\pi/4$ ($n = 3, \ldots, 9$) (b) Scaling property of $F(k,t)$. Each line in (a) is plotted versus $t/\tau$ where $\tau$ is defined by Eq. (9).

In Fig. 2, the imaginary part of complex susceptibility $\chi''(k,\omega)$ as a function of $\omega\tau$ is shown by taking the Fourier transformation of $F(k,t)$. The susceptibility for different $k$ values thus obtained naturally fit on a master curve as shown in Fig.2. A remarkable point is that the curve of $\chi''(k,\omega)$ is symmetric around its peak. This symmetric nature of $\chi''(k,\omega)$ is one of the most characteristic features of the Cole-Cole type relaxation.

Now we show that the empirical Cole-Cole form Eq. (1) for the complex susceptibility $\chi(k,\omega)$ is analytically derived in our model. From its definition, it follows that $\chi(k,\omega)$ in our model is rigorously described as

$$\chi(k,\omega) = 1 - i \frac{n_k \psi(-i\omega)(1 - \psi(-i\omega))}{1 - n_k \psi(-i\omega)}, \tag{7}$$

where $n_k = \cos(kd)$ and $\psi(-i\omega)$ is the Fourier-Laplace transform of $\psi(t)$.

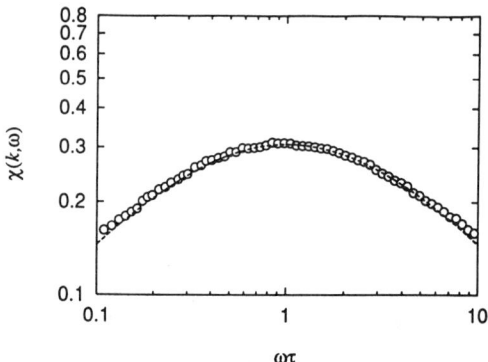

Figure 2: Imaginary part of complex susceptibility $\chi(k,\omega)$ vs. $\omega\tau$ in the case of $a = 0.7$ (open circle). The dashed line indicates the corresponding Cole-Cole form $1/(1 + (-i\omega\tau)^{0.7})$.

Since the relaxation under consideration is a phenomenon at large $t$ ($\gg t_0$) and at large $x$ ($\gg d$), we estimate Eq. (7) in the limit of $\omega \ll 1/t_0$ and $k \ll 1/d$. By retaining the leading terms in $k$ as well as in $\omega$, we obtain

$$\chi''(k,\omega) \sim \frac{(\omega\tau)^a \sin(a\pi/2)(1 - c(\omega\tau)^a \cos(a\pi/2) - c(\omega\tau)^{2a})}{1 + 2\cos(a\pi/2)(\omega\tau)^a + (\omega\tau)^{2a}}, \tag{8}$$

$$\tau^a(k) = \frac{n_k \Gamma(2-a)}{(1-n_k)(1-a)} t_0^a \sim k^{-2}, \tag{9}$$

where $\Gamma(\cdots)$ denotes the gamma function and $c \equiv (1-n_k)/n_k$. In the region of $\omega$ which satisfies $\omega^a \ll k$, the first term of the numerator of Eq. (8) becomes dominant and consequently the Cole-Cole susceptibility

$$\chi''(k,\omega) \sim \mathrm{Im} \frac{1}{1 + (-i\omega\tau)^a} \tag{10}$$

is derived.

It is readily seen from Eq. (9) that the peak position of $\chi''(k,\omega)$ is given by $\omega_{\max} = 1/\tau \sim k^2$, which is safely included in the region within the extent of our assumption, $\omega^a \ll k$. In other words, $\chi''(k,\omega)$ is expressed by Eq. (10) in the most of the interesting region of $\omega$ including $\omega_{\max}$. This result is extremely significant in the sense that the Cole-Cole type relaxation is evaluated analytically although this Cole-Cole form has been empirically known for a long time. Our formulation also ascertains that the parameter $a$ characterizing $\psi(t)$ becomes identical with the parameter $\alpha$ in Eq. (1).

Equation (8) shows $\chi''(k,\omega)$ depends on $k$ only through $\tau(k)$, which is consistent with the results attained from our simulation that the scaling law is fulfilled for $\chi''(k,\omega)$ of various $k$ values. This fact in turn lends support to the validity of our approximation used in the derivation of Eq. (10). Note the relaxation time $\tau$ has the power law dependences on $k$. Here we define an exponent $\gamma$ as $\tau \propto k^{-\gamma}$. As is clearly seen from above discussion, $\tau$ is equal to $2/a$. This exponent $\gamma$ is measurable in real experiments, such as neutron scattering.

The existence of scaling law for $F(k,t)$ and $\chi(k,\omega)$ suggests that there are some scaling relation in real $x$-space as well. In the following discussion, we are mainly concerned with the particle density $G(x,t)$ to see this property more closely. The definition of the exponent $\gamma$ implies that the scaling in the real space is realized as $x \propto k^{-1} \propto t^{1/\gamma}$. This argument is reinforced by the analysis below.

We first define the exponent $\theta$ through $\sigma^2(t) \propto t^\theta$. The long time behavior of transport quantities such as $\sigma^2(t)$ has been known analytically [e.g. Scher et al., 1991] and $\theta$ is given by $\theta = a$(for $a < 1$). Let us note that $\sigma(t)$ is expressed by using $G(x,t)$ or $F(k,t)$ as $\sigma^2(t) = \int x^2 G(x,t)dx = \frac{1}{2\pi} \int dx \int dk e^{-ikx} x^2 F(k,t)$. Through straight forward dimensional analysis, we obtain the relation between two parameters $\theta$ and $\gamma$:

$$\theta \times \gamma/2 = 1. \tag{11}$$

This equation is quite general and valid so far as the scaling law for $F(k,t)$ exists. Moreover Eq. (11) indicates the relation, $x \propto t^{1/\gamma} \propto t^{\theta/2} \propto \sigma$. This means the scaling variable in real space is $x/\sigma(t)$ which is obviously consistent with physical consideration. Equation (11) is also significant in the sense that it relates a parameter $\theta$ concerning *diffusion* to a parameter $\gamma$ concerning *relaxation*.

From our discussion concerning $\gamma$ and $\theta$, we see that Eq. (11) also holds in our model. In addition, results of simulations evidently support the scaling relation for $G(x,t)$. These facts undoubtedly show the consistency of our analyses.

D. Random walk on fractal structures

In this section, we study the structural relaxation on fractal structures by MC simulations of random walks [Fujiwara et al., 1993]. The algorithm of our simulations is the same as that for a random walk in a regular lattice except for the choice of structures. We choose two types of structures with fractal dimensions:
(1) structures with no dead ends,
(2) structures with dead ends.
As the first type, we choose the 2D Sierpinski Carpets (SC) with central cutout and the 3D Sierpinski Sponges (SS) whose surfaces are the 2D SC with central cutout. The fractal dimensions of the 2D SC and the 3D SS are respectively $D = \ln(b^2 - l^2)/\ln b$ and $D = \ln\{(b-l)^2(b+2l)\}/\ln b$, where $b$ is the system size and $l$ is the hole size. As the second type, we choose the 2D percolation clusters on a square lattice and the

3D percolation clusters on a cubic lattice at critical concentrations $p_c$ [Stauffer, 1985] ($p_c = 0.592745(2)$ for 2D and $p_c = 0.31161$). It is known that the fractal dimensions are $D = 91/48 \approx 1.896$ and $D = 2.524 \pm 0.008$, respectively [Nienhuis, 1982].

Our MC simulations are designed in such a way that, when the dice tells a random walker to walk into a wall, he or she remains unmoved for one time step. Periodic boundary conditions and mirror boundary conditions are respectively used for the Sierpinski structures and the percolation clusters to remove the influence of boundaries. The self-similarity of the system holds so far as the area where a random walker moves is much smaller than the size of the system. For the Sierpinski structures, the system sizes are taken between $243^2$ and $625^2$ for 2D SC and between $64^3$ and $125^3$ for the 3D SS. For the percolation structures, the system sizes are $400^2$ for the 2D and $50^3$ for the 3D percolation clusters. In each case, the number of samples is $2 \times 10^5$ and the time step of the MC simulations is $10^3$.

In Fig. 3, the imaginary parts of the complex susceptibility scaled by $\omega_p$ are shown in the case of the 2D SC with $b = 3, l = 1$, where $\omega_p$ is the frequency at which $\chi''(k,\omega)$ has a peak. From this figure, we find: (1) the imaginary parts of the complex susceptibility are symmetric with respect to $\omega_p$ in the log-log plot and are described by the Cole-Cole form (Eq. (1)) with the relaxation time $\tau = 1/\omega_p$ and (2) scaling law for the imaginary parts of the complex susceptibility for various wave numbers $k$ are fulfilled as in the case of fractal time random walk. This means the parameter $\alpha$ in Eq. (1) is independent of the wave numbers $k$.

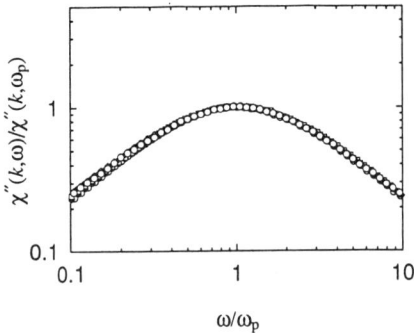
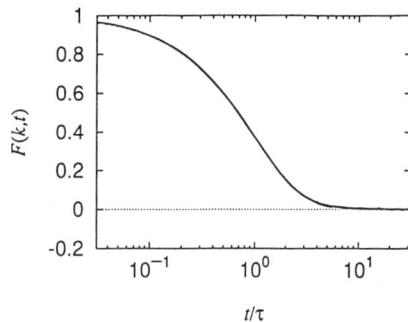

Figure 3: Scaling property of $\chi''(k,\omega)$. $\chi''(k,\omega)/\chi''(k,\omega_{max})$ is plotted vs. $\log(\omega/\omega_{max})$ for various values of $k$ ($k=n\pi/50$, $n = 4 \sim 10$). The results presented here are obtained from the 2D SC with $b = 3, l = 1$. The fractal dimension is $D \approx 1.893$.

Figure 4: $F(k,t)$ vs. $\log(t/\tau)$ for various values of $k$ ($k=n\pi/50$, $n = 3 \sim 11$). The results presented here are obtained from the 2D SC with $b = 3, l = 1$. The solid line is the relaxation function of the Cole-Cole type with the parameter $\alpha = 0.967$.

The scaling law in the relaxation function $F(k,t)$ is also found from the simulations of the 2D SC with $b = 3, l = 1$ as is expected from Eq. (4) when the scaling law in $\chi''(k,\omega)$ holds (Fig. 4). The solid line in Fig. 4 is the line obtained by the least-square-fitting of $F(k,t)$ to the Cole-Cole form.

The dependences of the parameter $\alpha$ on the normalized fractal dimension $D/D_0$ are presented in Fig. 5 for the Sierpinski structures, where $D_0$ is the Euclidean dimension of the space in which each fractal structure is placed. This figure shows that $\alpha$ decreases when $D/D_0$ is reduced, which reflects the fact that the smaller the fractal dimension, the slower the relaxation. The results for percolation clusters at $p_c$ are as follows: $\alpha = 0.748 \pm 0.005$ for 2D ($D/D_0 \approx 0.95$) and $\alpha = 0.666 \pm 0.004$ for 3D ($D/D_0 \approx 0.84$). The parameter $\alpha$ obtained from the percolation clusters is much smaller than that from the Sierpinski structures for the similar values of $D/D_0$. In fractal structures with no dead ends such as the Sierpinski structures, the slowing down of the relaxation is caused only by the blockings due to walls, so it is relatively easy for particles to find detours. On the other hand, in fractal structures such as the percolation clusters which are characterized by a number of dead ends at various stages of hierarchy, it takes quite a long time for particles to escape from dead ends once they are trapped therein, and accordingly the relaxation becomes slow. It is natural that $\alpha$ should depend not only on the fractal dimension but on other parameters like lacunarity [Gefen et al., 1984]. However it is interesting that $\alpha$ is roughly described by the normalized fractal dimension $D/D_0$ for Sierpinski structures with different Euclidean dimensions.

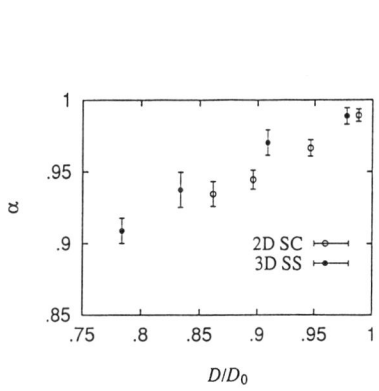

Figure 5: Parameter $\alpha$ vs. $D/D_0$ in the case of Sierpinski structures. In the figure, ○ and ●, respectively denote the results from the 2D SC and the 3D SS.

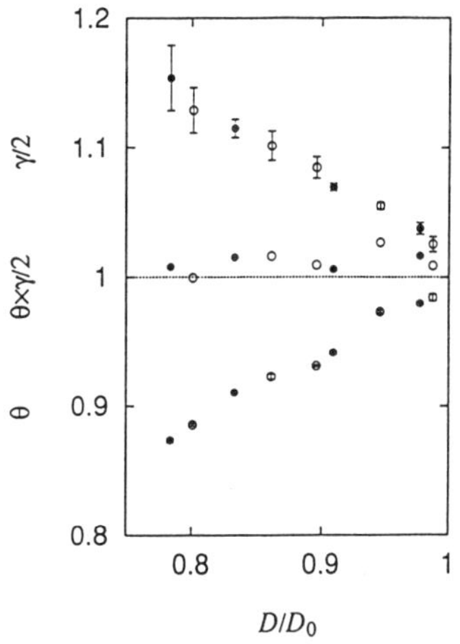

Figure 6: Parameters vs $D/D_0$: $\gamma/2$ (upper curves), $\theta$ (lower curves) and $\theta \times \gamma/2$ (center points). In the figure, ○ and ● respectively denote the results from the 2D SC and the 3D SS.

We also find that the $k$-dependence of $\tau$ is expressed by the same relation $\tau \propto k^{-\gamma}$ as fractal time random walk. The analysis of our simulations yields $\gamma \geq 2$, the equal sign being realized in the case of the Debye relaxation. The dependences of $\gamma$ on $D/D_0$ are given by the upper points in Fig. 6 for the Sierpinski structures, which shows that $\gamma$ increases when $D/D_0$ is reduced.

The parallel argument with fractal time random walk proceeds further. Our results of the simulations show the time dependence of the MSD is described by MSD $\propto t^\theta$ [Havlin and D. Ben-Avraham, 1987]. In Fig. 6, $\theta$ is presented for the Sierpinski structures. $\theta$ decreases as $D/D_0$ is reduced, showing that the diffusion is more delayed as the fractal dimension of the system becomes smaller. The parameter $\theta$ also is roughly described by $D/D_0$ for the different Euclidean dimensions. $\theta$ depends on the lacunarity as well as the fractal dimension and becomes smaller as the lacunarity increases for the 2D SC with the same fractal dimension [Kim et al., 1993]. The parameter $\theta$ obtained for the 2D SC gives the lower bound because the 2D SC with a central cutout has the largest lacunarity among those with the same fractal dimension. As seen by the center points in Fig. 6, the relation Eq. (11) holds. Correspondingly, the scaling law of $G(r,t)$ in the **r**-space is clearly shown in Fig. 7 for the 2D SC with $b=3, l=1$.

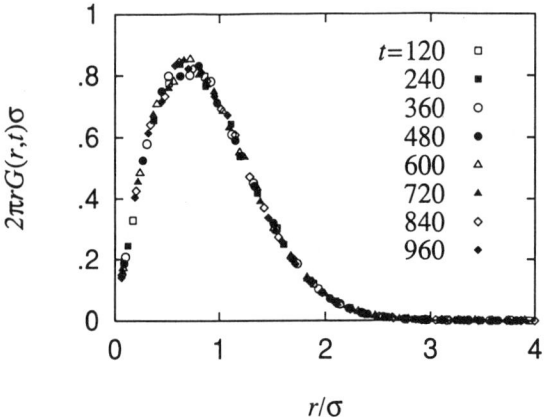

Figure 7: $2\pi r G(r,t)\sigma$ vs $r/\sigma$ for various times $t$. The results presented here are obtained from the 2D SC with $b=3, l=1$.

E. Summary and future work

In order to see how one-body picture is useful in describing the essential aspects of the anomalous relaxation, we make use of random walk models and take many-body effects into account through the temporal or spatial constraints for particles. In the above, we have shown that properties of fractal time random walk are quite similar to those of random walk on fractal structures. The following properties arising from the fractal nature are ascertained.

1. The anomalous relaxation is of the Cole-Cole type.
2. $\chi(k,\omega)$, $F(k,t)$ and $G(r,t)$ show scaling.
3. The relation between a parameter $\theta$ concerning diffusion and a parameter $\gamma$ concerning relaxation $\theta \times \gamma/2 = 1$ holds.

This work gives important clues to the understanding of the mechanisms of anomalous relaxation.

## II. DETERMINATION OF PERCOLATION THRESHOLDS WITH HIGH ACCURACY BY MONTE CARLO SIMULATIONS

Percolation theory has been studied for decades and applied to a large variety of physical systems [Stauffer, 1985] such as relaxation processes in supercooled fluids discussed in previous sections. Among various interests concerning percolation, the determination of percolation thresholds $p_c$ is a difficult problem to solve. They have been determined exactly in only a few kinds of lattices and others are remained unsettled in the following problems, the site problem of square and honeycomb lattice, and the bond problem of kagomé lattice. In order to determine the thresholds with high accuracy, we carried out Monte Carlo simulations, using Ziff's method.

Ziff et al. developed a new method to determine the percolation threshold for the site problem of square lattice [Ziff and Sapoval, 1986]. Their algorithm is rather quick and accurate because of no need of complicated cluster analysis and delicate extrapolations. We extended their method to the bond problem as well as to other kinds of lattice, and found more than four orders of the threshold of each lattice.

We performed Monte Carlo simulations in the following manner. We started with giving a gradient to the probability, with which sites or bonds are occupied, in one direction of the two dimensional lattice, and generated the percolating and non-percolating region (Fig. 8(a)). A walker walks along the surface between them, generating the hull of the surface. Then the percolation threshold can be calculated by

$$p_c = \frac{N_{\text{per}}}{N_{\text{per}} + N_{\text{non}}} \qquad (12)$$

where $N_{\text{per}}$ is the number of surface sites or bonds in the percolating region and $N_{\text{non}}$ is that of surface sites or bonds in the nonpercolating region. Counting $N_{\text{per}}$ and $N_{\text{non}}$, the threshold is calculated by the above relation.

Fig. 8(a) shows the structure of the surface in the case of site percolation of honeycomb lattice. Fractal property can be analyzed. Fig. 8(b) shows the distribution of the surface sites or bonds from which the mean position and deviation can be calculated. In the same way, a number of figures of these kinds are obtained in each kind of lattices.

In our simulations more than $10^8$ total surface sites or bonds were generated and probability gradient $\Delta p \sim 10^{-3}$. So statistical error is expected to be within $10^{-4}$ and we suppose four orders of $p_c$ are reliable. Percolation thresholds are summarized in Table 1. Comparing the previous and our results of the lattices in which the exact values are known, agreement of four orders is found and efficiency of $10^{-4}$ is supposed. For the site problem of honeycomb lattice, we found $p_c = 0.6970(9)$. This is a slightly different value from the previous one $p_c = 0.6962$ which was determined by Monte Carlo simulation [Ziff, 1992]. For the bond problem of kagomé lattice $p_c = 0.5244(3)$. This is in good agreement with the result from Potts model [Djordjevic, 1982], but there remains the possibility of deviation in higher orders.

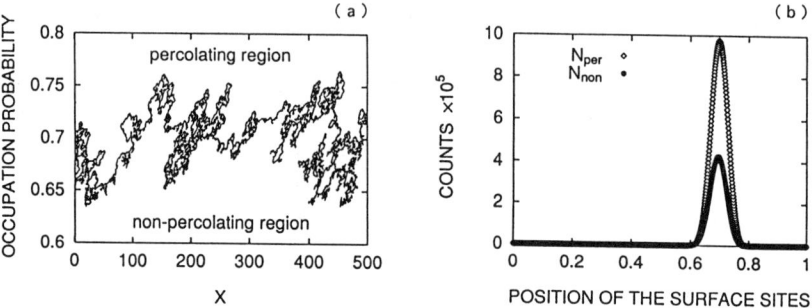

Figure 8: (a) The structure of the surface between the percolating and nonpercolating region in the case of site percolation of honeycomb lattice. (b) Distribution of the surface sites. It is found that the peak position is around $p_c$.

We found that Ziff's method provides good results for the determination of percolation thresholds, but it must be pointed out that there is no mathematical proof on the relation (12).

Previous Results

|  | | square | triangular | honeycomb | kagomé |
|---|---|---|---|---|---|
| site | $0.592746(0)^a$ | $1/2^*$ | $0.6962^b$ | $1 - 2\sin(\pi/18)^*$ | |
| bond | $1/2^*$ | $2\sin(\pi/18)^*$ | $1 - 2\sin(\pi/18)^*$ | $0.524429\cdots^c$ | |

Our Results

|  | square | triangular | honeycomb | kagomé |
|---|---|---|---|---|
| site | $0.5927(4)$ | $0.5000(6)$ | $0.6970(9)$ | $0.6526(9)$ |
| bond | $0.4999(7)$ | $0.3473(2)$ | | $0.5244(3)$ |

Table 1   * : Exact Solution   a : [Ziff and Sapoval, 1986]   b : [Ziff, 1992]   c : [Djordjevic, 1982]
$2\sin(\pi/18) = 0.347296\cdots$   $1 - 2\sin(\pi/18) = 0.652704\cdots$

## III. FINGER INSTABILITY ON THE WETTING CONTACT LINE

Studies of liquids interacting with solid surfaces, such as wetting, lubrication, adhesion and so on, are subjects of practical and technological importance. Recently they have been studied both experimentally and theoretically from a view point of physics of complex fluids[1]. In this section we report some topic on wetting.

---

[1] For example Physical Review E, containing a section of complex fluids, can be referred.

Wetting is an old but developing research area [de Gennes, 1985; Leger and Joanny, 1992]. We investigated by computers the motion of viscous and non-volatile fluids on a smooth and homogeneous solid surface. Treating the fluid as a continuum we considered fluid mechanics of the wetting and solved Navier-Stokes equations. With lubrication approximation which is applied when the flow is so slow that the inertial term can be neglected and the velocity gradient horizontal to the surface is small compared with the vertical one, we solved the following equation numerically

$$\frac{\partial h}{\partial t} + \nabla \frac{h^3}{3\eta} \left[ \gamma \nabla \nabla^2 h + \mathbf{F} \right] = 0 \tag{13}$$

where notations are defined in Fig. 9.

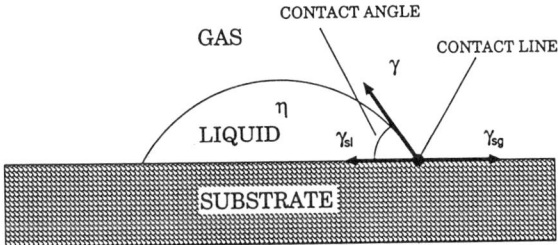

Figure 9: Wetting on a solid substrate. $\gamma$, $\gamma_{sl}$ and $\gamma_{sg}$ are surface tensions between gas and liquid, substrate and liquid, substrate and gas, respectively. $\eta$ is viscosity of liquid. The liquid height is defined by $h$ which is a function of position and time $t$.

It is known that under external forces, instability of the contact line appears and finger patterns are formed. Let us consider two cases one of which is liquid on an inclined surface under gravity [Huppert, 1982; Silvi and Dussan, 1985; Schwartz, 1989] and the other is a rotating drop under centrifugal force [Melo, Joanny, and Fauve, 1989].

Starting with a symmetrical shape of the fluid, we found that the instability did not appear. But giving fluctuations to the fluid height, they were reflected in fluctuations of the contact line and finger instabilities were observed Fig. 10.

In the case of liquid on an inclined surface, they are consistent with results obtained by Djordjevic [Djordjevic, 1982]. However the following facts are not in agreement with experiments in the case of a rotating drop. The shape of fingers is not so sharp that the lengths of them are short compared with the mean radius of the droplet, and the number of them is smaller than that observed in experiments. There are some explanations for these discrepancies. One possibility is that other factors besides surface tension and viscosity are needed or local fluctuations of surface tension must be taken into account. Another possibility is that lubrication approximation breaks down under strong external forces because the larger the radius of the drop develops the faster the fluid flows. In addition, microscopic dynamics at the contact line must be considered where long-range interaction with the surface plays an important role.

We are preparing for analytical and simulational approaches to explain these instabilities.

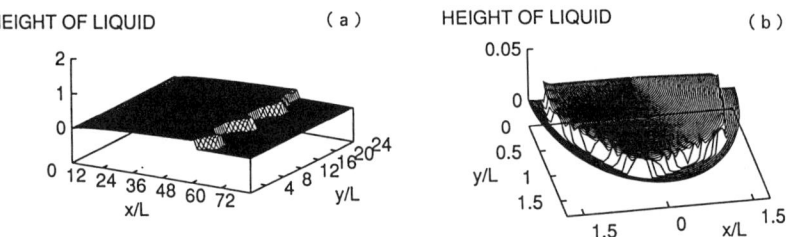

Figure 10: Results from our computer simulations in which finger instabilities are observed. Lengths are scaled by specific length L. (a) Liquid on an inclined plane. (b) Liquid on a rotating substrate.

**REFERENCES**
Binder, K. and A.P. Young, 1986, Rev. Mod. Phys., **58**, 801.
Cole, K.S. and R.H. Cole, 1941, J. Chem. Phys., **9**, 341.
Debye, P., 1929, *Polar Molecules*.
de Gennes, P. G., 1985, Rev. Mod. Phys. **57**, 827.
Djordjevic, Z. V., 1982, J. Phys. A, **15**, L405.
Fujiwara, S., S. Gomi, K. Morigaki and F. Yonezawa, 1993, J. Non-Cryst. Solids, **164-166**, 301.
Gefen, Y., A. Aharony and B.B. Madelbrot, 1984, J. Phys. A, **17**, 1277.
Gomi, S., S. Fujiwara and F. Yonezawa, 1993, J. Non-Cryst. Solids, **164-166**, 465.
Götze, W., 1991, *Liquids, Freezing and Glass Transition*, eds. J.-P. Hansen, D. Levesque and J. Zinn-Justin, (North-Holland, Amsterdam).
Havlin, S. and D. Ben-Avraham, 1987, Adv. Phys., **36**, 695.
Huppert, H. E., 1982, Nature (London), **300**, 427.
Kim, M.H., D.H. Yoon and I.M. Kim, 1993, J. Phys. A, **26**, 5655.
Klafter, J. and G. Zumofen, 1994, (preprint).
Kohlrausch, R., 1847, Ann. Phys. (Leipzig), **12**, 393.
Leger, L. & J. F. Joanny, 1992, Rep. Prog. Phys., **55**, 431.
Melo, F., J. F. Joanny and S. Fauve, 1989, Phys. Rev. Lett., **63**, 1959.
Nienhuis, B., 1982, J. Phys. A **15**, 199.
Scher, H. and E.W. Montroll, 1975, Phys. Rev. B, **12**, 245.
Scher, H., M.F. Shlesinger and J.T. Bendler, 1991, Physics Today, **44**, January, 26.
Schwartz, L. W., 1989, Phys. Fluids, **A 1**, 443.
Shlesinger, M.F., 1988, Ann. Rev. Phys. Chem., **39**, 269.
Silvi, N. and E. B. Dussan V, 1985, Phys. Fluids, **28**, 5.
Stauffer, D., 1985, *Introduction to Percolation Theory*, (Taylor & Francis, London and Philadelphia).
Williams, G. and D.C. Watts, 1970, Trans. Faraday Soc., **66**, 80.
Ziff, R.M. and B. Sapoval, 1986, J. Phys. A: Math. Gen., **19**, L1169.
Ziff, R. M., 1992, Phys. Rev. Lett., 2670.

# IX.2

# DENSITY FUNCTIONAL THEORY OF FREEZING OF SOFT-CORE SYSTEMS

M. Hasegawa

Department of Materials Science and Technology, Faculty of Engineering
Iwate University, Morioka 020, Japan

A density functional theory of freezing of soft-core systems is presented. The theory is based on the idea of the thermodynamic perturbation approach and may be viewed as a generalized version of the modified weighted-density approximation. The results of applications to the inverse-power systems and the classical one-component plasma are in good agreement with the simulation data.

## I. INTRODUCTION

In the so-called density functional theory (DFT) of freezing, the excess free energy functional, $F_{ex}[\rho]$, of the solid with a periodically modulated density distribution, $\rho(r)$, is described in terms of the free energy and structural functions of homogeneous fluids. Various type of theories have been developed and most of them have achieved ample success for the hard-sphere (HS) system (e.g., Evans, 1992). However, it has also been found that these theories more or less fail to predict freezing of systems interacting through soft-core or long-ranged potentials. This difficulty has been partly resolved by more complicated approaches: one is essentially based on the idea of the thermodynamic perturbation approach, which achieved great success in the studies of homogeneous fluids (e.g., Hansen and McDonald, 1986), and the other consists of extending the modified weighted-density approximation (MWDA) such that it is exact up to third order in the functional Taylor expansion of $F_{ex}[\rho]$ (Likos and Ashcroft, 1992, 1993).

In the present contribution we present a DFT of freezing which proved successful for the soft-core systems (Hasegawa, 1993, 1994). The theory consists of a generalization of the thermodynamic perturbation theory to inhomogeneous fluids and can be viewed as a generalized version of the MWDA. Various treatments within the present approach are investigated for the one-component plasma (OCP), for which the present theory is less successful.

## II. SUMMARY OF THE THEORY

We consider a system interacting through a pair potential, $\phi(r)$, and start with splitting $\phi(r)$ into two part, $\phi(r) = \phi_0 + \Delta\phi(r)$, where $\phi_0$ is a repulsive and short-ranged part and $\Delta\phi(r)$ is the remaining long-ranged part. The free energy of the system is then written, in accordance with this splitting, as

$$F[\rho] = F_{id}[\rho] + F_{0,ex}[\rho] + F_1[\rho], \tag{1}$$

where $F_{id}[\rho]$ is the ideal-gas contribution, $F_{0,ex}[\rho]$ the excess free energy of the reference system interacting through $\phi_0(r)$ and $F_1[\rho]$ the contribution due to $\Delta\phi(r)$. The functional form of $F_{id}[\rho]$ is known as

$$F_{id}[\rho] = \beta^{-1} \int dr \rho(r) \left\{ \ln\left[\rho(r)\Lambda^3\right] - 1 \right\}, \tag{2}$$

where $\beta = 1/k_B T$ and $\Lambda$ is the thermal de Broglie wavelength. We note that the explicit expressions for $F_{0,ex}[\rho]$ and $F_1[\rho]$ are irrelevant in the present theory.

Various versions have been developed based on the above approach and major differences between these theories consist in the choice of the reference system and the treatment of $F_1[\rho]$ [Curtin and Ashcroft, 1986; Lutsko and Baus, 1991; Mederos et al, 1993]. In the present approach [Hasegawa, 1993, 1994], we use the MWDA by Denton and Ashcroft [1989] to calculate both $F_{0,ex}[\rho]$ and $F_1[\rho]$:

$$F_0[\rho] \approx N f_{0,ex}(\hat{\rho}_0), \qquad (3a)$$

$$F_1[\rho] \approx N f_1(\hat{\rho}_1), \qquad (3b)$$

where $N$ is the number of particles in the system of volume $V$, $f_{0,ex}(\rho)$ the excess free energy per particle of the homogeneous reference fluid and $f_1(\rho) = f_{ex}(\rho) - f_{0,ex}(\rho)$, $f_{ex}(\rho)$ being the total excess free energy of the system. We assume that $\hat{\rho}_0$ and $\hat{\rho}_1$ are independent of each other and given as in the original MWDA by

$$\hat{\rho}_0 = \frac{1}{N} \int dr_1 \int dr_2 \, \rho(r_1)\rho(r_2) w_0(r_{12}; \hat{\rho}_0), \qquad (4a)$$

and

$$\hat{\rho}_1 = \frac{1}{N} \int dr_1 \int dr_2 \, \rho(r_1)\rho(r_2) w_1(r_{12}; \hat{\rho}_1), \qquad (4b)$$

with $r_{12} = |r_1 - r_2|$. The weight-functions, $w_0$ and $w_1$, are as yet unspecified.

The weight-functions are determined in the same way as in the original MWDA [Denton and Ashcroft, 1989]. More specifically, we first require that both $w_0$ and $w_1$ are normalized, which must be satisfied in the limit of a uniform density to ensure that the approximate free energies become exact in this limit. Then, we require that both approximate $F_{0,ex}[\rho]$ and $F_1[\rho]$ in Eqs. (3a) and (3b) exactly reproduce the corresponding two-body direct correlation functions (DCFs), $C_0^{(2)}(r;\rho)$ and $\Delta C^{(2)}(r;\rho) = C^{(2)}(r;p) - C_0^{(2)}(r;\rho)$, in the limit of a uniform density. Unique specifications of the weigh-functions follow from these requirements and the use of these results into Eqs. (4a) and (4b) leads to the equations,

$$(\hat{\rho}_0 - \bar{\rho})\beta f'_{0,ex}(\hat{\rho}_0) - \frac{1}{2N} \int dr_1 \int dr_2 \, \Delta\rho(r_1)\Delta\rho(r_2) C_0^{(2)}(r_{12}; \hat{\rho}_0), \qquad (5a)$$

and

$$(\hat{\rho}_1 - \bar{\rho})\beta f'_1(\hat{\rho}_1) - \frac{1}{2N} \int dr_1 \int dr_2 \, \Delta\rho(r_1)\Delta\rho(r_2) \Delta C^{(2)}(r_{12}; \hat{\rho}_1), \qquad (5b)$$

where $f'_{0,ex}(\rho) = \partial f_{0,ex}(\rho)/\partial \rho$, $f_1(\rho) = \partial f_1(\rho)/\partial \rho$, and $\Delta\rho(r) = \rho(r) - \bar{\rho}$, $\bar{\rho}$ being the average density,

$$\bar{\rho} = \frac{1}{V} \int dr \, \rho(r). \qquad (6)$$

We note that both $\hat{\rho}_0$ and $\hat{\rho}_1$ are functionals of $\rho(r)$ and determined by solving Eqs. (5a) and (5b), respectively, for a given $\rho(r)$. The approximate excess free energy per particle of the solid phase is then given by

$$\beta f_{ex}[\rho] \approx \beta f_{0,ex}(\hat{\rho}_0) + \beta f_1(\hat{\rho}_1). \qquad (7)$$

We denote the above theory as the generalized MWDA (GMWDA).

It is established that the nonperturbative treatment is essential for $F_{0,\text{ex}}[\rho]$, but the second-order perturbation theory (SPT) is expected to work well for $F_1[\rho]$. In this approximation, $\beta f_1[\rho]$ is given by

$$\beta f_1[\rho] = \beta f_1(\bar{\rho}) - \frac{1}{2N}\int \mathrm{d}r_1 \int \mathrm{d}r_2\, \Delta\rho(r_1)\Delta\rho(r_2)\Delta C^{(2)}(r_{12};\bar{\rho}). \tag{8}$$

This result is compared with $\beta f_1(\hat{\rho}_1)$ in Eq. (7) and the agreement between them was found to be quite good as expected.

## III. APPLICATIONS TO INVERSE-POWER SYSTEMS

We consider systems interacting through the potentials,

$$\phi(r) = \varepsilon(\sigma/r)^n. \tag{9}$$

The reduced excess thermodynamic quantities of these systems depend on a single parameter defined by

$$\gamma = (\bar{\rho}\sigma^3)(\beta\varepsilon)^{3/n}. \tag{10}$$

We adopted the Kang-Lee-Ree-Ree (KLRR) separation defined by [Kang et al, 1985]

$$\phi_0(r) = \left[\phi(r) - \phi(r_0) - (r - r_0)\phi'(r_0)\right]\theta(r_0 - r), \tag{11}$$

where $\phi'(r) = \mathrm{d}\phi(r)/\mathrm{d}r$. Following Lutsko and Baus [1991], we used the value, $r_0 = (\sqrt{2}/\bar{\rho})^{1/3} = 1.8094\bar{R}_s$, where $\bar{R}_s$ is the Wigner-Seitz sphere radius corresponding to $\bar{\rho}$. This value of $r_0$ is equal to the nearest-neighbor distance of a compact lattice structure.

All the calculations were performed using a variational method, in which the solid density was parametrized and given by the sum of the Gaussians peaked at each site of a periodic lattice, $\{R_i\}$:

$$\rho(r) = \left(\frac{\pi}{\alpha}\right)^{3/2}\sum_i \exp\left[-\alpha(r - R_i)^2\right]. \tag{12}$$

Then, the variational principle for the free energy, $F[\rho]$, reduces to a minimization of $F[\rho]$ with respect to $\alpha$.

As the input data necessary to solve the GMWDA equations, (5a) and (5b), we used the Monte Carlo (MC) results for $f_{\text{ex}}(\rho)$ [Hoover et al, 1971], the reference free energies $f_{0,\text{ex}}(\rho)$ calculated by the method of Andersen, Weeks and Chandler (AWC) [1971], and the DCFs obtained by the variational modified hypernetted-chain (VMHNC) theory [Rosenfeld, 1986].

The results of calculations for the freezing properties of inverse-power systems are summarized in Table I. The table also includes the results for the HS system for comparisons. We first note that the generalized effective-liquid approximation (GELA) [Lutsko and Baus, 1990] provides the best result for the HS system but completely fails for all other systems. While the MWDA and the GMWDA yield similar results for the freezing of systems with large $n$, the GMWDA greatly improves over the MWDA

**TABLE I.** Freezing parameters of inverse-power systems. The fcc structure is assumed for the solid phase. $\gamma_f$ and $\gamma_m$ represent the freezing and melting points, respectively, $\Delta\rho/\rho_m = (\gamma_m - \gamma_f)/\gamma_m$ is the fractional density change on melting, and $L$ the Lindemann ratio at $\gamma = \gamma_m$.

| $n$ | Method | $\gamma_f$ | $\gamma_m$ | $\Delta\rho/\rho_m$ | $L$ |
|---|---|---|---|---|---|
| $\infty$(HS) | MWDA[1] | 0.910 | 1.036 | 0.13 | 0.097 |
|  | MWDA[2] | 0.887 | 1.035 | 0.143 | 0.106 |
|  | GELA[3] | 0.945 | 1.041 | 0.10 | 0.095 |
|  | GELA[2] | 0.949 | 1.066 | 0.110 | 0.083 |
|  | MC[4] | 0.943 | 1.041 | 0.104 | 0.126 |
| 12 | MWDA[5] | 1.194 | 1.252 | 0.046 | 0.096 |
|  | GMWDA[6] | 1.097 | 1.159 | 0.053 | 0.113 |
|  | MC[7] | 1.15 | 1.19 | 0.037 | 0.15 |
| 9 | MWDA[6] | 1.488 | 1.514 | 0.018 | 0.077 |
|  | GMWDA[6] | 1.321 | 1.375 | 0.039 | 0.111 |
|  | MC[7] | 1.33 | 1.37 | 0.029 | 0.16 |
| 6 | MWDA[5] | 2.666 | 2.720 | 0.020 | 0.074 |
|  | GMWDA[6] | 2.130 | 2.171 | 0.019 | 0.111 |
|  | MC[7] | 2.18 | 2.21 | 0.013 | 0.17 |
| 4 | MWDA[5] | 8.176 | 8.238 | 0.0075 | 0.07 |
|  | PMWDA[8] | 6.68 | 6.73 | 0.007 |  |
|  | GMWDA[6] | 5.306 | 5.341 | 0.0066 | 0.111 |
|  | MC[7] | 5.54 | 5.57 | 0.0051 | 0.18 |

[1] Denton and Ashcroft [1989], [2] present work, [3] Lutsko and Baus [1990], [4] Hoover and Ree [1968], [5] Laird and Kroll [1990], [6] Hasegawa [1994], [7] Hoover et al [1971], [8] Salgi and Rajagopalan [1991].

for small $n$ and predicts $\gamma_f$ and $\gamma_m$ in good agreement with the MC results for all $n$ considered. As exemplified for the case of $n = 4$, the GMWDA also improves over the PMWDA [Salgi and Rajagopalan, 1991], a simplified version in which the HS system is used as the reference system and the SPT as given by Eq. (8) is used to calculate the long-range contribution [Curtin and Ashcroft, 1986]. This result suggests that the choice of the reference system is important and the present choice given by Eq. (11) is superior to the HS system. We note that the MWDA (Eqs. (3b) and (5b)) and the SPT as given by Eq. (8) yield similar results for $f_1[\rho]$ [Hasegawa, 1993, 1994]. Our final concern is the arbitrariness of the break point $r_0$ and we confirmed that freezing properties are stable against the variation of $r_0$ around the value, $r_0 = 1.8094\bar{R}_s$.

**TABLE II.** Freezing of the OCP into the bcc solid.

| Method | $r_0/\bar{R}_s$ | $\Gamma$ | $L$ |
|---|---|---|---|
| TPT[1] | | 140 | |
| EMWDA[2] | | 183 | 0.17 |
| GMWDA[3] | 1.8094 | 142 | 0.133 |
| | 1.75 | 147 | 0.135 |
| | 1.70 | 155 | 0.140 |
| MC[4] | | 178 | 0.17 |

[1]Iyetomi and Ichimaru [1988], [2] Likos and Ashcroft [1992, 1993], [3]present work, [4]Slattery et al [1980].

## IV. RESULTS FOR THE OCP

The GMWDA presented in the previous sections is less successful for the OCP. These results as well as those of other methods are summarized in Table II. For the OCP, we used the plasma parameter $\Gamma$ as the coupling constant, which is defined by $\Gamma = \beta e^2 / \bar{R}_s$, $e$ being the charge of the particle.

The third-order perturbation theory (TPT) was found to be successful in predicting freezing of the OCP into the bcc solid [Iyetomi and Ichimaru, 1988], and it seems essential to take into account terms up to third order in the functional Taylor expansion of $F_{\text{ex}}[\rho]$ for the OCP. Likos and Ashcroft [1992, 1993] extended the MWDA such that it is exact up to third order in the functional Taylor expansion of $F_{\text{ex}}[\rho]$, and found that the theory (EMWDA) yields freezing properties of the OCP in very good agreement with the MC data. However, these theories requires elaborate theoretical and numerical effort in treating the triplet DCFs and some uncertainties due to these input data are unavoidable.

We note that the GMWDA yields better results than the TPT, although the GMWDA results show some variation with varying break point $r_0$ of the reference system. For the reference system of the OCP, the packing fraction, $\eta = (\pi/6)\bar{\rho}\sigma^3$, of the underlying HS system determined by the AWC method is generally larger than those for inverse-power systems for the same value of $r_0$. In fact, we have $\eta = 0.5342$ for $r_0 = 1.8094 \bar{R}_s$ at $\Gamma = 150$, which seems too large to ensure that the structures of the reference and full systems are similar to each other. If we use smaller values of $r_0$, we have better results for the freezing properties of the OCP as shown in Table II. The value of $r_0 = 1.70 \bar{R}_s$ seems appropriate, which leads to $\eta = 0.4557$ at $\Gamma = 150$, but we have no rigorous criterion to specify the value of $r_0$ (or, more generally, reference system itself) other than the similarity of the structures mentioned in the above.

An accurate treatment of the reference free energy, $F_0[\rho]$, is probably the most crucial task in the GMWDA and similar approaches. We tried another method to gain insight into the nature of approximations involved in the calculations of $F_0[\rho]$. Similarity between the HS system and the present reference system interacting through

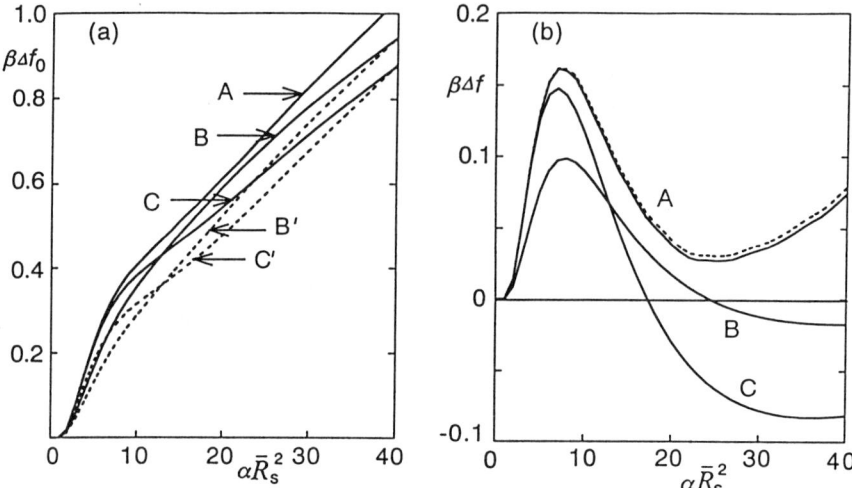

Fig. 1. (a) Free energies of the solid reference system relative to the liquid, $\beta\Delta f_0[\rho] = (F_0[\rho] - F_0(\bar{\rho}))/Nk_B T$, at $\Gamma = 150$ ($r_0/\bar{R}_s = 1.70, \eta = 0.4557$): curve A, full MWDA; curve B, GELA(HS)+SPT; curve C, MWDA(HS)+SPT; curves B' and C' are the HS contributions to the curves B and C, respectively. (b) Relative free energies of the OCP, $\beta\Delta f_0[\rho] = \beta\Delta f_0[\rho] + \beta\Delta f_1[\rho]$, with $\beta\Delta f_1[\rho] = (F_1[\rho] - F_1(\bar{\rho}))/Nk_B T$: each curve corresponds to that in (a), and $F_1[\rho] = Nf_1[\rho]$ was calculated either in the GMWDA as given by Eqs. (3b) and (5b) (full curves) or the SPT as given by Eq. (8) (dashed curve)

$\phi_0(r)$ naturally leads us to the idea of using the HS system as another reference system in calculating $F_0[\rho]$. In this method, $F_0[\rho]$ is given by

$$F_0[\rho] = F_{HS}[\rho] + F'_0[\rho], \qquad (13)$$

where $F_{HS}[\rho]$ is the HS free energy and $F'_0[\rho]$ is the remaining contribution given by the expression similar to that of $F_1[\rho]$. We used both the MWDA and the GELA to calculate $F_{HS}[\rho]$ and the SPT for $F'_0[\rho]$. We assumed $r_0 = 1.70\bar{R}_s$ for the break point, which seems approximate as discussed in the above. The results of these calculations for $F_0[\rho]$ at $\Gamma = 150$ are shown in Fig. 1(a).

The relative free energy, $\beta\Delta f_0$, obtained using the GELA for $F_{HS}[\rho]$ (denoted as GELA (HS) + SPT) is not much different (on the scale in Fig. 1(a)) from that obtained using the MWDA for the system with $\phi_0(r)$ (denoted as full MWDA) for realistic solids with $\alpha\bar{R}_s^2 \approx 20$. On the other hand, the result obtained using the MWDA for $F_{HS}[\rho]$ (denoted as MWDA (HS) + SPT) much differs from the above two results. Some important conclusions can be drawn from these results. Firstly, we may conclude that the GELA for $F_{HS}[\rho]$ and the MWDA for $F_0[\rho]$ are of similar nature and accuracy, provided that the SPT for $F'_0[\rho]$ is a good approximation as actually expected. This could be the reason why the GMWDA works rather well even for the OCP. This argument is based an on the expectation that the GELA is a nearly complete approximation scheme for the HS system. Secondly, the accuracy or the adequacy of the MWDA is quite different for $F_{HS}[\rho]$ and $F_0[\rho]$ in spite of the similarity of the two systems. More explicitly, the MWDA is not necessarily a good approximation for the HS system if it is used as the reference system. This argument is seemingly in contradiction to the fact that the MWDA predicts freezing of the HS system rather well (see Table I). Thirdly, we are concerned with this apparent contradiction. We found that the results for $F_{HS}[\rho]$ in the MWDA and the GELA show quite different behaviors as functions of the localization parameters, $\alpha \bar{R}_s^2$. The stabilized $F_{HS}[\rho]$ in these theories also behave quite differently as functions of the density (or $\eta$), although they yield similar freezing properties. One should be reminded of these results when one adopts the HS system as the reference system in the theory of freezing and related phenomena. Finally, we note that both schemes using the HS system (GELA (HS) + SPT and MWDA (HS) + SPT) produce much worse results for the freezing of the OCP, as suggested by the results in Fig. 1(b).

## V. CONCLUDING REMARKS

We have made systematic investigations of freezing of soft-core systems using a generalized version of the MWDA and confirmed that the theory works well for the inverse-power systems and the OCP. The theory (GMWDA) is based on the idea of the thermodynamic perturbation theory and we found that the choice of the reference system and its treatment is crucial, especially for the OCP, in predicting freezing properties. These findings suggest that it is still important to develop an improved DFT of freezing for system interacting through repulsive and short-ranged potentials, for which most theories work in predicting freezing itself but might have defects when used in the framework of the GMWDA-like approach.

## REFERENCES

Andersen, H. C., J. D. Weeks, and D. Chandler, 1971, Phys. Rev. A, **4**, 1597.
Curtin, W. A., and N. W. Ashcroft, 1986, Phys. Rev. Lett. **26**, 2775.
Denton, A. R., and N. W. Ashcroft, 1989, Phys. Rev. A, **39**, 4071.
DeWitt, H. E., 1979, *Strongly Coupled Plasmas*, ed. G. Kalman and P. Carini (New York, Plenum) p.81.
Evans, R., 1992, *Fundamentals of Inhomogeneous Fluids*, ed. D. Henderson (New York, Dekker) p.85.
Hasegawa, M., 1993, J. Phys. Soc. Jpn, **62**, 4316.
Hasegawa, M., 1994, J. Phys. Soc. Jpn, **63**, 2215.
Hansen, J.-P., and I. R. McDonald, 1986, *Theory of Simple Liquids*, 2nd ed., (London, Academic).
Hoover, W. G., S. G. Gray, and K. W. Johnson, 1971, J. Chem. Phys., **55**, 1128.
Hoover, W. G., and F. M. Ree, 1968, J. Chem. Phys., **49**, 3609.
Iyetomi, H., and S. Ichimaru, 1988, Phys. Rev. B, **38**, 6761.
Kang, H. S., C. S. Lee, T. Ree, and F. H. Ree, 1985, J. Chem. Phys., **82**, 414.

Laird, B. B., and D. M. Kroll, 1990, Phys. Rev. A, **42**, 4810.
Likos, C. N., and N. W. Ashcroft, 1992, Phys. Rev. Lett., **69**, 316.
Likos, C. N., and N. W. Ashcroft, 1993, J. Chem. Phys., **99**, 9090.
Lutsko, J. F., and M. Baus, 1990, Phys. Rev. A, **41**, 5547.
Lutsko, J. F., and M. Baus, 1991, J. Phys.: Condens. Matter, **3**, 6547.
Rosenfeld, Y., 1986, J. Stat. Phys., **42**, 437.
Salgi, P., and R. Rajagopalan, 1991, Phys. Rev. A, **44**, 5310.
Slatterly, W. L., G. D. Doolen, and H. E. DeWitt, 1980, Phys. Rev. A, **21**, 2087.

**IX.3**

# COMPUTER SIMULATION OF MATERIALS ON PARALLEL ARCHITECTURES: GLASSES, SOLID $C_{60}$, AND GRAPHITIC TUBULES

Priya Vashishta, Rajiv K. Kalia, Wei Jin, Jin Yu, and Aiichiro Nakano

Concurrent Computing Laboratory for Materials Simulations
Department of Physics & Astronomy
Department of Computer Science
Louisiana State University, Baton Rouge, LA 70803-4001, U. S. A.

A multiresolution algorithm is designed to carry out large-scale molecular dynamics (MD) simulations for systems with long-range Coulomb and three-body covalent interactions. Molecular dynamics simulations of porous silica, in the density range 2.2–0.1 g/cm$^3$, are carried out on 41,472-particle systems to study structural correlations. Pore interface growth and the roughness of internally fractured surfaces in silica glasses are investigated by MD simulations on a 1.12-million particle systems. The roughness exponent for fracture surfaces supports experimental claims about the universality. Structural transformation and dynamical behavior of SiO$_2$ glass at high pressures are investigated by MD simulations. The simulations reveal a tetrahedral to octahedral transformation, which was reported recently by Meade, Hemley, and Mao. Lattice dynamics of solid $C_{60}$ is studied using a unified interaction model which consists of a tight-binding potential for the intra-molecular interaction and a Lennard-Jones and bond charge model for the inter-molecular interaction. The effects of orientationally disordering and pressure on the inter- and intra-molecular phonons are investigated. Recently a new form of carbon—graphitic tubule—has been discovered. Using the tight binding molecular dynamics method, the structural and dynamical properties of graphitic tubules are studied.

## I. INTRODUCTION

Despite significant recent developments in materials-simulation techniques [1-11], the goal of reliably predicting the properties of new materials in advance of fabrication and measurement has not yet been achieved. The primary reason for this lack of success is the inability of sequential machines to handle large-scale simulations. For example, molecular dynamics (MD) simulations for long-range interactions scale as $N^2$ where N is the number of particles in the system. In many physical systems, the desired system sizes are in the range of $10^6$ particles. These are beyond the compute power of most sequential machines. However, the MD technique has considerable inherent parallelism. By exploiting this parallelism on emerging parallel architectures, it is possible to perform large-scale simulations for complex materials [3-6].

There has been a growing interest in porous materials because of their many technologically important applications [12-18]. Much of the recent work has focused on aerogel silica, a form of porous SiO$_2$ which is prepared by hypercritical drying of an alcoholic silica gel. It is an environmentally safe material with a large thermal resistance which makes it a suitable alternative to chlorofluorocarbon (CFC)-foamed plastic in thermal insulation of commercial and household refrigerators [12]. The application of porous glasses results from their unique selective separation capabilities, molecular transport, thermal resistance, and mechanical properties. All of these characteristics depend crucially on structural correlations such as the pore size, internal surface area, surface-to-volume ratio, and interface texture. We have studied the structural correlations in porous silica in the density range 2.2–0.1 g/cm$^3$ by MD simulations on 41,472-particle systems [18].

In recent years, a great deal of progress has been made in understanding the morphology of surfaces and interfaces. Scale-invariant surface fluctuations related to different growth processes have been observed in a wide variety of systems [19,20]: vapor deposition; fluid flow in porous media; sedimentation of granular materials; and thin-film growth. The root mean square surface fluctuations averaged over a distance $l$ obey the scaling relation [19,20], $W \sim l^\alpha$. Recent experiments on a wide variety of materials reveal that fracture surfaces exhibit the scaling properties with the roughness exponent $\alpha \sim 0.8$ [21,22]. This has led to the suggestion that the roughness exponent for fracture surfaces has a universal value. However the universality of the roughness exponent on the nanometer scale is still an unresolved issue [23]. We have investigated the roughness of internally fractured surfaces in $SiO_2$ glasses [24] by MD simulations on 1.12-million particle systems using parallel computer architectures.

In the past decade, numerous attempts have been made to investigate the structure and dynamics of crystalline and glassy states of $SiO_2$ at high pressures. Irreversible changes, indicating permanent densification, have been observed in the Brillouin and Raman spectra [25] of recovered $SiO_2$ glass samples. Infrared absorption measurements [26], however, indicate reversible changes in the $SiO_2$ glass at 20 GPa. Recently Meade, Hemley, and Mao have carried out in situ high-pressure x-ray diffraction experiments on $SiO_2$ glasses [27]. These measurements reveal significant changes in the intermediate range order. We have investigated the structural transformation of $SiO_2$ glass at high pressures by MD simulations [28].

Since the breakthrough in the synthesis of $C_{60}$ [29,30] and the discovery of superconductivity in $K_3C_{60}$ and $Rb_3C_{60}$ [31], tremendous effort has been made to understand the structure, orientational order, rotational dynamics, and inter- and intra-molecular interactions in fullerite and their compounds. By combining the inter-molecular (van der Waals and bond charge) and the intra-molecular (tight-binding) interaction models, we have calculated the inter-molecular and intra-molecular phonon density of states and dispersion for crystalline $C_{60}$ [32,33]. Effects of orientational disorder and pressure on the lattice dynamics have been studied.

Recently a new form of carbon - graphitic tubule was discovered in carbon rods under arc discharge by high-resolution transmission-electron microscopy [34]. It is the fourth member of the carbon family with dimension of 1 (diamond in 3D, graphite in 2D, and fullerene in 0D). A microtubule has the form of a rolled graphite sheet with a diameter of a few nanometers. The carbon-atom hexagons on the tubule are usually arranged in a helical fashion about the tubule axis. Due to crystalline perfection, various possible helical structures, and the dimensionality as well as the high efficiency of production, graphitic tubules may possess unusual mechanical, electronic, and optical properties which may find considerable technological applications (e.g. nanoscale devices, light-weight and high-strength composite materials, etc.). Using the tight-binding molecular-dynamics (TBMD) we have studied the structural and dynamical properties of the graphitic tubules [35].

In this paper, we describe our recent work on parallel algorithms for the MD approach and simulation results for structural and dynamical properties of materials. The outline of this paper is as follows. Parallel MD algorithms are discussed in Sec. II. Finite size effects on the intermediate-range order in amorphous $SiO_2$ are analyzed in Sec. III. Simulation results for structural properties of porous $SiO_2$ glasses and the roughness of fracture surfaces in $SiO_2$ glasses are discussed in Secs. IV and V, respectively. Section VI deals with the structural and vibrational properties of solid $C_{60}$ and graphitic tubules. Section VII describes the computing facilities in our laboratory. Concluding remarks are given in Sec. VIII.

## II. PARALLEL MOLECULAR DYNAMICS ALGORITHM

Molecular dynamics (MD) approach has played a key role in our understanding of physical systems [1]. For systems with a finite-range interparticle interaction, an efficient way to calculate the interactions is to use the linked-list method [1]. A major advantage of the linked-list technique is that the computation time is proportional to the number of particles, N. Furthermore, with the linked-list method the minimum-image convention can be implemented efficiently on parallel computers [4-6].

The computation of forces can be further reduced with the multiple time-step (MTS) approach [2]. The MTS approach exploits the fact that the force experienced by a particle can be separated into a rapidly varying primary component and a slowly varying secondary component. The primary interaction arises from nearest neighbors of a particle, whereas the secondary forces are due to other particles within its range of interaction. The primary component is calculated at every MD step. On the other hand, the secondary component is calculated at intervals of 5 to 15 steps. In between, the secondary component is extrapolated according to the Taylor series.

We have used the divide-and-conquer strategy based on domain decomposition [3] to implement the MTS-MD algorithm on distributed-memory MIMD (multiple instruction multiple data) machines [4,5]. The total volume of the system is divided into $p$ subsystems of equal volume, and each subsystem is assigned to a processor.

The three-body force calculation is a time consuming part of the MTS-MD algorithm. Speed-up of the three-body force calculation is achieved by decomposing the three-body potential into a separable form [5,7].

The most prohibitive computational problem is associated with the Coulomb potential. Because of its long range, each atom interacts with all the other atoms in the system. Therefore the evaluation of the Coulomb potential for an N-particle system requires $O(N^2)$ operations. Recently, we have implemented parallel MD simulations involving the Ewald summation [8] for Coulomb interaction [4]. The parallel algorithm for the Ewald summation reduces the computational complexity from $O(N^{3/2})$ to $O(N)$. This is achieved by ensuring that both the real-space and Fourier-space contributions scale linearly with the size of the system while maintaining the desired level of precision - 0.01% in the total potential energy.

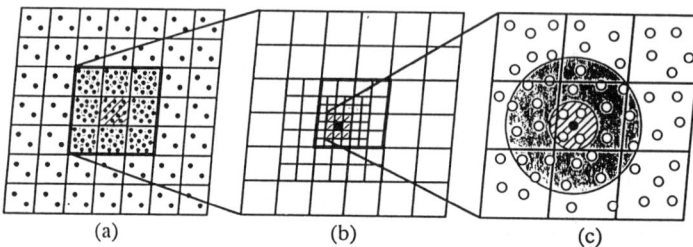

Figure 1: Multiresolution in space. (a) Periodically repeated images of the original MD box. Replacing far images by a small number of particles with the same multipole expansion up to a certain order reduces the computation enormously while maintaining the necessary accuracy. (b) A hierarchy of cells used in the fast multipole method. (c) The near-field force on a particle is due to primary, secondary, and tertiary neighbor particles.

Recent hierarchical algorithms [9,10] have revolutionized the computation of the Coulomb potential. The fast multipole method (FMM) uses the truncated multipole expansion and local Taylor expansion for the Coulomb potential field [10]. By computing both expansions recursively on a hierarchy of cells, the Coulomb potential is computed with O(N) operations. In many materials simulations, periodic boundary conditions are used to minimize surface effects. The summation over infinitely repeated image charges must be carried out to compute the Coulomb potential. Ding, Karasawa, and Goddard have developed the reduced cell multipole method (RCMM) which makes the computation of the Coulomb potential feasible for multimillion-particle systems with periodic boundary conditions [11]. In RCMM, distant images are replaced by a small number of fictitious particles with the same leading multipoles as the original system. With little computational effort, the Ewald summation is applied to these reduced images. We have developed a highly efficient MD algorithm based on multiresolutions in both space and time [6]. The Coulomb potentials in periodic systems are calculated with the RCMM and FMM, while non-Coulombic potentials are calculated by the MTS method, see Figure 1.

Performance of the multiresolution algorithm is tested for $SiO_2$ systems on the 512-node Intel Touchstone Delta machine at Caltech and the 128-node IBM SP-1 system at Argonne National Laboratory [6]. Figure 2 shows the execution time per MD step as a function of the number of processors, $p$. Number of particles is taken to be $8{,}232p$. For a 4.2-million particle system, the program requires only 4.84 seconds per MD step on the 512-node Delta. Communication accounts for only 8% of the total elapsed time. On the IBM SP-1, the computation part runs 4.8 times faster than on the Delta, while the communication performs at about the same speed.

## III. MD SIMULATIONS OF POROUS SILICA

Structural correlations in porous silica span many hierarchical regimes. The short-range ($< 4\text{Å}$) correlations arise from the structure of the $SiO_4$ tetrahedral unit [36]. The

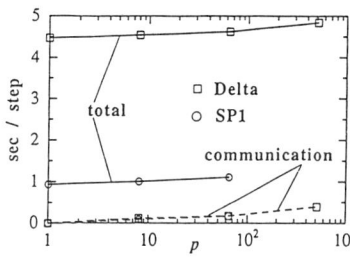

Figure 2: Execution time (solid curves) and communication time (dotted) per MD time step for $SiO_2$. Circles and squares represent the results on the IBM SP-1 and Intel Touchstone Delta, respectively. Here $p$ is the number of processors. The size of the system, N, increases as 8232 $p$.

intermediate-range (4–8Å) correlations, manifested as the first sharp diffraction peak (FSDP) in neutron- and x-ray diffraction experiments, arise from the connectivity of the tetrahedral units [36]. Beyond the intermediate range, small-angle neutron scattering (SANS) [13,14] and small-angle x-ray scattering (SAXS) [15,16] experiments on porous silica reveal a fractal structure.

The first MD simulation of 1,500-particle porous $SiO_2$ was performed by Kieffer and Angell [17]. We present the results of our MD simulations of porous $SiO_2$ at densities in the range of 2.2–0.1 g/cm$^3$ [18]. The system consists of 41,472 Si and O atoms. Even at the lowest density, 0.1 g/cm$^3$, the length of the MD box (240Å) covers all the hierarchical correlation regimes mentioned above. Simulations reported here took 1,200 hours on the 8-node iPSC/860 system at CCLMS.

The porous $SiO_2$ systems are prepared by successive expansions of a well-thermalized glass at the normal glass density 2.2 g/cm$^3$ and 300 K [17,18]. At the condensed amorphous phase above 1.6 g/cm$^3$, the amorphous system possesses only short- and intermediate-range correlations. However, as the density is lowered below 1.6 g/cm$^3$, density fluctuations that give rise to pores of various sizes set in. A close examination of snapshots reveals self-similarity at length scales between 5–25Å.

In Figure 3, we show a log-log plot of the pair distribution function $g(r)$ at various densities. Short-range correlations manifest themselves as peaks at distances less than 5Å. Some of the peaks split at lower densities, but the peak positions change very little. In the range of 5–25Å, a power-law decay is superimposed on the peak structures. From the power-law, the fractal dimension is calculated as $d_f = 3 + d\log[g(r)]/d\log(r)$ [17,18]. In real materials, the value of $d_f$ depends on the aggregation process and sample preparation conditions such as $pH$ value [13-16]. To investigate the effect of kinetic processes, we performed another set of MD simulations where the temperature was kept at 1,000 K instead of 300 K during the expansion process. At this high temperature, larger $d_f$ is observed. Kinetic processes during the expansion determine the structure of the resulting glass. For higher temperatures, larger diffusion overcomes the correlation in immediate neighbors and more global configuration space is searched. As a result,

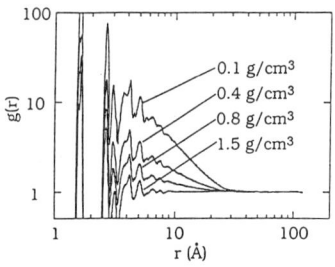

Figure 3: Log-log plot of pair distribution functions, $g(r)$, of 41,472-particle silica glasses for densities 0.1, 0.4, 0.8, and 1.5 g/cm$^3$ at 300 K.

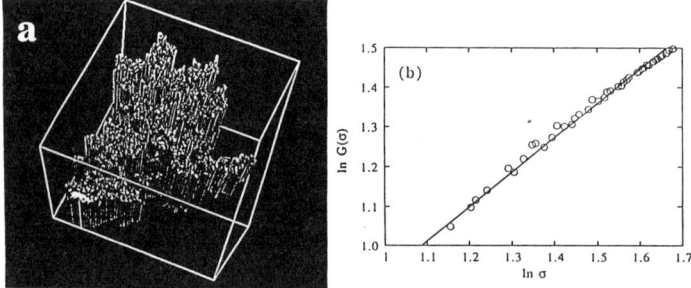

Figure 4: (a) Snapshot of an internally fractured surface resulting from a percolating pore in silica glass at a mass density of 1.4 g/cm$^3$. (b) Height-height correlation function (open circles) versus the in-plane distance, $\sigma$, for the fracture surface of a 1.12-million particle silica system computed on IBM SP-1. The solid curve is the best fit, $G(\sigma) \sim \sigma^\alpha$, with $\alpha = 0.87 \pm 0.02$ for $\sigma < 100$Å.

energetically favored packed networks with larger $d_f$ are formed. By controlling the balance between diffusion and correlation via temperature and expansion schedule, various dissimilar porous glasses with different $d_f$ can be produced in MD simulations.

## IV. ROUGHNESS OF FRACTURE SURFACE OF AMORPHOUS SILICA

We have performed large-scale MD calculations on 1.12-million particle amorphous silica systems to investigate the growth of pores with a decrease in the density [24]. The low-density MD glasses were obtained by uniformly expanding the normal-density glass. The pores begin to form when the density of the system is reduced to 1.8 g/cm$^3$. Further decrease in the density of the system causes an increase in the number of pores and also the pores coalesce to form larger entities. There is a dramatic increase in the size of pores when the mass density is reduced to the critical value, $\rho_c = 1.4$ g/cm$^3$. At that critical density, some pores percolate through the entire system by catastrophic growth. In Figure 4 (a) we show one of the surfaces of the percolating pore.

The roughness of this internally fractured surface is calculated from the height-height correlation function [24], $G(\sigma) = \langle [h(y + y_0, z + z_0) - h(y_0, z_0)]^2 \rangle^{1/2}$ where

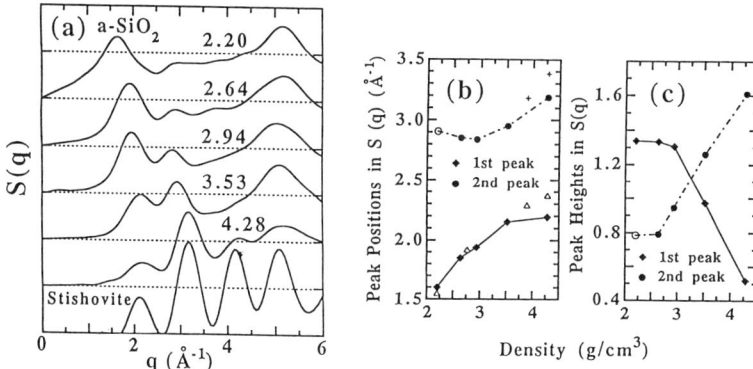

Figure 5: (a) MD results for the static structure factor, $S(q)$, of normal and high density SiO$_2$ glasses at 300 K; (b) Density dependencies of positions of the first two peaks in $S(q)$; (c) Density dependencies of heights of the first two peaks in $S(q)$. In (b) and (c), open circle at normal density are meant to indicate that the second peak is broad and has a small amplitude. Triangles and crosses in (b) represent the experimental data estimated from Ref. 27.

$\sigma = (y^2 + z^2)^{1/2}$ and $h(y,z)$ is the highest vertical coordinate at the point $(y,z)$. Figure 4 (b) shows that the MD results for $G(\sigma)$ are well-described by the relation, $G(\sigma) \sim \sigma^\alpha$ with $\alpha = 0.87 \pm 0.02$ for $\sigma < 100 \text{Å}$. Experimental measurements on bakelite, concrete, steel, and aluminum alloys indicate that the roughness exponent $\alpha$ has a universal value 0.8 [21,22]. The MD results for the roughness exponent agree with experimental measurements, thus lending further support to claims that the roughness exponent of fracture surfaces is a material-independent quantity. Furthermore, the MD results indicate that the universality of the roughness exponent may prevail even at length scales $\leq 10$ nm.

## V. SiO$_2$ GLASS AT HIGH PRESSURES

In Figure 5, we present the MD results for the density dependence of the static structure factor, $S(q)$. In the normal density glass, the FSDP is located at $1.6 \text{Å}^{-1}$. With an increase in the density, the height of the FSDP decreases, its width increases, and its position shifts to higher values of $q$. The high-pressure x-ray measurements by Meade, Hemley, and Mao [27] reveal similar behavior for the FSDP.

Figure 5 also shows a new peak in the static structure factor. It appears when the density of the normal system is increased by 20%. Located at $2.85 \text{Å}^{-1}$, the peak grows with further increase in the density. However, its position shifts only slightly at higher pressures [Fig. 5(b)]. These results are again well supported by the recent x-ray measurements [27].

Partial pair-distribution functions, $g_{\alpha\beta}(r)$, at the normal and the highest densities are shown in Figure 6. The position of the first peak in $g_{Si-O}(r)$ and the corresponding

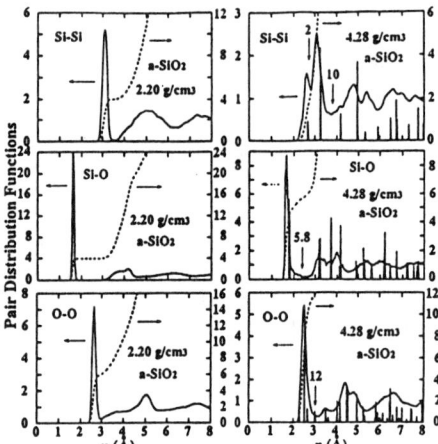

Figure 6: MD partial pair-distribution functions (solid lines) and coordination numbers (dotted lines) for $SiO_2$ glasses at normal and stishovite densities at 300 K. Sharp peaks in the figure at 4.28 g/cm$^3$ correspond to pair-distribution functions for crystalline stishovite.

Si-O coordination remain unchanged up to 3.53 g/cm$^3$. At a pressure of 42.3 GPa where the glass density (4.28 g/cm$^3$) reaches the stishovite density, the first peak in $g_{Si-O}(r)$ occurs at 1.67Å instead of 1.61Å and the Si-O coordination increases from 4 to 5.8. In stishovite, Si-O bond lengths are 1.76 and 1.81Å and the Si-O coordination is 6. In the glass at 4.28 g/cm$^3$, the second peak in $g_{Si-O}(r)$ is at 3.15Å, close to the next-nearest-neighbor (nnn) Si-O distance ($\sim$ 3.20Å) in the stishovite.

Figure 6 also shows how the Si-Si pair-distribution function changes upon densification. The first peak splits into two peaks when the density increases to 4.28 g/cm$^3$. One of these peaks is located at 2.59Å, close to the nn Si-Si distance (2.67Å) in the stishovite. The second peak appears at 3.07Å which is close to the nnn Si-Si distance (3.24Å) in the stishovite. The area under the first peak gives a coordination of 2 while the area under the first two peaks is 10. At normal density, the nn O-O coordination is 6. It increases to 10 at 3.53 g/cm$^3$ and to 12 at 4.28 g/cm$^3$. In stishovite crystal, the O-O coordination is 12.

We have also calculated O-Si-O and Si-O-Si bond-angle distributions in $SiO_2$ glasses at different densities. The results contain strong evidence for distorted $Si(O_{1/3})_6$ octahedra in the glass, joined at corners and sharing edges as well.

## VI. TBMD STUDY OF SOLID $C_{60}$ AND GRAPHITIC TUBULES

Carbon is unique among all the elements in that it can form strong covalent bonds with various coordination numbers. Owing to the quantum nature of the directional chemical bonding, the description of carbon system by a classical potential becomes inadequate. An alternative approach is the tight-binding approximation [37].

Since the minimum separation ($\sim$ 3Å) between any pair of atoms on different $C_{60}$ clusters is large compared with the C-C covalent bond length ($\sim$ 1.4Å), solid $C_{60}$ can be regarded as a typical molecular crystal. For a molecular crystal, the total energy is a sum of the intra- and intermolecular interaction energies. In our unified approach, we use a tight-binding model [37] to describe the intramolecular interaction and adapt a modified bond-bond charge model [38] for the intermolecular interaction.

A. Phonon dispersion and density of states of solid $C_{60}$

The phonon dispersion curves along some high symmetry directions in the first Brillouin zone are shown in Figure 7. Since there are four $C_{60}$ molecules per unit cell, there are 720 branches of dispersion curves. The lowest 24 branches correspond to the inter-molecular phonon modes and the remaining higher-energy branches correspond to the intra-molecular vibrational modes. Figure 7 shows that inter-molecular and intra-molecular phonon modes are well separated. This validates the rigid-molecule approximation. It is, however, surprising that not only the inter-molecular phonon modes but also the intra-molecular phonons show significant dispersions, especially those with energy below 70 meV. For comparison, we have plotted the vibrational spectrum of an isolated $C_{60}$ molecule together with the symmetry notation and degeneracy calculated with the tight-binding model. In the solid most of the low-energy vibrational modes are split into bands except for a few modes. The intra-molecular vibrational modes are also shifted upward due to the inter-molecular interaction. Note that most of the Raman active modes (with $H_g$ and $A_g$ symmetry) show strong dispersions. For example, the lowest $H_g$ Raman active mode splits into a broad band with a band-width of about 2.5 meV (20.1 cm$^{-1}$).

In Figure 8, the calculated intermolecular phonon dispersion and density of states are compared with the neutron scattering experiments [39]. The calculated intermolecular phonon density of states shows strong features around 2.3 and 3.7 meV. The highest intermolecular phonon mode is around 7.6 meV. The calculated phonon spectrum agrees very well with the experimental measurements.

We have calculated the phonon dispersion and density of states of solid $C_{60}$ at various pressures. The libron modes shift to higher frequencies at the rate of about 0.05 meV/kbar (0.4 cm$^{-1}$/kbar). Intramolecular phonon modes also show strong pressure dependence. When pressure is applied, the lower frequency modes are broadened into bands, and the higher frequency modes are split and shifted toward higher energy. Intramolecular phonon modes shift toward higher frequencies at a rate of up to 0.11 meV/kbar (0.88 cm$^{-1}$/kbar). Most Raman and IR active modes show strong pressure dependence. The high energy Raman ($H_g$) modes and the IR active ($T_{1u}$) modes are shifted upward and split into doublet or triplet. The splittings increase with increasing pressure. Our theoretical calculation agrees reasonably well with experimental high-pressure IR spectra [40].

B. Phonons in graphitic tubules

The structure of a graphitic tubule can be described by choosing two lattice points on a graphite sheet. Denoting a fixed lattice point by $O(0,0)$, a specification of another lattice point, $R_n(n_1, n_2)$, which will fold onto $O$ uniquely defines the tubule structure. The index $(n_1, n_2)$ can be used to specify the graphitic tubule. Here we denote it by $T(n_1, n_2)$ or $T(l_1, l_2)N$, where $n_1 = l_1 \cdot N, n_2 = l_2 \cdot N$ and $N$ is the largest common

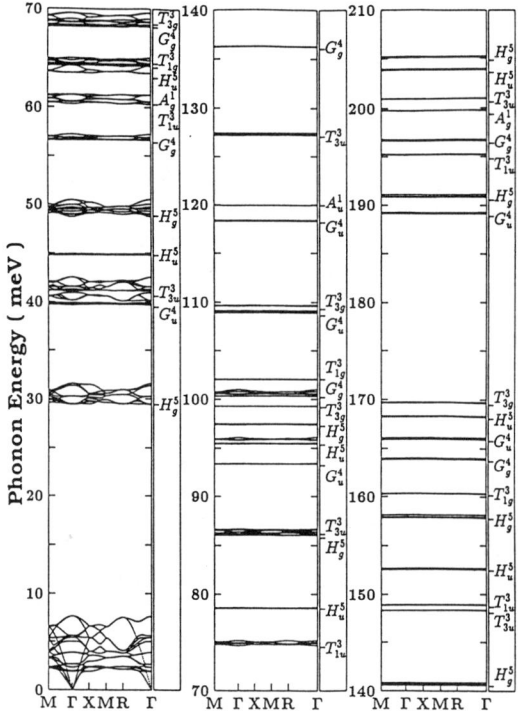

Figure 7: Phonon dispersion curves of solid $C_{60}$ at zero pressures and the vibrational frequencies of an isolated $C_{60}$ molecule. The symmetry of the $C_{60}$ molecule is indicated by the icosahedral $I_h$ group label, the number on the label's shoulder indicates the degeneracy of the corresponding group representation.

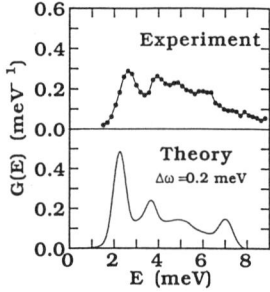

Figure 8: Comparison of the calculated intermolecular phonon dispersion and density of states of solid $C_{60}$ with experiment measurements [39].

divisor among $n_1$ and $n_2$. Two special cases are the tubules: $T(1,0)N$ and $T(1,1)N$. In

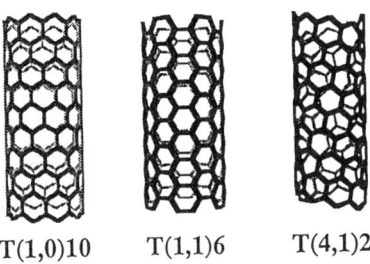

T(1,0)10  T(1,1)6  T(4,1)2

Figure 9: Examples of the structure of graphitic tubules: $T(1,0)10$; $T(1,1)$; and $T(2,1)$.

the tubule $T(1,0)N$, there exist C-C bonds which are parallel to the tubule axis, while in tubule $T(1,1)N$ there are bonds perpendicular to the tubule axis. Some examples of graphitic tubules with different helical structures are shown in Figure 9.

Using the TBMD method, we have calculated the phonon spectra of graphitic tubes with various helical structures and diameters [35]. Figure 10 shows the phonon densities of states of graphitic tubules $T(1,0)N$ and $T(1,1)N$ as a function of tubule diameters and a comparison with the spectrum of infinite graphite. The number of peaks in the low-frequency region of the spectra of graphitic tubules is proportional to the diameter. As the diameter increases, the tendency of the spectra toward that of graphite is evident. The high-frequency modes are softened by the curvature.

Recently, Jishi, et al. have studied the symmetry properties of chiral graphitic tubules [41]. We have calculated the phonon spectra for graphitic tubules with general chiral symmetries. The infrared and Raman active modes have been identified, and the symmetry of phonon modes has been analyzed. We have found that the lower-frequency part of the spectra corresponds to the radial modes with the number of peaks depending on the diameter of the tubule. The higher-frequency part of the spectra carries unique information about the tubule helical structures [35].

## VII. CONCURRENT COMPUTING LABORATORY FOR MATERIALS SIMULATIONS

In the past four years, we have received from the Louisiana Board of Regents three equipment enhancement grants totaling more than $ 2.4M. With the first grant we have purchased an 8,192-node MasPar MP-1 SIMD (single instruction multiple data) machine, an 8-node Intel iPSC/860 distributed-memory MIMD machine, a SPARCserver, and SPARC workstations. These facilities are housed in a laboratory in Nicholson Hall. The second grant has been used to purchase a 64-cell iWarp (a distributed-memory MIMD machine which supports fine-grain systolic communication), an 8-processor Silicon Graphics Power Center and associated visualization equipment, a 4-processor Sun as a front end for the iWarp, and X-terminals. These computing facilities are housed in a laboratory in an adjacent building (Coates Hall). The laboratory has a dedicated systems manager. From the third grant we plan to purchase a new parallel architecture. The overall setup of the Concurrent Computing Laboratory for Materials

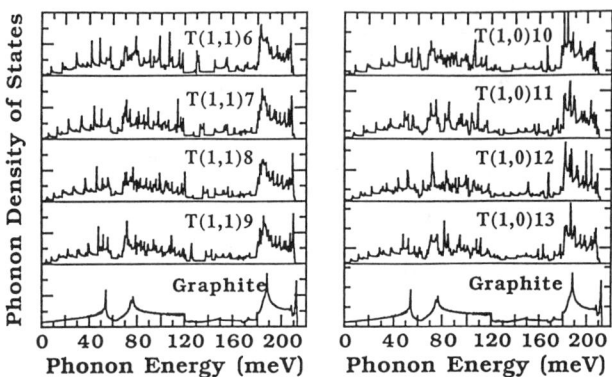

Figure 10: The phonon density of states of graphitic tubules $T(1,0)N$ and $T(1,1)N$ as a function of tubule diameters, and comparison with the density of states of an infinite graphite.

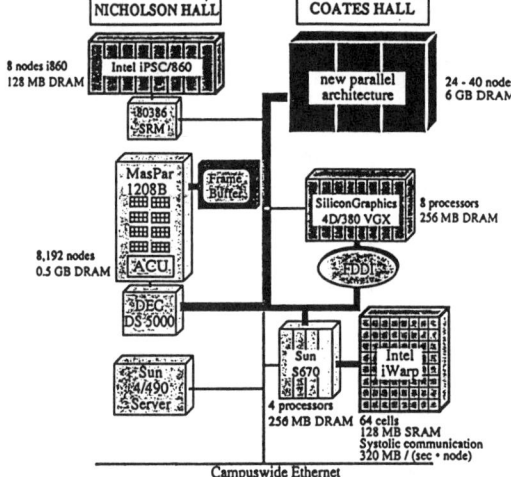

Figure 11: Computing facilities of the Concurrent Computing Laboratory for Materials Simulations (CCLMS) at Louisiana State University.

Simulations (CCLMS) is shown in Figure 11. Plans are in place to connect these machines by FDDI (Fiber Distributed Data Interface) to form a distributed multiparallel processing network.

## VIII. CONCLUSION

Algorithms have been designed to implement MD simulations on emerging concurrent architectures. Million particle MD simulations for materials have been performed on parallel computers. We believe that for the next ten years, there is an enormous opportunity to develop new and efficient algorithms for parallel computers to solve grand challenge problems in materials science.

## ACKNOWLEDGMENTS

This work was supported by the U.S. Department of Energy, Grant No. DE-FG05-92ER45477 and National Science Foundation Grants No. ASC-9109906 and ASC-9310314. The computations were performed using the facilities in the Concurrent Computing Laboratory for Materials Simulations (CCLMS) at Louisiana State University. The facilities in the CCLMS were acquired with Equipment Enhancement Grants awarded by the Louisiana Board of Regents through Louisiana Education Quality Support Fund (LEQSF). Computations were also performed on the Touchstone Delta and iPSC/860 machines operated by Caltech on behalf of the Concurrent Supercomputing Consortium and the 128-node IBM SP-1 at Argonne National Laboratory.

## REFERENCES

1) M. P. Allen and D. J. Tildesley, *Computer Simulation of Liquids* (Oxford University Press, Oxford, 1990).
2) W. B. Streett, D. J. Tildesley, and G. Saville, Mol. Phys. **35** (1978) 639.
3) D. C. Rapaport, Comput. Phys. Commun. **62** (1991) 217.
4) R. K. Kalia, S. W. de Leeuw, A. Nakano, P. Vashishta, Comp. Phys. Comm. **74** (1993) 316.
5) R. K. Kalia, S. W. de Leeuw, A. Nakano, D. L. Greenwell, and P. Vashishta, Supercomputer **54** (X-2) (1993) 11; A. Nakano, P. Vashishta, and R. K. Kalia, Comput. Phys. Commun. **77** (1993) 303.
6) A. Nakano, P. Vashishta, and R. K. Kalia, Comput. Phys. Commun., in press.
7) D. Frenkel, in *Simple Molecular Systems at Very High Density*, ed. A. Polian and P. Loubeyre (Plenum, New York, 1989).
8) S. W. de Leeuw, J. W. Perram, and E. R. Smith, Proc. R. Soc. London A **373** (1980) 27.
9) J. Barnes and P. Hut, Nature **324** (1986) 446.
10) L. Greengard and V. Rokhlin, J. Comput. Phys. **73** (1987) 325.
11) H.-Q. Ding, N. Karasawa, and W. A. Goddard, Chem. Phys. Lett. **196** (1992) 6.
12) J. Fricke, J. Non-Cryst. Solids **121** (1990) 188; ibid. **147** & **148** (1992) 356.
13) T. Freltoft et al., Phys. Rev. B **33** (1986) 269.
14) R. Vacher et al., Phys. Rev. B **37** (1988) 6500.
15) D. W. Shaefer and K. D. Keefer, Phys. Rev. Lett. **56** (1986) 2199.
16) T. Lours et al., J. Non-Cryst. Solids **121** (1990) 216.
17) J. Kieffer and C. A. Angell, J. Non-Cryst. Solids **106** (1988) 336.
18) A. Nakano, L. Bi, R. K. Kalia, and P. Vashishta, Phys. Rev. Lett. **71** (1993) 85; Phys. Rev. B **49** (1994) 9441.
19) F. Family and T. Vicsek, J. Phys. A **18** (1985) L75; M. Kardar, G. Parisi, and Y. C. Zhang, Phys. Rev. Lett. **56** (1986) 889; J. Villain, J. Phys. I (France) **1** (1991) 19.
20) F. Family and T. Vicsek, eds., *Dynamics of Fractal Surfaces* (World Scientific, Singapore, 1991).

21) E. Bouchaud, G. Lapasset, and J. Planes, Europhys. Lett. **13** (1990) 73.
22) K. J. Måløy et al., Phys. Rev. Lett. **68** (1992) 213.
23) V. Y. Milman et al., Phys. Rev. Lett. **71** (1993) 204.
24) A. Nakano, R. K. Kalia, and P. Vashishta, Phys. Rev. Lett., in press.
25) M. Grimsditch, Phys. Rev. Lett. **52** (1984) 2379.
26) Q. Williams and R. Jeanloz, Science **239** (1988) 902.
27) C. Meade, R. J. Hemley, and H. M. Mao, Phys. Rev. Lett. **57** (1986) 747.
28) W. Jin, R. K. Kalia, P. Vashishta, and J. P. Rino, Phys. Rev. Lett. **71** (1993) 3146.
29) H. W. Kroto et al., Nature **318** (1985) 162.
30) W. Kratschmer et al., Nature **347** (1990) 354.
31) H. Hebard et al., Nature **350** (1991) 600; M. Rosseinsky et al., Phys. Rev. Lett. **66** (1991) 2830; K. Holczer et al., Science **252** (1991) 1154.
32) J. Yu, R. K. Kalia, and P. Vashishta, Appl. Phys. Lett. **63** (1993) 3152.
33) J. Yu, R. K. Kalia, and P. Vashishta, Phys. Rev. B **49** (1994) 5008.
34) S. Iijima, Nature **354** (1991) 56.
35) J. Yu, R. K. Kalia, and P. Vashishta, to be published.
36) P. Vashishta, R. K. Kalia, J. P. Rino, and I. Ebbsjö, Phys. Rev. B **41**, 12197 (1990).
37) C. H. Xu, C. Z. Wang, C. T. Chan, and K. M. Ho, J. Phys.: Condens. Matter **4** (1992) 6047.
38) J. Lu, X. Li, and R. M. Martin, Phys. Rev. Lett. **68** (1992) 1551.
39) L. Pintschovius, et al., Phys. Rev. Lett. **69** (1992) 2662.
40) Y. Huang, D. F. R. Gilson, and I. S. Butler, J. Phys. Chem. **95** (1991) 5723.
41) R. A. Jishi and M. S. Dresselhaus, Phys. Rev. B **45** (1992) 11305.

# X. Laser and Shock Compressed Plasmas

# HIGH DENSITY PLASMA PHYSICS IN LASER PRODUCED PLASMAS

K. Mima, H. Takabe, Y. Kato, S. Miyamoto, and S. Kato

Institute of Laser Engineering, Osaka University
2-6 Yamada-oka, Suita, Osaka, Japan

Reviewed are the recent progresses of physics of high density plasmas produced by lasers. First of all, the parameter ranges for density and temperature of laser plasmas are described. In the laser implosion experiments at Osaka University, a mixture of carbon and deuteron is compressed to 600 g/cm$^3$ with a temperature of 300–500 eV. The degeneracy of the plasmas is found by measuring the ratio of the primary nuclear reaction (DD reaction) to the secondary reaction (DT reaction). The results indicate that (Fermi energy)/(thermal energy) $\sim 1$. Another recent progress is the technology of the ultra intense short pulse laser. By this new technology, the laser intensity reaches higher than $10^{20}$ W/cm$^2$. Such ultra intense lasers produce relativistic plasmas which generate MeV X-rays and electron-positron pairs which are discussed in the paper. Finally, the fusion burning in high density plasmas which will be demonstrated in near future is described. The hot dense burning plasmas may open up new research fields like the laboratory astrophysics in which the Coulomb shielding effect on nuclear reactions will be one of the important subjects.

## I. INTRODUCTION

Since laser was invented in the late '50s [1,2,3], high power lasers have been used to produce hot dense plasmas. In the early '70s, the laser implosion concept was proposed to ignite a micro fusion explosion. In the laser fusion research, the hot dense plasma physics is the main research subject.

The laser pulse energy has increased from J/nsec to 100 kJ/nsec in recent 20 years and the laser intensity reaches almost $10^{20}$ W/cm$^2$ recently. By using those high power lasers, extremely hot dense plasmas have been produced. As shown in Fig. 1, the electron densities of plasmas produced with the laser implosion reach higher than $10^{25}$ cm$^{-3}$ at Rochester University [4] and LLNL, which is approximately 200 times the density of solid hydrogen, and $10^{26}$ cm$^{-3}$, at Osaka University [5], which is 500 times the density of CD plastic. The Fermi energy of the plasma, $\varepsilon_F$ is about 0.5 keV for $10^{26}$ cm$^{-3}$, which is comparable to the electron thermal energy of compressed plasmas, $k_B T$ as shown in Fig. 1. The degeneracy parameter, $\theta = k_B T/\varepsilon_F$ is experimentally observed by using the $\theta$ dependence of the secondary nuclear reaction rate relative to the primary DD reaction rate in deuterium plasma, where 1 MeV tritons generated by DD reactions react with deuterons to produce 14 MeV neutrons.

The second topics of the recent laser plasma research is the atomic and radiation processes in hot dense plasmas. Since an expanding coronal plasma on a solid target surface is in the non-thermal equilibrium because of the spatial non-uniformities and the transient nature, the X-ray radiation processes are very unique. For an example, some excited states are populated higher than the lower energy states and the stimulated emissions in the soft X-ray regime are expected to occur and have been observed experimentally. The amplification processes of soft X-ray radiations are described in this paper as an example of the laser plasma atomic process.

The third topics is the ultra short pulse laser interaction with plasmas. Recently, the technologies of the intense ultra short pulse laser progresses very rapidly. Namely, it is now possible to use $10 \sim 100$ J/psec laser pulses for producing relativistic plasmas. In this case, the laser intensity can be the order of $10^{20}$ W/cm$^2$ which corresponds to 2 MeV of quivering electron energy. In such a case, the laser plasma density and temperature extend over the hatched area in Fig. 1. Since the solid target surface does not expand during the sub-picosecond laser heating, the short pulse lasers produce high temperature solid density plasmas, which emit intense hard X-rays and produced electron-positron pairs. The analysis of those processes are described in the text.

In summary, we describe the laser implosion processes in the sections II and III where hot dense plasma physics, and hydrodynamics are discussed. In the section IV, the radiation and atomic processes in laser coronal plasmas are described. Finally, in the section V we present the physics on ultra-short pulse laser plasma interactions.

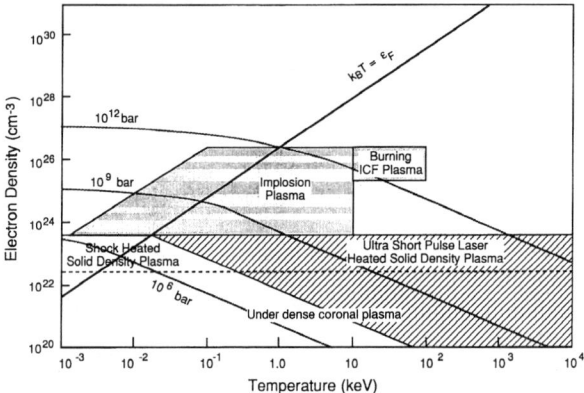

Fig. 1. Laser plasma parameters.

## II. LASER IMPLOSION AND DENSITY MEASUREMENT

When a spherical shell pellet is irradiated with may high power laser beams which are arranged spherically symmetric, spherically uniform hot dense plasmas are generated on the pellet surface. The surface plasma pressure reaches higher than 100 Mbar which implodes the pellet. The implosion dynamics has been measured with a X-ray framing camera which has 10 mm and 80 psec space and time resolution. The figure 2 shows an example of a series of X-ray framing photographs for the implosion of a plastic shell target by the Gekko XII laser system at Osaka University, where the time intervals between two frames are 170 psec. The X-ray images indicate shapes of pellet surface plasmas which are produced by laser ablation. Since the X-ray opacity is high for compressed dense plasmas, it is not possible to diagnose the internal structure of the compressed plasmas by X-rays. The 1D spherically symmetric implosion dynamics has been simulated with 1D hydrodynamic computer codes. The radial motions of fluid (plasma as a continuum) elements obtained by the computer simulation are compared with the results of X-ray framing Camera measurement in Fig. 2. This figure shows that the motion of ablation front for the simulation agrees well with the measured X-ray emission surface positions indicated by squares in Fig. 3.

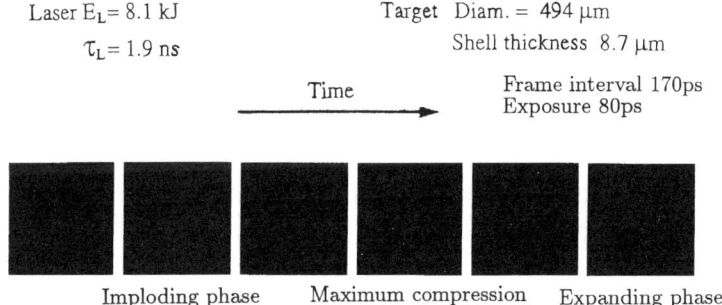

Fig. 2. Imploding pellet X-ray framing photographs.

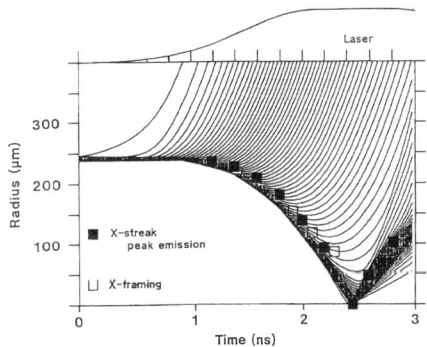

Fig. 3. Flow diagram of laser implosion simulation and comparison with X-ray image measurement.

In order to measure the highly compressed plasma parameters, we have to use higher energy X-ray and/or fusion products such as neutrons. As an example, the density measurement by observing the electron degeneracy on which the secondary fusion reaction rate depends is described in this section. In the experiments, CD plastic shell pellets have been used. In the imploded plasmas, the primary fusion reactions;

$$D + D \rightarrow \begin{cases} {}^3He\ (1\ MeV) + n\ (3 MeV) \\ T\ (1\ MeV) + p\ (3\ MeV) \end{cases} \quad (1)$$

take place in a mixture of carbon and deuterium. The branching ratio of the above two reactions is about 50%. The 1 MeV tritium produced by the DD reactions reacts with the plasma deuterium to produce 14 MeV neutrons. By measuring both the 3 MeV neutron yield and the 14 MeV neutron yield, we obtain a ratio of the secondary reaction rate to the primary reaction rate.

When compressed plasma area density is sufficiently large, the reaction ratio depends only on the stopping power of 1 MeV tritium in a dense CD plasma. Namely, the reaction ratio is estimated by the following formula;

$$\frac{Y_{2nd}}{Y_{1st}} = n_D \int_0^{E0} \sigma_{DT} \left(\frac{dE}{dl}\right)^{-1} dE \qquad (2)$$

which assumes that the secondary DT reactions occurs before the thermalization of the tritium and the DT reaction after the thermalization of tritons is negligible. This assumption is justified when the plasma area density is smaller than 0.3 g/cm$^2$, because the ratio given by Eq. (2) is the order of $10^{-3}$. Namely, when $\rho R \leq 0.3$ g/cm$^2$, and the plasma temperature is lower than 1 keV,

$$\frac{\langle \sigma v \rangle_{DT} n_D n_T}{\langle \sigma v \rangle_{DD} n_D n_D} \approx \frac{1}{4} \langle \sigma v \rangle_{DT} n_D \tau \ll 10^{-3}$$

is met, where $\langle \sigma v \rangle_{DT}$, $\langle \sigma v \rangle_{DD}$ and $\tau$ are the reaction rates averaged over a Maxwellian distribution for DT and DD, and reaction duration respectively.

The stopping power $dE/dl$ in Eq. (2) due to electron is inversely proportional to the cubic of the electron average velocity approximately, when the tritium velocity is less than the electron average velocity, $v_e$. For an example; $k_B T = 500$ eV, $v_e \approx 10^9$ cm/sec which corresponds to 2 MeV triton velocity. In this case, the stopping power is roughly scaled as follows;

$$\frac{dE}{dl} \propto \frac{n_e}{T^{3/2}} \quad \text{for a non degenerate plasma} \qquad (3)$$

and

$$\frac{dE}{dl} \propto \frac{n_e}{\varepsilon_F^{3/2}} \quad \text{for a degenerate plasma}, \qquad (4)$$

where $\varepsilon_F$ is the Fermi energy which is proportional to $n_e^{2/3}$. Therefore, for a given temperature, the stopping power is proportional to the plasma density when it is not degenerate and independent of the density when degenerate. From Eqs. (3) and (4) together with Eq. (2), the reaction ratio is found to increase with plasma density when the plasma is degenerate, while it is constant with respect to the density change when non degenerate. For partially degenerate plasmas, it is necessary to evaluate the stopping power more precisely. We use the following formula to obtain the temperature and density dependencies of the reaction ratio which is shown in Fig. 4;

$$\frac{dE}{dl} = \frac{e^2}{2\pi^2} \int d\vec{k} \, \frac{\vec{k} \cdot \vec{v}}{k^2 |\vec{v}|} \text{Im} \left[\frac{1}{\epsilon(k, \vec{k} \cdot \vec{v})}\right]. \qquad (5)$$

Here, $\epsilon(k, \omega)$ is the dielectric function and $v$ is the velocity of a triton.

Using Fig. 4 with the experimentally measured reaction ratio $Y_{2nd}/Y_{1st}$, we obtain the target shell thickness dependencies of degree of degeneracy $\theta = k_B T / \varepsilon_F$ and the compressed plasma density. The degeneracy q is related to the plasma electron density by

$$n_e \approx 2 \times 10^{26} (k_B T / 1 \text{KeV})^{3/2} / \theta^{3/2} \quad (\text{cm}^{-3}) \ .$$

As shown in Fig. 5, typically, $\theta = 0.7$ and $k_B T \approx 0.4$ keV yield $n_e \approx 8 \times 10^{25}$ cm$^{-3}$ which is approximately 300 times solid density. Note here that the plasma ion temperature was measured with the Doppler broadening of the neutron spectrum and the electron temperature is assumed equal to the ion temperature. According to the above plasma parameter, the average Coulomb interaction energy between Carbon ions is about 2 keV which is 5 times greater than the thermal energy. Namely, the hydrogen and carbon plasmas produced by the laser implosion will give us high temperature strongly coupled plasmas. The more detail measurements of this plasma will become possible in near future and provide fruitful experimental data which disclose the physical properties of hot dense strongly coupled plasmas.

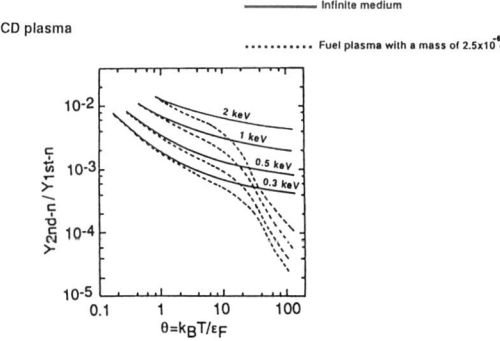

Fig. 4. Degeneracy dependences of secondary reaction ratio for CD plasmas.

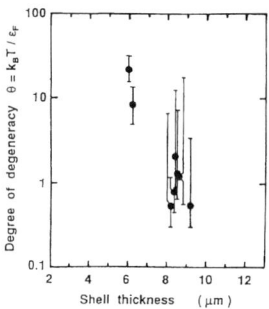

Fig. 5. Experimental data of degeneracy for various pellet shell thickness.

## III. TURBULENT MIXING IN LASER DRIVEN IMPLOSION

A. Introduction

In laser driven implosion, hydrodynamic stability is the most important key issue. Many implosion experiments have been carried out with high power laser system such as Gekko XII; however, the experimental results have not been explained without taking account of hydrodynamic instability regardless whether it is direct [6,7] or indirect [8] schemes. Long-wavelength nonuniformity induces a deformation from implosion sphericity, while short-wavelength nonuniformity becomes a source of fluid turbulence with a help of instability and its nonlinear process. In most of the experiments carried out so far, we have used beam profiles not uniform enough to eliminate the generation of short-scale turbulent fluid motion. In order to analyze the experimental results, therefore, it is essential to model a capsule mixing by the turbulent convection.

Relating to the direct-drive glass-micro-balloon implosion experiment, it has been concluded that a systematic disagreement between the experimental neutron yields and the conventional one-dimensional implosion code results stem from the Rayleigh-Taylor instability in the stagnation phase. However, we can enumerate the following three scenarios for explaining the disagreement.

(1) If hot electrons or tail component of heated electron preheat the glass shell in the acceleration phase and the density of the glass reduces drastically compared to the conventional code result, the glass pusher can not reflect the shock wave coming from the DT plasmas, consequently no stagnation is observed.

(2) If the hydrodynamic instability grows substantially in the acceleration phase, the shell-breakup takes place in this phase. Then, averaged density of the glass shell reduced due to mixing and dynamical pressure ($\rho u^2$) may reduce and no stagnation occurs due to the same reason as case (1).

(3) Even if the shell does not break up in the acceleration phase, the hydrodynamic instability in the stagnation phase prevents the heating and compression in the stagnation phase.

Actually, the above three may occurs in the same time, while in the present paper we focusing our attention to the mixing in the stagnation phase. In the gas-filled GMB implosion, glass plasma penetrates into the DT fuel, consequently neutron yield reduces due to cooling of the fuel. We have obtained a systematic agreement with the experimental neutron yield, when we assume that 10–20% of the implosion kinetic energy is converted to the turbulent kinetic energy in the acceleration phase. It is also found that the same relative fraction of the turbulent energy can explain neutron yield, core temperature, and density-radius product for the high density compression experiment with CD-shell targets as reported in Ref. 9.

B. Turbulent mixing in GMB target implosion

The turbulent mixing model has been applied to a single shock tube problem [10,11] and also applied to the laser implosion [12]. The numerical constants relating to the turbulent diffusion coefficients are adjusted so that the experimental results is well reproduced in the shock tube experiments [11]. Although a trial to apply the mixing model to the laser implosion has been repented is a short articles [12], it is not clear how the model works to explain some experimental results.

We have applied the turbulent mixing model to the case of high neutron yield experiments reported in Ref. 6. In the experiment, so-called LHART-type targets characterized with large diameter ($\sim$1mm), thin shell ($\sim 1\mu$m) glass micro balloon are used to contain deuterium tritium (DT) mixture fuel gas of about 10 atmosphere pressure. Most of target irradiations have been done with about 10 kjoule, green light ($\lambda = 0.53\ \mu$ m) of Gekko XII laser system. The neutron yield of about $10^{10} \sim 10^{13}$ has been observed.

Implosion dynamics has been compared with conventional one-dimensional simulation. It is reported that a systematic agreement is obtained for the implosion timing by adjusting an electron heat flux limiter [13]. Implosion parameters such as neutron yield, compressed ion temperature and density, and so on have also been compared with one-dimensional simulation for many shots. It is reported that the neutron yields are not well reproduced [6]. The same kind of comparison on neutron yield has been carried out in several institutes and the same kind of disagreement has been reported. The disagreement becomes remarkable with the increase of importance of the stagnation dynamics in neutron production. It has been phenomenologically concluded in Ref. 6 that because of high level nonuniformity in irradiated laser intensity profile, uneven ablation pressure play a role in generating perturbation source of the Rayleigh-Taylor instability. Since the Rayleigh-Taylor instability grows explosively in the stagnation phase [14], we may not observe the neutrons generated in this phase in the conventional one-dimensional codes.

In the present report, we have included the turbulent mixing in the stagnation phase. For the case of highest neutron shot (# 3826), the details of radius-time diagram of trajectories of each fluid element (numerical grid of Lagrangian description) is plotted in Fig. 6(a). It is typical that a shock wave collides at the center and a reflected shock wave in generated and propagates outward. Then, the shock wave collides at the fuel-pusher interface (25-th flow line from the center) then reflected again toward the center. After same time the pusher (glass material )tend to be decelerated due to the pressure of the fuel. These dynamics is called "stagnation dynamics" and shows the same evolution as in the shock tube experiments [e.g., 10]. In such situation, the contact surface between DT fuel and glass ($SiO_2$) pusher is unstable for Richtmyer-Meshkov and Rayleigh-Taylor instabilities.

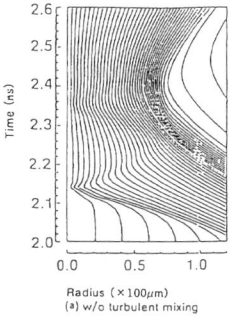
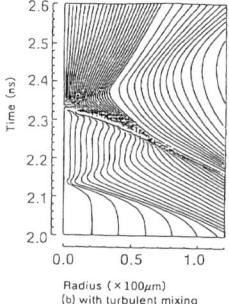

Fig. 6. Flow diagram of gas-filled GMB implosion.

In Fig. 7(b), the flow-diagram is shown for the case when the turbulent mixing model is coupled. In this simulation, the turbulent mixing starts to be solved at the time ($t = t_s$) when the shock wave collides at the center ($t \approx 2.1$ nsec) and the turbulence energy of the value of $\xi \equiv 2k/u_o^2$ at $t = t_s$) = 20% is assumed to be generated in the acceleration phase ($0 < t < t_s$). It is clearly seen in Fig. 7(b) that the glass pusher can not be decelerated by the reflected shock wave and freely penetrates toward the center. Since the material can diffuse among Lagrangian zones, this flow diagram does not tell how the glass penetrates into the fuel region.

In Figs. 7(a) and (b), the equi-contours of glass concentration are plotted for the cases without and with the turbulent mixing model, respectively. In Fig.7(a), the trajectory of decelerating contact surface is recognized. However, by including the mixing effect, the glass material is not decelerated by the fuel pressure and the material font characterized with sharp front of the glass concentration penetrates toward the center. As the results, the fuel is cooled down and the neutron production which is very sensitive to the temperature of the fuel is suppressed. In the case, the neutron yield without the mixing was $Y_N = 2 \times 10^{13}$, while it reduced to $Y_N = 1 \times 10^{13}$ by including the mixing. It is noted that the corresponding experimental neutron yield is $Y_N = 1 \times 10^{13}$. What we can conclude in such study is that about 20% of the kinetic energy given by the rocket effect in the acceleration phase may be converted to the energy of disordered fluid motion due to uneven ablation pressure acceleration and/or hydrodynamic instability. It is found that in this case the fraction of the turbulence energy grows up to almost 10% of the total kinetic and thermal energy in the fuel at the maximum compression ($t = 2.4$ nsec).

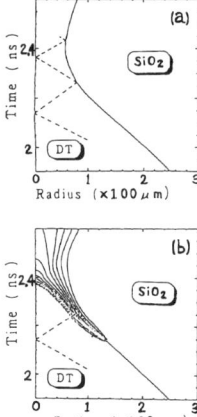

Fig. 7. Equi-contour of glass concentration for cases without and with mixing.

The same kind of simulations have been carried out for a variety of experimental shots. In Fig. 8, the ratio of the neutron yields with the turbulent mixing model ($Y_N^*$) and those without the turbulent mixing model ($Y_N(0)$) is plotted as a function of the radius convergence ratio ($R_0/R_f$). The radial convergence ratio is obtained from a conventional one-dimensional simulation without the mixing model which yields the fusion neutrons $Y_N(0)$. The radius $R_0$ is the initial radius of a glass-fuel interface and $R_f$ is the minimum value of the radius of the interface. In the simulation, we have varying the radius convergence ratio by varying the initial pressure of the DT gas without raring the parameters of the glass shell and the irradiated laser. In Fig. 8, the marks ×, ◇, ○, ⊞, □, and △ represent the cases when the levels of turbulence are assumed to be $\xi = 0, 2, 6, 10, 20,$ and 40%, respectively. Here, $\xi = 2k/u_0^2$ is the ratio of the turbulence energy $k$ and the kinetic energy of implosion $1/2u_0^2$ at the time where the shock wave generated in the gas fuel collides at the center.

In Fig. 8, the solid circles are the the experimental neutron yields divided by these obtained with the conventional 1-D code without the mixing model [6]. As reported in Ref. 6, with the increase of the contribution of the stagnation dynamics to the neutron yields, the radial convergence ratio increases due to the compression by this dynamics and the discrepancy between the experimental yields and the corresponding 1-D simulation ones increases. By comparing the neutron yields obtained in the experiment and in the 1-D simulation with the mixing model, we can conclude that the turbulent energy of 2-20% out of the kinetic energy of implosion is generated before the stagnation phase. In the stagnation phase, the turbulence energy is seen to be enhance by the instability to finally terminate the predominant production of the neutrons due to the cooling of the fuel through the penetration of pusher material (glass) by the turbulent diffusion process.

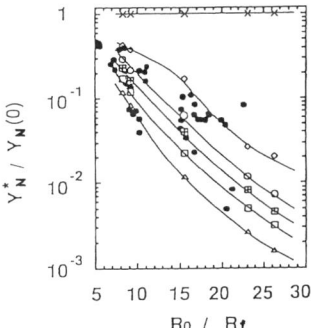

Fig. 8. Reduced neutron yield by turbulent mixing.

C. Summary and conclusion

In order to treat the effect of small scale turbulence in laser driven implosion, we have installed the mixing model into the one dimensional implosion code ILESTA-1D and used the code to analyze the experimental results obtained by imploding DT gas-filled $R_f$ glass micro balloon with Gekko XII laser system at Osaka. The $k$-$\epsilon$ type mixing model is introduce to couple with the one-fluid, two-temperature fluid equations.

We have focused on the turbulent mixing phenomena in the final compression called stagnation phase and studied how high level of the turbulence energy has been generated in the acceleration phase. As the results, it is found that about $2 \sim 20\%$ out of the kinetic energy for implosion goes to the energy of turbulent mention in this phase.

## IV. ATOMIC PROCESSES IN X-RAY LASERS: RECENT EXPERIMENTS

Atomic processes are manifested, often quite explicitly, in the performances of X-ray lasers. In this paper, we present some of recent experiments on X-ray lasers where there are intriguing issues on atomic physics in plasmas and laser interaction with matter.

Followings are the major topics covered in this presentation.

A. Collisional excitation X-ray lasers

1. "$J = 0$–1 anomaly" problem

In the first successful demonstration of soft X-ray amplification in neon-like selenium ions at LLNL [5] the $J = 0$–1 transition showed only a small gain among 5 lasing lines. This was in contradiction to the theoretical prediction and led to reexamination of atomic physics modeling. However the $J = 0$–1 line has been shown to have the highest gain in our recent experiments where curved slab targets were irradiated with short laser pulses, resolving the long-standing "$J = 0$–1 anomaly" problem.

2. Amplification in nickel-like ions by double pulse pumping

The short pulses pumping of curved targets are also effective in improving the pumping efficiency of the collisional excitation X-ray lasers. Amplification has been observed at around 70 A in nickel-like ions of various lanthanides elements with the efficiency approximately 10 times higher in comparison to previous experiments using single, long pulse irradiation pumping.

3. Polarization of soft X-ray lasers

Soft X-ray lasers generated in laser-produced plasmas as amplified spontaneous emission is generally considered to be unpolarized. However our recent measurement suggests that the $J = 0$–1 line in neon-like germanium laser is polarized. Implication of this finding is discussed.

B. Lyman-$\alpha$ ground state laser

Amplification in the Lyman-$\alpha$ transition in hydrogenic Li at 135 A has been demonstrated by Midorikawa et al. [16]. Our recent collaborative experiment on measurement of the time dependence of the Lyman-$\alpha$ line shows that the electron temperature is approximately 1 eV, consistent with generation of population inversion in this ground state lasing scheme.

## V. ULTRA INTENSE SHORT PULSE LASER INTERACTION WITH PLASMAS

Recently, a new laser technology for generating an intense short laser pulse progressed extensively. As the result, relativistic plasmas whose temperature is in the MeV range can be produced by the intense laser irradiation. The relativistic laser plasmas will provide us the opportunities of carrying out laboratory experiments related to astrophysical plasmas. For an example, the intense electromagnetic field interactions with charged particles leads to the electron-position pair production.

The electron motion in the laser field is described by

$$P_L = mv_E \left[\vec{e}_y \sin\phi - \vec{e}_x(v_E/4\gamma c)\cos 2\phi\right], \qquad (6)$$

when a laser vector potential $\vec{A}$ is given by

$$\vec{A} = A\vec{e}_y \sin\phi,$$

where

$$v_E = eA/m, \quad \gamma = \sqrt{1 + v_E^2/c^2}$$

and $\phi = kx - \omega t$. The quivering kinetic energy

$$\epsilon_{os}(P_L) = c\sqrt{P_L^2 + m^2c^2} - mc^2$$

is approximately evaluated to be

$$\epsilon_{os}/mc^2 \approx \sqrt{1 + 40 I_L \lambda^2/10^{20} \text{ W}\mu\text{m}^2/\text{cm}^2} - 1, \qquad (7)$$

where $I_L$ and $\lambda$ are the laser intensity and wavelength respectively. The quiver motion becomes relativistic when the laser intensity is higher than $10^{18}$ W/cm$^2$ for $\lambda = 1$ $\mu$m. Such laser intensity has been achieved by so called Charped Pulse Amplification (CPA) method with a relatively small laser system. When the ultra intense laser pulse is irradiated on a solid target, electrons on a surface are strongly expelled from the focal region. The following two forces will act on the electrons; the ponder motive potential is given by

$$\Phi_p = \epsilon_{os} \approx 2.5 \left(I_L \lambda^2/10^{20} \text{ W}\mu\text{m}^2/\text{cm}^2\right)^{1/2} \text{ MeV},$$

and the radiation pressure is given by

$$P_r = 30 I_L/(10^{20} \text{ W/cm}^2) \text{ Gbar}.$$

In a short pulse laser where the pulse width is shorter than 1 psec, electrons are separated from ions by the above forces to produce a strong electromagnetic field. For the laser intensity of $10^{20}$ W/cm$^2$, the induced electric field is the order of $10^{12}$ Volts/m. The magnetic field can be estimated from the electron current which will be generated by the strong local heating. The electron heating will be mainly attributed to the electron plasma wave breaking on the solid target surface. By this process, heated electrons have the energy in the range of MeV and those electrons will penetrate deeply into the solid target. The electric current density carried by those high energy electrons is estimated by,

$$j = I_{ab} e/E_h,$$

for the absorbed laser intensity $I_{ab}$ and the electron energy $E_h$. Therefore the magnetic field on the surface of the target is approximately evaluated to be

$$B_{\text{theta}} = \frac{2\pi r_0}{c} \frac{I_{ab}}{E_h} e ,\qquad(8)$$

for a laser focal spot radius $r_0$. This magnetic field can be the order of $10^{10}$ gauss for $10^{20}$ W/cm$^2$ and $r_0 = 10$ mm. Those extremely high quasi-static electro-magnetic fields will introduce new atomic physics and other electromagnetic plasma phenomena.

The electron momentum distribution for a laser irradiated relativistic plasma is approximated by the oscillating Maxwellian distribution. By the Lorentz transform of the Maxwellian to the oscillating form, we will get

$$F(P) = \frac{D}{m^3 c^3} \left[ \left(1 + \frac{\epsilon_{os}}{mc^2}\right) \left(\frac{mc^2}{k}\right)^3 - \left(\frac{mc^2}{k}\right)^2 \right]$$
$$\times \exp\left[ -\frac{mc^2}{k_B T} \sqrt{\frac{1+(P-P_L)^2}{k}} - 1 \right] . \qquad(9)$$

For $I_L = 10^{20}$ W/cm$^2$ and $\lambda = 0.53$ μm, the momentum distribution given by Eq. (9) is given in Fig. 9, where the temperature $T$ is assumed to be 100 keV. Because of the high energy quivering motion, the electron momentum extends to higher than 10 MeV.

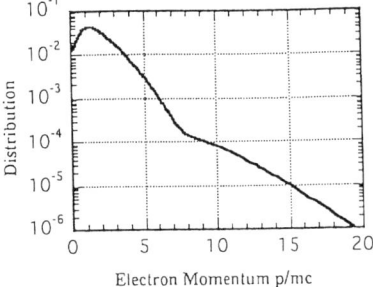

Fig. 9. Momentum distribution of laser heating plasma for the laser intensity $10^{20}$ W/cm$^2$.

The X-ray spectrum and angular distribution for the bremsstrahlung of this relativistic plasma are evaluated. The angular distribution is shown in Fig. 10. Since the radiation from relativistic electrons are strongly collimated toward the direction of motion, the angular distribution of radiation is anisotropic. Because of the 8-figure motion, the radiation direction is declined from the transverse direction to the laser propagation direction. The X-ray energy spectrum is shown in Fig. 10. As indicated by the shadow, significant amount of MeV γ-rays are generated for the laser intensity of $10^{20}$ W/cm$^2$. Because of those MeV $\gamma$-rays, the electron-position pair is generated.

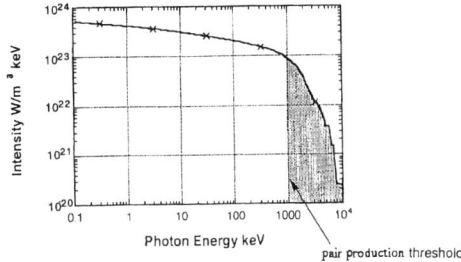

Fig. 10. X-ray spectrum of a relativistic plasma generated by $10^{20}$ W/cm$^2$ laser irradiation.

There are several processes for producing positrons in relativistic plasmas. They are particle-particle collisions;

$$e + e \rightarrow 2e + e + e^+ ,$$

$$e + p \rightarrow e + e + e^+ ,$$

and $\gamma$-ray-particle collisions;

$$p + \gamma \rightarrow p + e + e^+ ,$$

$$e + \gamma \rightarrow e + e + e^+ .$$

The lowest threshold for the required energy is the (P, $\gamma$) process, which is 1 MeV. The pair production cross section for (P, $\gamma$) process is given by

$$\frac{\Phi_{\text{pair}}}{\Phi} = 0.112 \times \left( \frac{\hbar \omega}{mc^2} - 2 \right)^{3.55} ,$$

where $\Phi = z^2 \alpha a_0^2$ for ion charge $z$, hyperfine constant $\alpha$ and the electron classical radius $a_0$. Using the $\gamma$-ray spectrum given by Fig. 10, the positron density is evaluated as shown in Fig. 11 for a Aluminum target irradiated with $10^{20}$ W/cm$^2$, $\lambda = 0.53$ $\mu$m laser. Note here that the positron-electron annihilation has been neglected because the annihilation time for a positron in the solid target is about 1 nsec which is much longer than the laser pulse. The positron density becomes significant, namely $10^{-6}$ times solid density when the laser intensity is higher than $10^{20}$ W/cm$^2$. In this case, the annihilation 0.5 MeV $\gamma$-ray intensity will be comparable to or stronger than the bremsstrahlung $\gamma$-ray. From the active galactic center (Sgr A) [17,18] the annihilation $\gamma$-ray is considered to be observed as shown in Fig. 11 when it is in the active phase.

The plasma dynamics of such a relativistic plasma can be investigated in the near future in the laboratory.

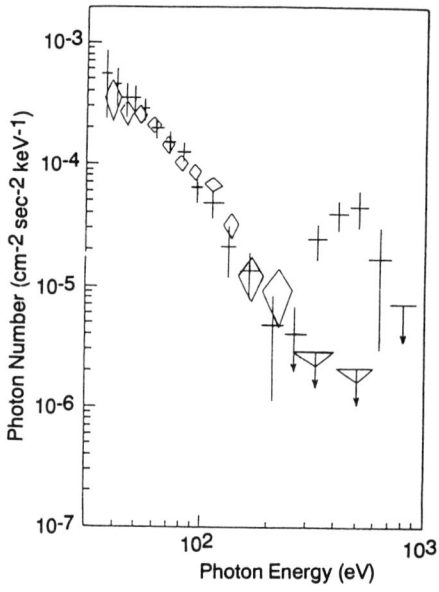

Fig. 11. $\gamma$-ray spectrum from Sgr A galactic center during March $\sim$ April '90 for $\Diamond$ and October '90 for $+$. Note that 511 keV peak corresponds to electron-positron annihilation.

**REFERENCES**
[1] A.L. Shawlow and C.H. Townes, Phys. Rev., **112**, 1940 (1958).
[2] N.G. Basov and A.M. Prokhorov, Zh. Exsp. Teor. Fiz., **28**, 249 (1955).
[3] T.H. Maiman, Nature, **187**, 493 (1960).
[4] R.L. McCrory, et al., Nature, **335**, 225 (1988).
[5] H. Azechi et al., Laser and Particle Beams, **9**, 193 (1991).
[6] H. Takabe et al., Physics of Fluids, **31**, 2884 (1988).
[7] H. Azechi et al., Laser and Particle Beams, **9**, 193 (1991).
[8] H. Nishimura et al., to be published in Physics of Plasma (1994).
[9] H. Takabe et al., Proc. 15-th Int. Conf. on Plasma Physics and Controlled Nuclear Fusion Research, Seville, 26 September – 1 October 1994.
[10] V.A. Andronov, Sov. Phys. JETP, **44**, 424 (1977).
[11] S. Gauthier and M. Bonnet, Phys. Fluids A, **2**, 1685 (1990).
[12] V.A. Andronov et al., JETP Lett., **29**, 56 (1979).
[13] H. Azechi et al., Appl. Phys. Lett., **55**, 945 (1989).
[14] F. Hattori et al., Phys. Fluids, **29**, 1917 (1986).
[15] D.L. Mathews et al., Phys. Rev. Lett., **54**, 110 (1985).
[16] Midorikawa et al., Phys. Rev. Lett., **71**, 3774 (1994).
[17] S. Fukue, Parity, **8**, No. 2, 50 (1993).
[18] L. Bouchet et al., Astrophys. J., **383**, L45 (1991).

# X.2

# SHOCK COMPRESSED NONIDEAL PLASMAS

V.E. Fortov

High Energy Density Research Center
Russian Academy of Sciences, Moscow, Russia

Review is given of recent experimental advances in the study of physical properties of strongly nonideal plasmas generated by compression and irreversible heating of the matter by intense shock and rarefaction waves. Experiments have been performed in the proximity of chemical explosives employing geometrical and gradient cumulation and by nuclear explosion. Porous metal and aerogel samples were used to increase the temperature of shock compressed plasmas. High-speed diagnostic technique and interpretation of dynamic experiments are described. Experimental results on the thermodynamic and optical properties of strongly coupled plasmas up to extremely high energy densities $\leq$ GJ/cc are discussed. Measurements of the hydrodynamic and optical emission from adiabatically expanded previously compressed metal plasmas give the information on thermodynamics and opacity of the strongly nonideal plasmas in near critical region of the metals. A comparison of the experimental data with the nonideal plasma theoretical models is made. The role of the discrete spectrum, strong Coulomb interaction and phase transitions in strongly coupled plasmas are analyzed.

## I. INTRODUCTION

Nonideal plasmas, the most widely spread state of matter occurring in the nature, have always attracted the attention of physicists due to a great variety of physical properties [Ichimaru, 1993; Fortov et al., 1984; Hubbard et al., 1991] and practical applications in astrophysics and in some advanced high energy installations. Magnetohydrodynamic (MHD) and flux compression generators, gas-core fusion reactor, laser- and beam-driven inertial confinement fusion projects are based on high energy density cumulation in strongly coupled plasmas. In these designs high pressure plasmas played the role of a "working medium" similar to steam in turbines and other heat engines.

The advent of intense sources of highly directional forms of energy, such as laser beams, electron, ion, and neutron beams, shock and electromagnetic waves, and so on, has in turn led to laboratory studies and to technology applications of ultradense states of matter at previously unattainable extreme pressures and temperatures [Avrorin et al., 1993]. Numerical simulations, based on computer solutions to the equations of the mechanics of continuous compressible media in which different complicated physico-chemical transformations take place at high temperatures and pressures, have become an essential component of physical analyses and optimizations of impulse processes involving dense plasmas. Studies of these processes constitute an exceedingly difficult and interesting problem in the science of extreme states of matter.

Along with the pragmatic interest in high-pressure plasma pure scientific interest has grown appreciably because the nonideality of plasma is responsible for behaviour of matter in a wide region of the phase diagram, from solid and liquid to neutral gas, including the high temperature phase boundaries of boiling and melting, as well as the metal-dielectric transition region [Hensel et al., 1990].

The plasma physical properties are greatly simplified at extremely high pressures and temperatures, when the kinetic energy of particles considerably exceeds that of interparticle interaction, such that models of ideal homogeneous degenerate (or Boltzmann) plasmas can be applied with assurance. A weak interparticle interaction can then be taken into account with the perturbation theory methods in the framework of classical (Debye-Huckel) [Ebeling et al., 1991] or quasiclassical (Thomas-Fermi) [Kirzhnits et al., 1975] self-consistent field methods. In strongly compressed plasmas the interaction energy is comparable to or exceeds the kinetic energy of particle motion, which hinders the application of perturbation theory to such systems. Parameterless numerical simulation methods (Monte Carlo, molecular dynamics) [Hansen, 1973; DeWitt et al., 1978, Zamalin et al, 1977] provides comprehensive information about the simplest models beyond the framework of the perturbation theory, e.g., the one-component plasma [Hansen, 1973; DeWitt et al., 1978] and the pseudopotential model of multycomponent plasma [Zamalin et al., 1977]. However, for the second model great difficulties arise when one tries to choose a qualitatively correct electron-ion pseudopotential, while it is difficult to apply the one-component plasma results to real plasmas. Therefore, for a qualitative analysis of the physical properties of strongly compressed plasmas there are heuristic models in use now, based on extrapolations of general ideas concerning the role of collective and quantum effects by the Coulomb interaction. These models predict physical effects that are new in principle, e.g., metalization and clusterization of plasma as well as the formation of yet unknown exotic plasma phases [Ebeling et al., 1991; Zamalin et al., 1977; Alekseev et al., 1983]. The significant point here is that several theoretical models predict highly nonmonotonic behaviour of thermodynamic functions, due to thermal ionization and (or) ionization by pressure, deformation, and rearrangement of the energy spectrum of atoms and ions on compression, and also strong Coulomb interactions in plasmas. This phenomena can often lead to hypothetical phase transitions that can substantially distort the usual form of the phase diagram of matter and can complicate the qualitative description of time-dependent hydrodynamic phenomena. Naturally, all these theoretical predictions need verification in experiments with real plasmas at high pressure.

In spite of the fact that the major part of matter in the Universe is the state of a strongly compressed plasma, our experimental knowledge about such a plasma has been quite limited until now because of great difficulties in generation and diagnostics of the high pressure plasma under laboratory or semilaborarory conditions [Fortov et al., 1984; Avrorin et al., 1993; Hensel, 1990; Alekseev et al., 1983; Fortov, 1982; Bushman et al., 1983; Zel'dovich et al., 1966].

The main difficulties in the production of a nonideal plasma are to make considerable local energy concentrations, which produce high pressures and temperatures above the thermostrength limits of the device structural materials. Consequently, it is necessary to carry out experiments in a forced pulsing regime at high power levels. In this case a serious problem is the diagnostics of a strongly compressed plasma which is very often opaque to the light.

In the experimental physics of strongly compressed plasmas the most widely used are dynamical methods [Fortov, 1982; Zel'dovich et al., 1966] which employ intense shock wave techniques for the compression and irreversible heating of matter due to the viscous dissipation of energy in the shock front. In this way physical measurements have been carried out over a wide range unaccessible to traditional plasma experimental methods; in particular aluminum superdense plasma of extremely high concentration 0.7 GJ/cc and pressures near 4 Gbar has been created [Simonenko et al., 1985].

The aim of this review is to discuss factual information about the thermodynamics and opacity of condensed matter at extreme thermal-energy concentrations. Since a considerable number of specialist reviews and monographs is already available on this topic (see Refs mentioned above and references therein) we shall concentrate our attention on studies of ultraintense shock waves and expansion adiabats of metals in shock-compressed and heated states, published during the last decade. In the final analysis, it was these studies that have resulted in a considerable expansion of the range of parameter values accessible to physical experimentation. They have also led to the exploration of an exceedingly wide portion of the phase diagram of metals, which now consist of seven orders of magnitude in pressure and five orders in density. Very different, complicated, and little known physical processes occur in plasma in this region. They include multiple thermal ionization, deformation of the energy spectrum of bound electrons ("shell effects"), the lifting of the degeneracy of electrons, the overcoming of the strong Coulomb interaction, the metal dielectric transition, and high temperature boiling.

## II. GENERAL DESCRIPTION OF PLASMA EXTREME STATES

Qualitative analysis of extreme states of the electron subsystem can be based on the consideration of Fig. 1 which summarizes the characteristic dimensionless parameters of plasmas, their technological applications, and also the typical pressures (in the atomic system of units in which $P_a = e^2/a_b^4 \sim e^{10}m^4/\hbar^8 \sim 300$ Mbar) that are encountered in cosmic and nuclear objects. The relative strength of the interaction between particles in a Coulomb system is characterized by the dimensionless parameter $\Gamma = E_e/E_k$, i.e., the ratio of the mean Coulomb energy ($E_e = Ze^2/r_e$) to the kinetic energy ($r_e$ is the screening length).

Below the $\Gamma \sim 1$ curve, the interaction in the classical plasma is weak and can be described by the chemical equilibrium model with perturbation theory [Ebeling et al., 1991] used to calculate corrections for nonideal plasma behavior. Such calculations are, strictly speaking, asymptotic in character ($\Gamma \to 0$). However, dynamic experiments with cesium and inert gas plasma [Fortov et al., 1984; Alekseev et al., 1983; Fortov, 1982] show that several of the most successful descriptions allow extrapolations of these corrections to the region right up to $\Gamma \to 1$.

The statistics of the electron component is determined by the degeneracy parameter $n_e \lambda_e^3$, were $\lambda_e = \sqrt{\hbar^2/2\pi m_c kT}$ is the thermal de Broglie wavelength. The characteristic scale for the kinetic energy $E_k$ in such plasmas is the Fermi energy $E_F = e^2 n^{2/3}/2m_e$, so that the compression of plasmas beyond the quantum nonideality limit $E_F \sim e^2 n_e^{1/3}$ leads to a simplification of its thermodynamic properties as

$\Gamma \to 0$. In real plasmas, the presence of positively charged nuclei results in the formation of manyelectron atoms and ions which must be described quantum-mechanically.

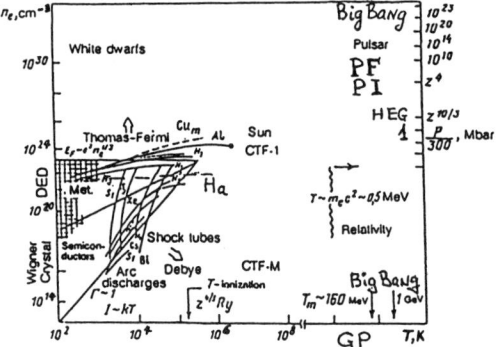

Fig. 1. Phase diagram of plasma. $S_1$ – adiabatic compression of saturated cesium vapor, $H_1'$, $H_2$; $H_2'$, $H_1$ – compression of saturated cesium vapor and inert gases by incident and reflected shock waves; shock compression of solid ($H_3$) and porous ($H_m$) metals; $H_a$ – shock compression of aerogel; $Al$ and $Cu_m$ – shock adiabats aluminum and porous copper, $S_2$ – adiabatic expansion of shock compressed metals. $Bi$ – expansion isentropes of bismuth. $GP$ – gluon plasmas, $PF$ – pesofusion, $PI$ – pressure ionization, $HEG$ – homogeneous electron gas.

One of the simplest model in this field is the Thomas-Fermi model [Kirzhnits et al., 1975] which is based on the quasiclassical approximation to the self-consistent field approximation and is valid for ultrahigh pressures $P \geq P_a \sim 300$ Mbar. However, comparison of the results obtained in this way with the more accurate band models shows that there are considerable discrepancies for pressures $P \geq P_a$ for which quasiclassical considerations suggest the electron shells become 'crushed' and a quasiuniform electron-density distribution is produced. One of the basic problems in the study of the superextreme states of plasmas is the elucidation of the physical role of shell effects and the determination of the range of validity of the quasiclassical description.

It is clear from Fig. 1 that only the periphery of the phase diagram is accessible to rigorous theoretical analysis. The nonideal region of plasmas is of the greater practical interest but, at the same time, presents the greatest difficulties for the theory. Use of various energy sources, vis. compressed and electroheated gas, chemical and nuclear explosives, powerful laser and electron beams, pneumatic and electrodynamic guns (see reviews [Fortov et al., 1984; Avrorin et al., 1993; Fortov, 1982]) has made it possible to create strongly compressed plasmas of different elements over a wide region in the phase diagram.

The Boltzmann nonideal gas-like plasma was obtained by the dynamic compression of high pressure gases, the initial states of which were in the neighborhood of the saturation curve (cesium, noble gases [Fortov et al., 1984; Fortov, 1982]) or even under supercritical conditions [Mintsev et al., 1979; Zaporoshets et al., 1984]. While registering the states of single ($H$) and double ($H'$) compressions, one manages to obtain plasmas with supercritical parameters at pressures up to 170 kbar, temperatures up to $10^5$ K and electron densities up to $10^{22} \mathrm{cm}^{-3}$ and to gain access from the side of the "gas" phase to the region of a condensed state. In the case of xenon plasma [Mintsev et al., 1979; Zaporoshets et al., 1984] a maximum density of 4.5 g/cc has been obtained which is 1.5 times larger than the xenon crystallographic density and near to the solid aluminum density. Adiabatic compression of saturated vapors of cesium and potassium [Isakov et al., 1984] (adiabat $s_1$ in Fig. 1) leads to less intensive heating of plasmas when charge-neutral interaction prevails. It is essential that the shock wave and adiabatic compression product not only a nondegenerate plasma with extremely high energy concentrations over a wide range of the phase diagram, but also under these conditions to perform detailed thermodynamical, electrophysical and optical [Fortov et al., 1984; Fortov, 1982] measurements as well as those of the laser beam reflection [Zaporoshets et al., 1984]. Very interesting experimental information can be obtained by the shock compression of low density ultrafine aerogel samples [Holmes et al., 1984].

The compression of metals by intense shock waves enables one to create strongly compressed plasmas with the electron's component either degenerate or partially degenerate (states $H_3$ on Fig. 1). For this purpose explosion and cumulation methods [Fortovet et al., 1984; Glushak et al., 1989; Bazanov et al., 1985], light gas gun [Nellis, 1985], laser [Cauble et al., 1993; Lower et al., 1993] and electron [Akkerman et al., 1992] beams, powerful underground explosions are applied successfully. The main problem in laser and beam generated shock waves [Anisimov et al., 1984] was solved recently by conversion of directed energy fluxes into soft X-ray which was then used for shock generation. The successful experiments with conversion of laser beam [Lower et al., 1993] and Z-pinch [Grabovski et al., 1994] into soft X-ray have shown very good symmetry and uniformity of generated shock waves.

Of special interest for plasma physics are experiments [Simonenko et al., 1985; Zubarev et al., 1978; Trunin et al., 1989, 1993] in which, by using porous samples and ultraintense shock waves, the plasma states at record high temperatures and concentrations of heat energy with nongenerate electron component at densities $n_t \sim 10^{23} \mathrm{cm}^{-3}$ have been obtained. These experiments make it possible to investigate the thermodynamics of metal plasmas over the entire range of condensed state and to penetrate into the region of quasiclassical [Kirzhnits et al., 1975] description up to exotic conditions, where the pressure and the energy of the equilibrium radiation become important.

The method of adiabatic expansion (curves $S_2$) of metals, compressed and heated by powerful shock fronts ($H_3$), is quite effective for the generation of plasmas with densities below those of the normal solid. This technique makes it possible to explore a wide region of the phase diagram from the strongly compressed metallic liquid to the ideal gas, including the nonideal degenerate, Boltzmann plasma region, and the critical point region [Fortov et al., 1984; Alekseev et al., 1983; Fortov, 1982]. These results serve as a base for wide-ranged semiempirical equations of state [Bushaman et al., 1983, 1992] and, in combination with the data from static [Hensel, 1990] and electroexplosive [Hixson et al., 1985] experiments, enable one to draw more definite conclusions as to the form of the phase diagram of metals.

## III. SHOCK WAVE COMPRESSION OF NONIDEAL PLASMA

The dynamic experimental diagnostic methods are based on utilization of the relationship between the thermodynamic properties of the medium under investigation and the experimentally observed hydrodynamic phenomena occurring upon cumulation of high energy densities in the matter [Zel'dovich et al., 1966]. In the dynamic investigations one tends to use self-similar solutions of the type of stationary shock wave or centered Riemann expansion wave which express the conservation laws in a simple algebraic or integral form.

Upon propagation through the material of a stationary shock-wave discontinuity, the laws of conservation of mass, momentum and energy are satisfied in the front of the latter [Zel'dovich et al., 1966]

$$\frac{V}{V_0} = \frac{D-u}{D} \; ; \quad p = p + \frac{Du}{V} \; ; \quad E - E_0 = \frac{1}{2}(P + P_0)(V - V_0) \; , \tag{1}$$

which permits of finding the hydrodynamic and thermodynamic characteristics of the material from the recording of any two of five parameters characterizing the shock wave discontinuity $(E, p, V, D, u)$. The shock velocity $D$ is measured most readily and accurately using baseline techniques.

The choice of the second parameter to be measured depends on the actual experimental conditions. In the case of highly compressible ("gas") media, it is practical to perform the recording of density, $\rho = V^{-1}$, of shock compressed material. There has been presently developed a procedure for such measurements based on the registration of absorption by cesium, argon and air plasma of "soft" X-ray radiation [Fortov et al., 1984; Fortov, 1982]. In the case of lower compressibility of the system (condensed media), tolerable accuracies are ensured [Zel'dovitch et al., 1966] by recording the mass velocity of travel, $u$. Found in this manner were the states of degenerate metal plasma and of a dense Boltzmann plasma of argon and xenon.

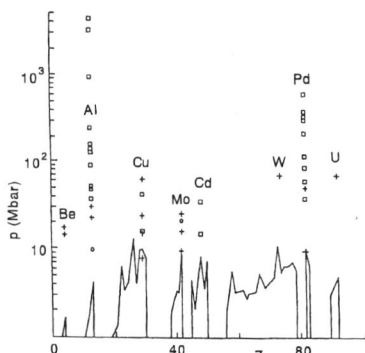

Fig. 2. Pressure ranges for experiments with intense shock waves, in which shock compressibility of the elements has been measured. Regions at the bottom (with sharp peaks) – experiments with explosives and guns [Al'tshuler, 1992; Fortov, 1982; Zel'dovitsh et al., 1966] underground nuclear explosions: circles – absolute measurements [Avrorin et al., 1993], crosses – American data, squares – Soviet data.

There is now a very considerable volume of experimental data on shock compressibility. Existing experimental data on the thermodynamic properties of dense materials have been obtained in dynamic experiments and have been reviewed in the literature

[Van Thiel, 1977; Marsh, 1980; Al'tshuler, 1992]. The amount of published information on shock compressibility obtained in dynamic studies is illustrated in Fig. 2.

Measurements of shock compressibility and isentropic expansion are often accompanied by measurements of adiabatic compressibility (velocity of sound), opacity, high and low frequency electrical conductivity, and stopping power of fast particles in the plasmas [Fortov et al., 1984; Fortov, 1982].

First shock wave experiments with cesium, argon, and xenon plasmas by Fortov [1984,1982] and also shock compressed liquid argon and xenon [Nellis et al., 1979] show that the particle interaction in highly compressed partially ionized plasmas produces a deformation and splitting of energy levels.

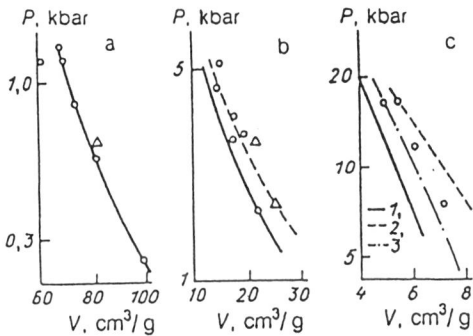

Fig. 3. Equation of state of argon plasma [Bushman et al., 1983]: a–c – calculated adiabats from states which initial pressure of 1, 5, and 20 bar, respectively; 1 – Debye approximation in grand canonical ensemble, 2 – confined atom model, 3 – pseudopotential model of plasma [Fortov et al., 1984; Ebeling et al., 1991]; points – experiment.

This effect can not be described by the perturbation theory and requires a complete solution of the many-electron quantum- mechanical problem. In the confined-atom model [Fortov et al., 1984; Ebeling et al., 1991] internal atomic or ionic electrons are contained in a spherical cell of radius $r_0$ in which the electron-ion interaction potential has the Coulomb form

$$v_{ei} = \begin{cases} \frac{Ze^2}{r}, & \text{for } r < r_0 \\ \infty, & \text{for } r > r_0 . \end{cases}$$

The radial wave function of each electron and the electron energy levels are calculated in this model by numerical integration of the Hartree-Fock equations with a nonlocal (exchange) potential. The resulting excitation energies are then used to calculate the partition functions for the chemical plasma model into which additional corrections are introduced for the Coulomb interaction, electron degeneracy, and the interaction between hard spheres of radius $r_0$, calculated from molecular dynamics. The variational principle $\delta F/\delta r_0 = 0$ then gives the equilibrium value of $r_0$. This is of course, different from the value widely used in the cell model with dense packing [Ashcroft et al., 1976]. In contrast to cell models, 'the confined atom' approximation is introduced within the framework of the quasichemical description in which the translation motion of the individual particles is explicitly taken into account together with the collective Coulomb interaction. Fig. 3 compares this plasma model with shock compression data for the nonideal cesium and argon plasmas.

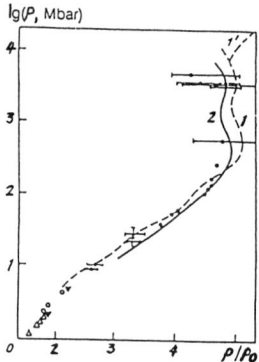

Fig. 4. Shock adiabat of aluminum: points – experiment, 1 – confined atom plasma model, 1' – contribution of radiation, 2 – SCF calculation [Avrorin et al., 1993].

Very recently the new experiments were carried out on shock compression of aerogel ($SiO_2$) samples with extremely low $\sim 0.01$ g/cc density. The characteristic pressure $\sim 100$ kbar, temperature – 1–2 ev, electron concentration $n_e \sim 10^{21} cc^{-1}$ and plasma coupling parameter $\Gamma \sim 2$–3 were achieved.

A very interesting and new region for applying the chemical model of the plasma is the multi-megabar pressure range, realized by compression of solids by shock waves of extreme intensities. Fig. 4 shows a comparison between theoretical models and experimental data at ultrahigh pressures [Avrorin et al., 1986; Vladimirov et al., 1984]. At the record pressure value achieved in these works, i.e., $P = 4$ Gbar, temperatures $T \sim 7 \times 10^6$ K and energy density of about $6 \times 10^4$ KJ/g, the plasma is fully ionized with $n_c \sim 3 \times 10^{24} cm^{-3}$, and is weakly degenerate ($\lambda_e^3 n_e \sim 0.06$). As the pressure is reduced, the nonideality parameter $\Gamma$ rises on the shock adiabat from 0.05 to 8 (at $P \sim 10$ Mbar) and the short-range repulsion remains appreciable which justifies the utilization of the Debye approximation and the solid-sphere model [Fortov et al., 1984; Avrorin et al., 1993; Ebeling et al., 1991]. Additional modification of plasma confined atom model was done to take into account fixture of the ions of different ionic charges. One of the recent examples of chemical model applications is the interpretation of the experiments on the shock compression of porous metal samples by intense shocks [Zubarev et al., 1978; Trunin et al., 1989, 1993] which make it possible to obtain extremely high concentrations of thermal ($\sim 1$ MJ/cc) energy.

The Hugoniots calculated by the plasma chemical model are represented in Fig. 5 for initial porosities of sample $m = \rho_{00}/\rho_0$. One can see the satisfactory agreement with the experiment. From the chemical model effects under consideration the most essential is the short-range repulsion, the absence of which leads calculated plasma densities to be 1.5 times more than measured. The results of calculation are less depended upon the Coulomb correction and exited states, inclusion of both leads to an increase of calculated density. The least significant is the degeneration of electrons.

The chemical model of plasmas [DeWitt et al., 1978] make it possible to analyze qualitatively some effects of the equation of state including the so-colled electron-shell structure effects. In the plasma equation of state of copper the Coulomb corrections, electron degeneration and the short-range repulsion of ions were taken into account. The parameter Ri of the short-range repulsion is evaluated from the Hartree-Fock calculations. Partition functions of atoms and ions were calculated using energy levels and ionization energies calculated by Hartree-Fock method. Degrees of ionization up to five were taken into consideration.

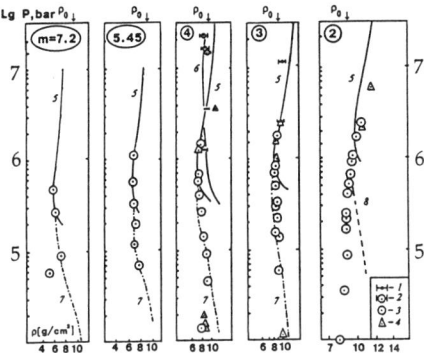

Fig. 5. Hugoniot of porous copper. Experiments: 1 − [Kormer et al., 1962], 2 − [Zubarev et al., 1978], 3 − [Trunin et al., 1989 (1)], 4 − [Trunin et al., 1989 (2)]. Confined atom model: 5 − hard sphere radius $R_0$ =2 a.u. and ionic radius $R_i$ calculated from ionizations potentials. 6 − $R_i = R_0$. Particles repulsion energy $\Delta F \sim -V^{-\delta}$: 7 − $\delta = 1/3$, 8 − $\delta = 1$.

## IV. ISENTROPIC EXPANSION OF SHOCK-COMPRESSED METALS

Intense shock waves can be used to produce high pressures and temperatures in shock-compressed plasma, but the region of reduced pressure that is occupied by the dense hot liquid and partially degenerate plasma is inaccessible to this techniques [Fortov, 1982; Zel'dovich et al., 1966]. The situation can be remedied by using adiabatic expansion of condensed material after preliminary compression and heating by powerful shock waves [Fortov et al, 1984; Avrorin et al., 1993; Fortov, 1982].

From the physical point of view there is particular interest in situations in which near- or post-critical states are produced in the rarefaction wave.

The thermodynamic parameters of the material are determined from hydrodynamic measurements of pressure $P$, particle velocity $u$, by evaluating the Riemann integrals expressing the conservation laws for the particular self-similar flow:

$$V = V_A + \int_P^{P_A} \left(\frac{du}{dP}\right)^2 dP,$$
$$E = E_A - \int_P^{P_A} \left(\frac{du}{dP}\right)^2 dP. \tag{2}$$

Mechanical measurements have now been augmented by optical temperature determinations through a transparent helium barrier.

To generate high shock waves, intensity of which allow to reach supercritical metal plasma in rarefaction wave, the multistage cumulative ballistic and conical Mach shock wave generators fed by chemical explosives were used. To increase entropy in shock wave front the low density (porous) targets were used.

Figure 6 shows the $P, u$ diagram for bismuth [Glushak et al., 1989]. Calculations of thermodynamic parameters based on Reamann integrals (2) show that the experiments were performed in a wide range lying between states on the shock adiabat with $P \approx 6.7$ Mbar, $\rho \approx 2.6$, $\rho_0$ and the low-pressure metal vapor with $\rho \sim 10^{-2}\rho_0$.

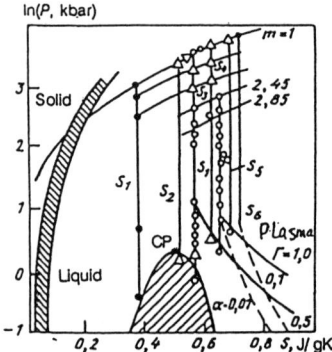

Fig. 6. Entropy diagram of bismuth: $m$ – initial porosity on shock adiabats, $S$ – release isentropes, $\alpha$ and $\Gamma$ – ionization and non-ideality. CP – critical point; mixed-phase regions are shaded.

In this range of parameters adiabatic expansion of metals results in a complex range of physical processes that have not been adequately investigated. Degeneracy is thus lifted, electrons recombine, the energy spectrum of the material undergoes a radial change, a metal-dielectric transition takes place in the electronically disordered structure, and the dense plasma produced in the process exhibits non-ideal properties for different form of particle interaction.

There is particular interest in the strong expansion of metals, which takes them to the region of highly postcritical conditions. Fig. 6 plots the nonideality parameter $\Gamma$ and the degree of ionization $\alpha = n_e/(n_e + n_0)$, calculated from the chemical models of plasma.

The experiments performed so far have revealed neither appreciable discontinuities in thermodynamic functions nor some other hydrodynamic anomalies that could be interpreted as plasma phase transitions [Ebeling et al., 1991; Alekseev et al., 1983]. We emphasize that such phase transitions are particular probable precisely in this part of the phase diagram because a rise and a reduction in the density of Boltzmann plasma, and also an increase in pressure in the degenerate plasma should lead to a relative reduction in the interaction between the particles.

The detection of optical phenomena accompanying the arrival of powerful shock waves on the free surface provides interesting data on optical properties that are closely related to the structure, composition, and energy spectrum of exploding plasma. The first experimental work on the emission of radiation by nonideal bismuth plasma when a powerful shock wave producing a pressure of a few million atmospheres emerged on the free surface was reported in [Kvitov et al., 1991; Fortov et al., 1992]. Intense shock waves were generated by explosive cumulative propulsion devices in which thin (0.2 and 0.1 mm) molybdenum liners were accelerated to speeds of 7.0 and 8.3 km/s by using the gradient cumulation effect [Glushak et al., 1989]. The impact of these liners on a bismuth target produced shock waves with amplitude pressure of 2.8 and 3.6 Mbar, respectively. The emergence of these shocks on the free surface of the sample was accompanied by the adiabatic expansion of the dense plasma whose optical emission was recorded by photodetectors converters.

The nonideal plasma states produced in these experiments are shown in Fig. 7 on the $T, V$ plane ($T$ is the temperature and $V$ is the volume). The figure shows the boundary of the two-phase region (curve 1) and also the isentropes $S_1(P_s = 2.8$ Mbar) and $S_2(P_s = 3.6$ Mbar) for particular segments identified experimentally. Curves of constant ionization $\alpha = n_e/m_i + n_a$. Coulomb nonideaity $\Gamma = \sqrt{8\pi n_e}/(kT)^{3/2}$, and ion-atom interaction strength $\gamma_i a = (2\pi a e^2 N)(r_a kT)^{-1} \approx 1$ [$\alpha \approx 50 a_0^3$] is the polarizability of neutrals and $r_a \approx 3 a_0 = 3 h^2/m_e (2\pi e)^2$ were calculated from a plasma model that included degeneracy effects, the interaction between charges and between charge and neutrals, and also the atomic and ionic radii [Ebeling et al., 1991].

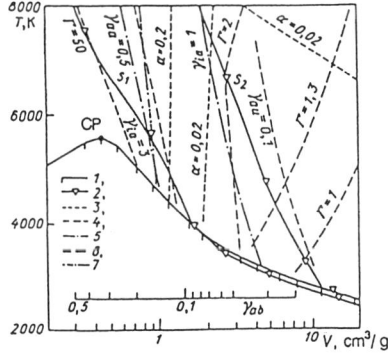

Fig. 7. The $T, V$ diagram of bismuth: 1 – boundary of two-phase region, 2 – $S_1$, 3 – $\alpha$ = const, 4 – $\Gamma$ = const, 5 – $\gamma$ = const, 6 – $E_F/kT = 1$.

It is seen that the states detected in adiabatic rarefaction waves correspond to dense low-temperature plasma with a strong Coulomb interaction ($\Gamma \approx 1.3$–30), a change in the type of statistics near the $\epsilon_F/kT = (3 n_e/\pi)^{2/3} h^2/8 m_e kT = 1$ curve, and appreciable polarization interaction between charges and neutrals: $\gamma_{ia} = 0.5$–3. Since the states produced in the rarefaction wave are close to the boiling curve, the Van der Waals interaction between neutrals is also significant, its parameters $a$ and $b$ being given by $\gamma_{aa} = Na/kT \approx 0.2$–1, $\gamma_{ab} = 3Nb \approx 0.03$–0.3. The optical radiation intensity leaving the expanding plasma is described by the simultaneous solution of the hydrodynamic and radiation transfer equations. The corresponding calculations are shown in Fig. 8 and confirm the estimates made in [Zel'dovich et al., 1966] according to which the effective layer of plasma that is responsible for the emission of radiation lies at a distance corresponding to unit optical path length to the interface with vacuum, i.e.

$$\int_0^x r_\nu dx = 1 \; .$$

This condition enables us to estimate the magnitude of the plasma absorption coefficient $k_\nu$ from the semiempirical equation of state [Bushman et al., 1992] and the measured function $T_{\text{eff}}(t)$. Since $k_\nu$ is an exceedingly rapidly-varying (exponential in $T$) function of the parameters of state, the temperature $T$ determined in this type of experiment is actually dictated by the local value of $k_\nu$ in the effective radiating layer.

The strong particle interaction illustrated in Fig. 1 complicates the systematic theoretical analysis of the composition and optical properties of the plasma and forces us to consider simplified methods. Curve 2 (Fig. 8) shows $k_\nu$ calculated from the Kramers-Unsold formula [Ebeling et al., 1991; Zaporoshets et al., 1984] with terms representing the photoionization of the upper excited states and bremsstralung in the field of ions and neutrons for a given degree of ionization of the plasma ($\alpha = 1$). Curves 3 and 4 show the absorption coefficient calculated from a hydrogen-like model of the bismuth atom [Ebeling et al., 1991]. For low densities and low nonidealities $S_2$, detailed allowance for ionization processes results in reasonable agreement with experiment even for a very approximate description of the contribution of bremsstrahlung and photoionization processes to $k_\nu$. The plasma density on the adiabat $S_1$ (Fig. 8) is greater by approximately an order of magnitude than $S_2$, which reduces the argument between experiment and simple models and indicates that the contribution of photoionization processes is similar in compressed plasma. It is possible that, as in [Fortov et al., 1992], we are dealing here with a plasma transmission effect in which some of the high-lying energy levels are transferred to the continuum as the plasma density increases.

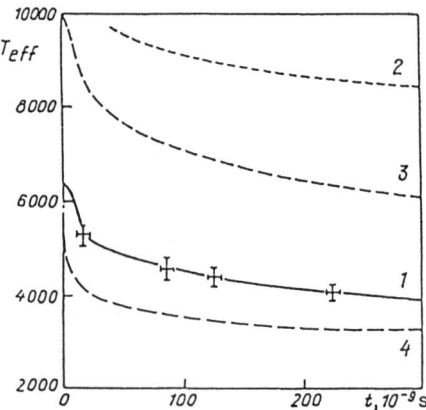

Fig. 8. Temperature $T_{\text{eff}}(t)$ of bismuth plasmas in rarefaction wave: 1 – experimental data (crosses indicate the ranges of uncertainty), 2 – Kramers-Unsold, 3 – calculations based on [Ebeling et al., 1991] including only the bremsstrahlung absorption channel, 4 – calculations based on [ibid.] with the inclusion of bremsstrahlung and photoionization.

## V. CONCLUSION

The attempt was made in this review to provide a systematic review of an extensive range of experimental data obtained by the dynamical methods in the last few years in the physics of extreme states of nonideal plasma. Experiments involving cumulative high explosive and intense underground explosions, which were being carried out until quite recently, have produced record thermal energy densities ($10^9 J/cm^2$) that were comparable with the specific energy release of nuclear explosives. The upper limit of the pressure range in 'hot' plasmas is now about 4 billion atmospheres, which is comparable with the pressures in the interior of the sun and other stars. Above this limit, equilibrium radiation provides the dominant contribution, so that the thermodynamics of these exotic states is determined by radiation and is not very dependent on the structure of the material itself. The adiabatic expansion of a metal after it has been traversed by powerful shock waves takes it to a state that can be described as low temperature nonideal plasma with a complex and extensive spectrum of particle interactions. It was precisely in this way that it has been possible to take metals to their near-critical state. This has yielded information about the high temperature part of their boiling curves, the properties of boiling and condensation, and the radiative properties of nonideal plasma.

It seems that these expensive and labour-intensive experiments are of considerable contribution to our understanding of properties of plasmas at extreme energy densities.

We believe that further advances in this part of physics will depend on the application of new experimental techniques for the concentration of energy, which will reply on powerful laser, charged particle beams, and electrodynamic accelerators. The aim of this paper has been to attract the attention of specialists to this interesting problem.

To finish this review we clearly see that in spite of the recent progress in strongly coupled plasma our knowledge in this field is very limited. As Voltaire [1976] declared: "Principalities of some german and italian princes, which one can pass through without difficulty in a halfe an hour, are in comparison with Turkish, Moscovitic or Chinese empires only extreme feeble representation of those wanderful contrasts which the Nature shows in all its creations".

## REFERENCES

Akkerman, A., et al., 1992, *Application of High-Current Charged Particle Beams in Dynamic High Pressure Physics in Thermal Physics Reviews*, eds A. Sheindlin and V. Fortov (Harwood Academic Publishers).

Alekseev, V.A., V.E. Fortov, and I.T. Yakubov, 1983, Usp. Fiz. Nauk, **139**, 193.

Al'tshuler, L.V., 1992, *Shock Compression of Condensed Matter*, (Amsterdam, North-Holland), p 3.

Anisimov, S.I., A.M. Prohorov, and V.E. Fortov, 1984, Sov. Phys. Usp., **27**, 181.

Ashcroft, N.W., and N.D. Mermin, 1976, *Solid State Physics*, (N.Y., Holt, Rinehart, and Winston).

Avrorin, E.N., B.K. Vodolaga et al., 1986, JETP Letter, **43**, 308.

Avrorin, E.N., et al., 1993, Sov. Phys. Usp., **36**, 337.

Bazanov, O.V., V.E. Bespalov, A.P. Zharkov, et al., 1985, Teplofiz, **23**, 976.

Bushman, A.V., and V.E. Fortov, 1983, Usp. Fiz. Nauk., **140**, 177.

Bushman, A.V., I.V. Lomonosov, V.E. Fortov, 1992, *Equation of State of Metal at High Energy Density*, (Chernogolovka); J. Noncryst. Solids, 1993, **156**, 631.

Cauble, R., et al., 1993, Phys. Rev. Lett., **70**, 2102.
DeWitt, H.E., G. Kalman, and P. Carini, 1978, *Strongly Coupled Plasma*, (N.Y., London, Plenum), p.83.
Ebeling, W., et al., 1991, *Thermophysical Properties of Hot Dense Plasmas*, (Stuttgart-Leipzig, Teubner-Texte zur Physik).
Fortov, V.E., 1982, Usp. Fiz. Nauk., **138**, 361.
Fortov, V.E., and I.T. Yakubov, 1984, *Physics of Nonideal Plasma*, (Chernogolovka).
Fortov, V.E., et al., 1992, *Shock Compression of Condensed Matter*, (Elsevier Science Pulishers), p. 745.
Glushak, B.L. et al., 1989, Sov. Phys. JETP, **69**, 739.
Grabovski, E.V., et al., 1994, JETP Pisma.
Hubbard, W.B., et al., 1991, Science, **253**, 648.
Hansen, J.P., 1973, Phys. Rev. A, **8**, 3096.
Hensel, F., 1990, J. Phys.: Condensed Matter, **2**, SA33.
Holmes, N.S., Radovsky et al., 1984, Appl. Phys. Lett., **45**, 626.
Hixson, R.S., M.A. Wincler, J.W. Shaner, 1985, High Temperatures – High Pressures, **17**, 267.
Isakov, I.M., A.A. Likal'ter, B.N. Lomakin, A.D. Lopatin, and V.E.Fortov, 1984, Sov. Phys. JETP, **87**, 832.
Ichimaru, S., 1993, Rev. Mod. Phys., **65**, 255.
Kirzhnits, D.A., Yu.E. Lozovik, and G.V. Shpatakovskaya, 1975, Usp. Fiz. Nauk., **117**, 3.
Kormer, S.B., et al., 1962, JETP, **43**, 686.
Kvitov, S.V., A.V. Bushman, et al., 1991, Pisma Zh. Eksp. Teor. Fiz., **53**, 338, 353.
Lower, T., and R. Sigel, 1993, Contributions to Plasma Physics, **33**, 355.
Marsh, S.P., 1980, *LASL Shock Data*, (Berkely, University of California Press).
Mintsev, V.B., and V.E. Fortov, 1979, Pisma Zh. Eksp. Teor. Fiz., **30**, 401.
Nellis, W.J., 1985, *High Pressure Measurement Techniques*, (London, Appl. Sci. Pubblishers), p.68.
Nellis, W.J., et al., 1979, Phys. Rev. Lett., **48**, 816.
Nellis, W.J., et al., 1980, J. Chem. Phys., **73**, 6137.
Simonenko, V.A., N.P. Voloshin, A.S. Vladimirov, et al., 1985, Zh. Eksp. Teor. Fiz., **88**, 1452.
Thiel, M.Van, J.M. Shaner, and E. Salinas, 1977, LLNL Rep., VCRL-50-108, 1-3.
Trunin, R., et al., 1989, JETP, **96**, 1024 (1).
Trunin, R., et al., 1989, JETP, **95**, 631 (2).
Trunin, R., et al., 1993, JETP, **103**, 2180.
Vladimirov, A.S., and N.P. Voloshin, 1984, Pisma Zh.Eksp. Teor. Fiz., **39**, 69. (1984, JETP Lett., **39**, 82).
Voltaire, F.M., 1976, *Micromegas in Samtliche Romane*, (Frankfurt, Insel).
Zamalin, V.M., G.E. Norman, and V.S. Filinov, 1977, *Monte Carlo Method in Statistical Thermodynamics*, (Moscow, Nauka).
Zaproroshets, Yu.B., V.B. Mintsev, V.E. Fortov, and O.M.Batovskii, 1984, Pisma Zh. Tekhn. Fiz., **10**, 1139.
Zel'dovich, Ya.B., and Yu.P. Raizer, 1966, *Physics of Shock Waves and High-Temperature Hydrodynamic Phenomena*, (Moscow, Nauka).

# X.3

# SPECTROSCOPIC ANALYSIS OF HOT DENSE LASER PRODUCED PLASMAS: REVIEW AND UPDATE

C.F. Hooper, Jr.,[†] R.C. Mancini,[‡] D.A. Haynes, Jr.,[†] and D.T. Garber[†]

[†]University of Florida, Gainesville, FL 32611, U. S. A.
[‡]University of Nevada, Reno, NV 89557, U. S. A.

Brief discussion of the standard theory of line broadening is presented together with an analysis of laser driven implosion experiments. The effect of improved theoretical procedures on experimental analysis is discussed. The experiments discussed were performed at the University of Rochester Laboratory for Laser Energetics (LLE). The results presented in this paper illustrate the usefulness of plasma line broadening in diagnosing laser-driven implosions and in understanding fundamental plasmas processes.

## I. INTRODUCTION

For almost two decades, high power lasers have been used to implode microballoons filled with gases such as neon, argon, deuterium, or a mixture of deuterium and argon. These implosions have generated high-temperature (1keV) and high density ($10^{23}$cm$^{-3}$–$10^{25}$cm$^{-3}$) plasmas.

As a results of these experiments we are able to observe the radiative properties of highly charged ions in the presence of a variety of strongly coupled plasmas. Spectral radiation observed from these experiments is frequently in the X-ray region and the radiative properties are greatly influenced by plasma effects. In section II of this paper we will discuss the theoretical techniques employed to interpret these spectra and will describe three sets of implosion experiments. In section III we will list some conclusions.

## II. THEORY AND EXPERIMENTS

A. Preliminary considerations

1. Plasma microfield effects and Stark splitting

If an isolated hydrogen atom or hydrogenic ion is raised to its first excited electronic state it will subsequently return to its ground state with the emission of a photon. A collection of non-interacting excited atoms or ions will radiate one photon per radiator and will produce an intensity distribution $I(\omega)$, as illustrated below.

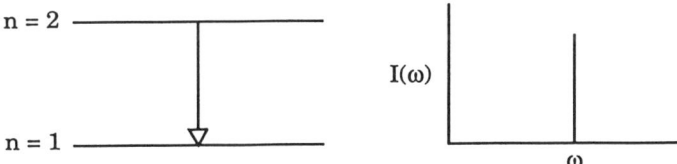

Figure 1. Diagram of an unperturbed 2→1 transition in a one-electron radiator and the spectrum emitted.

If these excited hydrogenic atoms/ions are placed in external electric field, the energy-level structure and resulting intensity will be altered by the familiar Stark effect.

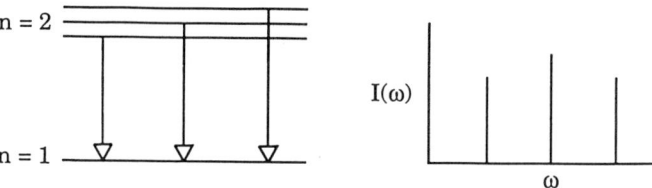

Figure 2. Diagram of an 2→1 transition in a one-electron radiator perturbed by an external electric field and the spectrum emitted.

Such an electric field could be generated by a static configuration of ions in a plasma, in which case it is called an ion microfield.

Next suppose that we place our radiating ions in a plasma where the ions are assumed to create a static microfield and where each transition is broadened by dynamic electrons perturbing the individual states; then, if the radiating ion is $Ar^{+17}$, we find, for two different microfield values, the line profiles are as pictured.

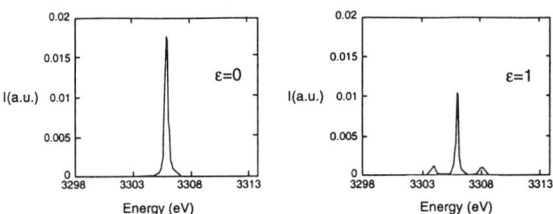

Figure 3. Ar Lyman-$\alpha$ transition under the influence of two different microfields. $n_e = 5 \times 10^{23}$ cm$^{-3}$, $kT_e = 800$ eV.

$\epsilon$ is defined to be $E/E_0$ where $E_0$ is defined by the expression

$$E_0 = \frac{e}{R_0^2}$$

and

$$R_0 = \left(\frac{3}{4\pi n_e}\right)^{1/3} .$$

These figures illustrate the effect of a given ion microfield on the intensity pattern. Of course in a real plasma each radiator would generally experience a different static ion microfield and hence the final line profile would involve a weighted sum over all values of $\epsilon$. Plasma line broadening theory has developed to calculate spectral line profiles characteristic of radiators immersed in the plasma environment [see e.g., Griem, 1974].

2. Plasma line broadening theory

The "standard theory" of line broadening which is usually employed in practical calculations uses the static ion approximation, in which the plasma ions are treated as static during the radiative lifetime, while the plasma electrons are considered dynamically. The ions produce a statistical distribution of ion microfields at the position of the radiators in the plasma. The calculation of such a distribution is detailed in the literature [Hooper, 1966; Hooper, 1968; Iglesias et al., 1983]. The character of microfield distribution is illustrated in the following figure.

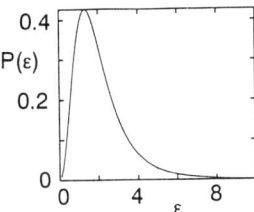

Figure 4. Microfield distribution function, $P(\epsilon)$, for a pure hydrogen-like Ar plasmas, $n_e = 5 \times 10^{23} \text{cm}^{-3}$, $kT_e = 800\text{eV}$.

The dynamic plasma electrons are included using a second order quantum mechanical perturbation theory. For both the perturbing ions and electrons the radiator-perturber interaction potentials are treated in the dipole approximation. The expression for the lineshape is given by

$$I(\omega) = \int_0^\infty P(\epsilon) J(\omega, \epsilon) d\epsilon ,$$

where $P(\epsilon)$ is the ion microfield distribution function and $J(\omega, \epsilon)$ represents the effect of electron broadening in the presence of an ion microfield $\epsilon$. The lineshape depends on the atomic physics, in particular on whether or not relativistic effects are included. If relativistic atomic physics is used for the Ar Lyman-$\alpha$ line, a much different line shape is obtained. Relativistic effects produce both fine structure splitting and a shift to lower transition energies. For the $2 \to 1$ transition in $\text{Ar}^{+17}$, the Ar Lyman-$\alpha$ line, with $n_e = 5 \times 10^{23} \text{cm}^{-3}$ and $kT_e = 800\text{eV}$ we arrive at the following results for $I(\omega)$.

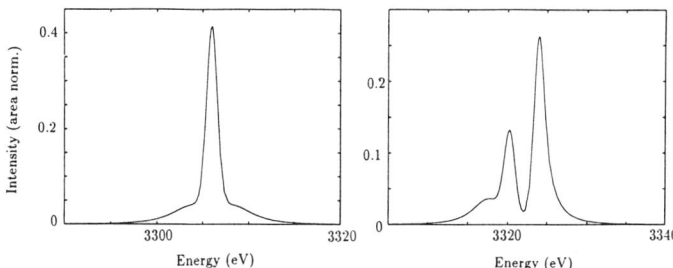

Figure 5. Ar Lyman-$\alpha$ without (left) and with (right) relativistic atomic physics. $n_e = 5 \times 10^{23} \text{cm}^{-3}$ and $kT_e = 800\text{eV}$.

The use of Stark broadening of spectral features as a plasma density diagonostic is particularly effective since many such features are very sensitive to changes in density while not particularly sensitive to changes in temperature. However, where several isolated spectral features are observed simultaneously, integrated intensity ratios can supply temperature inferences as well. The density sensitivity of the lineshape is illustrated in the following figure, for two different transitions in hydrogen-like Ar.

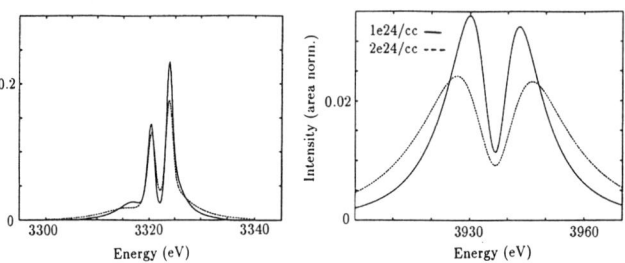

Figure 6. Density sensitivity of the Ar Lyman-$\alpha$ (left) and -$\beta$ (right). $kT$ =800eV.

It is clear from a comparison of the plots that for the given temperature and density conditions the lineshape of the $3 \rightarrow 1$ transition, the Ar Lyman-$\beta$, is far more density sensitive than the Lyman-$\alpha$ line. This is due to the fact that the $n = 3$ level is more sensitive to perturbation than is the $n = 2$ level.

To analyze radiative plasma properties realistically, it is also necessary to include the influence of opacity and Doppler effects on the lineshape. Photons emitted in the core of such a plasma may well be absorbed and reemitted one or more times before escaping the plasma and being observed spectroscopically. Hence the observed photon can have an energy different from the core photons [see, e.g., Mihalas, 1978]. The degree to which the lineshape is effected by this process is a function of the optical depth at line center, $\tau_0$, which is defined as the ratio of plasma source size to photon mean free path. Figure 7 shows two Ar Lyman-$\alpha$ profiles, one for $\tau_0 = 0$ and one for $\tau_0 = 5$.

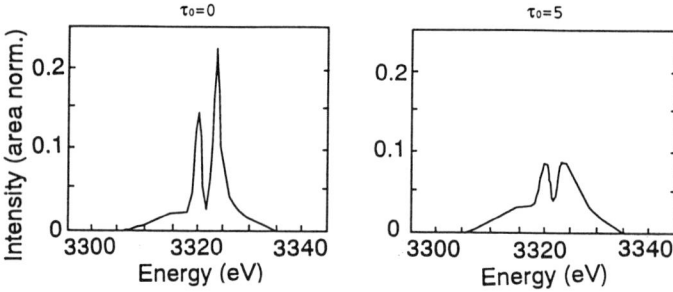

Figure 7. Effect of opacity on the Ar Lyman-$\alpha$ lineshape. $n_e = 1 \times 10^{24} \text{cm}^{-3}$, $kT$ =800eV.

In our analysis of experimental data we always include this effect.

In the following analyses of experimental data the theoretical requirements are demanding, especially the need to include perhaps thousands of atomic states [Cowan, 1981]. Our procedure [Woltz and Hooper, 1988; Mancini et al., 1991] for dealing with these complicated physical situations has the modular form as indicated below, where specialized codes produce the input for the line broadening code.

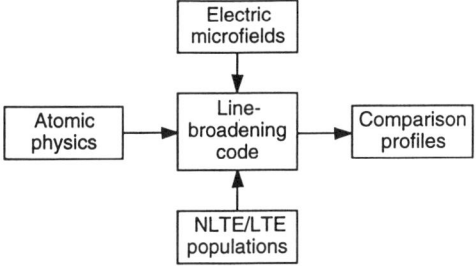

Figure 8. Diagram illustrating the inputs for MERL, the multi-electron radiator line broadening code.

## B. Experiments and analysis

All of the experiments that we will discuss in this section involve implosions of gas-filled microballoons performed at the University of Rochester Laboratory for Laser Energetics (LLE). Although these experiments generally involve a variety of diagnostics, our discussion will concentrate on theoretical analysis of time-integrated and time-resolved X-ray spectra.

Before we compare our theoretical spectra to experimental data, we convolve our theoretical profiles with an instrumental response function. Also, since our theory treats only bound-bound transitions, we must evaluate and subtract the background emission due to free-free and free-bound transitions from the experimental data. In addition to simple estimates, spectral analysis programs, such as ROBFIT [Coldwell and Bamford, 1991], exist which can be used to more accurately determine the background.

### 1. 1985 LLE experiments

The primary purpose of these experiments was to study the feasibility of using L-shell X-ray line emission as a density diagonostic. The laser used to perform these experiments was the 24-beam $\Omega$ system which has a wavelength of 3500Å, delivering 2000J in approximately 600ps. The targets imploded were plastic microballoons filled with an argon/krypton mixture (50/50); coated with a 500Å aluminum sealing layer. Time integrated spectra were recorded on two spectrographs, one dedicated to recording L-shell krypton spectra ($3 \to 2$) and the other to recording argon K-shell spectra. Theoretical analysis was performed using multi-electron radiator line-broadening theory. As a result of our analysis of the $n = 3$ to $n = 2$ transitions in the krypton spectra, we determined that these lines were not density sensitive in the $10^{23} \text{cm}^{-3}$ and $10^{24} \text{cm}^{-3}$ density range and hence were not useful as a density diagonostic.

However, in Figure 9, we see that the spectograph which was calibrated to record spectra from hydrogenic and helium-like argon lines also detected L-shell krypton line spectra due 4 → 2 transitions. These lines did prove to be density sensitive.

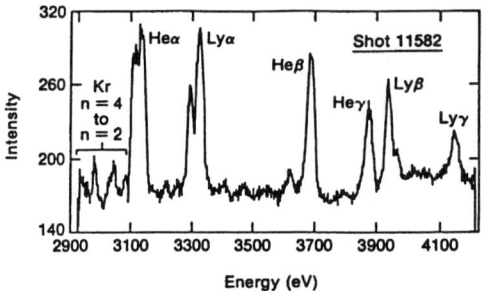

Figure 9. Experimental spectrum displaying both Ar K-shell lines and 4→2 Kr L-shell lines.

Figure 10 illustrates this density sensitivity for $n = 4$ to $n = 2$ transitions in Li-like krypton.

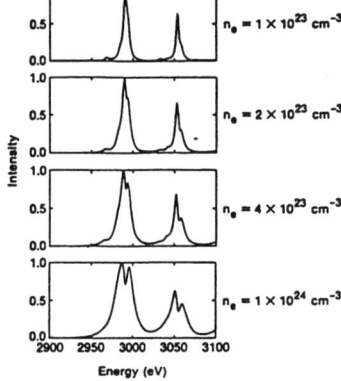

Figure 10. Theoretical results for Kr 4→2 transitions at various densities.

A comparison of our theoretical fit to the part of the spectrum in Figure 9 corresponding to transitions in Li-like and Be-like Ar is shown in Figure 11. The agreement with experiment is quite good except for some differences in intensity, especially in the higher energy range of the spectrum shown. The temperature (1.1keV) and electron density ($1.5 \times 10^{23} \text{cm}^{-3}$) inferences from this fit were consistent with those from our analysis of the Ar K-shell spectra also displayed in Figure 10. Hence we conclude that Stark broadening of L-shell X-ray spectra did indeed offer promise as a plasma-density diagnostic.

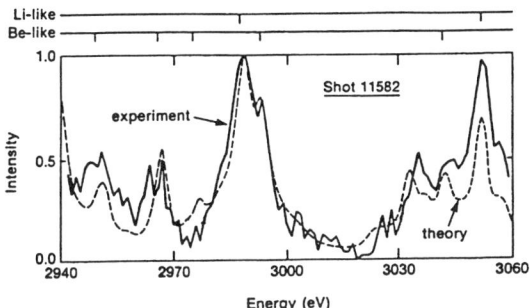

Figure 11. Comparison of theoretical spectrum to experimental spectrum of Figure 9.

2. 1987 LLE experiments

In this set of experiments, we attempted to extend and refine our 1985 experiments. The driver was again the 24-beam $\Omega$ system with a wavelength of 3500Å and an energy of 2000J delivered in approximately 600ps. One difference in these experiments was the use of a new beam-smoothing technique.

The targets were plastic microballoons each with nominal dimensions of 450 microns (outer diameter) × 6 microns (wall thickness) and a 1000Å aluminum sealing layer; gas fills of 2 and 10 atmospheres were employed. Some of the targets contained all argon gas fills while others included argon/krypton mixtures. Empty shells were also imploded. When we examine the recorded spectra resulting from the implosions we found that there was little or no line radiation coming from those shells containing argon/krypton mixtures. Spectra from shells containing only argon, however, did show both the hydrogenic and helium-like $\alpha$ lines which appeared to be greatly broadened; no higher series members were noted. We concluded that plasma densities resulting from the implosions of these targets were greater than those previously observed. The effect of the new beam-smoothing technique was apparently to produce cooler implosions. In addition, radiative cooling from highly ionized krypton lines prevented the temperatures in those experiments involving argon/krypton-filled microballoons from rising to a point where appreciable line radiation could take place. Analysis of spectra from the all-argon filled microballoons was of considerable interest from the point of view of Stark broadening, especially since for these experiments time resolved spectra were available.

Our discussion here will concentrate on three of the implosions involving argon-filled microballoons. We observe the He-$\alpha$ line of Ar XVII together with its Li-like satellites as well as Lyman $\alpha$ line of Ar XVIII plus its He-like satellites. At early times, the spectra observed from each of the three shots was similar to those from the 1985 experiments. In the first two shots the time-evolution of the spectra gave rise to line emission. In the third shot, which was produced under nominally the same conditions as the second shot, the time-evolution of the spectrum leads to reduced emission and a strong absorption feature on the low-energy side of the He-$\alpha$ line. Time resolved spectra from these three experiments are shown in Figure 12.

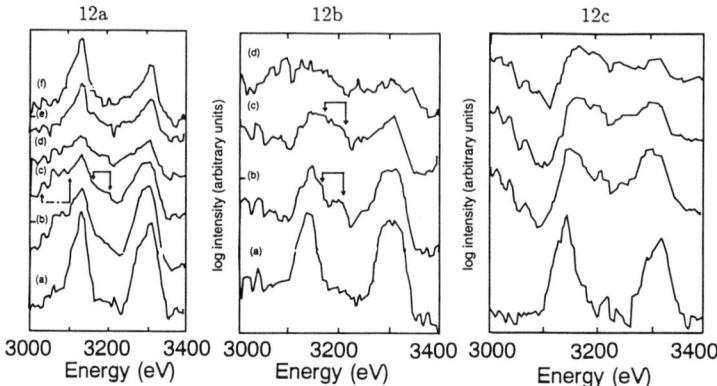

Figure 12. Three series of time-resolved lineouts from the 1987 LLE experiments. The lineouts have been shifted by an arbitrary amount along the y-axis for the sake of clarity, with the higher lineouts corresponding to later times in the shot.

In the first two shots, analysis of the spectra were complicated by the fact that the satellite lines merge with the resonance lines creating a composite spectral feature. Thus, our theoretical analysis involved fitting composite satellite/resonance features. As a result of our analysis of the first shot, we determined the time sequence of temperatures and densities indicated in Table I.

Table I. Temperature and density inferences from analysis of Figure 12a.

| Fig. 13a spectrum | Time interval (ps) | Electron density (cm$^{-3}$) | Temperature (eV) |
|---|---|---|---|
| (a) | 312–375 | $3-6 \times 10^{24}$ | 600–900 |
| (b) | 348–405 | $5-7 \times 10^{24}$ | 600–800 |
| (c) | 375–427 | $6-8 \times 10^{24}$ | 500–800 |

These temperatures and densities are consistent with results from modeling done using the LILAC code [Delettrez et al., 1991]. The electron densities that were achieved in this series of experiments were the highest that we have achieved at LLE. Figure 13 illustrates an example of our line fitting calculations.

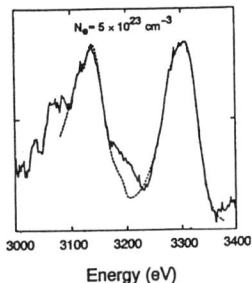

Figure 13. Comparison of theoretical spectrum to experimental spectrum of Figure 12a. The density is $5 \times 10^{24}$ cm$^{-3}$ and the temperature is 800eV.

Figure 14 illustrate the results of our analysis of the absorption features noted in the third shot where time-resolved diagnostics were employed to analyse the implosion of a microballoon containing an all argon gas fill.

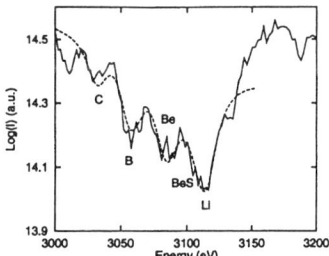

Figure 14. Three layer theoretical fir to experimental spectrum from Figure 12c.

Figure 14 illustrates the fitting of spectra using a three-layer absorption model, each layer characterized by a different temperature, density and layer thickness. One- and two-layer absorption models could not reproduce the details of the experimental spectrum, indicating the presence of temperature and density gradients in the absorbing medium [Mancini et al., 1990; Haynes et al., 1993; Mancini et al., 1994].

In summary, this set of experiments illustrated the attainment of high electron densities and our ability to analyze the resulting composite line structures in emission and absorption. The results of the third shot (Figure 14) indicated the existence of a hot plasma core surrounded by a cooler region which was responsible for the absorption. This picture was consistent with LILAC modeling predictions.

Finally, the blue satellite features of the He-$\alpha$ noted in Figure 12 are as yet unexplained. We speculate, however, that they may be due to enhanced and broadened two electron-one photon satellite transitions.

3. 1993 LLE experiments

In the fall of 1992, a series of experiments was performed at LLE with the objective of studying the relative importance of ion dynamics and opacity on observed lineshapes.

By varying the composition of the filled gas from 20 atm of DD with a trace of Ar to 20 atm of pure Ar, we were able to alter systematically the mix of perturbing ions. For cases where the concentration of Ar was small, the static ion approximation employed in the "standard model" becomes suspect, and we implemented the BID formalism [Boercker et al., 1982] for the inclusion of ion dynamic effects. When the concentration of Ar in DD is small, ion dynamic effects significantly modify the static-peak structure. As the relative concentration of Ar rises, the ion dynamics effects become less important, and the effects of opacity begin to significantly alter the lineshape. Since the effects of ion dynamics and opacity broadening primarily modify the peak lineshape structure, careful and consistent modeling of these two effects will be required for the analysis of the data from these experiments.

The 24-beam $\Omega$-laser system was used for these experiments, operating at 3500 Å, delivering approximately 1000J in 600ps and, again, employing beam smoothing techniques. The targets were plastic microballoons with nominal dimensions of 250 microns (diameter) × 6 microns (wall thickness) with a 1000Å aluminum sealing layer. The fill gas was a mixture of Ar and DD with a total pressure of 20 atm. The partial pressure of Ar varied from 0.01 atm to 20 atm (pure Ar).

The effects of changing the relative concentration of Ar to DD, and thus the effect of ion dynamics, is illustrated by comparing a Stark broadened Ar resonance transition occurring in a pure Ar plasma to the same Ar transition taking place in a DD plasma with only a trace of Ar. See Figure 15.

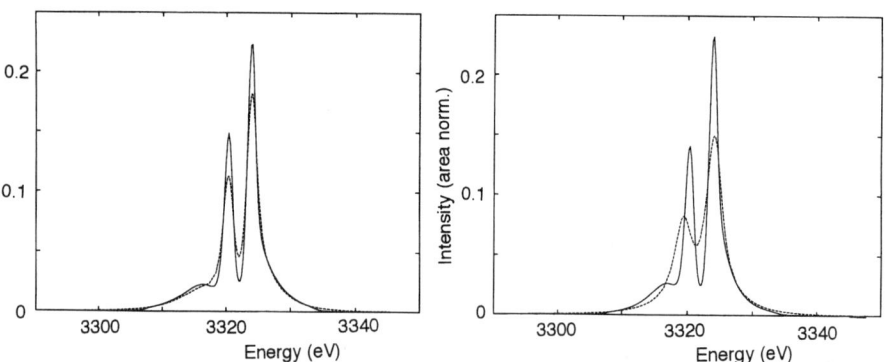

Figure 15. The effect of ion dynamics on the Ar Lyman-$\alpha$ lineshape. Pure Ar plasmas (left), 0.25% Ar in DD (right). $n_e = 1 \times 10^{24}$cm$^{-3}$, $kT$ =800eV; static calculation, solid line; dynamic calculation, dotted line.

Time resolved spectra were again recorded and multi-electron radiator line broadening theory was employed in the analysis of the resulting spectra [Haynes et al., 1995]. In Figure 16 we demonstrate the interplay of ion dynamics and opacity in our fit to spectral data from a shot with 0.1 atm Ar in 20 atm of DD. It is important to note that the inclusion of ion dynamics effects substantially decreases the inferred optical depth of the plasma source. The optical depth of the plasma inferred using the static ion approximation is not consistent with X-ray microscopic data, while that inferred using dynamic ion broadening can is.

The development of theoretical spectra to compare to this experimental data requires the use of NLTE population information (including the effect of radiative transfer), opacity effects, and ion dynamic Stark broadening calculations. The results presented here are only preliminary, we are currently performing a systematic analysis of the data.

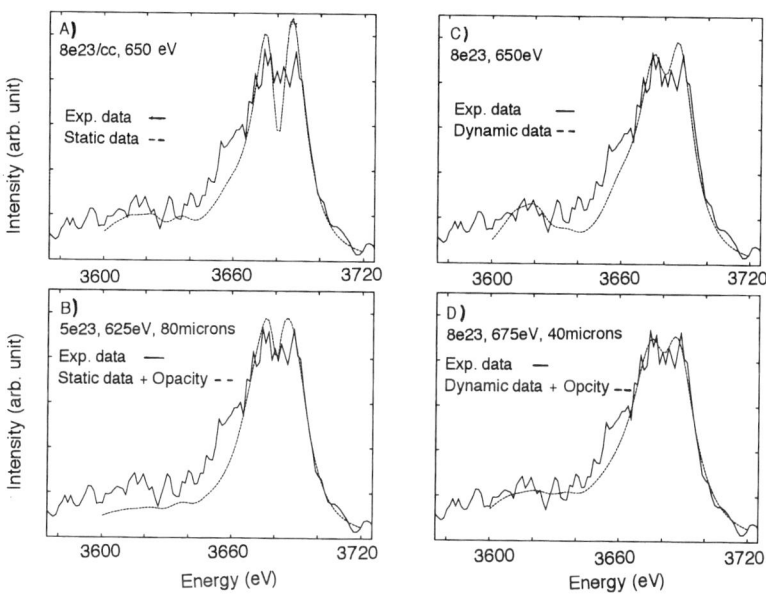

Figure 16. Experimental spectra and various theoretical fits to the Ar He-$\beta$ line and its Li-like satellites.

## III. CONCLUSION

The results of our comparisons of experimental X-ray spectra and theoretical calculations verify that plasma spectroscopy provides useful diagnostic tools to determine the plasma temperatures and densities that occur in these high power laser implosions. Spectroscopy also allows us to understand how the physical processes that occur in these plasmas influence the resulting spectra. With the new laser upgrades currently anticipated, or in progress, an even greater range of plasma densities and temperatures will be achieved in the future. It is highly probable that these plasmas will lead to new physics. With the continuing improvement of experimental diagnostics and theoretical tools, we anticipate that plasma spectroscopy will continue to be an indispensable tool in understanding such plasmas.

**REFERENCES**

Boercker, D.B. *et al.*, 1982, Phys. Rev. A., **36**, 2254.

Coldwell, R.L., and Bamford, G.J., 1991, *The Theory and Operation of Spectral Analysis using ROBFIT*, (New York, American Institute of Physics).

Cowan R.D., 1981, *The Theory of Atomic Structure and Spectra*, (Berkley and Los Angels, University of California Press).

Delettrez, J. *et al.*, 1991, in *Radiative Properties of Hot Dense Matter*, edt. Goldstein *et al.*, (Singapore, World Scientific), p. 309.

Griem, H.R., 1974, *Spectral Line Broadening By Plasmas*, (New York, Academic Press).

Haynes, D.A. *et al.*, 1993, Laser and Particle Beams, **11**, 205.

Haynes, D.A. *et al.*, scheduled for Jan 1995, Rev. Sci. Instrum.

Hooper, C.F., 1966, Phys. Rev., **149**, 177.

Hooper, C.F., 1968, Phys. Rev., **165**, 215.

Iglesias, C.A. *et al.*, 1983, Phys. Rev. A, **28**, 1667.

Mancini, R.C. *et al.*, 1990, Rec. Sci. Instrum, **61**, 2378.

Mancini, R.C. *et al.*, 1991, Comp. Phys. Comm., **63**, 314.

Mancini, R.C. *et al.*, 1994, J. Quant. Spectrosc. Radiat. Transfer, **51**, 201.

Mihalas, D., 1978, *Stellar Atmosphere*, (New York, W.H. Freeman).

Woltz, L., and Hooper, C.F., 1988, Phys. Rev. A, **38**, 4766.

# X.4

# ELECTRON TRANSPORT PHENOMENA AND DENSE PLASMAS PRODUCED BY ULTRA-SHORT-PULSE LASER INTERACTION

Richard M. More

Lawrence Livermore National Laboratory
Livermore, California 94550, U. S. A.

Recent experiments with femtosecond lasers provide a test bed for theoretical ideas about electron processes in hot dense plasmas. We briefly review aspects of electron conduction theory likely to prove relevant to femtosecond laser absorption. We show that the Mott-Ioffe-Regel limit implies a maximum inverse bremsstrahlung absorption of about 50% at temperatures near the Fermi temperature. We also propose that sheath inverse bremsstrahlung leads to a minimum absorption of 7–10% at high laser intensity.

## I. FEMTOSECOND LASER AND TRANSPORT PROCESSES

A new generation of high power short-pulse lasers was made possible by the technique of chirped pulse amplification [Strickland and Mourou, 1985]. These lasers generate pulses in the femtosecond range, typically 100 fsec, and can be used to prepare a well-characterized solid-density plasma. Recent femtosecond laser experiments give data concerning electron transport in hot dense matter, data valuable precisely because of the simplicity of the experimental conditions.

Electron transport, which we take to include AC electrical conduction as well as heat conduction, is a key to understanding the laser interaction. The AC conductivity determines the absorption and controls how much energy is deposited in the target. Heat conduction determines how far the energy spreads and this determines the temperature.

Typical parameters for a femtosecond pulse laser are: total laser energy, 0.1 Joule per pulse; pulse length, 100 fsec; power, 1 terawatt; focal spot diameter as low as 10 microns; laser intensity, $10^{12}$–$10^{18}$ Watts/cm$^2$; wavelength in the visible and repetition rate above one shot per second. Most crucial to control of the target interaction is the contrast between peak power in the pulse and the background laser emission ("prepulse"); for a well-adjusted laser this contrast can be better than $10^4$, and after harmonic conversion the contrast improves to $10^8$.

As we understand it, the laser interacts with the target over a skin depth of 100 Å. The absorbed energy is carried into the target to a depth of $\sim$ 500 Å during the laser pulse by an electron thermal wave. At high intensity, the target expands as much as 100 Å during the laser pulse, a velocity of 100 km/sec. At low intensity there is virtually no expansion during the heat pulse. The peak electron temperature ranges from 1 eV at low intensities to 1000 eV at the highest intensities. The absorbed energy is transformed into electron thermal energy, ionization energy, ion thermal energy, kinetic energy of hydrodynamic motion, magnetostatic energy and x-rays emission.

Femtosecond laser targets are not in thermal equilibrium. The electron temperature is higher than the ion temperature. The ionization is usually less than equilibrium (LTE) or even than a steady collisional-radiative non-LTE ionization state and the radiation is much less than an equilibrium (black-body) distribution. However the collision rates probably suffice to bring free electrons to a local equilibrium distribution.

Femtosecond laser experiments explore the transition from solid state to plasma and the transport phenomena must be examined from viewpoints traditionally associated with these two distinct phases of matter.

## II. HOW DOES SOLID MATTER BECOME A PLASMA?

The ideal solid is translationally invariant and its elementary excitations are labelled by wave-vector $k$ or crystal momentum $p$. The atomic vibrations are travelling sound waves or phonons. Electron excitations are described in terms of energy bands and Bloch functions. Inner-shell electrons are associated with narrow bands of propagating states, and even electrons far above the Fermi level can be Bragg-reflected and therefore have energy gaps at the Brillouin zone boundaries. Bragg-reflection gives the electrons an effective mass $m^*(k)$ which can be significantly different from the usual electron mass.

The dense plasma is disordered; there is short-range ion correlation but no long-range order. Because of this phonons no longer propagate very well. Free electrons have energies defined in $k$-space, but bound electrons are localized on specific ions. The ions have definite charge states whose probabilities are determined by a Saha equation. The electrons make transitions corresponding to excitation and ionization processes, but the zero-order Hamiltonian is diagonal in a representation which assigns integer charge to each ion rather than one which assigns a definite momentum to each electron.

How do solid-state energy levels evolve into those of the plasma? The effect of temperature, aside from disordering the ion positions, is to excite electrons to states of higher energies. Quite generally, excited electrons (bound or free) are less able to screen the nuclear charge than they were in states of lower energy. This means the self-consistent electrostatic potential grows stronger as electrons begin to be thermally excited. In turn, this increases the spacing of energy levels and moves lines to shorter wave-lengths. What is the change in the electron band-gaps?

It seems clear that as temperature rises the core bands become very narrow, both as a result of the stronger potential and as a consequence of disorder, which spoils the resonant hopping required for propagating states. However, band gaps at energies above the original Fermi energy evidently decrease with loss of Bragg scattering even if the potential ultimately becomes stronger. This is seen in the optical properties of liquid metals [Faber, 1972]. In particular, conduction electrons in liquid metals (and presumably in hotter plasmas) respond with essentially the free-electron mass.

There remain questions how ionic and covalent materials behave as their temperature is raised. To a certain extent one can make predictions based on the room-temperature band-structure. On the high-temperature side one has the usual plasma picture. In the range between , there are various possibilities of charge transfer, excitonic phases, etc. For example, room temperature glass consists of $Si(4+)O(2-)$ ions.

Partway to the plasma state, there will doubtless be some free electrons, i.e., electrons in 3s, 3p conduction bands, but we do not know how the core charge states vary.

## III. ELECTRON CONDUCTION THEORY

The classical theory of Brownian motion shows the connection between diffusion and frictional drag forces which determine the mobility ($\mu$ = velocity/force) for an electron. Already this simple theory gives an expression for the AC conductivity of nondegenerate electrons,

$$\sigma(\omega) = \frac{ne^2}{m} \left\langle \frac{\tau}{1 - i\omega\tau} \right\rangle \tag{1}$$

where $n$ = electron number density, $e$ = electron charge, $m$ = electron mass, and $\tau = m\mu$ = electron collision time. The average is taken over the velocity distribution when $\tau$ depends on electron energy.

In the 1930's Sommerfeld and Frank studied electron transport in simple metals using the Boltzmann equation with Fermi-Dirac statistics for the unperturbed distribution function. This theory also gives Eq. (1) for degenerate or nondegenerate electrons (the average is weighted by $Edf/dE$, the derivative of the equilibrium distribution function).

When we use Eq. (1) to calculate absorption, we must solve Maxwell's equations for the evanescent laser wave penetrating into the target over a skin depth $\delta = c\omega_p$. This calculation gives a useful qualitative formula for the absorbed energy fraction [More et al., 1988]:

$$A = 2/\omega_p \tau . \tag{2}$$

Here $\omega_p$ is the electron plasma frequency and $\tau$ is an average collision time (strictly, $1/\tau = \langle 1/\tau(E) \rangle$ is the appropriate average). Equation (2) describes steady-state absorption of a homogeneous laser beam by an idealized homogeneous target, so $\omega_p$ and $\tau$ do not depend on position, and applies in the high-frequency limit $\omega\tau \gg 1$ (this is the case at high target temperatures). The target is assumed to have a sharp interface and to be overdense to the laser ($\omega \ll \omega_p$). Equation (2) should be corrected for hydrodynamic expansion of the target, temperature gradients, the spatial beam profile and the time-dependence of laser energy and target temperature. These corrections are straightforward aspects of a numerical simulation.

Another theory of laser absorption is given by the radiative kinetic equation which describes inverse bremsstrahlung transitions (absorption and stimulated emission). It is not difficult to show, using the Kramers cross-section for radiative rates, that the radiative kinetic equation gives the same result as Eq. (1) for conditions $\omega\tau \gg 1$ when $\hbar\omega \ll kT$ (non-degenerate electrons) or $\hbar\omega \ll E_f$ (degenerate electrons, discussed further below). The agreement with Eq. (1) requires that we use a high-frequency Coulomb logarithm, $\ln \Lambda = \pi/\sqrt{3}$ in the evaluation of Eq. (1). For lower frequencies ($\omega\tau < 1$) one must correct the radiative kinetic equation to include interference in electron-ion multiple-scattering, [Zel'dovich et al., 1966] and then one obtains again Eq. (1). At high frequencies ($\hbar\omega \gg kT$ or $\hbar\omega \gg E_f$), the radiative kinetic equation gives a different answer discussed below.

Starting with a quantum equation of motion for the density matrix, Kohn and Luttinger [Kohn and Luttinger, 1958] derived a transport theory for nondegenerate electrons. The equation of motion is expanded with respect to a DC applied electric field (linear response) and also with respect to the perturbation due to randomly located ions. When averaged over ion positions, the diagonal elements of the density matrix correspond to the Boltzmann distribution function and are larger, in a certain sense, than the off-diagonal elements. The diagonal elements are governed by an equation essentially equivalent to the Boltzmann equation.

Kubo recast the density-matrix equation of motion to obtain a general formal expression for the electrical conductivity [e.g., Kubo, 1957; Kubo, 1959], relating the dissipation produced by Joule heating to fluctuations in the spontaneous currents existing in thermal equilibrium in the absence of the applied field ("Fluctuation-dissipation theorem").

Another systematic theory of conduction phenomena is given by the Green's function perturbation theory [e.g., Edwards, 1958; Langer, 1960, Abrikosov et al., 1963]. This theory reproduces all the previous results and is readily extended to the BCS superconducting state. Disordered materials with strong scattering pose a problem of high-order perturbation theory; one method for treating this case is the coherent potential approximation [Soven, 1967].

In research on heavily doped semiconductors a rule has emerged, known as the Mott-Ioffe-Regel limit, to the effect that the minimum electron mean free path is approximately the atomic spacing [e.g. Mott, 1966; Mott and Davis, 1971]. With this rule goes a minimum electrical conductivity and an explanation for any smaller values of the conductivity: that would require localized electrons unable to propagate. Lee and More [1984] propose this limit applies to laser plasmas at temperatures of $kT \cong E_f$ because these are the conditions of maximum electron-ion scattering. (Localized conduction electrons are very unlikely in a high-temperature system.)

The condensed-matter theories mainly discuss electron scattering by screened ion potentials. In plasma physics the Coulomb potential gives a long-range interaction which is traditionally treated by the Fokker-Planck approximation, applied by Spitzer and Harm [1953], Rosenbluth et al. [1957], and others. For dense plasmas the Coulomb collisions can also be handled by the method of correlation functions, i. e., the Ziman formula.

The Ziman formula shows how interference between scattering on different centers limits the electron mean free path. In particular the Ziman formula gives the small-angle cutoff required for a finite Coulomb logarithm [Hubbard and Lampe, 1969]. The simplest version of this formula is

$$\frac{1}{\tau} = n_i v \frac{\pi}{k^4} \int_0^{2k} q^3 \frac{d\sigma}{d\Omega} S(q) dq \ . \tag{3}$$

Here $\hbar k$ = electron momentum, $\hbar q$ = momentum transferred in the collision, $S(q)$ = ion structure factor, $d\sigma/d\Omega$ = electron-ion differential scattering cross-section. The formula is derived with the Born approximation for this cross-section, but the Born approximation is rarely justified except for fully-ionized or low-$Z$ plasmas. When a higher-order cross-section is used one has kept some terms in a multiple-scattering series (and ignored others). For many years it has been known that an expression like Eq. (3) should be used for inverse bremsstrahlung absorption [Ichimaru, 1973].

Much of the recent literature on transport in strongly coupled plasmas consists of evaluating the Ziman formula with various approximations for the structure factor $S(q)$, and various treatments of the electron screening of the ion potential. In this context we mention work of Ashcroft and Schaich [1970], Hansen et al. [e.g., Minoo et al., 1976; Bernu and Hansen, 1982; Baus et al., 1981], DeWitt et al. [Rogers et al., 1981], Ichimaru et al. [e.g. Ichimaru and Tanaka, 1985; Iyetomi et al., 1992], and Perrot and Dharma-Wardana [e.g. Perrot and Dharma-Wardana, 1987; Dharma-Wardana and Perrot, 1992]. Because most of these authors are present to describe their own work, we do not need to summarize these important developments further.

## IV. SOME POINTS TO DISCUSS

A. Momentum transfer cross-section

To obtain the collision time $\tau$ which appears in transport theory the electron-ion differential cross-section must be weighted by $(1 - \cos\theta)$, where $\theta$ is the scattering angle. The importance of collisions is determined by the momentum-loss they cause.

The factor $(1 - \cos\theta) = q^2/2k^2$ appears explicitly in the Ziman formula. For Brownian motion this effect emerges from a treatment of correlated random walks. In the Green's function theory, it appears as the difference between the average of a product of Green's functions and the product of the averages [e.g. Edwards, 1958; Langer, 1960; Abrikosov et al., 1963]. Kubo derives this effect from a memory-function in his Colorado lectures [Kubo, 1957; 1959]. All transport theories include this effect, so it is not a difficult question.

B. Landau-Peierls' effect

The intuitive derivation of the Boltzmann equation seems to fail under conditions where $\hbar/\tau > kT$, because in that case one cannot defend the picture of a small wave-packet traveling between collisions. The inequality is satisfied in low-intensity femtosecond laser interactions for many materials. Theoretical analyses from various viewpoints [e.g., Kohn and Luttinger, 1957; Kohn and Luttinger, 1959] have shown there is no Landau-Peierls' effect, i.e., no quantum correction to Eq. (1) when $\hbar/\tau > kT$. The nature of the transport in these circumstances is close to Brownian motion of large electron wave-packets and we began with the observation that Eq. (1) governs this case.

C. Quantum AC transport

The question here concerns high frequencies, $\hbar\omega > kT$. The Kubo formula and the radiative kinetic equation contain a factor $1 - \exp(-\hbar\omega/kT)$ in which the first term (unity) describes absorption and the second term describes stimulated emission. Equation (1) has no such factor. Does this mean the Boltzmann equation fails?

For nondegenerate electrons the quantum theories give a conductivity equal to a factor $(kT/\hbar\omega)[1 - \exp(-\hbar\omega/kT)]$ times the result in Eq. (1). When $\hbar\omega \ll kT$ this factor reduces to unity and we have again Eq. (1).

For degenerate electrons, the quantum theories give a more complicated result, and it is quite remarkable that this agrees with Eq. (1) up to photon energies of the order of the Fermi energy, much larger than $kT$. It is well-known that Eq. (1) gives a good description of liquid metals interacting with visible light [Faber, 1972]. Thus the quantum correction only affects the ultra-violet or soft x-ray absorption.

D. Electron-electron collisions

The greatest technical difficulties of conduction theory are associated with electron-electron collisions. In the Boltzmann transport theory one has a nonlinear integro-differential equation for the perturbed part of the distribution function. The equation simplifies to a Fokker-Planck approximation, expressed by the Lenard-Balescu collision operator, but still requires a nontrivial numerical calculation.

Fortunately we are mainly interested in target plasmas having high ion charge states—for example, aluminum has $Z^* = 11$ for intense laser pulses. In this case the relative importance of electron-electron scattering is small, proportional to $1/Z^*$, and we simply neglect electron-electron collisions in view of the uncertainties in other aspects of the modelling.

Is the electron-electron interaction even a $1/Z^*$ correction? In degenerate matter it is well-known that e-e collisions are inhibited by the exclusion principle (two electrons within $kT$ of the Fermi level must find final states also in this narrow energy range).

There is also the subtlety that when electrons carry a given current and a pair of electrons collide, their individual momenta change, and the distribution function $f(v)$ changes, but there is no change in the total current of the electrons. Electron-electron collisions have an indirect effect, for example, moving an energetic electron down to lower energies where its next electron-ion collision will happen sooner. Even this effect is partly cancelled because, while reducing the energy of one electron, the electron-electron collision raises the energy of the collision partner.

Thus the high-energy part of the distribution is affected by electron-electron collisions, and while this plays an important role in the DC conductivity and/or the heat conduction, Eq. (1) shows that the high-frequency AC conductivity is less sensitive to high-energy electrons and this is another reason to omit e-e collisions from a laser absorption model.

E. Minimum mean free path

The range of temperatures near the Fermi energy represents the boundary between solid-state and plasma physics. In this case the electron mean free path must be very short because one has strong lattice disorder and the conduction electrons have enough excitation that they can lose energy by exciting other electrons from the Fermi sea. Even in cold matter, electrons with energies a few eV above $E_f$ have mean free paths no more than a few atomic diameters. At higher temperatures electrons have reduced Coulomb cross-sections but this effect is not yet significant at temperatures near $E_f$.

It is also evident that despite the strong disorder electrons are not localized or trapped because they are constantly exposed to small energy transfers from other free electrons. Thus we expect the collision time to be essentially that associated with a mean free path of $2R_o$ (i.e., one atomic diameter) [Lee and More 1984]

$$\tau = 2R_o/v_F . \tag{4}$$

If we combine this approximation with Eq. (2) for the absorption, we find a maximum inverse bremsstrahlung absorption,

$$A_{\max} = 1.407(a_0^3 n)^{1/6}(Z^*)^{-1/3} \tag{5}$$

where $a_o = 0.529 \times 10^{-8}$ cm is the Bohr radius, and the numerical value for aluminum is $A_{\max} = 53.3\%$ (assuming a valence of 3).

The weak dependence on electron density is the most striking feature of Eq. (5). If there is no other absorption mechanism, Eq. (5) predicts a maximum absorption of 50% at temperatures near the Fermi temperature for all target materials. (It is assumed that insulators break down to a metallic state with a well-defined electron density and Fermi temperature.) At higher and lower temperatures the mean free path is longer and the absorption is lower.

Equation (5) is not a rigorous limit on the inverse bremsstrahlung absorption because of the corrections discussed in connection with Eq. (2). However any material showing absorption significantly above Eq. (5) is probably showing an additional absorption mechanism (such as line absorption).

F. Nonlinear phenomena

Whereas the Kubo formula claims to give the exact linear response conductivity, when we think about laser plasmas we rapidly leave the linear regime. The AC electric field of the laser is of the order of a billion volts per centimeter and the temperature gradient is more than a kilovolt per micron or $10^{11}$ Kelvin per centimeter. Even referred to voltage or temperature changes over one mean free path these numbers are large enough to cast doubt on the application of linear-response theory.

One specific nonlinear-response phenomenon is the Silin effect: at high laser intensity, the electron quiver velocity is larger than the thermal velocity and should replace it in the collision cross-sections (leading to a smaller collision rate). There is every reason to believe this effect occurs in high-power interaction with low-density plasmas. For femtosecond laser interactions, however, the incident laser field is partly cancelled by a reflected wave leading to a smaller electric field, and smaller quiver velocity, and the high target density implies a high collision rate, so the quiver energy is thermalized and the thermal velocity keeps up with the quiver velocity.

The Fourier law for heat conduction breaks down in very high temperature gradients, another nonlinear response phenomenon. It is replaced by a nonlocal conduction process which usually implies a maximum heat current, $q = fnvkT$, where $n$ = electron density, $v$ = electron thermal velocity, $T$ = electron temperature and $f$ is a constant which should be 0.6 on the Knudsen model (free streaming electrons). There is an extensive literature based on modeling laser-plasma experiments with inhibited heat conduction, meaning values of $f$ as low as 0.05. For the moment the femtosecond laser experiments seem to be compatible with the theoretical value, $f = 0.6$.

A related issue is the possible existence of non-Maxwellian electron distributions. These can be caused by atomic processes (depletion of energetic electrons by ionization) as well as by the laser absorption itself or by escape of the most energetic electrons. Because the Coulomb cross-section depends strongly on electron energy, a depletion of the high-energy portion of the distribution would strongly affect the heat transport coefficients. However in femtosecond laser plasmas, the laser absorption pushes the distribution out of equilibrium only over the skin depth while thermalization can occur throughout the plasma heated by the electron thermal wave.

Finally we mention instability effects, for example, a mechanism by which electrons generate ion sound waves when they carry a large enough current to have a supersonic drift velocity. Such effects have a threshold and are not included in the linear-response formalism.

G. Geometric effects

It has been suggested that the anomalous skin effect could occur in femtosecond laser plasmas if the electron mean free path were to exceed the skin depth. This could, hypothetically, occur at high laser intensities where the target temperature reaches 1 keV in aluminum. However the criterion for the anomalous skin effect is two-fold: the ratio of mean free path to skin depth must exceed unity but also must exceed $\omega\tau$. This second condition is very hard to satisfy in a femtosecond laser-heated target, and for this reason we are skeptical that the anomalous skin effect occurs in these plasmas.

The author has proposed a different surface-related absorption mechanism, Sheath Inverse Bremsstrahlung, which generalizes the surface-assisted absorption mechanism of Holstein. This mechanism applies for very high temperatures (1 keV) on targets which still have a sharp interface with the vacuum, and which may again correspond to conditions difficult to achieve. For the ideal sharp interface, the theory gives an extra absorption

$$A_{\text{sib}} = 1.6v/c . \tag{6}$$

Here $v = (kT/m)^{1/2}$ is the average electron thermal velocity. The formula predicts as much as 10% absorption at keV temperatures.

Finally, it should be mentioned that the laser focal spot is almost certainly surrounded by a large toroidal magnetic field. While preliminary estimates show this field does not greatly alter the conductivity responsible for absorption of the laser, more thorough study of magnetic effects is certainly required.

## V. PRACTICAL THEORY OF CONDUCTION PHENOMENA

The comprehensive conductivity model of Lee and More [1984] combines a semi-empirical solid/liquid conductivity, the high-temperature Spitzer result, and a minimum conductivity obtained from Eq. (4). It is in general agreement with several laser reflectivity experiments, when combined with appropriate modelling of the hydrodynamics [e.g. Ng *et al.*, 1986; Milchberg *et al.*, 1988; Ng *et al.*, 1994]. It is essential that the AC conductivity be calculated from the DC model by Eq. (1) as written to obtain good agreement with experiment. The new experiments with shorter laser pulses and higher intensities will give much more conclusive tests of all these ideas about conduction phenomena, just because of the reduced importance of the hydrodynamic corrections to the raw data.

It remains to mention several other comprehensive practical models for conduction coefficients developed by Rinker [1985; 1988], Drska and Vondrasek [1989], and Kalitkin and Ermakov [1979].

## ACKNOWLEDGEMENT

The theoretical ideas presented here have been greatly strengthened by an inspection of preliminary experimental data obtained by D. Price and R. Shepherd at the Livermore Ultra-Short Pulse laser facility, and by LASNEX calculations performed in collaboration with R. Walling and E. Alley. The author has also benefitted from discussions with W. Rozmus concerning Eq. 1. This work was supported under the auspices of the United States Department of Energy through the Lawrence Livermore National Laboratory under contract number W-7405-Eng-48.

## REFERENCES

Abrikosov, A., L. Gorkov, and I. Dzyaloshinski, 1963, *Methods of Quantum Field Theory in Statistical Physics*, (Prentice-Hall, Engelwood Cliffs).
Ashcroft, N. and W. Schaich, 1970, Phys. Rev. **B1**, 1370.
Baus, M., J.-P. Hansen, and L. Sjogren, 1981, Phys. Lett. **82A**, 180.
Bernu, B., and J.-P. Hansen, 1982, Phys. Rev. Lett. **48**, 1375.
Catto, P. and R. More, 1977, Phys. Fluids **20**, 704.
Dharma-Wardana, M. and F. Perrot, 1992, Phys. Letters A **163**, 223.
Drska, L. and J. Vondrasek, 1989, Laser and Particle Beams **7**, 237.
Edwards, S., 1958, Phil. Mag. **3**, 33, 1020
Ermakov, V. and N. Kalitkin, 1979, Sov J. Plasma Phys. **5**, 365.
Faber, T. E., 1972, *Theory of Liquid Metals*, (Cambridge Univ. Press, Cambridge).
Hubbard, W. and M. Lampe, 1969, Ap. J. Suppl. **18**, 297.
Ichimaru, S., 1973, *Basic Principles of Plasma Physics*, (Benjamin, Reading, Mass.).
Ichimaru, S. and S. Tanaka, 1985, Phys. Rev. **A32**, 1790;
Iyetomi, H., S. Ogata, and S. Ichimaru, 1992, Phys. Rev. **A46**, 1051.
Kohn, W. and J. Luttinger, 1957, Phys. Rev. **108**, 590; 1958, Phys. Rev. **109**, 1892.
Kubo, R., J., 1957, Phys. Soc. Japan **12**, 570.
Kubo, R., 1959, in *Lectures in Theoretical Physics, vol. I*, Ed. by W. E. Brittin and L. G. Dunham, (Interscience Publishers, New York).
Langer, 1960, J., Phys. Rev. **120**, 714
Lee, Y. and R. More, 1984, Phys. Fluids **27**, 1273.
Minoo, H., C. Deutsch, and J.-P. Hansen, 1976, Phys. Rev. **A14**, 840
More, R. M., Z. Zinamon, K. H. Warren, R. Falcone, and M. Murnane, 1988, Journal de Physique **49**, C7-43.
Mott, N. F., 1966, Phil. Mag. **13**, 989.

Mott, N. F., and A. Davis, 1971, *Electronic Processes in Non-Crystalline Materials*, (Oxford Univ. Press, Oxford).

Milchberg, H., R. Freeman, S. Davey, and R. More, 1988, Phys. Rev. Lett. **61**, 2364.

Ng, A., P. Celliers, A. Forsman, R. More, Y.-T. Lee, F. Perrot, M.W.C. Dharma-Wardana, and G. Rinker, 1994, Phys. Rev. Lett. **72**, 3351.

Ng., A., D. Parfeniuk, P. Cellies, L. DaSilva, R. More, and Y.-T. Lee, 1986, Phys. Rev. Lett. **57**, 1595.

Perrot, F. and M. Dharma-Wardana, 1987, Phys. Rev. A **36**, 238.

Rinker, G., 1985, Phys. Rev. **B31**, 4207.

Rinker, G., 1988, Phys. Rev. **A37**, 1284.

Rogers, F., H. DeWitt, and D. Boercker, 1981, Phys. Lett. **82A**, 331.

Rosenbluth, M., W. MacDonald, and D. Judd, 1957, Phys. Rev. **107**, 1.

Soven, P., 1967, Phys. Rev. **156**, 809.

Spitzer, L. and R. Harm, 1953, Phys. Rev. **89**, 977.

Strickland, D. and G. Mourou, 1985, Opt. Commun. **56**, 219.

Zel'dovich, Ya. B. and Yu. Raizer, 1966, *Physics of Shock Waves and High-Temperature Hydrodynamic Phenomena*, (Academic Press, New York).

# X.5

# INTERNAL STRUCTURE OF A COMPRESSED ATOM AND MOLECULE IN DENSE PLASMAS

Y. Furutani and A. Fukuyama

Department of Electrical and Electronic Engineering, Okayama University
Okayama 700, Japan

Internal structure of an atom and of a diatomic molecule compressed by a surrounding dense plasma is investigated within the framework of the statistical model of atoms. We also propose an elementary two-temperature model of atoms and compare it with a one-temperature model.

## I. INTRODUCTION

History of the statistical model of atoms began at the second half of 1920's [e.g., Thomas, 1927; Fermi, 1927]. For long, the field of atomic physics seemed to be less attractive than many other fields of physics, e.g., solid-state physics, nuclear and plasma physics, astrophysics and quantum electronics. Since the advent of inertially confined fusion plasmas produced by high-power laser and/or high-energy particle beams, vivid interest was renewed on the problem of interaction of radiation fields with a dense matter. In this problem, a precise, quantitative knowledge of the internal structure of a compressed, partially ionized high-Z atom or molecule is required to estimate the opacity, the photo-ionization cross section, the line spectra and so forth. To carry out a self-consistent calculation of these quantities, we should pursue a first-principles approach via the many-electron Schrödinger or Hartree-Fock equation. This is a herculean task, even with the aid of a most powerful modern supercomputer, in view of the intricate correlations existing among electrons. For this reason, a series of our works relies mainly upon the Thomas-Fermi-Dirac-Weizsäcker (TFDW) statistical model, based on the density functional theory (DFT) [Hohenberg and Kohn, 1964; Kohn and Sham, 1965; Mermin, 1965]. The internal structure of a compressed single atom and diatomic molecule is thus entirely determined by boundary conditions we impose from one case to the other. We outline below three models studied by Okayama group.

We use the atomic units throughout this article, unless otherwise specified.

## II. SINGLE-CENTER MODEL

By single center model (SCM), we mean an isolated or a compressed atom carrying an effective charge $z$, where $z = Z - N$, with $Z$ a charge of the nucleus and $N$ a number of bound electrons confined within the ion. The time-honored TFDW statistical model includes the electron exchange and correlation effects in the local density approximation and the Weizsäcker gradient correction to the Thomas-Fermi-Dirac (TFD) form of the kinetic energy.

The grand canonical thermodynamic potential $\Omega[n]$ as a functional of the electron density is given by [Perrot, 1979]

$$\Omega[n] = F_{\text{KG}}[n] + F_{\text{XC}}[n] - \int d\vec{r}\, \frac{Z}{r} n(r) + \frac{1}{2} \iint d\vec{r}d\vec{r}\,'\, \frac{n(r)\,n(r')}{|\vec{r} - \vec{r}\,'|} - \mu \int d\vec{r}\, n(r) \quad (1)$$

where the suffixes K, G, X and C denote, respectively, kinetic, gradient correction, exchange and correlation. $\mu$ is the Lagrange's multiplier. We associate to each $F$ the energy density $\mathcal{F}[n]$ per unit volume. Following the DFT, a correct density profile $n(r)$ satisfying appropriate boundary conditions minimizes $\Omega[n]$, with temperature $T$ kept constant. Performing the functional differentiation, we obtain from eq.(1)

$$\frac{\delta \mathcal{F}_K}{\delta n} - \frac{\delta}{\delta n}\left(\frac{h}{n}\right)(\vec{\nabla} n)^2 - 2h\frac{\nabla^2 n}{n} + \frac{\delta \mathcal{F}_{XC}}{\delta n} - U(r) = \mu \, . \tag{2}$$

Eq.(2) was derived, using the boundary condition: $(dn/dr)_{r=R} = 0$, with $R$ the ion radius. The kinetic energy density $\mathcal{F}_K$ including the entropy contribution is given by

$$\mathcal{F}_K[n] = \frac{\sqrt{2}}{\pi^2}\beta^{-5/2}\left[-\frac{2}{3}I_{3/2}(\alpha) + \alpha I_{1/2}(\alpha)\right] \tag{3}$$

where the parameter $\alpha$ is related to a local density $n(r)$ through the relation

$$I_{1/2}(\alpha) = \frac{\pi^2}{\sqrt{2}}n\beta^{3/2} \tag{4}$$

with $I_\nu(\alpha)$ the Fermi function defined as

$$I_\nu(\alpha) = \int_0^\infty dx \frac{x^\nu}{e^{x-\alpha}+1} \, . \tag{5}$$

$U(r)$ is the Hartree potential (aside from the sign) defined as

$$U(r) = \frac{Z}{r} - \int d\vec{r}' \frac{n(r')}{|\vec{r}-\vec{r}'|} \, . \tag{6}$$

The gradient correction factor $h[n]$, due to Perrot [1979] and often criticized as not to be a unique choice, is explicitly given by

$$h[n] = -\frac{\sqrt{2}}{24}\pi^2 n\beta^{3/2}\frac{\partial}{\partial \alpha}\left[\frac{1}{I_{-1/2}(\alpha)}\right], \quad \beta = \frac{1}{k_B T} \, . \tag{7}$$

Since there are two unknown functions $n(r)$ and $U(r)$, we require another equation independent of eq.(2), which is the Poisson's equation

$$\nabla^2 U(r) = 4\pi n(r) \, . \tag{8}$$

As for the exchange-correlation energy density $\mathcal{F}_{XC}[n]$, we borrow the fitting formula, devised by Tanaka, Mitake and Ichimaru [1985], which is arranged in the form of the Padé approximant so as to reproduce the Singwi-Tosi-Land-Sjölander scheme [1968]. Their formula is believed to go beyond the RPA expression. Eqs.(2) and (8) constitute the basic set of equations for the SCM. When they are applied to an isolated ion, we require *five* boundary conditions, since elimination of $n(r)$ between them gives rise to the differential equation of fourth order for the screening function defined as

$$\frac{Z}{r} \phi(r) = U(r) + \mu + \frac{1}{2\pi^2} . \qquad (9)$$

They are quoted as

$$\phi(0) = 1, \quad n(R) = \bar{n}, \quad n'(0) = -18Zn(0), \quad n'(R) = 0 \qquad (10\text{a} \sim \text{d})$$

and

$$\frac{z}{Z} = \phi(R) - R\phi'(R) . \qquad (10\text{e})$$

The prime means the derivative with respect to the argument. $\bar{n}$ is a constant density of a surrounding plasma. At the ground state $(T = 0)$, the TFDW equation (2) reduces to

$$\frac{1}{2} (3\pi^2 n)^{2/3} + \left.\frac{\delta \mathcal{F}_{\text{XC}}}{\delta n}\right|_{T=0} + \frac{1}{72} \left[ \left(\frac{\vec{\nabla} n}{n}\right)^2 - 2\frac{\nabla^2 n}{n} \right] - U(r) = \mu . \qquad (11)$$

To explore the fitting formula for $\phi(r)$ and $n(r)$, Tu [1991] studied the structure of the Thomas-Fermi (TF) and TFDW equations at $T = 0$. One of his main results is a precise determination of $n(0)$ in terms of $Z$, given by

$$n(0) = \alpha(Z) Z^3 \qquad (12)$$

with

$$\alpha(Z) = 6.8401 \frac{Z + 18.145}{Z + 19.371} . \qquad (13)$$

The exact slope of the density at the origin: $n'(0) = -18Zn(0)$ was proven by Steiner [1963]. Having analyzed the asymptotic behavior of $\phi(x)$ and $n(x)$, Tu established the Padé approximant valid both near the origin and in the asymptotic domain. Though his approximate expressions for $\phi(x)$ and $n(x)$ are of great use as an initial input to the iterative solution of the set of equations (8) and (11) to accelerate convergence, they are called only once at the first iteration. In this respect, close examination of his approximate solution to the TF and TFDW equations is important from a mathematical, rather than a computational, viewpoint.

To finish this section, we quote our complete DFT calculation [Furutani, Totsuji, Komaki, and Tanabe, 1988] which consists in a self-consistent solution of the one-electron Schrödinger equation, using as an initial input the effective potential obtained from the TFDW model. In this work, a difficulty pertinent to a counting of the number of bound electrons in shallow (Rydberg) states was remedied with recourse to the Planck-Larkin sum [Larkin, 1960]. One of the results is quoted in Fig. 1.

## III. TWO-CENTER MODEL

There exists a situation in which, with increasing ion density, a wave function of adjacent ions overlaps so strongly that an electron should be considered to belong not to

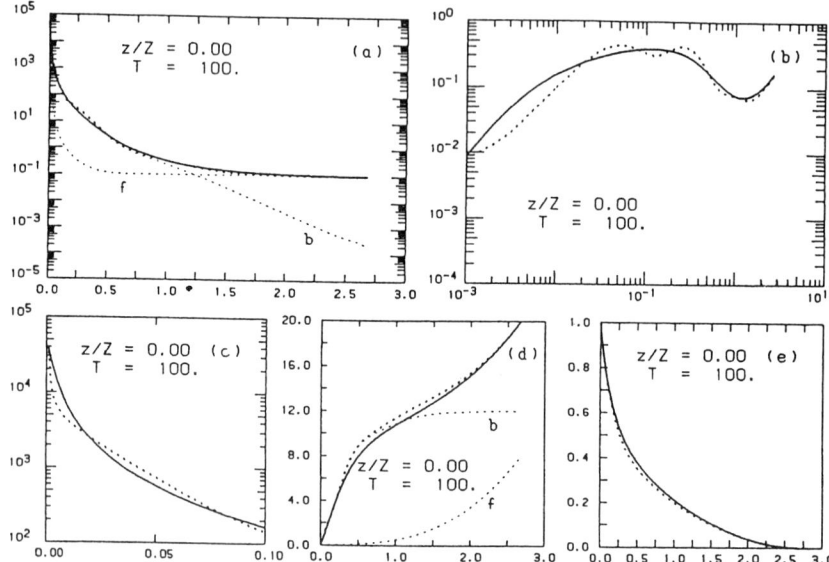

Fig. 1. Comparison of the TFDW results (solid line) with the iterative solution of the DFT [dotted (initial) and solid (final) lines] for $T = 100$eV and $z/Z = 0$. (a) $n(r)$ vs $r$, (b) $4\pi r^2 n(r)/Z^{4/3}$ vs $r$, (c) $n(r)$ vs $r$ in an enlarged scale, (d) integrated number of electrons within $r$, and (e) $rU(r)/Z$ vs $r$. The dotted lines in (a) to (d) include both bound and free electrons. In (a) and (d), contributions from bound (b) and free (f) electrons are shown separately by thin dotted lines.

a single ion but to a molecule or an ion cluster. This case was analyzed by specializing to a diatomic molecule, to fill a gap between the SCM and the spherical cell model (multicenter model) proposed a decade ago by Perrot and Dharma-wardana and explored extensively since then [e.g., Dharma-wardana and Perrot, 1982; Perrot and Dharma-wardana, 1984; Perrot, 1987; Perrot, Furutani, and Dharma-wardana, 1990]. We now construct the two-center model (TCM) describing a compressed, neutral, diatomic molecule. It consists of two nuclei kept fixed a given distance apart and electrons, so that the whole system be electrically neutral [Yonei, 1971; Gross and Dreizler, 1979]. A shape of the boundary surface, rotationally symmetric about the nuclear axis but *a priori* unknown, has to be determined by appropriate boundary conditions. Just as in the case of the SCM, we have analyzed a spatial profile of the electron density and of the Hartree potential inside a molecule, together with the Helmholtz excess free energy to examine a possibility of the molecular binding.

The formulation which follows is a simple extension to the two dimensions of the one already presented in Sec. II and we skip it to avoid redundancy. In connection with one of the boundary conditions, adopted in our work, that the normal derivative of the electron density on the boundary surface vanishes, one comment is in order. In their study of the TCM to examine the effect of neighboring ions on the ionic

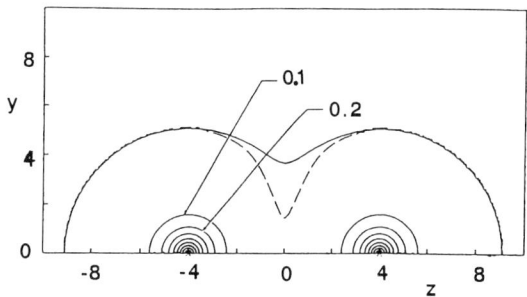

Fig. 2(a). Contour of constant $\phi(\sigma,\tau)$ for the H-H pair at $T=0$, $\bar{n}=1.0\times10^{-4}$. The dashed curve represents the inner $\phi(\sigma,\tau)=0$ curve.

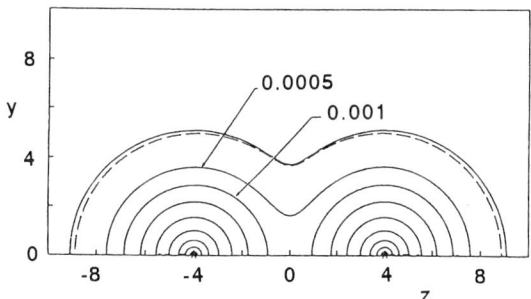

Fig. 2(b). Contour of constant $n(\sigma,\tau)$. The dashed curve represents the inner $n(\sigma,\tau)=\bar{n}$ curve.

levels, Salzmann *et al* [1991] assumed that, given $\bar{n}$, the electric field on the boundary vanishes. Along the same line of reasoning, Balàzs [1967] proved the Teller-Balàzs' theorem on an impossibility of the molecular binding within the TF and TFD models. In view of this, we believe that, even if the electrical neutrality were ensured within the boundary, the lines of force of electric dipolar field could cross the boundary, such that they are only summed to zero over the entire boundary surface. We depict in Fig. 2 the contour of constant $\phi(\sigma,\tau)$ and $n(\sigma,\tau)$ for the H-H pair [Furutani, Ohashi, Shimizu, and Fukuyama, 1993]. The set of variables $(\sigma,\tau)$ is that of the prolate spheroidal coordinate system. Fig. 2(b) clearly shows a deficit of electrons in the peripheral region, which produces a sink of the electric dipolar field.

## IV. TWO-TEMPERATURE SCM

Motivated by a very naive argument that bound electrons lying in deep levels be likely to be in their ground state, we shall construct a monatomic two-temperature model (TTM) composed of an ion core in its ground state ($T=0$) with $Q_b$ electrons and a surrounding (clad) electron gas at finite temperatures ($T>0$). Since at the interface separating the latter from the former there exists a discontinuity of temperature, two electron gases are no longer in thermal equilibrium. When we intend to work out according to the DFT, it is legitimate to introduce the internal energy for the ion core and Helmholtz's free energy for the clad electron gas. The DFT is then applied to two gases separately, giving rise to two TFDW equations, which are further coupled with the respective Poisson's equation. The formulation for the TTM parallels closely that of Sec. II for the SCM at uniform temperature. For bound electrons inside the ion core, we replace the free energy $F[n_f]$ and its density $\mathcal{F}[n_f]$ by the internal energy $E[n_b]$ and its density $\mathcal{E}[n_b]$, where $n_b$ and $n_f$ are, respectively, the bound and free electron density. The free energy of the clad electron gas

$$F[n_f] = F_{KG}[n_f] + F_{XC}[n_f] - \int_{\Omega_>} d\vec{r}\, \frac{Z}{r} n_f(r) + \int_{\Omega_>} d\vec{r}\, \frac{Q_b}{r} n_f(r) + \frac{1}{2} \int_{\Omega_>} d\vec{r}\, d\vec{r}'\, \frac{n_f(r)\, n_f(r')}{|\vec{r} - \vec{r}'|} \quad (14)$$

is minimized according to the recipe: $\delta F[n_f]/\delta n_f = \mu_f$. On the other hand, the internal energy of the ion core can be given by the replacement of $F$, $\Omega_>$, $n_f$, $Q_b/r$, and $\mu_f$ by $E$, $\Omega_<$, $n_b$, $V_0$, and $\mu_b$, respectively. $\Omega$ is a volume of bounded region. $V_0$, given below, is a constant potential in the ion core produced by free electrons in $\Omega_>$. Application of the DFT to eq.(14) yields, as in Sec. II, the TFDW equation

$$\frac{\delta \mathcal{F}_K}{\delta n_f} + \frac{\delta \mathcal{F}_{XC}}{\delta n_f} - \left\{ \frac{\delta}{\delta n_f}\left[\frac{h[n_f]}{n_f}\right] [n_f'(r)]^2 + 2h[n_f]\frac{\nabla^2 n_f}{n_f}\right\} - U_>(r) = \mu_f. \quad (15)$$

The equation for the ion core can be readily obtained by replacing $\mathcal{F}$, $h[n_f]$, $U_>(r)$, and $\mu_f$ by $\mathcal{E}$, $1/72$, $U_<(r)$, and $\mu_b$. The Hartree potential $U_>(r)$ is defined by

$$U_>(r) = \frac{Z - Q_b}{r} - \int_{\Omega_>} d\vec{r}'\, \frac{n_f(r')}{|\vec{r} - \vec{r}'|}, \quad (16)$$

while $U_<(r)$ follows immediately from the above, remembering the replacement of $Q_b/r$ by $V_0$, where $Q_b$ is the number of electrons within the ion core and

$$V_0 = \int_{r_c}^{R} 4\pi r'\, dr'\, n_f(r'), \quad (17)$$

with $r_c$ and $R$ the radius of the ion core and of the whole atom, respectively. Inspection of eqs.(16) and (17) indicates the continuity of $U(r)$ and $U'(r)$ at the interface ($r = r_c$). To solve eq.(15) and its counterpart for the ion core, it is convenient to introduce the same screening function $\phi(r)$ as that of eq.(9), which satisfies the Poisson's equation

$$\frac{Z}{r}\phi''(r) = 4\pi\, n(r). \quad (18)$$

Combination of eqs.(15) and (18) and their counterpart for the ion core yields two nonlinear equations of fourth order in two regions. To solve them, we should impose *ten* boundary conditions as listed below. At the origin, we have

$$\phi_<(0) = 1, \qquad n_b'(0) = -18\, Z\, n_b(0). \quad (19\text{a, b})$$

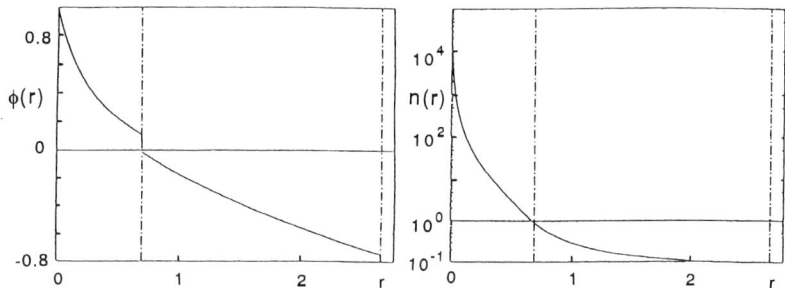

Fig. 3. Behavior of (a) $\phi(r)$ vs $r$ and (b) $\ln n(r)$ vs $r$. Two vertical lines indicate the position of $r_c$ and $R$, respectively. External parameters are: $Z=20$, $Q_b = Q_f = 10$, $T=100$eV, and $\bar{n}=0.1$. Main results are: $r_c = 0.70905$, $R = 2.6396$, $\mu_b = -4.7783$, and $\mu_f = -8.2375$.

At the interface $(r = r_c)$, we can establish

$$n_b(r_c) = n_f(r_c), \qquad n'_b(r_c) = n'_f(r_c) \qquad (20a,b)$$

$$U_<(r_c) = U_>(r_c), \qquad (21a)$$

$$\phi_<(r_c) - r_c \phi'_<(r_c) = \phi_>(r_c) - r_c \phi'_>(r_c) = 1 - \frac{Q_b}{Z}. \qquad (21b)$$

At the outer boundary, we impose

$$n_f(R) = \bar{n}, \qquad n'_f(R) = 0, \qquad (22a,b)$$

$$U_>(R) = \frac{Z_i}{R}, \qquad Z_i = Z - (Q_b + Q_f).$$

We have analyzed the two sets of equations in the inner and outer region as eigenvalue problem. The discontinuity of $\phi_<(r_c)$ and $\phi_>(r_c)$ which naturally stems from that of $\mu_b$ and $\mu_f$ can be accounted for with recourse to eq.(21b). To solve the above four equations, we convert them into a set of algebraic equations by means of the finite difference method, details of which will be published elsewhere [Furutani and Fukuyama, 1995]. One of the preliminary results of $\phi(r)$ and $\ln n(r)$ versus $r$ are depicted in Fig. 3, where we observe the jump of $\phi(r)$ which reflects the discontinuity of $\mu$ at the interface as described previously.

## V. CONCLUDING REMARKS

In all cases discussed above, several problems underlying the TFDW model may be summarized thus.

(1) Difference between an isolated and a compressed atom may be accounted for only through the boundary condition imposed on the electron density at the ionic surface. The other boundary conditions remain the same.

(2) The effect of continuum lowering (or of pressure ionization) can only be examined by means of the full DFT theory, which provides reference data of energy eigenvalues and associated wave functions (instead of hydrogen-like wave functions). Their interpretation and justification are still open to future discussion.

(3) A number of boundary conditions, five in the SCM and ten in the TTM used in our works is not always convincing. Specifically in the TTM, external parameters we specify are: nuclear charge $Z$, density $\bar{n}$ and temperature $T$ of a surrounding plasma and number $Q_b$ of bound electrons inside the ion core. The fact that, given any set of these parameters, there is always a converged solution may suggest a lack of one more boundary condition, which will allow to determine, for instance, a ratio $Q_b/Z$ as a function of $T$. We ignore for the moment if it were a definition of the bound electron within the TFDW theory.

One of the authors, Y. F., deeply acknowledges Dr. F. Perrot (CEL-V, CEA, France) for valuable discussions on the TTM through private communication. He also thanks Dr. R.M. More (LLNL, USA) for critical reading of the manuscript and Professor S. Ichimaru for his interest in this work.

**REFERENCES**
Balàzs, N. L., 1967, Phys. Rev., **156**, 42.
Dharma-wardana, M. W. C., and F. Perrot, 1982, Phys. Rev., **26**, 2096.
Fermi, E., 1928, Z. Phys., **48**, 73.
Furutani, Y., K. Ohashi, M. Shimizu, and A. Fukuyama, 1993, J. Phys. Soc. Jpn., **62**, 3413.
Furutani, Y., H. Totsuji, K. Komaki, and M. Tanabe, 1988, J. Phys. Soc. Jpn., **57**, 3778.
Furutani, Y., A. Fukuyama, to be published.
Gross, E. K. U., and R. M. Dreizler, 1979, Phys. Rev., **A20**, 1798.
Hohenberg, P., and W. Kohn, 1964, Phys. Rev., **B136**, 864.
Kohn, W., and L. J. Sham, 1965, Phys. Rev., **A140**, 1133.
Larkin, A. I., 1960, Sov. Phys. JETP., **11**, 1363.
Mermin, N. D., 1965, Phys. Rev., **A137**, 1441.
Perrot, F., 1979, Phys. Rev., **A20**, 586.
Perrot, F., 1987, Phys. Rev., **A35**, 1235.
Perrot, F., and M. W. C. Dharma-wardana, 1984, Phys. Rev., **A30**, 2619.
Perrot, F., Y. Furutani, and M. W. C. Dharma-wardana, 1990, Phys. Rev., **A41**, 1096.
Salzmann, D., J. Stein, I. B. Goldberg, and R. H. Pratt, 1991, Phys. Rev., **A44**, 1270.
Singwi, K. S., M. P. Tosi, R. H. Land, and A. Sjölander, 1968, Phys. Rev., **176**, 589.
Steiner, E., 1963, J. Chem. Phys., **39**, 2365.
Tanaka, S., S. Mitake, and S. Ichimaru, 1985, Phys. Rev., **A32**, 1896.
Thomas, L. H., 1927, Proc. Camb. Phil. Soc., **23**, 542.
Tu, K., 1991, J. Math. Phys., **32**, 2250.
Yonei, K., 1971, J. Phys. Soc. Jpn., **31**, 882.

# XI. Magnetic Reconnection

# XI.1

# RECENT EXPERIMENTS ON MAGNETIC RECON-NECTION IN LABORATORY PLASMAS: A REVIEW

Masaaki Yamada

Princeton Plasma Physics Laboratory, Princeton University
P.O.Box 451, Princeton, New Jersey 08543, U. S. A.

The present paper reviews recent laboratory experiments on magnetic reconnection including the merging experiments of two toroidal plasmas and the most recent result from the high temperature tokamak plasmas with Magnetic Reynolds numbers exceeding $10^7$. These recent laboratory experiments create an environment which satisfies the criteria for MHD plasma in which the global MHD boundary condition can be controlled externally. Experiments with fully three dimensional reconnection are now possible. Examples will be drawn from electron current sheet experiments, merging spheromaks, and tokamaks. Particularly, in the recent TFTR tokamak discharges. Motional Stark effect (MSE) data have verified the existence of a partial reconnection. In the experiments of spheromak merging, a new plasma acceleration parallel to the neutral line has been indicated. Together with the relationship of these observations to the analysis of magnetic reconnection in space and the solar flares, important physics issues such as global boundary condition, local plasma parameters, merging angle of the field lines, and the 3-D aspect of the reconnection are discussed.

## I. INTRODUCTION

Magnetic reconnection is one of the most important processes in magnetized plasmas [1-4]. It is considered to be a key process in the description of the evolution mechanisms [1] for solar flares and the earth's magnetosphere. Magnetic reconnection also occurs as one of the relaxation processes in fusion research plasmas; it often plays a dominant role in determining the confinement characteristics of the high temperature fusion plasmas.

The study of the evolution and interaction of the solar flares has been intensified because of the soft X-ray pictures of the sun recently taken by the Yohkoh satellite [5]. Many large solar flares were observed to be interacting actively and changing their topology rapidly, on a short time scale of a few minutes, much faster than the value predicted by classical theory. Magnetic reconnection is attributed to the observed activities [1,5].

In highly conductive plasmas for fusion research [6-8], magnetic reconnection plays an essential role in the dynamic changes of magnetic configuration. The major and minor disruptions of tokamak plasmas for magnetic fusion research are caused by the reconnection of magnetic field lines, resulting in degradation of confinement characteristics [6-8]. Magnetic reconnection always occurs during plasma formation and/or configuration change and is regarded as the most important self-organization phenomena in plasmas.

A few recent laboratory experiments have created an environment which satisfies the criteria for MHD plasma ($V_A/c \ll 1$, $S \gg 1$, $\rho_i/a \ll 1$) to study magnetic reconnection, in which the global boundary condition can be externally controlled. Experiments with fully three dimensional reconnection are now possible. In the most recent experiment, an investigation of three-dimensional MHD effects of magnetic reconnection dynamics has been carried out by the use of axially colliding spheromaks [9-11]. The toroidal shape spheromak plasmas were forced to merge in an external equilibrium field. The reconnection angle $\theta$ between the merging field lines was varied by changing the magnitude and polarity of an externally applied toroidal field. It was observed that the speed of counter-helicity merging with $\theta \approx 180°$ was 3-5 times faster than that of co-helicity merging with $\theta \approx 90°$, suggesting the significance of the third component of the magnetic field line vector and of 3-D MHD effects on the reconnection process. The reconnection speed increased proportionally with the initial approaching speed of the spheromaks. The observations suggest that a plasma compressibility plays an important role in the driven reconnection experiments [9,11]. The local features of the neutral sheet current density and the induced electric field were also measured together with the anomalous resistivity. A careful analysis of magnetic field line evolution indicated a new plasma acceleration mechanism parallel to the neutral sheet [11]. A significant ion heating was also measured. The recent numerical MHD simulations [12,13] also recognized the similar 3-D effect as well as an observation that the large component of the magnetic field vector parallel to the reconnection line shows down the speed of the magnetic reconnection.

To describe the motion of magnetic field lines in a plasma, Eq. (1) cam be derived by combining Maxwell equations and Ohm's law,

$$\frac{\partial \vec{B}}{\partial t} = \nabla \times (\vec{V} \times \vec{B}) + \frac{\eta}{\mu_0} \nabla^2 \vec{B} \tag{1}$$

where

$$\tau_D = \frac{\mu_0 a^2}{\eta} : \quad \text{Diffusion time},$$

$$\tau_A = \frac{a}{V_A} : \quad \text{Alfven time, and}$$

$$S = \frac{\tau_D}{\tau_A} \gg 1 \quad \text{for MHD plasmas}.$$

The first term of the right hand side represents a motion of plasma moving with field lines [1,14-16] (plasma is frozen to the field lines) and the second term describes a diffusion of magnetic fields with the diffusion coefficient proportional to the plasma resistivity. $S$ is called the Magnetic Reynolds Number and has to be significantly larger than unity in order for plasma to be treated as MHD fluids. For typical examples of the MHD plasma, solar flares have very large $S$, i.e., $S > 10^{10}$, for tokamaks [7] $S > 10^7$, and for the MRX/TS-3 experiments [9-11] $S \sim 10^2$–$10^3$. In the past decades, a few laboratory experiments have been carried out to investigate the magnetic reconnection. Three typical examples can be mentioned.

A. Electron current sheet experiments

Stenzel and Gekelman [17-19] investigated the magnetic reconnection, using parallel conductor plate currents in a linear plasma. In their elaborate series of experiments, in which electrons were magnetized but the ion gyro-radius was as large as the plasma size, many important local features of magnetic reconnection were investigated. In particular, the typical 2-D feature of particle acceleration was verified [18].

B. Sawtooth relaxation of tokamak plasma

A sawtooth relaxation of tokamak plasma can provide a good example of magnetic reconnection [6,7]. If an initial plasma state has a certain $q$ value ($q$ = safety factor < 1) and is unstable to 1/1 kink mode, then the plasma goes through an unstable regime and settles in another state with a different $q$ while field lines have to rearrange themselves by breaking their initial topological configuration. Thus the sawtooth relaxation (crash) is a typical magnetic reconnection phenomenon.

C. MRX/TS-3 experiments

In the TS-3 and MRX experiments [9-11], two spheromaks collide to form a new plasma configuration and field lines reconnect through a topological reconfiguration. An evidence of three dimensional effects is seen together with a new plasma acceleration mechanism.

The present paper summarizes these three recent laboratory experiments in which good progress has been made for documenting important aspects of magnetic reconnection.

## II. TWO-DIMENSIONAL RECONNECTION MODEL AND EXPERIMENTAL OBSERVATIONS

Fig. 1 presents the most commonly used 2-D description of magnetic reconnection [1,14-16], in which two sets of field lines are directed oppositely in the top and bottom of the separatrix. As magnetized plasmas move from both sides toward the separatrix, a strong sheet current develops perpendicular to the page sheet plane due to a strong curl $\vec{B}$ effect. This sheet current diffuses due to plasma resistivity and a reconnection of field lines occurs. This region is often called "diffusion region". After the reconnection of field lines, acceleration of plasma particles (up to Alfven velocity) occurs parallel to the plane.

The Sweet-Parker model and the Petschek model are based on this 2-D picture [14-16] and evaluate the reconnection rate of field lines.

A series of well diagnosed experiments were carried out by Stenzel and Gekelman using a linear plasma device [17-19]. In their experiments, a reconnection regime was created by driving currents in the parallel sheet conductors shown in Fig. 2a, and a detailed local study of magnetic reconnection was made. Although, the ion gyro-radius was too large to treat the experiment as full MHD phenomena, precise local measurements were made to identifying microscopic physics issues associated with reconnection (Fig. 2b–d). They carried out spatial and temporal measurements of plasma pressure $n_e T_e$, magnetic force density $\vec{J} \times \vec{B}$ and ion velocity $\vec{v}$.

In particular they identified the classical ion acceleration pattern associated with reconnection by using different particle detectors. Figs. 2c and 2d depict typical 2D ion flows drifting from diffusion regime to outside in perpendicular to the neutral line. The local force

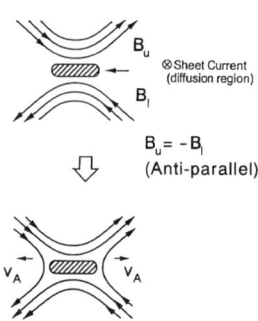

Fig. 1. A typical 2-D description of magnetic reconnection.

on the plasma, $(\vec{J} \times \vec{V} - \mathrm{grad}p)$ was compared with the measured particle acceleration. It was found that the ion acceleration was strongly modified by scattering of wave turbulence. After several Alfven times, the plasma was observed to develop the classic flow pattern, jetting from the neutral sheet with velocities close to the Aflven speed. Furthermore, Gekelman *et al.* has extended their experiments for a 3D study [19].

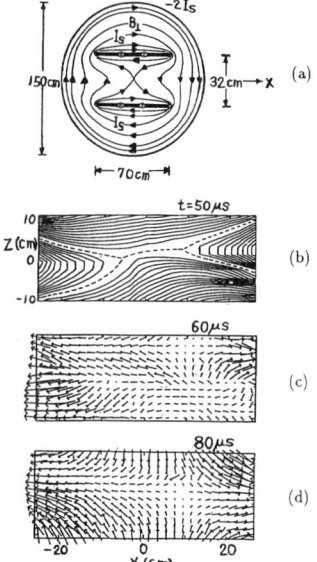

Fig. 2. Results from Stenzel and Gekelman's experiment. a) Cross-sectional view of the device, showing the magnetic field geometry without plasma. b) Transverse field lines contours at $t = 50\mu s$. c) and d) Measured ion velocity vectors at $t = 60$ and $80\mu s$ at axial position of $z = 87$cm.

## III. TOKAMAK SAWTOOTH OSCILLATION AND MAGNETIC RECONNECTION

A sawtooth crash in a tokamak is a typical example of magnetic reconnection in high temperature plasma. A sawtooth oscillation [6,7,20] is characterized by a periodic repetition of $T_e$ peaking and crash as shown in Fig. 3. The crash phase of the sawtooth can manifest important physics mechanisms for magnetic reconnection, since field line re-arrangement is expected during this period. A highly peaked $T_e$ profile often can lead to a peaked current profile with central $q < 1$, where $q$ denotes inverse of rotational transform and is called a safety factor. Due to this current profile, $m = 1/n = 1$ MHD instability develops near the $q = 1$ flux surface and this instability can induce magnetic reconnection. According to Kadomtsev, the reconnection event (crash) makes central $q$ rise to unity, and similar cyclic evolution are repeated.

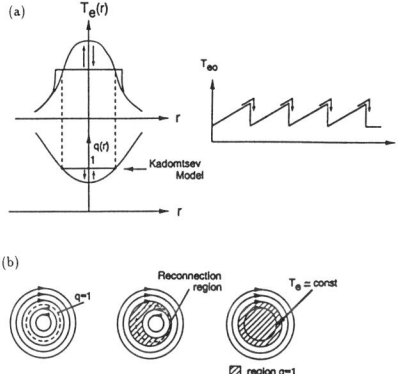

Fig. 3. a) Sawtooth oscillation of a tokamak plasma. A peaked current profile associated with a highly peaked $T_e$ profile leads to $q <1$ ($q$ = safety factor) and the plasma becomes unstable against 1/1 kink. b) Kadomtsev model; $m = 1/n =1$ MHD instability develops near $q =1$ flux surface and induce magnetic reconnection. After the reconnection event (crash), central $q$ rises to unity.

A. Electron temperature profile evolution

If we regard that a tokamak consists of nested flux surfaces on which $T_e$ is constant, electron cyclotron emission (ECE) radiation spectrum can provide $T_e$ profile thus flux surface evolution except for a short crash period in which plasma parameters can change in a time scale less than 50$\mu$sec.

The electron temperature profile on the poloidal plane of the plasma is measured by ECE with a 2$\mu$sec time step. This measurement can provide local electron temperature evolution utilizing rotation of the plasma [22,23]. The Fig. 4 demonstrates a typical $T_e(R)$ profile evolution during the sawtooth crash. It is observed that electron heat transfer occurs from the inside to just outside of the inversion radius ($\sim r_a = 1$) through a reconnection region [24].

Overlaying color contours of $\Delta T_e$ on the 3D-$T_e(r,\theta)$ profile allows one to visualize a transfer of an excess electron heat during the crash period as shown in Ref. 23. The entire crash phase takes 150–800$\mu$sec, while a sawtooth period is 100-200msec. Just before the crash, a shrinking circular hot peak shows up and a crescent-shaped flat island grows inside the inversion radius. A fast heat transfer can be seen in the area of inversion radius. It is expected that magnetic reconnection causes this fast heat flow. The $T_e$ profile inside the inversion radius becomes flat after the crash, which is consistent with the prediction made by Kadomtsev [6].

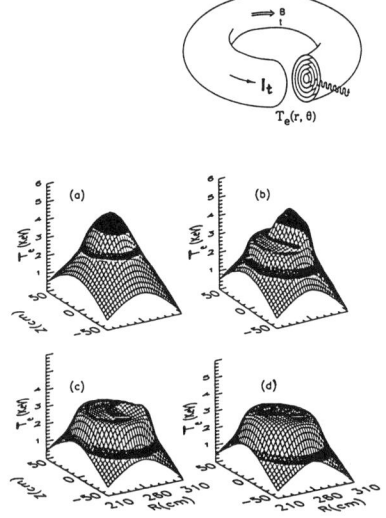

Fig. 4. ECE schematic display and $T_e(r,\theta)$ profiles in 3-D display during the crash period of sawtooth. Time step between each figure is $\sim 120\mu$sec.

B. Verification of magnetic reconnection by measuring $q$ profile evolution

We employ the motional Stark effect (MSE) diagnostic to obtain the profile of magnetic pitch angle, and hence the $q$ profile ($q(R)$ = local safety factor), using polarimetry measurements of the Doppler shifted $D_\alpha$ emission from a neutral deuterium-beam (NBI) heating line [21]. An important advantage of this technique is that this non invasive and nonperturbative measurement of the field-line pitch is localized to the geometric intersection of the field of view with the neutral beamlines, which leads to good spatial resolution of $\delta r = 3 \sim 5$ cm. If the plasma is considered to have good flux surfaces, the measured field line pitch can be translated into a radial profile of rotational transform, or $q(R)$ profile, based on tokamak equilibrium calculations [24]. The $q$ profiles measured by the MSE diagnostics indicate that central $q$ values increase by 5–10%, typically from 0.7 to 0.75, during the sawtooth crash phase but do not relax to unity even while the pressure gradient diminishes inside the $q = 1$ region. In this example, as well as most tokamak sawtoothing discharges, $q_0$ is below one, and remains below unity throughout the sawtooth cycle. Because only field-line breaking and re-arrangement can make a $q(R)$ change on such a short time scale, this verifies a magnetic field-line reconnection, although it is small.

C. A heuristic model for sawtooth crash

The observations raise an important question as to why the magnetic field lines inside the $q = 1$ region do not form a flat $q \sim 1$ region after the crash as suggested by Kadomtsev [6], while the temperature gradient diminishes to zero as predicted by him for full reconnection. So a simulation analysis of $T_e(r,\theta)$ and $q(R)$ profile evolution

[24] was made in TFTR tokamak. Based on the experimental result, a heuristic model has been proposed for a sawtooth crash [24]. In this model, the plasma is viewed as concentric toroidal regions separated by the $q = 1$ surface. The $m/n = 1/1$ kink instability is primarily confined to the inner region while the outer region plays the role of a stable conducting shell. The kink mode displaces the pressure contours on an ideal MHD time scale with a helical (1/1) structure. In the final stage, the pressure contours will be displaced more than the magnetic flux surfaces (the plasma pressure is <u>not</u> constant on a flux surface in this crash period). The external conducting region reacts to suppress the flux displacement by inducing image currents near the interface of the two regions, often called an "X-point". A fast reconfiguration of magnetic field lines occurs connecting the inner region of the plasma to the outside through the X-point region. A rapid efflux of thermal energy occurs through the X-point region along newly connected field lines. The precipitous drop of the pressure gradient, which occurs within a short period of 100–200$\mu$sec ($\ll t_{\text{Sweet-Parker}}$ [14,15]), removes the free energy to drive the $m/n = 1/1$ instability, inhibiting further progress of the mode and thus full reconnection.

The central $q$ values have been measured in the sawtooth plasmas by several groups for the past ten years [28-32]. There is agreement among all experiments that the relative change of the central $q$ values during sawtooth crash is small ($\Delta q/q \leq 0.1$). However, there is a major difference among them on the final values of the central $q$ after the crash. The cause of their apparent disagreement with the above results is yet to be determined. Since all reported $\Delta q < 0.1$ during sawteeth and the results can be be explained by the proposed model [24], the inconsistency should be regarded as small.

The most important finding to date is that the magnetic reconnection in tokamak plasma is determined by and/or strongly influenced by <u>the three dimensional boundary conditions</u> of the sawtoothing plasmas as well as the local plasma parameters. The reconnection rate is determined by the plasma's MHD stability condition (for 1/1 mode) based on the plasma parameters ($n_e(R)$, $T_e(R)$, and $T_i(R)$), boundaries, and current profiles ($q$ profiles).

## IV. THREE DIMENSIONAL STUDY OF MAGNETIC RECONNECTION IN COLLIDING SPHEROMAK EXPERIMENTS

Three dimensional effects of magnetic reconnection has been recently investigated extensively by use of axially merging two spheromaks [9-11]. In these experiments, two toroidal plasma rings, with equal or opposite helicity, are formed and brought together, contacting along a toroidally (azimuthally) symmetric line. It has been found that plasmas of opposite helicity reconnect appreciably faster than those of similar helicity [9], and the direction of toroidal field plays an important role in the reconnection process.

A. Experimental profile

Now we consider an experiment in which two plasma toroids of spheromak-type are merging together. The spheromaks with a flux hole in the center are easy to be translated. They can be generated with different helicities. Two spheromaks carry identical toroidal current with the same or the opposite toroidal field (let us call them a co-helicity merging or a counter-helicity merging, respectively) are made to merge to induce reconnection by controlling external coil currents.

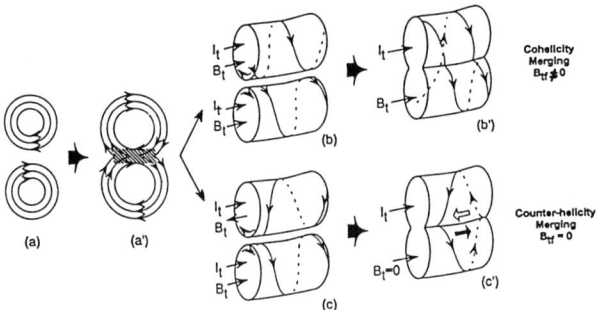

Fig. 5. Three dimensional description of magnetic reconnection. a) and a'), evolution of 2D poloidal picture of magnetic field line at the reconnection point; b) and b'), 3D description of evolution for merging two toroidal plasmas with equal helicity, before and after reconnection; c) and c'), 3D description of evolution for two plasmas with opposite helicity, before and after reconnection.

In both cases, identical amount of toroidal currents are flowing parallel to each other. Thus, the 2-D pictures of these 2 cases are exactly the same. However, they are quite different in the 3 dimensional picture as seen in Figs. 5b and 5c. In the co-helicity case, the field lines merge with large angles at the reconnection region, while in the counter-helicity case, they merge exactly with anti-parallel symmetry.

There is another important difference in the reconnection pattern. For co-helicity merging, the transition of the configuration should be globally smooth. But in counter-helicity merging, the pitch of the field lines changes abruptly at the reconnection point as seen in Fig. 5c. We expect a violent plasma acceleration in the toroidal direction as the field lines contract after the merging of two toroidal plasmas of the opposite helicities. This acceleration mechanisms and direction is importantly different from that conjectured in the typical 2-D models [14-16]. In the 2-D picture a plasma acceleration occurs perpendicular to the neutral sheet as shown in Fig. 1. But in the 3-D picture shown in Fig. 5, plasma acceleration can occur parallel to the neutral sheet.

B. Experimental results

To identify critical physics issues for reconnection, experiments have been carried out at the University of Tokyo in collaboration with PPPL. Fig. 6a shows the set up of the TS-3 experiment in which 2 spheromaks of toroidal shape are created and allowed to merge together. The toroidal flux of each spheromak is generated by the Z-discharges between electrodes. To document the internal magnetic structure of the reconnection on a single shot, 2 dimensional magnetic probe array is placed on an

$r$–$z$ plane or toroidal cut-off plane as shown in this figure. Plasma parameters are as follows; $B \sim 0.5$–$1$kG, $T_e \sim 10$eV, and $n_e \sim 2$–$5 \times 10^{14}$cm$^{-3}$. Fig. 6b shows the time evolution of the poloidal flux contours derived experimentally from internal probe signals for the merging of co-helicities and counter-helicities. Other plasma parameters were held identical for each discharges.

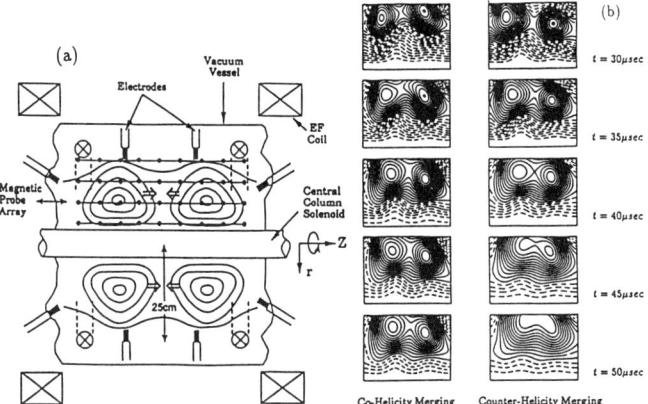

Fig. 6. a) Experimental set-up in TS-3 device. The central column provides stability effects for spheromaks with a flux-hole (currentless region) at the major axis. b) Evolution of poloidal flux contours for co- and counter-helicity merging. The other plasma parameters are kept identical for the cases shown. Each step for flux contours ($\Delta\Psi$) corresponds to $2 \times 10^{-4}$ Webers ($2 \times 10^4$Gr. cm$^2$).

A merging of spheromaks of opposite helicity is shown to be more efficient compared to merging of the same helicity. During the initial phase, reconnection progresses with the same speed for both. In the case of co-helicity merging, the reconnection rate is seen to slow down significantly in midway, while for counter-helicity merging, reconnection continues until they merge completely.

Here let us consider a possible cause of the observed faster reconnection for counter-helicity merging. When two plasmas of parallel toroidal fields are brought together, a new equilibrium is formed among the toroidal field pressure (outward), poloidal-field pressure (attracting force), and the plasma pressure (toward). For the merging of the plasmas of two anti-parallel toroidal fields, the central toroidal field is quickly reduced to zero and the attracting force becomes so dominant that reconnection is accelerated.

Based on the magnetic field evolution, toroidal current contours were deduced for the same sequence of shots [11], which verified an important 2D feature of magnetic reconnection. It was measured that the plasma resistivity based on the observed current decay was enhanced by more than factor of 10.

Another important observation was a strong dependence of the reconnection speed on global forcing, i.e., in this case, colliding velocity of the two plasmas. Fig. 7 presents reconnection rate $\gamma$ (defined by a flux transfer rate, $(1/\Psi)\delta\Psi/\delta t$, versus initial colliding velocity of two plasmas for co- and counter-helicity merging cases. It is observed that $\gamma$ increases almost proportionally to $v_i$. This result is not consistent with the leading 2-D theories of Sweet and Parker and/or Petschek, which suggest no dependence on external forcing. The experiment clearly suggests importance of an external driving force in determining reconnection rate, and support an important aspect of a driven-reconnection model [33]. More qualitative analysis is underway.

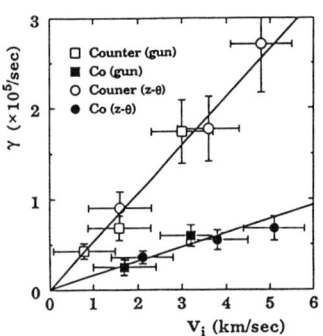

Fig. 7. Evidence of driven reconnection; reconnection rate $\gamma$ versus colliding velocity $v_i$ for co-helicity and counter-helicity merging.

An evidence of such mechanism was observed in the recent experiment by Ono et al. [11]. Fig. 8a depicts time evolution of profile of toroidal field, $B_t$, versus $z$ (axial) direction for counter-helicity merging. This result was obtained by a $B_t$ probe array inserted at $r$ = 14cm, which penetrated through the magnetic axis. Initially, the merging plasmas formed the $B_t$ profile shown in the figure, positive in the left and negative in the right side. As reconnection progressed, the value of $B_t$ decreased as expected but then the $B_t$ profile flipped (changing its polarity) between $t$ = 20 and 30$\mu$sec. This overshoot is regarded as an evidence of the sling shot effect [9] discussed in the earlier section, because the figures shown in Fig. 8b describe schematically a dynamic (3D) evolution of magnetic field lines during and after the reconnection. An energy transfer from magnetic to plasma thermal energy has been expected in this dynamic reconnection process of toroidal field annihilation. A strong ion heating has been documented during the counter-helicity heating and reported elsewhere.

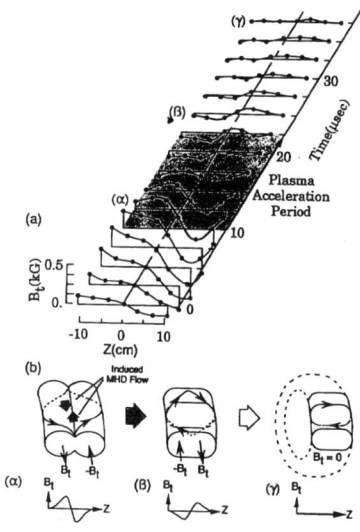

Fig. 8. Observation of sling-shot effect. a) Time evolution of axial profile of toroidal magnetic field $B_t$. b) Schematic description of overshoot phenomena of toroidal field.

## V. SUMMARY

A number of recent laboratory experiments on magnetic reconnection have been reviewed with special focus on the merging experiments of two toroidal plasmas and the observations from the sawtoothing tokamak plasmas with Magnetic Reynolds numbers exceeding $10^7$. These recent laboratory experiments created an environment which satisfied the criteria for MHD plasma ($V_A c \ll 1$, $S \gg 1$, $\rho_i/a \ll 1$) and in which the global MHD boundary condition could be controlled externally. Experiments with fully three dimensional reconnection are now possible. Examples were drawn from electron current sheet experiments by Stenzel and Gekelman, merging spheromaks, and sawtooth crash in tokamaks. It was found that the magnetic reconnection was influenced by global boundary condition, merging angle of the field lines, and the 3-D aspect of the reconnection, as well as local plasma parameters.

During a sawtooth crash, a magnetic reconnection has been verified as partial mixing of field lines, and the resulting changes of $q$ profile are documented. It is shown that the $q$ values stay substantially below 1, despite that $T_e$ contour evolution from ECE diagnostics show apparently the Kadomtsev-like full reconnection patterns. A simultaneous analysis of poloidal $T_e(r,\theta)$ and $q(R)$ profile evolutions has led us to propose a model in which 3 dimensional MHD effects play an essential role. A fast parallel transport along the newly connected field lines can cause a rapid efflux of internal energy through the X-point region and thus creating a fast crash of the central plasma pressure. The observed small change of $q$ values, an indication of partial reconnection of field lines, is attributed to the precipitous drop of the pressure gradients which drives the instability while being constrained by a flux conservation principle.

The most important finding is that the magnetic reconnection in tokamak plasma is strongly influenced by the three dimensional boundary conditions of the sawtoothing plasmas as well as the plasma transport. The reconnection rate is determined by the plasma's MHD stability condition for often non-axisymmetric modes (typically, $m = 1/n = 1$ mode), the local plasma parameters ($n_e(R)$, $T_e(R)$, and $T_i(R)$), boundaries, and by current profiles ($q$ profiles).

In colliding spheromak experiments, local and global MHD physics issues for magnetic reconnection have been extensively investigated in a 3-D geometry. The three-dimensional features of magnetic reconnection were found to be quite different from the two-dimensional features depending on whether the plasma toroids have co-helicity or counter-helicity configurations. Evidences of driven reconnection have been observed and a quantitative dependence of reconnection rate on external force was documented ($\gamma_{Rec} \propto v_i$), verified during the 3D reconnection process [34]. The results have proved that a double spheromak geometry is a well suited configuration for a basic study of magnetic reconnection. Further comprehensive study to determine the dependence on local and global characteristics will give a full picture of magnetic reconnection in three dimensions to address issues such as (a) What are the most critical factors, both global and local, to determine the rate of magnetic reconnection? (2) How does directional flow energy thermalize and/or could be utilized in the fusion devices? (3) Do deviations from axisymmetric spontaneously arise? And how do they occur?

## ACKNOWLEDGMENT

The author acknowledges many valuable inputs from Drs. W. Gekelman, M. Ono, Y. Ono, N. Pomphrey, and T. Tajima for writing this review. The work was supported by NSF Grant #ATM-9114924 and US. DoE Contract No. DE-AC02-76-CHO3073.

## REFERENCES

[1] Parker, E. N., 1979, *Cosmical Magnetic Fields*, (Clarendon, Oxford).
[2] Vasyliunas, V. M., 1975, Rev. Geophys. Space Phys., **13**, 303; Shi, Y., C. C. Wu, L. C. Lee, 1988, Geophys. Res. Lett., **15**, 295.
[3] Taylor, J. B., 1986, Rev. Mod. Phys., **28**, 243.
[4] Priest, E. R., *Solar Magnetohydrodynamics*, 1984, (Reidel, Dordrecht), Chap. 10; Parker, E. N., 1973, Astrophys. J., **180**, 247.
[5] Tsuneta, S., *et al.*, 1991, Solar Phys., **136**, 37.
[6] Kadomtsev, B. B., 1975, Sov. J. Plasma Phys., **1**, 389.
[7] Wesson, J. A., 1987, *Tokamaks*, (Clarendon, Oxford), pp. 167–183.
[8] Taylor, J. B., 1986, Rev. Mod. Phys., **58**, 741; Finn, J. M., and T. Anotonsen, 1985, Comments on Plasma Phys. Contr. Fusion, **9**, 111.
[9] Yamada, M., Y. Ono, A. Hayakawa, M. Katsurai, and F. W. Perkins, 1990, Phys. Rev. Lett., **65**, 721.
[10] Yamada, M., *et al.*, 1991, Phys. Fluids B, **3**, 2379.
[11] Ono, Y., A. Morita, K. Katsrai, and M. Yamada, 1993, Phys. Fluids B, **5**, 3691.
[12] Matsumoto, R., *et al.*, 1993, Astrophys. J., **414**, 357.
[13] Hawkins, J. G., *et al.*, 1994, J. Geophys. Res., **99**, 5869.
[14] Sweet, P. A., 1958, in *Electromagnetic Phenomena in Cosmical Physics*, edited by B. Lehnert (Cambridge Press, New York), p. 123.
[15] Parker, E. N., 1957, J. Geophys. Res., **62**, 509.
[16] Petschek, H. E., 1964, *Magnetic Field Annihilation*, NASA Spec. Pub. SP-50, 425.
[17] Stenzel, R. L., and W. Gekelman, 1981, J. Geophys. Res., **86**, 649.
[18] Gekelman, W., *et al.*, 1982, J. Geophys. Res., **87**, 101.
[19] Gekelman, W., and H. Pfister, 1988, Phys. Fluids, **31**, 2017.
[20] Wesson, J. A., 1986, Plasma Phys. Contr. Fusion, **28**, 243.
[21] Levinton, F. M., *et al.*, 1993, Phys. Fluids B, **5**, 2554.
[22] Nagayama, Y., *et al.*, 1990, Rev. Sci. Instrum, **61**, 3265; Nagayama, Y., *et al.*, 1991, Phys. Rev. Lett., **67**, 3257.
[23] Yamada, M., *et al.*, 1992, Rev. Sci. Instrum, **63**, 4623.
[24] Yamada, M., *et al.*, 1994, PPPL Report #3004 to be published in Phys. Plasmas (1994).
[25] Levinton, F. M., 1992, Rev. Sci. Instrum, **63**, 5157; Levinton, F. M., 1989, Rev. Sci. Instrum, **63**, 2060.
[26] Greene, J. M., *et al.*, 1971, Phys. Fluids, **14**, 671.
[27] Lichtenberg, A. J., 1984, Nucl. Fusion, **24**, 1277; Baty, H., *et al.*, 1993, Phys. Fluids B, **5**, 1213.
[28] Osborne, T. H., *et al.*, 1982, Phys. Rev. Lett., **49**, 734.
[29] West, W. P., *et al.*, 1987, Phys. Rev. Lett., **58**, 2758.
[30] Weisen, H., *et al.*, 1989, Phys. Rev. Lett., **62**, 434.
[31] Wolf, R. C., *et al.*, 1993, Nucl. Fusion, **33**, 663.
[32] Wroblewski, D., and R. Snider, 1993, Phys. Rev. Lett., **71**, 859.
[33] Sato, T., 1985, Phys. Rev. Lett., **54**, 1502.
[34] Ono, Y., M. Katsurai, and M. Yamada, 1993, *Proc. US-Japan Workshop on Effects of Finite Gyro-radius on Plasmas*, Irvin, Calif.

# XI.2

# DYNAMICS OF THE SOLAR CORONA OBSERVED WITH YOHKOH

Saku Tsuneta

Institute of Astronomy, University of Tokyo, Mitaka, Tokyo 181, Japan

The highlights from the solar observation satellite *Yohkoh* (meaning *sun beam* or *sun light*) are presented with emphasis on the results from the Soft X-ray Telescope (SXT). The highly dynamical and transient nature of the coronal magnetic fields revealed by SXT is drastically changing our understanding of the solar corona and behavior of magnetized plasma in general. This paper focuses on the new observations as well as on the new problems raised by the *Yohkoh*.

## I. INTRODUCTION

*Yohkoh* satellite was launched on 1991 August 30 for the observations of solar flares and the solar corona in X-ray and gamma-ray wavelengths. *Yohkoh* carries 2 X-ray telescopes. With the hard X-ray telescope, it is possible for the first time to take the images of the solar flares above 30 keV with a moderate spatial (7 arcsec) resolution. With these hard X-ray images, we are able to obtain a spatial distribution of purely non-thermal electrons produced in solar flares. The soft X-ray telescope (Tsuneta *et al.* 1991) is the grazing incidence X-ray (5-50 Å) telescope to observe the solar corona and solar flares with a high spatial (2.5 arcsec) and time (2 sec) resolution. The soft X-ray telescope is sensitive to plasma with temperatures ranging from $10^6$K to $10^8$K. Two other instruments onboard are the Bragg crystal spectrometer to observe iron, calcium, and sulfur lines from flare plasma, and the wide band spectrometer to observe the flare spectra from 5 keV to 10 MeV. For details on *Yohkoh* onboard instrumentation, the readers are referred to *Solar Physics* **136**, 1991.

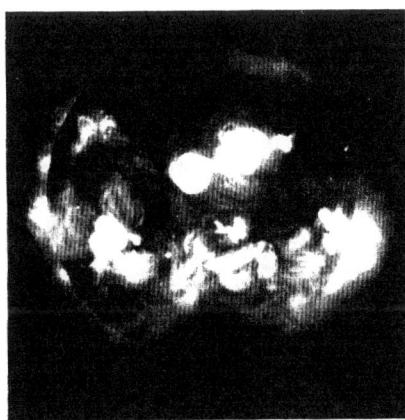

Figure 1. X-ray corona taken by the Soft X-ray Telescope aboard *Yohkoh* on 1991 November 12. North is up. East is to the left.

447

The new discoveries obtained with *Yohkoh* cover wide area in solar physics. To mention several of them with emphasis on the work done in my group:
(1) Global magnetic restructuring (Tsuneta *et al.* 1992b, Tsuneta 1993a): The X-ray movie shows frequent topological changes of coronal magnetic fields on global and local scales. This restructuring is associated with X-ray brightening (heating), suggesting that the magnetic reconnection is a primary device for the restructuring.
(2) Solar flares as an on-going magnetic reconnection (Tsuneta *et al.* 1992a, Tsuneta 1993b): *Yohkoh* observations clearly show that X-type neutral sheet structure is created in association with MHD instability involving global active region fields. Magnetic reconnection takes place at the neutral sheet formed by the instability, and is responsible for flare energy release.
(3) Microflares and coronal heating (Shimizu *et al.* 1992, 1994, Tsuneta 1994a): Microflares are frequently seen in active regions. Multiple loop interaction (reconnection) appears to be involved in some of the brightenings. Although the energy supply from the observed microflares is not enough to heat the active region corona, there is a possibility that the transient energy release may significantly contribute to the coronal heating.
(4) High temperature plasma of active regions (Hara *et al.* 1992): Active region temperatures obtained with SXT are 4-5 MK, considerably higher than those obtained by *Skylab*. This high temperature component appears to coexist with the known 1-2 MK component.
(5) Higher coronal hole temperature (Hara *et al.* 1994): The temperatures of the coronal holes obtained with SXT reach around 2 MK, which is considerably higher than those obtained with the previous observations. The temperature is as high as the temperatures of the quiet region.
(6) Strong X-ray emission from buoyant emerging fluxes (Shibata *et al.* 1994): Magnetic fluxes emerged from below the photosphere are very bright in X-rays. Strong heating occurs through reconnection between the emerging fluxes and the existing coronal fields.
(7) Rigidly rotating coronal holes (Tsuneta 1993a): We find that the thin long (along N-S direction) coronal holes often seen in SXT images are rigidly rotating with rate close to the equatorial rotation rate. It is well known, however, that the photospheric plasma has the differential rotation: The equatorial region rotates faster. The probable explanation for the discrepancy can be found in Wang, Nash and Sheeley (1991).
(8) Expansion of the activer region magnetic fields (Uchida *et al.* 1992): Active region magnetic fields sometimes show almost continuous expansion. It appears that they eject mass and magnetic fluxes into the interplanetary space.

This paper is concerned with some of the new findings and the theoretical problems raised by those discoveries.

## II. X-RAY CORONA AND SOLAR MAGNETIC FIELDS

Figure 1 is a soft X-ray image of the quiet sun. We can see the complex structures; active regions, large scale magnetic structures, coronal holes on both polar regions, and an X-ray jet (Shibata *et al.* 1992) from a bright region located on the North-West side of the Sun. These complex structures essentially represent the magnetic structures of the corona, because the collision frequency in the solar corona is generally much smaller than the gyro-frequency for the typical magnetic field strength (10-1000 G). High resolution soft X-ray imaging is the only means to "see" the coronal magnetic structures.

Figure 2 illustrates the generation and dissipation of the solar magnetic fields. The toroidal magnetic fields are believed to be generated at the base of the convection zone or below it, where the intense toroidal flux tubes are intensified from the poloidal fields possibly due to the rotational shearing ($\omega$ mechanism). When the field strength reaches the critical value determined by the local superadiabaticity and the plasma $\beta$ (several 10 KG), it becomes buoyant (Parker buoyancy) and emerges to the photosphere with duration much shorter than the solar cycle of 11 years (Spruit and van Ballegooijen 1981, van Ballegooijen 1982). A wonderful numerical simulation on the dynamics of the rising thin flux tubes in the solar convection zone has been recently done by Fan, Fisher and DeLuca (1993). The sunspots and the plage regions seen on the photosphere are the cross-sections of the distinct flux tubes. The magnetic fields emerged to the photosphere then heat the corona, and produce solar flares in the corona, where the plasma $\beta$ is generally much lower than 1. The reversed poloidal fields, which are the seeds of the toroidal fields in the next solar cycle, are generated from the toroidal fields ($\alpha$ effect, *e.g.* Cowling 1981, Leighton, 1964, 1969).

Our understanding of the solar cycle and the dynamo process is, however, far from complete. The most serious challenge is that the internal rotation profile of the Sun expected from the various dynamo theories is not consistent with the observed rotation profile obtained with the helioseismology (Shibahashi 1994). Also, it is not yet well understood how the toroidal fields are ultimately removed from the Sun for the solar cycle. Wang, Sheeley and Nash (1991) propose an attractive new idea on the solar cycle and the dynamo action.

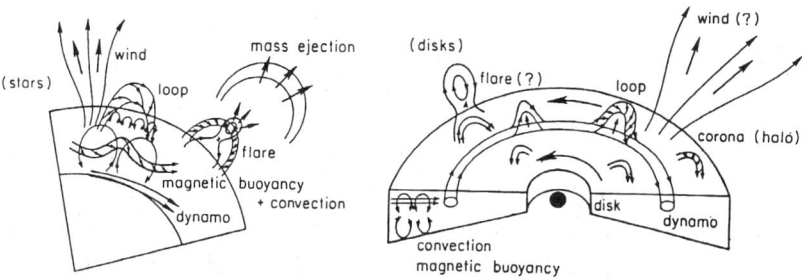

Figure 2. Solar magnetic fields: The toroidal magnetic fields are generated at the base of the convection zone and then emerge to the photosphere, and dissipate. Coronal heating, flares and other dynamical phenomena occur in the corona as a part of the dissipation process. Similar dynamical processes would take place in the accretion disk of AGN etc. (Makishima 1994).

In this situation, we expect the soft X-ray images to serve as a diagnosis tool of the internal flux tubes and the dynamo action. For instance, the spatial distribution of the X-ray loops sometimes appears to be part of "sea-serpents (Zwaan 1987)" of the main flux tubes located in the convection zone.

Figure 3 shows an X-ray image with the corresponding optical by *Yohkoh*, Hα and magnetogram images. The bright regions in X-rays (active regions) correspond very well with the strong magnetic pairs. Furthermore, most of the enhancements seen in X-rays also correspond to small magnetic pairs on the photosphere. This correspondence shows that the energy input to the corona is closely related to the magnetic field strength of the photosphere (see section IV for exceptions).

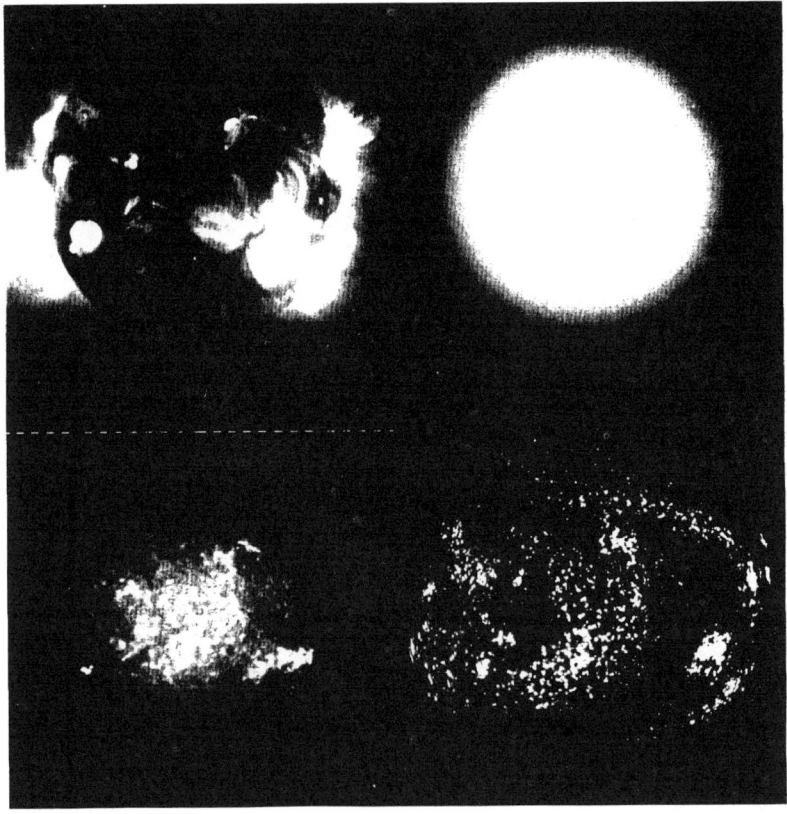

Figure 3. X-ray (top left) and white light (top right) images both taken by *Yohkoh* soft X-ray telescope together with Hα (bottom left) and magnetogram (bottom right) images. The gray scale of the magnetogram image indicates the polarity and magnitude of the longitudinal component of the photospheric magnetic fields.

## III. GLOBAL TOPOLOGY CHANGE OF CORONAL MAGNETIC FIELDS

A. Example of restructuring

So far, based on the low cadence *Skylab* X-ray images, it was thought that the coronal magnetic field structures are robust, and keep the same configurations for hours, days and sometimes weeks. This is theoretically interpreted as due to the extremely high electrical conductivity of the coronal plasma, which impedes the interactions (reconnection) of coronal magnetic fields. The magnetic Reynolds number is about $10^{12}$ for the classical coronal plasma [T = 2 - 40 $\times$ $10^6$ K, n = $10^{8-11}$ cm$^{-3}$, L = $10^{3-5}$ km]. The plasma $\beta$ range from $10^{-5}$ (sunspots) to around $10^{-1}$ (flares) in the corona, whereas the $\beta$ on the photosphere is much larger than 1.

As shown in the high cadence *Yohkoh* movie, however, the solar corona is full of highly transient phenomena, both global and local on many time scales ranging from a few minutes to several hours. Solar coronal magnetic fields restructure themselves, while dissipating the magnetic energy. If there is no reconnection, the coronal magnetic fields should have had more and more complex configuration owing to the random and systematic photospheric plasma flow, and should have built up higher and higher energy without relaxation.

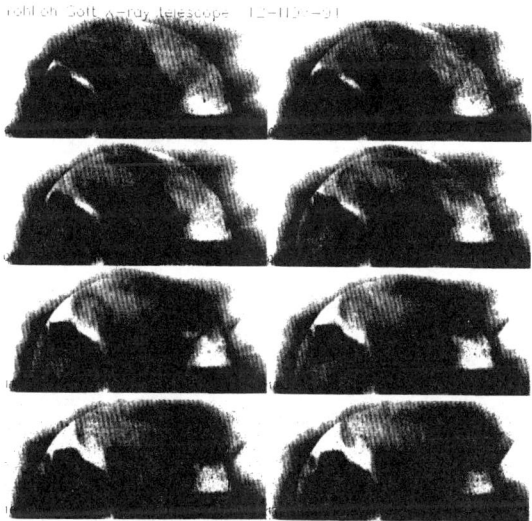

Figure 4. Large scale coronal restructuring that occurred on 1991 November 12. The closed loop or cusp structure is formed over 10 hours owing to magnetic reconnection at the neutral sheet. The cusp structure seen on the west limb (to the right) indicates the separatrix of the reconnected field lines. Note that the height and the separation of the footpoints increase as a result of reconnection.

Figure 4 shows a spectacular formation of the closed field lines from the open field lines over 10 hours. Figure 5 shows the schematic evolution of the coronal magnetic structure. First, the open field structure with the neutral sheet is formed by a global MHD instability. Then, the closed field structure is formed from the open field structure due to magnetic reconnection. The closed loops are heated, and then emit X-rays. Thus, the height and the footpoint separation of the closed loops increase. The cusp seen in X-rays would identify the separatrix of the reconnected magnetic fields. The formation of the open field structure with the neutral sheet, which has higher magnetic energy than the closed field configuration for the given boundary condition on the photosphere, is an important theoretical question raised by the *Yohkoh* observation.

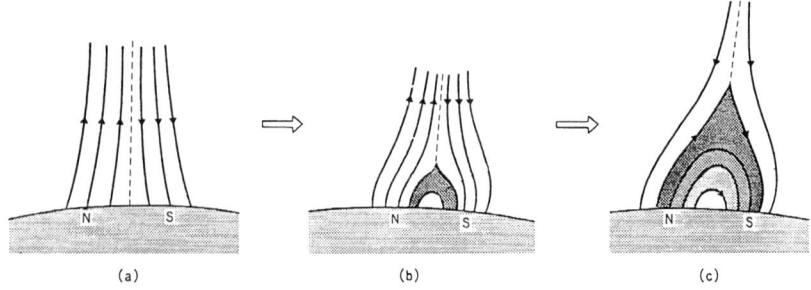

Figure 5. Evolution of the magnetic fields for the global restructuring shown in Figure 4.

B. Solar corona as self-organized critical system

These observations indicate that the global topological changes do occur in the solar corona. Actually, much more complex restructurings were frequently observed: The *Yohkoh* soft X-ray movie shows that the solar corona is full of such global restructurings. Magnetic reconnection itself is a local phenomena. But, the global topological change of the coronal magnetic fields occur through single X-point as a result of reconnection.

The movie also shows many examples of successive propagation of the restructuring. When the local restructuring occurs somewhere on the Sun, it appears to trigger another restructuring at another place. This chain sometimes propagates far away from the first restructuring. A restructuring would create new magnetic neutral sheet, which may be the location of next global restructuring. Magnetic reconnection and the associated topology changes may go on like avalanche.

How is the magnetic energy built up and how is the neutral sheet formed as a result of a MHD instability? The coronal magnetic fields are constantly agitated through their footpoints by random velocity fields due to granulation and super-granulation motions as well as some systematic velocity fields on the photosphere, where the plasma $\beta$ is much higher than 1. Solar corona may be regarded as a statistical self-organized critical system (Bak, Tang and Wiesenfeld 1988) against such disturbances (noise) applied from the photosphere, from which new magnetic fluxes are also constantly supplied. The solar corona may continuously adjusts (relaxes) itself toward the lower energy state through reconnection process.

## IV. STEADY (DC) MAGNETIC LOOPS AND TRANSIENTS

A. Steady (DC) magnetic loops

Figure 6 is the X-ray and optical images of an active region both taken by *Yohkoh*. Sunspots and strong field regions surrounding the sunspots, called penumbra, are seen in the optical image. We notice two kinds of X-ray structures in the active region; diffuse dark potential-like (current-free) large loops and thin small loops which are much brighter than the diffuse loops. Although active region corona is filled with the magnetic fields, it is apparent that these small loops are selectively heated by unknown mechanism. It is extremely important to investigate why these loops are heated, and the others are not.

We also notice that the corona just above the sunspots is very dark. We do not know if there is normal hot corona above the sunspots. It appears that organized *unipolar* magnetic fields (like sunspots) are not associated with the coronal heating, whereas the *complex* magnetic structures with plus and minus polarities appear more closely related to the heating of the corona.

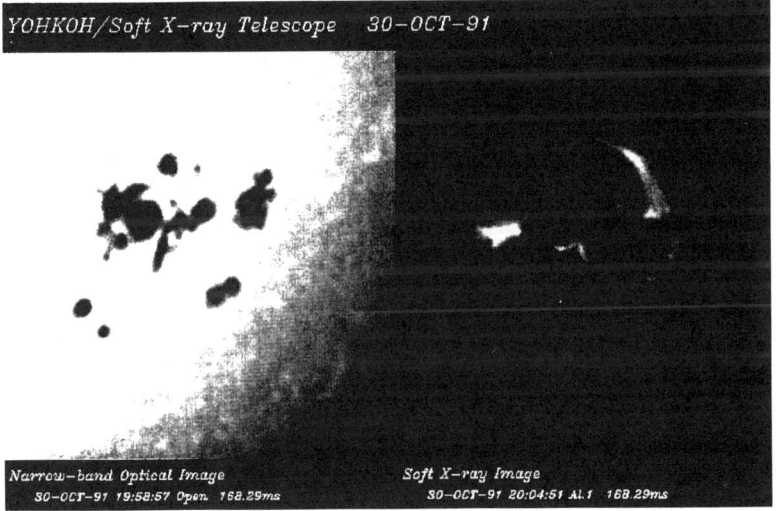

Figure 6. X-ray and optical images of an active region both taken by *Yohkoh*. Note the dark hole in X-rays seen just above the sunspot.

B. Transients: Microflares in active regions

SXT discovered numerous transient brightenings in active regions (Shimizu et al. 1992, 1994). The microflares, like flares, typically have a duration from a few minutes to tens of minutes, but their energy scale is much (several orders of magnitude) smaller than that of the smallest flares. Energetic active regions which produce intense flares tend to produce more microflares. Figure 7 shows active region X-ray images with a high time resolution. Microflaring loops are shown by arrows.

In the normal state, these compact loops are faint and are obscured by the diffuse component of the active region. Figure 9 shows an example of the fine structure of the microflares. Multiple footpoints brighten first, followed by the brightenings of the multiple loops. It is evident from this image that the multiple loops are involved in microflares, implying that magnetic reconnection of multiple loops is the energy source of microflares.

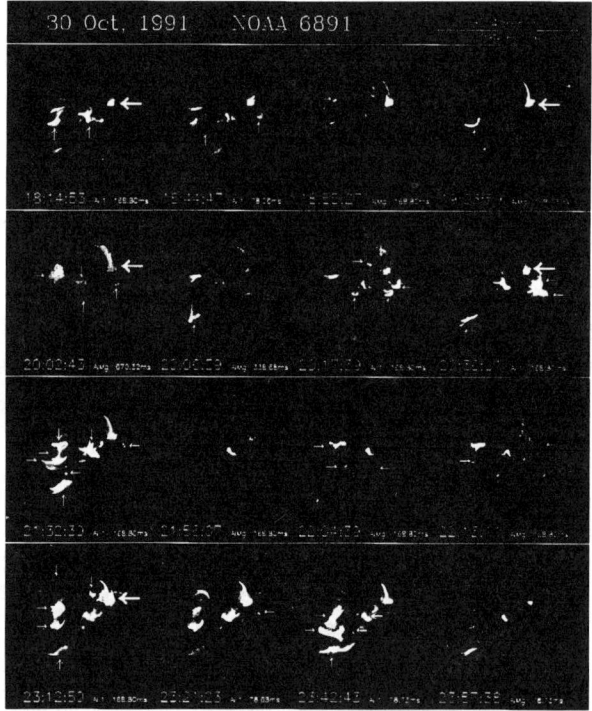

Figure 7. X-ray images of an active region. Frequent microflaring are seen as indicated by arrows (Shimizu et al. 1992). The FOV is 8 arcmin, and the pixel size is 2.5 arcsec.

Coronal heating problem has been one of the central problems in solar physics: Among various candidates, heating by coronal DC currents and Alfven wave dissipation may be viable. A new conjecture from the *Yohkoh* observation is that the solar corona is heated by the microflares. Figure 8 shows the frequency distribution of the microflares observed by *Yohkoh*.

According to Shimizu (1994), the frequency distribution of the microflares is a power-law from $10^{26.5}$ ergs to $10^{29}$ ergs with index of -1.4 to -1.6. The frequency distribution of the hard X-ray flares observed by *SMM*, which are more intense than microflares, has a similar power law (power index of -1.53) from $10^{28.5}$ ergs to $10^{32}$ ergs (total non-thermal electron energy above 25 keV) (Dennis 1985, Crosby et al. 1993).

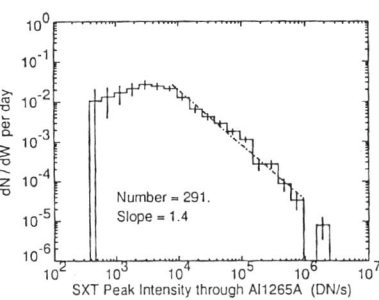

Figure 8. Frequency distribution of the microflares as a function of the soft X-ray peak flux. The error bars represent $\pm 1\sigma$ uncertainties. The distribution can be fitted by a power law with an index of 1.4 (dash-dotted line). The deviation from the power law in weaker events is due to the telescope sensitivity limit (Shimizu 1994).

The similarity of the power indices implies that the microflares and the flares have the same basic physics. Although these energy estimates contain uncertainty, and are not calibrated with each other, the combined histogram may indicate the power-law distribution over 5 orders of magnitude.

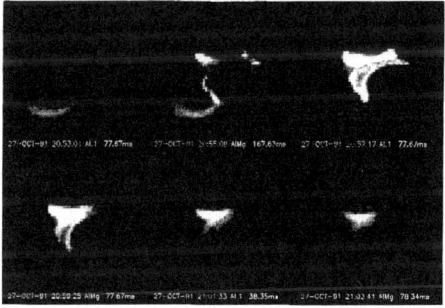

Figure 9. X-ray evolution of a microflare that occurred on 1991 October 27. The footpoint brightenings are followed by brightening of the entire loops.

Shimizu (1994) concludes that the microflares observed by *Yohkoh* alone cannot heat the active-region corona, assuming that the same power-law continues to weaker undetectable events. Do these two observations rule out Parker's nano flare hypothesis (Parker 1988)? X-ray morphology and fairly flat frequency distribution of microflares appear not to be consistent with the hypothesis. There is, however, a possibility that we observe only the tip of the iceberg because of the high X-ray background (diffuse loops) of the active regions, and that there is a steep increase of the number of weaker microflares undetectable with *Yohkoh*. We need a higher sensitivity (higher spatial resolution and a relevant temperature range) to give the definite result.

## V. TEMPERATURE OF CORONAL PLASMAS

A. Temperature of the active regions

SXT is equipped with 5 broad band metal filters, each of which has different temperature response (Tsuneta et al. 1991). Images taken with the thin aluminum filter is sensitive to plasma above 1 MK, while images with the thick beryllium filter is sensitive to plasma only above 4-5 MK, which is mainly prepared for flare observations. Unexpectedly, the coronal structure, especially active regions, are clearly seen in the beryllium images. An immediate implication is that there is a steady temperature component higher than 4 MK in the active regions. *Skylab* and other observations, however, give a temperature of about 2 MK for active regions. The temperatures obtained with the various filter pairs are concentrated around 5-6 MK for active regions (Hara et al. 1992). Some filter pairs, however, give much lower temperature around 2-3 MK, which is consistent with the previous observations. This indicates that the emission measure distribution (as a function of temperature) of active regions generally extends (from 1-2 MK) up to 5-6 MK, considerably higher than those obtained by *Skylab*. The discovery of the high temperature component may suggest a higher energy input to the corona, because the dominant conductive energy loss of the corona sharply depends on the temperature.

B. Temperature of coronal holes

The temperatures of the coronal holes obtained with *Yohkoh* reaches around 2 MK, which is considerably higher than those obtained with the previous observations of 1 MK or below (Hara et al. 1994). This again suggests the presence of the hot component in the coronal hole. The low X-ray brightness of coronal holes is due to its low density and not low temperature. The absolute temperature obtained with the broadband filters has to be carefully examined, when the plasma has the multiple temperature components. Although the absolute temperature of the coronal holes is subject to this uncertainty, it is certain that the coronal holes have the component with temperature as high as the temperatures in the quiet region.

Coronal holes are identified as the source of the high speed solar wind. A number of mechanisms has been proposed so far to explain the fast wind from cool coronal holes. *Yohkoh* observations, however, imply that the thermal (Parker) wind may explain the high speed solar wind without (or with less) additional non-thermal heat or momentum inputs.

These surprising results will seriously affect our current understanding of the overall energetics of the solar corona and the coronal heating problem.

## VI. SOLAR FLARES

A. Magnetic Reconnection

Solar flares are the most violent and spectacular phenomena. It produces $10^{29-32}$ ergs/flare in the forms of non-thermal electrons [ $10^{33-35}$ electrons/sec are accelerated to above 20 keV.], a high temperature plasma with temperatures of $2 - 4 \times 10^7$ K, and a mass ejection. It is now certain from *Yohkoh* observations that magnetic reconnection is involved in the flare energy release (Tsuneta et al. 1992a, Tsuneta 1993a, 1993b).

We have several pieces of strong evidence that magnetic reconnection of the neutral sheet formed at the top of the flare loop participates in the flare energy release (see also Figure 5):

(1) The flare loop increases its height and footpoint separation in soft X-rays.
(2) The outer X-ray arches generally have higher temperatures. This is consistent with the flare loop resulting from the reconnected magnetic fields.
(3) An X-type neutral sheet configuration is formed near the peak of hard X-rays in association with the eruption of the entire magnetic structure of the active region. Solar flares themselves are a local phenomena, but are the result of an instability of the global magnetic structure in the active region.

In the quiet sun observation, we suggested that magnetic reconnection plays an essential role for non-explosive restructuring of the coronal magnetic fields. Present observation demonstrates that similar process is responsible for flares with smaller spatial dimension but with much larger energy scale. The only difference is difference of the magnetic field strength: the magnetic field strength of the quiet sun restructuring is much smaller than that of flares, because latter occur in the general corona with larger scale size. Nevertheless, magnetic reconnection appears to play a fundamental role both for flares and non-explosive restructuring of the quiet sun. Again, we should stress the unresolved problem: How is the neutral sheet (open field) structure created as a result of MHD instability? (Note that the magnetic fields emerged from the photosphere have a loop structure.)

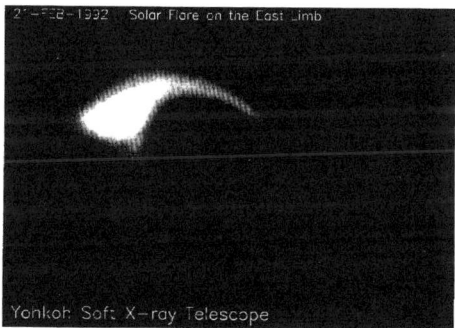

Figure 10. X-ray flare that occurred on 1992 February 21. The flare occurred on the east limb. The field of view is 2.6 arcmin, and the pixel size is 2.5 arcsec.

B. Nonthermal Particle Acceleration

We have, therefore, substantial pieces of evidence from *Yohkoh* observations that magnetic reconnection is involved in a flare energy release. It is, however, not clear how much energy can be released by magnetic reconnection. In particular, non-thermal electrons carry substantial amount of energy released by solar flares. Can magnetic reconnection alone accelerate the electrons upto 100 keV or more, and thus supply the entire flare energy in the form of non-thermal electrons?

Sakao (1994) obtains the following general conclusions from the *Yohkoh* hard X-ray observations:

(1) Hard X-ray flares essentially have double source structures, which are located at the footpoints of the soft X-ray loop.

(2) The intensities of the double sources correlate in time within the accuracy of 100 msec.

The double source structure and the intensity correlation indicate the bi-directional particle acceleration toward footpoints. Particle acceleration in the neutral sheet located above the soft X-ray loop is not consistent with such a concentrated double-source structure. If particle acceleration is driven by the uni-directional field-aligned electric fields, we should have seen a single source structure on either end of the soft X-ray loops. If the double source structure is due to the magnetic mirroring with uni-directional acceleration, systematic time delay of the hard X-ray emission from one of the footpoints with respect to the other should have been detected.

These simple considerations lead to the conclusion that the electrons are accelerated by bi-directional electric fields in the soft X-ray loop. *Yohkoh* observations support the conjecture from *Hinotori* observations that there are many (fine scale) anti-parallel sub-Dreicer electric fields along the loop over the macroscopic distances (Tsuneta 1985, 1987, Holman 1985). It has been long argued how to directly convert the magnetic energy to particle acceleration. Tsuneta (1994a) points out that the hydromagnetic reconnection outflows with speed of Alfven speed ($\sim 10^3$ km/sec) has energy comparable to the total flare energy, and that the bi-directional field-aligned currents are driven by the vortices at the loop top created by the reconnection out-flow colliding with the loop structure.

## VII. CONCLUDING REMARK

Lifetimes of the transient events observed by *Yohkoh* range from tens of seconds to several 10 hours. When the time scales are expressed by the Alfven transit times, the time scales of the transients range from 10 to 100 Alfven transit times, whereas the lifetime of the DC type loops may reach more than 1000 Alfven transit time, suggesting different mechanisms for heating (Table I, Tsuneta 1994b). Recent laboratory experiments (Yamada *et al.* 1990, 1991) on magnetic reconnection gives similar Alfvenic transit time scales, suggesting the common basic physics of magnetic reconnection between the solar transients and the laboratory plasma.

The author would like to thank Mr. Victor Shvetsky of the University of California, Irvine for his help in preparing the manuscript.

TABLE I. Time Scales of Transient Phenomena Observed by *Yohkoh*.

| | B(gauss) | Length | n(cm$^{-3}$) | Time scale |
|---|---|---|---|---|
| Quiet Sun Restructuring | 10 | $1 \times 10^5$ km | $10^9$ | 2 hr $\sim 50\tau_A$ |
| Transient Brightings | 100 | $5 \times 10^3$ km | $10^{10}$ | 2 min $\sim 40\tau_A$ |
| Flares | 100-500 | $10^4$ km | $10^{11}$ | 10 min $\sim 60\tau_A$ |
| Steady loops | 100 | $5 \times 10^3$ km | $10^{10}$ | $1 hr \sim 10^{3-4}\tau_A$ |
| Laboratory experiment (Yamada *et al.* 1990, 1991) | 500 | 15 cm | $3 \times 10^{14}$ | $30\mu sec \sim 10 - 60\tau_A$ |

Note: $\tau_A$ is the Alfven transit time obtained from the tabulated parameters. The time scales should be regarded as order-of-magnitude estimates to demonstrate the gross similarity in the time scales of the various transient phenomena (except steady coronal loops).

## REFERENCES

Bak, P., Tang, C., Wiesenfeld, K., 1988, Phys. Rev. A, **38**, 364.
Cowling, T. G., 1981, Ann. Rev. Astron. Astrophys., **19**, 115.
Crosby, N., Aschwanden, M., Dennis, B., 1993, Solar Phys., **143**, 275.
Dennis, B., 1985, Solar Phys., **100**, 465.
Fan, Y., Fisher, G. H., DeLuca, E. E., 1993, Astrophys. J., **405**, 390.
Hara, H., Tsuneta, S., Acton, L. W., Lemen, J. R., McTiernan, J. M., 1992, Publ. Astron. Soc. Japan, **44**, 135.
Hara, H., Tsuneta, S., Acton, L. W., Lemen, J. R., Bruner, M., Ogawara, Y., 1994, Publ. Astron. Soc. Japan, in press.
Holman, G. 1985, Astrophys. J., **293**, 584.
Leighton, R. B., 1964, Astrophys. J., **140**, 1547.
Leighton, R. B., 1969, Astrophys. J., **156**, 1.
Makishima, K., 1994, these proceedings.
Ogawara, Y., Takano, T., Kato, T., Kosugi, T., Tsuneta, S., Watanabe, T., Kondo, I., Uchida, Y., 1991, Solar Phys., **136**, 1.
Parker, E., 1988, Astrophys. J., **330**, 474.
Sakao, T., 1994, Ph. D. Thesis, University of Tokyo.
Sakao, T., Kosugi, T., Masuda, S., Yaji, K., Inda-Koide, M., Makishima, K. 1994, *X-ray Solar Physics from Yohkoh*, eds. Uchida, Y. *et al.*, Universal Academy Press, Tokyo, p 91.
Shibahashi, H., 1994, these proceedings.

Shibata, K., Ishido, Y., Acton, L., Strong, K., Hirayama, T., Uchida, Y., McAllister, A., Matsumoto, R., Tsuneta, S., Shimizu, T., Hara, H., Sakurai, T., Ichimoto, K., Nishino, Y. and Ogawara, Y. Shibata, K., 1992, Publ. Astron. Soc. Japan, **44**, L173.

Shibata, K., Nitta, N., Matsumoto, R., Tajima, T., Yokoyama, T., Hirayama, T., Hudson, H., 1994, *X-ray Solar Physics from Yohkoh*, eds. Uchida, Y. *et al.*, Universal Academy Press, Tokyo, p. 29.

Shimizu, T., Tsuneta, S., Acton, L., Lemen, J., Uchida, Y., 1992, Publ. Astron. Soc. Japan, **44**, L147.

Shimizu, T., Tsuneta, S., Acton, L., Lemen, J., Ogawara, Y., Uchida, Y., 1994, Astrophys. J., **422**, 906.

Shimizu, T., 1994, Proc. *Kofu* symposium, eds. Enome, S., Hirayama, T., National Astronomical Observatory, Japan, in press.

Spruit, H. C., van Ballegooijen, A. A., 1981, Astron. Astrophy., **106**, 58.

Tsuneta, S., 1985, Astrophys. J, **290**, 353.

Tsuneta, S., 1987, Solar Phys., **113**, 35.

Tsuneta, S., Acton, L., Bruner, M., Lemen, J., Brown, W., Caravalho, R., Catura, R., Freeland, S., Jurcevich, B., Morrison, M., Ogawara, Y., Hirayama, T., Owens, J., 1991, Solar Phys., **136**, 37.

Tsuneta, S., Hara, H., Shimizu, T., Acton, L., Strong, K., Hudson, H., Ogawara, Y., 1992a, Publ. Astron. Soc. Japan, **44**, L63.

Tsuneta, S., Takahashi, T., Acton, L., Bruner, M., Harvey, K., Ogawara, Y., 1992b, Publ. Astron. Soc. Japan, **44**, L211.

Tsuneta, S., 1993a, *Advances in Stellar and Solar Coronal Physics*, eds. Linsky, J. and Serio, S., Kluwer, p. 113.

Tsuneta, S., 1993b, *The magnetic and velocity fields of solar active regions*, eds. Zirin, H. and Ai, G., IAU colloquium No. **141**, Astronomical Society of the Pacific, p. 239.

Tsuneta, S., 1994a, *X-ray Solar Physics from Yohkoh*, eds. Uchida, Y. *et al.*, Universal Academy Press, Tokyo, p. 115.

Tsuneta, S., 1994b, *Solar Active Region Evolution*, eds. Balasubramanian, K.S., and Simon, G., Astronomical Society of the Pacific, in press.

Uchida, Y. *et al.*, 1992, Publ. Astron. Soc. Japan, **44**, L155.

van Ballegooijen, A. A., 1982, Astron. Astrophy. J., **113**, 99.

Wang, Y.-M., Sheeley, N. R., Nash, A. G., 1991, Astrophys. J., **383**, 431.

Wang, Y.-M., Nash, A. G., Sheeley, N. R., 1991, Science, **245**, 712.

Yamada, M., Ono, Y., Hayakawa, A., Katsurai, M., 1990, Physical Review Letters, **65**, 721.

Yamada, M., Perkins, F.W., MacAulay, A.K., Ono, Y., Katsurai, M., 1991, Phys. Fluids B, **3** (8), 2379.

Zwaan, C., 1987, Ann. Rev. Astron. Astrophys., **25**, 83.

# XII. Positron Plasmas

# XII.1

# PHYSICS WITH TRAPPED POSITRON GASES AND PLASMAS

C.M. Surko and R.G. Greaves

Department of Physics, University of California, San Diego
La Jolla, CA 92093-0319, U. S. A.

Techniques are described to capture positrons efficiently from a radioactive source in a modified Penning trap, using a nitrogen buffer gas. Cold positron gases and plasmas are created, and methods are described to monitor the plasma density and temperature, and the shape of the charge clouds by non-perturbatively studying their modes of oscillation. The resulting positron plasmas and gases are being used for a range of physics experiments, and several examples are discussed. One topic is the study of the physics of electron-positron plasmas, in which the mass of both charge species are equal. Both species can also be both equally and strongly magnetized. Another topic is the study of the interaction of slow positrons with various forms of ordinary matter, including atoms, molecules, atomic clusters and dust grains. These experiments can address important issues in atomic and molecular physics, condensed matter physics, and gamma-ray astrophysics.

## I. INTRODUCTION

In this paper, we review recent progress in developing techniques to manipulate low energy antimatter, in the form of positron gases and plasmas, and we describe the utilization of these collections of antimatter for physics research in a number of areas.

The confinement of small numbers of particles in Penning traps was developed by H. Dehmelt and co-workers at the University of Washington. They demonstrated that single particles can be confined for months, and they used the trapped particles for such experiments as the high precision measurement of the $g$-factor of the electron [Dehmelt, 1990].

J. H. Malmberg and co-workers at the University of California, San Diego, modified the Penning trap to study clouds of electrons that were sufficiently cold and dense to form plasmas [Malmberg and de Grassie, 1975]. They demonstrated that these pure electron plasmas have remarkably good confinement properties [Malmberg and Driscoll, 1980; O'Neil, 1980], and they have explored a wide range of physics issues including transport [Driscoll et al., 1988], collective modes [White et al., 1982], and simulations of fluid vortices [Driscoll and Fine, 1990].

The excellent confinement properties of Penning traps make them a natural choice for the accumulation of antiparticles. We have modified the Malmberg-Penning trap to accumulate positrons from a radioactive source by scattering from a buffer gas [Surko et al., 1988] and we are now able to accumulate more than $10^8$ positrons with a lifetime of about half an hour. Small numbers of positrons were earlier trapped by Schwinberg et al. [1981], who obtained an accurate measurement of the positron $g$-factor. Antiprotons from the low energy antiproton ring (LEAR) at CERN have been trapped in Penning traps [Gabrielse et al., 1989; Holzscheiter, 1993], and have provided the most accurate measurement of the antiproton mass available [Gabrielse et al., 1990].

Trapped positrons provide many opportunities for physics research. The positron is a qualitatively new plasma particle, in that it has the mass of the electron and a positive charge. Thus, in principle, one can study the physics of an equal-mass, neutral plasma. While equal-mass, neutral, ion plasmas have been studied previously [Sheehan and Rynn, 1988; Cooney et al., 1993], they have had sufficient densities of free electrons to significantly modify the plasma dynamics from that expected for a pure equal-mass plasma. The electron-positron plasma system is also of interest in that both charge species can be magnetized to the same degree. Such strongly magnetized neutral plasmas have not been investigated previously, and it is possible that studies of confinement in such systems can lead to a better understanding of the confinement of neutral plasmas in general. There is considerable history of using positrons for plasma research, dating from the work of Gibson et al. [1963], who used positrons as test particles to investigate confinement in a magnetic mirror. However, classical, single-component positron plasmas have only recently been created in the laboratory [Surko et al., 1989], and it has only been in the last year that our group has been able to introduce an electron beam into a positron plasma to begin to investigate the electron-positron plasma system [Greaves et al., 1994a].

Trapped positrons also have technological applications. For example, they could be applied to produce pulsed beams of positronium atoms, which have been proposed as a transport diagnostic in tokamak fusion plasmas [Surko et al., 1986; Murphy, 1987]. They are also being developed as a source of low-emittance positron beams [Wineland et al., 1993], which are important for such applications as surface studies [Suzuki et al., 1992]. The production of ions by positron annihilation is a qualitatively different physical process from ionization by electron-impact ionization and is being investigated as a source of ions for mass spectrometry [Passner et al., 1989; Glish et al., 1994].

At this point, the research has several facets: one is to refine the trapping techniques to confine larger quantities of antimatter for longer periods of time. A second is to develop techniques to monitor, cool and manipulate collections of antimatter in a non-destructive manner. Finally, experiments have begun to use these collections of antiparticles to address outstanding physics issues.

This paper is organized in the following way. We first review the positron trapping techniques that we have developed, including discussion of capture efficiency and storage times. We then go on to describe methods to cool, heat and manipulate these trapped positron plasmas. Following that, we discuss physics experiments in progress and other possible areas of research using cold accumulations of stored positrons.

## II. THE POSITRON TRAP

Here, we present a brief overview of the operation of our positron trap. A more detailed description can be found elsewhere [Murphy and Surko, 1992]. The positron source is typically about 100 mCi of $^{22}$Na ($\tau_{1/2} = 2.6$ years). Positrons from the source with energies in the range from zero to 540 keV pass through a 0.1 $\mu$m thick Ti window.

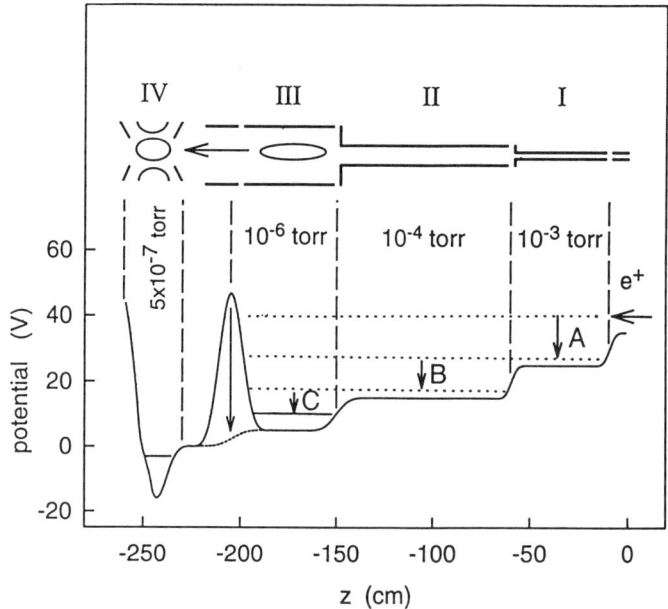

Figure 1: Layout of the four-stage positron trap. Positrons from the moderator enter the trap at $z=0$ cm. The potential barrier at $z=-210$ cm is lowered periodically to transfer trapped positrons from stage III to stage IV.

In order to use the trapping techniques described here, the positrons must be slowed to energies of a few electron volts. In our previous experiments, this was accomplished using a single-crystal tungsten foil approximately 1 $\mu$m thick [Gramsch et al., 1987]. We have recently installed a rare gas solid moderator, following the design developed by Mills and co-workers [Mills, Jr. and Gullikson, 1986; Khatri et al., 1990]. In this design, the source is located near the apex of a cone or paraboloid, and all surfaces are coated with a layer of a solid rare gas, usually krypton or neon, at temperatures below 10 K. Fast positrons lose energy in the solid rare gas and then diffuse to the surface, where they are ejected with energies of a few eV. Tungsten moderators typically have efficiencies $\epsilon \sim 2 \times 10^{-4}$, defined as slow positrons produced, per fast positron emitted from the source. The efficiency of the rare gas solid moderators is an order of magnitude larger, (i.e., $\epsilon \sim 2 \times 10^{-3}$). The energy spread of the positrons from the tungsten is about 0.3 eV, while the corresponding energy spread associated with rare gas solids about one order of magnitude larger (e.g., a few electron volts).

As shown in Fig. 1, moderated positrons are guided by a magnetic field into a potential well created by a series of cylindrical electrodes. A nitrogen buffer gas is introduced into the electrode structure, and inelastic excitation of electronic transitions in the nitrogen molecules is used as an energy loss mechanism to trap the positrons in the potential well. The present version of our trap has four stages, each with successively lower pressure. The pressure in stage I is adjusted so that, on average, a positron makes one inelastic collision (denoted 'A' in Fig. 1) from an $N_2$ molecule in one round trip through the trap and is thereby confined in the trap. After further transits in the trap, the positron makes two more inelastic collisions (denoted 'B' and 'C' in Fig. 1) and is then near the bottom of the potential well in a sufficiently low pressure of buffer gas so that diffusion across the magnetic field is slow. The potential drop between the stages is adjusted to be between 8 and 9 eV. At these energies, there is an appreciable cross-section for electronic excitation of the $N_2$, but their energy is low enough that formation of positronium by $e^+$–$N_2$ collisions is small. This later process is the dominant loss process in this trapping scheme, since the neutral positronium atoms can exit the trap and the positrons annihilate with electrons at the walls of the trap. However, positronium atom formation has an energy threshold of about 8.6 eV in $N_2$, so that this process is forbidden once the positrons are in stage III. At typical operating pressures, incident positrons are trapped in stage III in about 30 ms. At a buffer gas pressure of $1 \times 10^{-6}$ torr, they cool to room temperature by subsequent vibrational, rotational and momentum transfer collisions with the $N_2$ in about 1 s.

For positrons from a tungsten moderator, the efficiency of the trap is approximately 40%, defined as the fraction of incident positrons trapped in stage III, per moderated positron entering the trap. For positrons from rare gas moderators, the larger energy spread results in a lower trapping efficiency, typically 25%, but this is more that compensated by the larger moderation efficiency which at present is about 0.3%. We expect that the trapping efficiency can be further improved.

Confinement in stage III is limited by "direct" annihilation on the $N_2$ molecules, in which the positron annihilates with an electron on the molecule in the course of an elastic collision. For a buffer gas pressure of $2 \times 10^{-6}$ torr, the annihilation time is 40 s. Shown in Fig. 2 is a summary of positron storage and confinement results, comparing positron confinement for the usual buffer gas pressure in stage III with that obtained when the buffer gas is pumped out, and also for electrons in the same situation. The positron confinement is still limited by annihilation with residual gas molecules in the vacuum system, even at pressure of $6 \times 10^{-10}$ torr. Our present vacuum system has viton o-rings and is bakable only to 60 °C, so we expect that the confinement can still be greatly improved.

## III. PLASMA CHARACTERIZATION AND MANIPULATION

We have now added a fourth stage to the trap, placed about 50 cm from stage III and operating at a pressure of $1 \times 10^{-6}$ torr. Using it, we have shown that we can periodically (e.g., every 5 s or so) shuttle bunches of positrons from stage III to IV [Greaves et al., 1994a]. Having demonstrated this stacking capability, we envision that it will not be difficult to arrange a much lower pressure confinement region, separated from III by a differentially pumped region or by two fast gate valves and a stage of differential pumping. In this way, one could achieve high accumulation efficiencies, long accumulation times, and long-term storage in a UHV environment. With these improvements, we expect to be able to increase the total number of trapped positrons by another two orders of magnitude. The ability to produce such dense clouds of positrons in a UHV environment can also be expected to be useful for the production of antihydrogen.

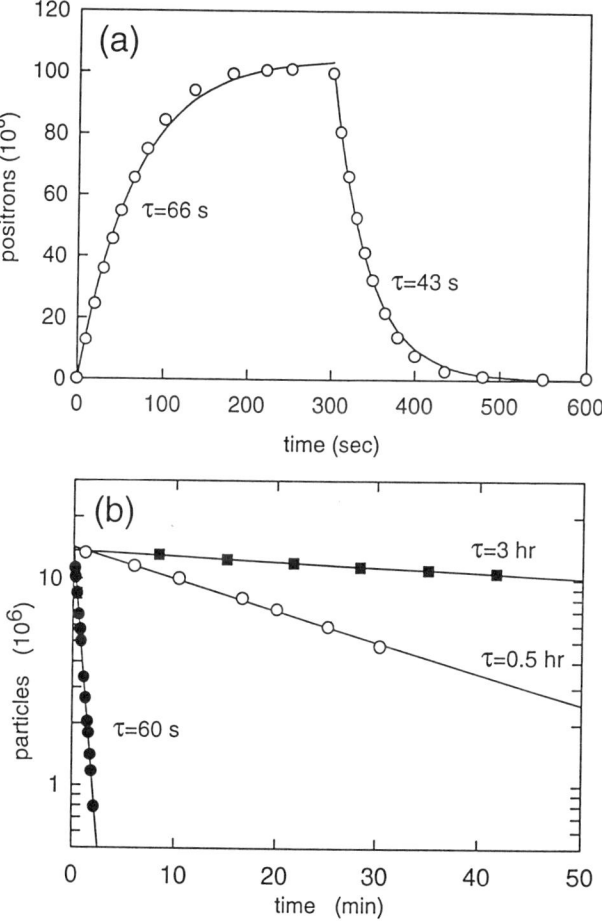

Figure 2: (a) Positron filling and storage from a 70 mCi $^{22}$Na radioactive source in the presence of buffer gas at a pressure of $6 \times 10^{-7}$ torr. The source is switched on at $t = 0$ and switched off at $t = 300$ s. The apparent difference in the two lifetimes shown arises from changes to the filling rate caused by the positron space charge. (b) Positron (o) and electron (■) storage when the buffer gas feed is switched off after loading particles ($p = 6 \times 10^{-10}$ torr); positron (•) storage in the presence of the buffer gas ($p = 6 \times 10^{-7}$ torr).

Density, temperature and radial profiles of the positron plasmas can be measured by dumping them [Greaves et al., 1994a]. Line-integrated density at a given radius is measured by dumping the plasma onto a set of annular collectors. Temperature is measured by dumping the plasma onto a "magnetic beach" energy analyzer [Boyd et al., 1973]. Knowing the line-integrated density and the temperature, the plasma shape and the local density can be obtained by solving the Poisson equation. These techniques have the disadvantage that a plasma must be destroyed for each measurement. We have now developed non-destructive, remote sensing diagnostics of plasma parameters, by studying the collective modes [Dubin, 1991; Dubin, 1993] of these spheroidal plasmas [Tinkle et al., 1994]. These modes also been studied in cryogenic pure electron plasmas [Weimer et al., 1993] and in pure ion plasmas [Greaves et al., 1994b; Bollinger et al., 1994; Bollinger et al., 1993].

For convenience, the work on modes thus far has emphasized electron plasmas. Excitation and detection of the center-of-mass mode of plasma motion in the external potential well gives a measure of total particle number. This measurement can be calibrated by dumping the plasma once. The lowest-order compressional mode (the quadrupole mode) of the plasma in the direction along the magnetic field is strongly dependent on the temperature and aspect ratio, $\alpha = L/2r_p$, where $L$ is the length and $r_p$ is the plasma radius, and the frequency of this mode can be used to measure either of these quantities if the other is known. Measuring the frequency of a third mode provides enough information to obtain the plasma temperature, shape and total particle number, thereby characterizing completely these single component plasmas.

We have investigated the temperature and aspect ratio dependence of these modes in electron plasmas using an rf heating technique that we have developed [Tinkle et al., 1994]. A typical result is presented in Fig. 3(a) which shows how the plasma temperature increases as a short rf pulse is applied, and then falls as the plasma cools on the buffer gas [Greaves et al., 1994a]. The frequency shift of the quadrupole mode is shown in Fig. 3(b). We are also able to change the aspect ratio of the plasma by changing the magnetic field after the plasma is loaded, and so we were able to investigate the temperature dependence of the quadrupole mode for a range a aspect ratios. A typical set of data is shown in Fig. 4, together with the results of a numerical simulation of the same plasma.

## IV. POSITRON PLASMA EXPERIMENTS

A. Pure positron plasmas

The maximum density positron gas that we have achieved thus far is approximately $2.3 \times 10^6$ cm$^{-3}$, using a 70 mCi $^{22}$Na source and a solid neon moderator. The positron temperature is 300 K, which gives a Debye length, $\lambda_D$ of 0.8 mm. The characteristic size of the spheroidal charge cloud is $L \simeq 20$ cm and $r_p \simeq 0.8$ cm. For these parameters, $N_D \sim 5 \times 10^3$, and $\omega_p \tau_{pn} \gg 1$, where $\tau_{pn}$ is the positron-neutral collision time. Thus, all the classical criteria for plasma behavior are comfortably satisfied, and we are able to create, for the first time, robust pure positron plasmas.

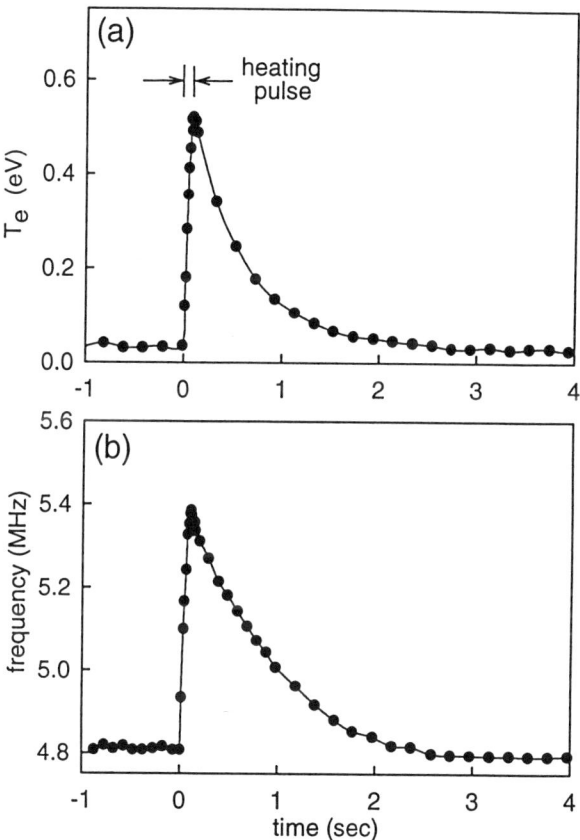

Figure 3: (a) Temperature of a pure-electron plasma during a cycle of heating by broadband rf noise and buffer gas cooling. Note that the temperature is close to 300 K before and after the heating cycle. (b) Frequency of the quadrupole mode which is a measure of the plasma temperature.

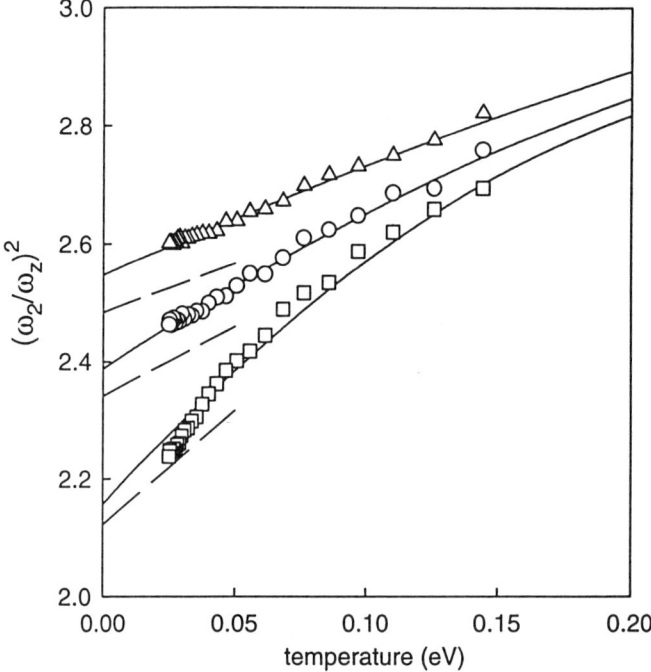

Figure 4: Dependence of quadrupole mode frequency on electron temperature for plasmas with various aspect ratios; (□) $\alpha =2.2$, (○) $\alpha =4.4$, (△) $\alpha =7.8$. The solid lines are from a numerical simulation, and the dashed lines are from theory [Dubin, 1991].

B. Electron-positron plasmas

One of the goals of the experiment is to explore the physics of electron-positron plasmas. Such plasmas belong to the larger class of equal-mass plasmas where the dynamical symmetry between the electrons and positrons leads to properties that differ from electron-ion plasmas [Tsytovich and Wharton, 1978; Iwamoto, 1993; Stewart and Laing, 1992; Stewart, 1993; Zank and Greaves, 1994]. The electrostatic confinement schemes used currently for pure positron plasmas are not suitable for plasmas with oppositely charged species. However, magnetic mirror devices [Tsytovich and Wharton, 1978] and Paul traps [Schermann and Major, 1978] offer the possibility of studying equal-mass plasmas experimentally.

Electron-positron plasmas can also be studied as a beam-plasma system where only one species needs to be confined. This can be accomplished by transmitting an electron beam through a positron plasma confined in the trap using the potential profile illustrated in Fig. 5. These experiments are now in progress in our laboratory. Information on the nature of the interaction is obtained by detecting modulation on the beam, and signals excited on the confining electrodes. The plasma can also be dumped after transmission of the beam to detect modifications to the plasma. A number of physics issues are planned for investigation. One is the excitation of beam-plasma instabilities which have substantially larger growth rates than in conventional

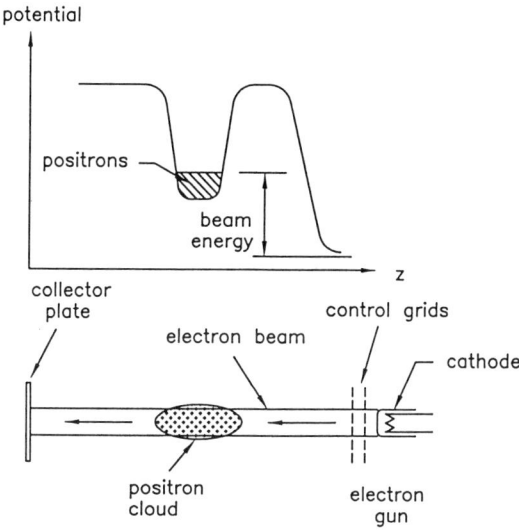

Figure 5: The electron-beam positron plasma experiment.

electron-ion plasmas [Zank and Greaves, 1994]. A second issue is the confinement properties of electron-positron plasmas, in which both species are equally and highly magnetized.

## V. POSITRON ANNIHILATION EXPERIMENTS

A. Positron interactions with large molecules

While the interaction of positrons with ordinary matter and the annihilation of positrons with electrons have been studied extensively ever since the discovery of the positron, such interactions are understood in relatively few cases. For example, first-principles calculations of positron annihilation rates are successful only for the positronium atom and for a few small atoms and molecules. Early in the development of our positron trapping scheme, we discovered that small numbers of impurity molecules (e.g., large hydrocarbons, such as oils and greases) could severely limit the achievable lifetime in the trap [Surko et al., 1988]. While this effect had been previously recognized for the case of molecules such as butane [Paul and Saint-Pierre, 1963; Heyland et al., 1982], we have been able to extend these measurements to much larger molecules and to study the effects of chemical composition, molecular symmetry, and the molecular vibrational modes on these interactions [Iwata et al., 1994].

The positron trap provides a way of isolating two-body interactions between the molecules and positrons at a known temperature. Shown in Fig. 6(a) is the annihilation rate per molecule for a variety of molecular species containing only single bonds [Murphy and Surko, 1991; Iwata et al., 1994]. In Fig. 6(b), the annihilation rate, $\Gamma$, is given in terms of $Z_{\text{eff}} = \Gamma/\Gamma_e$, where $\Gamma_e = \pi r_0^2 cn$ is the annihilation rate of a positron in an uncorrelated electron gas with a density, $n$, equal to the density of molecules, $r_0$ is the classical radius of the electron, and $c$ is the speed of light [Dirac et al., 1930]. If the positron-molecule interaction occurs via elastic collisions, one would expect $Z_{\text{eff}} \lesssim Z$, where $Z$ is the total charge on the molecule. The fact that $Z_{\text{eff}}$ can be as much as four or five orders of magnitude larger than $Z$ leads us to conclude that the interaction results in long-lived resonances between the positron and molecule. In these resonances, the positron spends a much longer time in the vicinity of the molecule, and the probability of annihilation per collision is correspondingly larger.

Such resonances can occur only if there is a positive binding energy between the positron and molecule (i.e., a positron affinity, $\epsilon_A > 0$). We have found empirically that the annihilation rate is a function of the difference between the molecular ionization potential, $E_I$, and the binding energy of a positronium atom, $E_{Ps}$, which is 6.8 eV [Murphy and Surko, 1991]. In particular, as shown in Fig. 6(b), $\log Z_{\text{eff}}$ is proportional to $(E_I - 6.8\text{eV})^{-1}$. We have speculated that this dependence might indicate a model of positron binding in which a highly correlated electron and positron (i.e., a "pseudo-positronium" atom) moves in the field of the positive molecular ion [Murphy and Surko, 1992]. Further considerations regarding $\epsilon_A$ are discussed in Sec. B, below.

Given that $\epsilon_A > 0$, a plausible model of the lifetime of the positron-molecule resonance can be formulated in terms of the Marcus-Rice modification of the Rice-Ramsperger-Kassel theory (i.e., RRKM theory) of unimolecular reactions [Surko et al., 1988; Iwata et al., 1994]. In this model, an incoming positron excites vibrational modes in the molecule and it is effectively trapped, to escape only when the excess vibrational energy again happens to be concentrated in the kinetic energy of the positron. Then, for a large molecule with many degrees of freedom, the lifetime of such a resonance will be long compared to the time of an elastic collision. Recent data on the dependence of the annihilation rate on the vibrational mode frequencies, studied by comparing deuterated and undeuterated hydrocarbons, lends support to this model [Iwata et al., 1994].

B. Annihilation gamma-ray spectra

An important feature of the interaction of positrons with matter is the energy spectrum of annihilation gamma rays. When a positron annihilates with an electron, the fastest mode of annihilation is via two gamma rays with equal and opposite momenta, each having an energy of 511 keV. If the center-of-mass of the electron and positron is moving, the gamma rays will be Doppler-shifted. In the case where a positron annihilates with a bound electron, there is a contribution to the center-of-mass momentum due to the momentum distribution of the bound electron [Brown and Leventhal, 1986]. Thus, measurement of the gamma-ray spectra gives information about the site of the annihilation [Tang et al., 1992]. The energy shift of the gamma rays will be $\Delta E \sim 511(v/c)$ keV, where $v$ is the typical velocity of the electron in the atom. For annihilation with electrons in molecular bonds or in the outer shell of atoms, $\Delta E \sim 1 - 2$ keV. Shown in Fig. 7 are spectra for two molecular species. Comparison

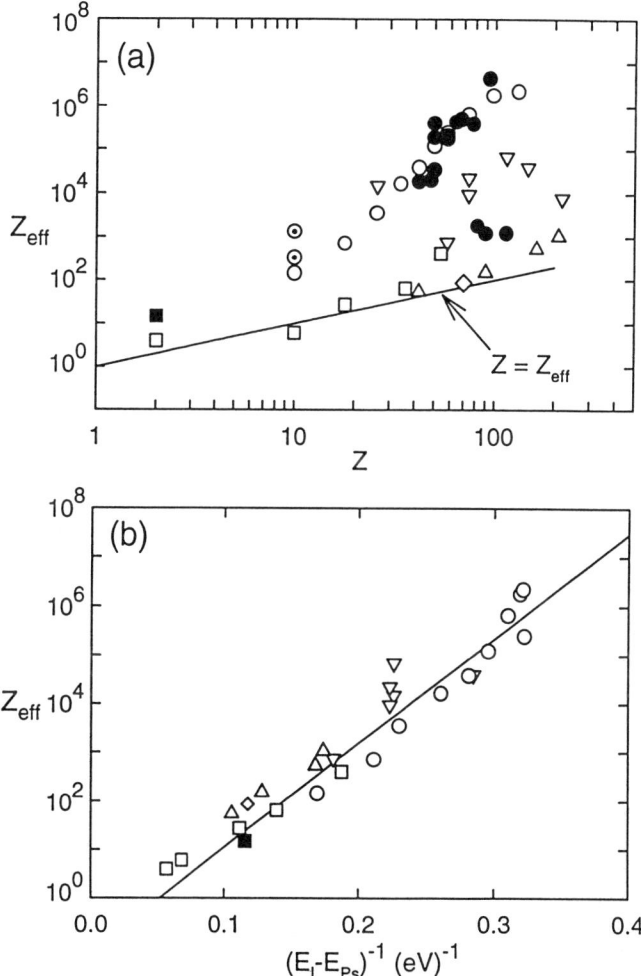

Figure 6: Positron annihilation rates in units of $Z_{eff}$ for a selection of molecules (○) alkanes; (□) rare gases; (△) perfluorocarbons; (▽) halocarbons; (■) hydrogen; (◇) small non-polar molecules; (●) ring compounds; (⊙) small polar molecules. (a) $Z_{eff}$ vs $Z$ (b) $Z_{eff}$ vs $(E_I - E_{Ps})^{-1}$ eV$^{-1}$, where $E_I$ is the ionization energy and $E_{Ps} = 6.8$ eV is the binding energy of the positronium atom.

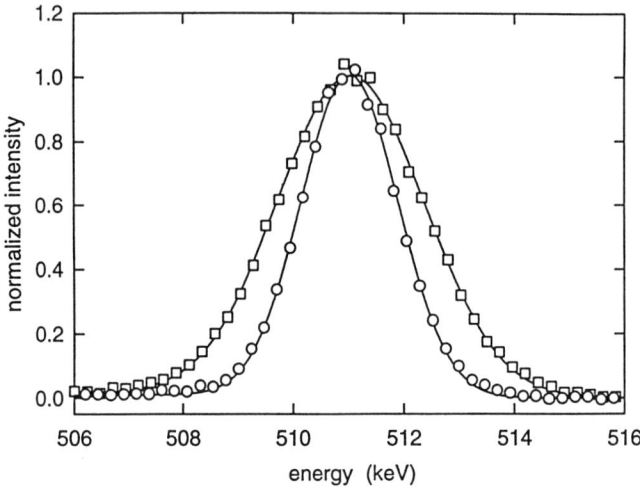

Figure 7: Annihilation gamma-ray spectra for hydrogen (○) and carbon tetrafluoride (□).

of the measured Doppler widths for hydrocarbons with calculations of the momentum distribution of electrons in the C–C and C–H bonds indicate that the positron annihilates predominantly with electrons in the C–H bonds [Surko et al., 1993]. We expect that this fact will be useful in guiding theoretical calculations of $E_A$ for these molecules. At present, progress in understanding the nature of positron interactions with large molecules is hindered significantly by the lack of such calculations.

C. Gamma-ray astrophysics

The positron annihilation line at 511 keV is the strongest gamma-ray line of astrophysical origin [Schönfelder, 1990; Leventhal et al., 1993; Gehrels et al., 1991]. Large aromatic molecules, such as the four-ring compound pyrene, are estimated to be present in the interstellar medium at molecular abundances of about $10^{-7}$ that of hydrogen. On the other hand, our measurements indicate that the annihilation rate per molecule is approximately a factor of $10^{+7}$ greater for pyrene as compared with hydrogen. Thus, we have concluded that annihilation on such large aromatics could be a significant contribution to the gamma-ray radiation from the interstellar medium, and this will be reflected in the measured gamma-ray linewidths [Surko et al., 1993].

This is one example of a number of experiments that can be done in the laboratory, using positron trapping techniques to model astrophysical processes [Surko et al., 1993]. Other possible experiments include measurement of the annihilation rates and energy spectra when slow positrons annihilate on atomic clusters and dust grains. We expect that these particles will behave similarly to the large molecules in that long-lived resonances are likely to occur, thereby enhancing the annihilation rates.

The process by which relativistic positrons slow down in partially ionized gases and plasmas is also an important issue. As fast positrons slow in neutral matter, they lose energy by atomic excitation and ionization processes. At energies below about 100 eV and above the threshold for positronium formation (which is of the order of 10 eV), there is an appreciable probability of forming fast positronium atoms. This results in a broad feature in the gamma ray spectrum and a smaller fraction of thermal positrons. If the medium is appreciably ionized, the positrons can lose energy in collisions with free electrons and thermalize without forming positronium. The fractional ionization above which there is no appreciable positronium formation is an outstanding issue. This parameter could be measured in a suitably designed laboratory experiment [Surko *et al.*, 1993].

## VI. CONCLUDING REMARKS

The ability to efficiently accumulate and store large numbers of positrons and to heat, cool and otherwise manipulate the resulting positron gases and plasmas offers new and important research capabilities. In the near future, it is likely that such collections of antimatter will be used to pursue research opportunities in many areas, including astrophysics, plasma, atomic and molecular and condensed matter physics, as well as for a range of technological applications.

## ACKNOWLEDGMENTS

We would like to acknowledge the collaboration of K. Iwata, T. J. Murphy, A. Passner, S. Tang, M. D. Tinkle, and F. J. Wysocki and the technical assistance of E. A. Jerzewski on many of the experiments described here. This work is supported by the Office of Naval Research and by the National Science Foundation under grant PHY 9221283.

## REFERENCES
Bollinger, J. J., D. J. Heinzen, F. L. Moore, W. M. Itano, D. J. Wineland, and D. H. E. Dubin, 1993, Phys. Rev. A, **48**, 525.
Bollinger, J. J., D. J. Wineland, and D. H. E. Dubin, 1994, Phys. Plasmas, **1**, 1403.
Boyd, D., W. Carr, R. Jones, and M. Seidl, 1973, Phys. Lett., **45A**, 421.
Brown, B. L., and M. Leventhal, 1986, Phys. Rev. Lett., **57**, 1651.
Cooney, J. L., D. W. Aossey, J. E. Williams, and K. E. Lonngren, 1993, Phys. Rev. E, **47**, 564.
Dehmelt, H., 1990, Rev. Mod. Phys., **62**, 525.
Dirac, P. A. M., 1930, Proc. Cambridge Phil. Soc., **26**, 361.
Driscoll, C. F., and K. S. Fine, 1990, Phys. Fluids B, **2**, 1359.
Driscoll, C. F., J. H. Malmberg, and K. S. Fine, 1988, Phys. Rev. Lett., **60**, 1290.
Dubin, D. H. E., 1991, Phys. Rev. Lett., **66**, 2076.
Dubin, D. H. E., 1993, Phys. Fluids B, **5**, 295.
Gabrielse, G., X. Fei, L. A. Orozco, R. L. Tjoelker, J. Haas, H. Kalinowsky, T. A. Trainor, and W. Kells, 1989, Phys. Rev. Lett., **63**, 1360.
Gabrielse, G., X. Fei, L. A. Orozco, R. L. Tjoelker, J. Haas, H. Kalinowsky, T. A. Trainor, and W. Kells, 1990, Phys. Rev. Lett., **65**, 1317.
Gehrels, N., S. D. Barthelmy, B. J. Teegarden, J. Tueller, M. Leventhal, and C. J. MacCallum, 1991, Ap. J., Letters., **375**, L13.
Gibson, G., W. C. Jordan, and E. J. Lauer, 1963, Phys. Fluids, **6**, 116.

Glish, G. L., R. G. Greaves, S. A. McLuckey, L. D. Hulett, C. M. Surko, J. Xu, and D. L. Donohue, 1994, Phys. Rev. A, **49**, 2389.

Gramsch, E., J. Throwe, and K. G. Lynn, 1987, Appl. Phys. Lett., **51**, 1862.

Greaves, R. G., M. D. Tinkle, and C. M. Surko, 1994a, Phys. Plasmas, **1**, 1439.

Greaves, R. G., M. D. Tinkle, and C. M. Surko, "Modes in a pure ion plasma at the Brillouin limit", submitted to Phys. Rev. Lett., 1994b.

Heyland, G. R., M. Charlton, T. C. Griffith, and G. L. Wright, 1982, Can. J. Phys., **60**, 503.

Holzscheiter, M. H., 1993, Hyperfine Interactions, **81**, 71.

Iwamoto, N., 1993, Phys. Rev. E, **47**, 604.

Iwata, K., R. G. Greaves, T. J. Murphy, M. D. Tinkle, and C. M. Surko, "Measurements of positron annihilation rates on molecules", submitted to Phys. Rev. A, 1994.

Khatri, R., M. Charlton, P. Sferlazzo, K. G. Lynn, A. P. Mills, Jr., and L. O. Roellig, 1990, Appl. Phys. Lett., **57**, 2374.

Leventhal, M., S. D. Barthelmy, N. Gehrels, B. J. Teegarden, J. Tueller, and L. M. Bartlett, 1993, Ap. J., Letters, **405**, L25.

Malmberg, J. H., and J. S. de Grassie, 1975, Phys. Rev. Lett., **35**, 577.

Malmberg, J. H., and C. F. Driscoll, 1980, Phys. Rev. Lett., **44**, 654.

Mills, Jr., A. P., and E. M. Gullikson, 1986, Appl. Phys. Lett., **49**, 1121.

Murphy, T. J., 1987, Plasma Phys. Contr. Fusion, **29**, 549.

Murphy, T. J., and C. M. Surko, 1991, Phys. Rev. Lett., **67**, 2954.

Murphy, T. J., and C. M. Surko, 1992, Phys. Rev. A, **46**, 5696.

O'Neil, T. M., 1980, Comments Plasma Phys. Contr. Fusion, **5**, 213.

Passner, A., C. M. Surko, M. Leventhal, and A. P. Mills, Jr., 1989, Phys. Rev. A, **39**, 3706.

Paul, D. A. L., and L. Saint-Pierre, 1963, Phys. Rev. Lett., **11**, 493.

Schermann, J. P., and F. G. Major, 1978, Appl. Phys., **16**, 225.

Schönfelder, V., 1990, Adv. Space Res., **10**, 2, 243.

Schwinberg, P. B., R. S. Van Dyck, Jr., and H. G. Dehmelt, 1981, Phys. Lett., **81A**, 119.

Sheehan, D. P., and N. Rynn, 1988, Rev. of Sci. Instrum., **59**, 1369.

Stewart, G. A., 1993, J. Plasma Phys., **50**, 521.

Stewart, G. A., and E. W. Laing, 1992, J. Plasma Phys., **47**, 295.

Surko, C. M., M. Leventhal, W. S. Crane, A. Passner, and F. Wysocki, 1986, Rev. of Sci. Instrum., **57**, 1862.

Surko, C. M., A. Passner, M. Leventhal, and F. J. Wysocki, 1988, Phys. Rev. Lett., **61**, 1831.

Surko, C. M., M. Leventhal, and A. Passner, 1989, Phys. Rev. Lett., **62**, 901.

Surko, C. M., R. G. Greaves, and M. Leventhal, 1993, Hyperfine Interactions, **81**, 239.

Suzuki, R., Y. Kobayashi, T. Mikado, H. Ohgaki, et al., 1992, Jpn. J. Appl. Phys., **31**, 2237.

Tang, S., M. D. Tinkle, R. G. Greaves, and C. M. Surko, 1992, Phys. Rev. Lett., **68**, 3793.

Tinkle, M. D., R. G. Greaves, C. M. Surko, R. L. Spencer, and G. W. Mason, 1994, Phys. Rev. Lett., **72**, 352.

Tsytovich, V., and C. B. Wharton, 1978, Comments Plasma Phys. Contr. Fusion, **4**, 91.

Weimer, C. S., J. J. Bollinger, F. L. Moore, and D. J. Wineland, 1994, Phys. Rev. A, **49**, 3842.

White, W. D., J. H. Malmberg, and C. F. Driscoll, 1982, Phys. Rev. Lett., **49**, 1822.

Wineland, D. J., C. S. Weimer, and J. J. Bollinger, 1993, Hyperfine Interactions, **76**, 115.

Zank, G. P., and R. G. Greaves, "Linear and nonlinear modes in nonrelativistic electron-positron plasmas", to be submitted to Phys. Rev. E, 1994.

# XII.2

# PRODUCTION OF SLOW POSITRONS WITH A LINAC AS A SOURCE OF POSITRON PLASMA FORMATION

A. Mohri,[†] H. Tanaka,[†] T. Michishita,[†] Y. Yuyama,[†] Y. Kawase,[‡] and T. Takami[‡]

[†] Department of Fundamental Sciences, Kyoto University, Kyoto 606-01, Japan
[‡] Reactor Research Institute, Kyoto University, Kumatori 590-01, Japan

Several improvements on slow positron production with a LINAC are pointed out to serve positron plasma formation, and are examined experimentally using 30 MeV LINAC. Points of practical improvement are effective utilization of photons produced for pair creation, enlargement of positron emerging surface area, and effective simultaneous collection of thermal and nonthermal positrons from the surface. The obtained production rate of positrons with 3 keV energy spread is $3 \times 10^{13}$ [e$^+$/s] during the LINAC pulse. Moderation of the collected beam with Ar, Xe and N$_2$ solid films produces slow positrons ($< 6$ eV) at $\sim 0.5$ % efficiency, where the rate during the LINAC pulse is $1.2 \times 10^{11}$ [e$^+$/s]. A preliminary experiment on stacking of the slow positrons in an electrostatic well is described and confinement in toroidal systems is discussed.

## I. INTRODUCTION

For a formation of positron plasma in laboratories, it is necessary to have a large number of positrons in an easily controllable energy range. These positrons are obtained either from $\beta^+$ emissive isotopes or from pair-production appliances attached to electron linear accelerators (LINACs). Positrons from a $^{22}$Na isotope were moderated through a thin tungsten film and gradually accumulated in an electrostatic well to a plasma state (Surko, Leventhal and Passner, 1989; Murphy and Surko, 1992; Greaves, Tinkle and Surko, 1994). On the other hand, the use of LINACs has a potential of increasing the production rate of positrons and a positron plasma of a larger size is expected to be formed easily. Positron beams produced in this way are pulsive and the stacking of them are now being examined experimentally (Cowan et al., 1993; Mohri et al., 1994). Here, improvement in the positron production efficiency is requisite for realizing a practically useful source of positrons. This article mainly describes methods of efficiency enhancements, developed through experiments using a 30 MeV LINAC of Kyoto University (KURRI-LINAC), and a preliminary experiment on stacking of pulsed positron beams which are moderated with rare gas solid films.

## II. EFFICIENT PRODUCTION OF POSITRONS WITH LINAC

Production of slow positrons by the use of LINACs is easy to raise the production rate and extensive experiments have been carried out at laboratories having LINACs with output electron energies ranging from 20 MeV to 300 MeV (reviewed by Dahn et.al., 1988; Ley, 1992; Charlton and Laricchia, 1992) or 2 GeV (Shidara et al., 1994). Figure 1 shows a schematic drawing of such a usual system. Electron beams bombard a tantalum target. Shower-enhanced photons and positrons emerging from the target irradiate tungsten foils, in which both pair-created positrons *in situ* and incident ones slowdown their energies and emerge from the surfaces. Here, thermalized ones diffuse to the surfaces and come out by the help of the negative work function of tungsten. These positrons are transported through a magnetic duct either to an experimental apparatus or to a remoderator.

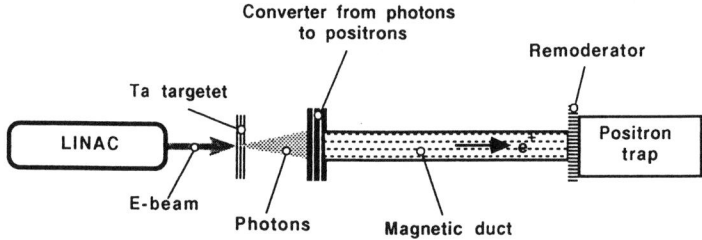

Fig. 1 Schematic drawing of a positron production system using a LINAC.

There are several points for further improvement in the production efficiency as follows:
(1) The number density of positrons born and implanted in the tungsten foils should be maximized by the effective use of the photon flux from the target. This optimization can be made with a Monte Carlo simulation program: the EGS4 code (Nelson et al., 1985) where all of the surrounding materials and geometrical factors can be taken into account.
(2) The flux of emerging positrons can be made higher by providing a large area of the emission surface without deteriorating the extraction efficiency (Mohri et al., 1991; Tanaka et al., 1992).
(3) Nonthermalized positrons with a wide energy spread mix together with thermalized ones in the emerging flux. Additive use of their low energy components raises the total yield when appropriate remoderation should be applied (Mohri et al., 1994). Here, the transport of positrons over a wide energy spread becomes necessary.
(4) Remoderator or cooling means, if adopted, should work effectively.

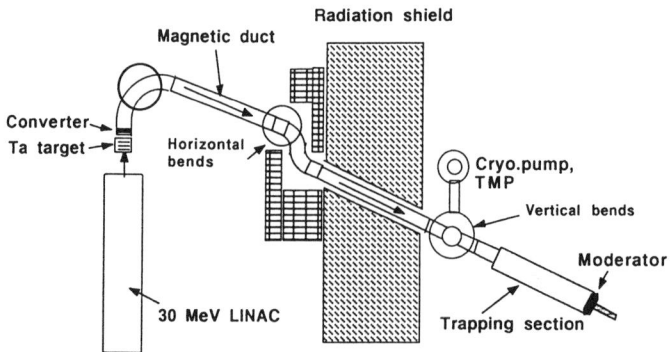

Fig. 2 Experimental setup for positron production in Kyoto University.

Hereafter, experiments reflecting upon above points are to be explained. The experiments were performed with KURRI-LINAC and its experimental setup is shown in Fig. 2. KURRI-LINAC delivers 30 MeV, 0.4A electron beams with 4 μs pulse width at 120 pps repetition rate. Three 2 mm thick tantalum plates are housed in a water-cooling Al jucket and the Al end plate of the vacuum vessel is 5 mm thick.

A. Optimal arrangement of the target and the converter

Figure 3(a) illustrates the arrangement of the tantalum target and the converter of a tungsten disk. Forward-directed bremsstrahlung X-rays caused when the target is struck by LINAC beams penetrate the jucket and vessel plates and then pass through the converter set in a vacuum vessel. Maximization of the number of pair-creation events in the converter gives the arrangement suitable for obtaining the highest efficiency of positron production. Using EGS4 code, we can determine the thicknesses and diameters of the components. The arrangement in Fig. 3 (a) was found in this way, where the electron-positron conversion efficiency inside the tungsten per 1mm thickness per 30 MeV LINAC electron was estimated to be 0.03.

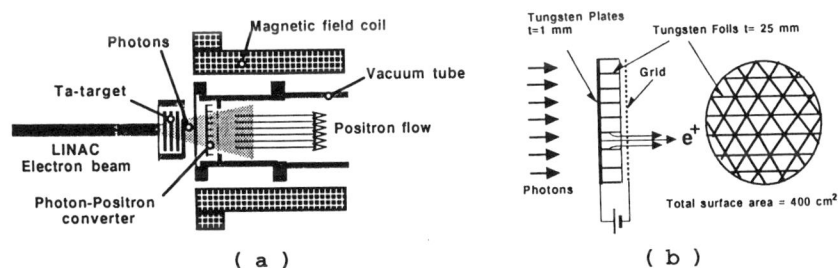

Fig. 3 (a) Arrangement of the tantalum target and the photon-positron converter. (b) Photon-positron converter to extract nonthermal positrons.

B. Enlargement of the emerging surface

The region in a tungsten foil contributing to the thermalized positron emission is restricted within the depth of diffusion length ($\sim 10^{-5}$ cm) from the surface. Therefore, the yield of slow positrons can be increased by enlarging the emission surface area so long as the collection efficiency is kept high. Figure 4 shows the arrangement developed to meet the above principle, i.e., so called modular photon-positron converter (MPPC). Many thin tungsten foil strips of width $w$ are aligned parallel to each other with separation $g$ and assembled into a module. Many modules are stacked perpendicular to the LINAC beam direction. X-rays from the target pass through the stacked foils and slow positrons emerging from thus formed wide foil surfaces are directed by applied extraction electric fields to flow out through spaces present among the foils. Setting of highly transparent metal meshes on each module under the condition as $w/g \sim 1$ allows the permeation of the collection field into deep inside the gaps between neighboring foils. There appears a favorable focusing effect on positron motions, so that annihilation losses of positrons due to collisions with foil surfaces are suppressed.

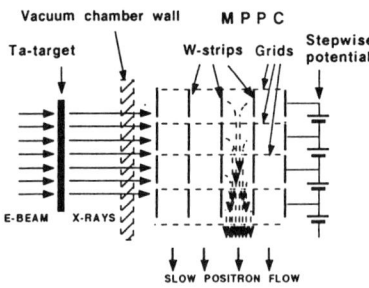
Fig. 4 Principle of MPPC.

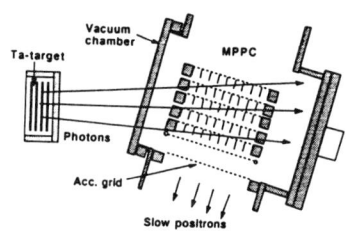
Fig. 5 MPPC used for POP.

Figure 5 shows the setting of the MPPC which was used for proof of principle. The MPPC consists of five modules, each of which holds ten tungsten foil strips 25 $\mu$m thick, 60 mm long and 8 mm wide. The stepwise potential of an equal step is applied to the modules. The extracted slow positrons are transferred through the magnetic duct 9 m long to the diagnostic room. The maximum positron yield by a 3 $\mu$s LINAC pulse is $3.7 \times 10^5$, corresponding to the time averaged production rate of $4.4 \times 10^7$ [e$^+$/s]. The obtained production rate during a pulse ($= 1.2 \times 10^{11}$ [e$^+$/s]) is compared with numerical estimations: the production rate found with the EGS4 code and the amount of losses in the MPPC and in the duct, obtained from orbit calculations. The estimated rate coincides with the experimental one if the emission efficiency of positrons from foil surfaces is set to be 0.13. This experiment suggests that more efficient production becomes possible by making more tight coupling of the photon flux to the converter which have a larger emission surface area.

C. Use of nonthermalized positrons

Pair-creation occurs nearly uniformly in X-ray irradiated tungsten foils, while positrons present in moderators used for radioactive $\beta^+$ sources are all implanted ones. An energy spectrum of positrons emerging from a X-ray irradiated tungsten surface, found with the EGS4 code, is shown in Fig. 6. Here, 10 keV is the lower energy limit of positrons dealt in the code. Practically, the spectrum must rise up as going lower energy to the work function. Addition of nonthermalized positrons to thermalized ones increases the total yield if the magnetic duct has a function to transport positrons in a wide energy range. Despite of worse remoderation efficiency for higher energy positrons, the total yield after remoderation will become higher. Figures 3 (a) and (b) show converters designed to extract nonthermalized positrons. Nested 25 $\mu$m thick tungsten foils are attached to a 1 mm thick tungsten disk, when the total surface area becomes 400 cm$^2$. The numbers of positrons reaching the duct end through retarding electric potential $V_r$ are shown in Fig. 7 for different field strengths of the duct $B_d$. At higher $B_d$, positrons of higher energy components increase in number, and, as shown in Fig. 7, the total transported number during LINAC pulse becomes larger because losses of higher energy positrons in the duct are reduced more. The obtained yield in a LINAC pulse is $1.3 \times 10^8$ which corresponds to the production rate of $3 \times 10^{13}$ [e$^+$/s] and the time-averaged rate for 120 pps of $1.5 \times 10^{10}$ [e$^+$/s]. The e$^-$ → e$^+$ conversion efficiency in this case is as high as $5 \times 10^{-5}$. The obtained positron flow has a large diameter and a wide energy spread up to 3 keV, so that suitable remoderation of them is indispensable.

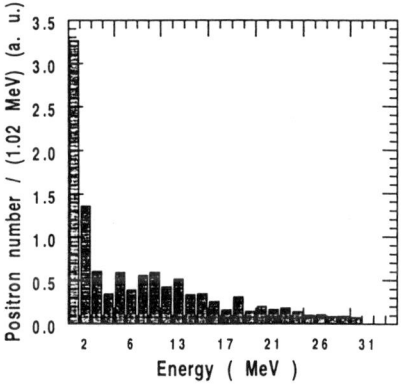 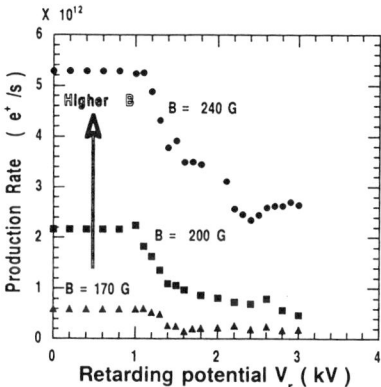

Fig. 6 Energy spectrum of positrons emerging from a tungsten foil, calculated with the EGS4 code.

Fig. 7 Transported positron numbers through retarding potential $V_r$ for various field strength $B_d$.

## III. REMODERATION WITH GAS SOLID FILMS

Solid films of frozen gases such as Xe, Ar (Gullikson and Mills, Jr., 1986; Mills, Jr. and Gullikson, 1986) and N$_2$ are adopted as moderators of reflection type. Evolutions of implanted positrons were examined with a new Monte Carlo simulation code including elastic and inelastic scattering with cored- or valence-electrons and phonons. From obtained implantation profiles, rare gas solid films 1000 Å thick is found sufficient to moderate positrons with energies less than 5 keV.

The diameter of such a film can be made large compared with that of the positron flow. Frozen layers are formed on a gold-plated copper disk cooled by a refrigerator at 7–9 K. Argon solid films exhibits to have properties fit to the purpose, e.g., fairly high moderation efficiency and low vapor pressure which is essential for positron trap. Moderated positrons are once trapped in an electric potential and, 50 μs after, exhausted for annihilation measurement. The potential of the well bottom is biased by $V_b$ against the moderator to give a retardation to the moderated positrons. The moderated positron energies extend to 6 eV and the energy distribution has its peak at ~ 1.0 eV as shown in Fig. 8. The yield of moderated positrons per pulse $Y_+$ and the moderation efficiency $\eta_M$ for the Xe solid film are plotted in Fig. 9 as a function of the duct field $B_d$. At higher $B_d$, $\eta_M$ becomes lower due to the inclusion of higher energy positrons while the yield $Y_+$ continuously increases overcoming the lowering of $\eta_M$. The maximum value obtained in the case of Ar solid films is $Y_+ = 5 \times 10^5$ at $B_d$ = 320 G, corresponding to the production rate during pulse of $1.2 \times 10^{11}$ [e$^+$/s] and the time-averaged rate of $6 \times 10^7$ [e$^+$/s].

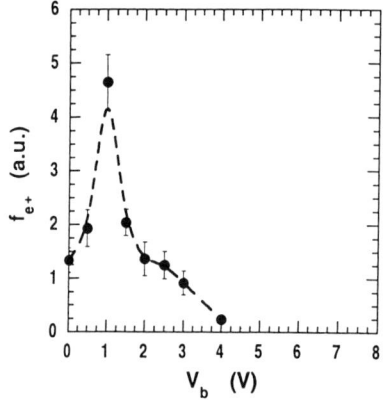
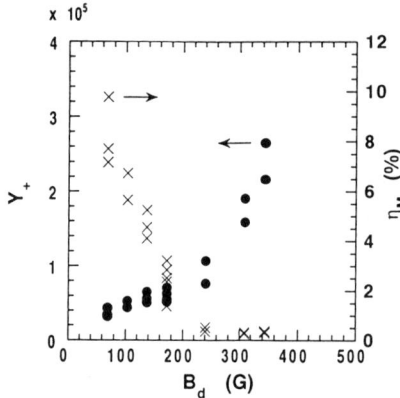

Fig. 8  Energy distribution $f_{e+}(E)$ of positrons moderated with an Ar solid film.

Fig. 9 Slow positron yield per pulse, $Y_+$, and moderation efficiency, $\eta_M$, versus $B_d$.

## IV. STACKING

To form positron plasma, slow positrons should be accumulated in a confinement system. Penning traps of axisymmetric configuration have been adopted at many laboratories to trap charged particles with advantages of long confinement, easy injection and ejection of positrons along the axis and controllability to transport the plasma as a bunch. Formation of antihydrogen in this type of trap is now under consideration. Slow positron beams generated with LINACs are pulsive in contrast with radioisotope sources. There need be some contrivances to stack pulsed positron beams in a Penning trap.

Here consider an electric axial potential well which is $\ell_p$ in length and equipped with a gate on one side. A single pulsed beam is injected into the well by opening the gate and the gate is closed immediately after the pulse end. To trap all of the beam positrons, the beam length $\ell_b = (pulse\ width : \tau_b) \times (velocity : v_b)$ should be shorter than $2 \cdot \ell_p$. Therefore, either slow beams (i.e., moderated beams), or short-pulsed beams are required to reduce the axial size of the well. Accumulation of positrons in the well may become possible by repeating the gate opening, synchronized with each beam pulse, if a cooling down mechanism is present there. Inelastic collisions with molecules (Surko et al., 1989) and cyclotron radiation (e.g., Shlyapnikov et al., 1993; Cowan et al., 1993; Haarsma et al., 1993) are available loss mechnisms to the pulsed beam injection.

A. Trapping of a single pulse

Figure 10 shows the potential well made with segmented cylindrical electrodes and its axial potential distribution having slopes falling to the well bottom. The total length of the well is $\ell_p = 103$ cm so that a slow positron beam with 2.4 $\mu$s pulse width and energies less than 2 eV can be trapped. A pulsed positron beam transferred through the magnetic duct passes over the well region and reaches the thin solid film moderator of argon. Slow positrons reemitted from the film are introduced into the well by opening the potential gate and trapped by closing it. Figure 11 shows the decay of the trapped positron number in the case that the applied magnetic field is 320 G and the potential well depth is 10 V. The trapped number just after the injection is $N_s^+ = 4 \times 10^5$. The confinement time is about 15 ms which is much shorter than the time observed in the case of electron trap under the same condition. However, the ratio of the positron number trapped in the sloping potential region to the total number increases with time, suggesting the presence of a relaxation mechanism.

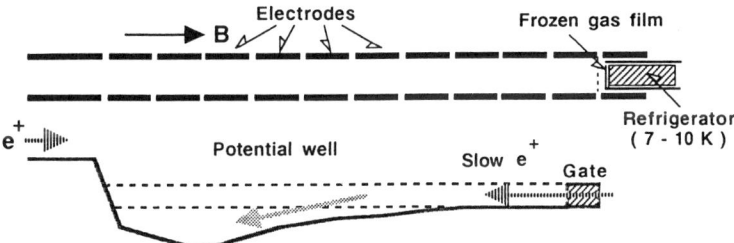

Fig. 10 Axial potential distribution used to trap moderated positrons.

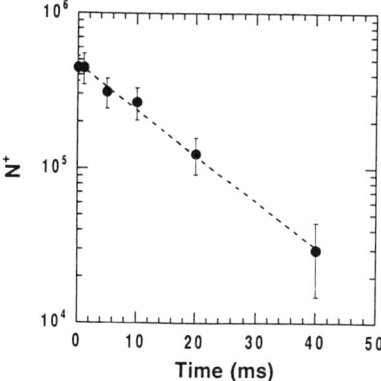 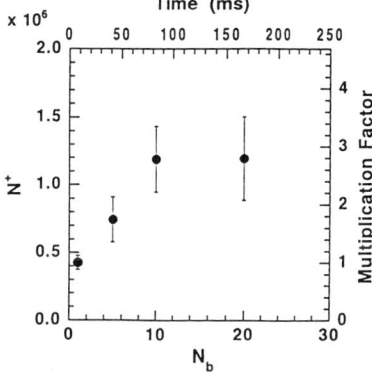

Fig. 11 Decay of the trapped positron number.

Fig. 12 Stacked positron number, $N^+$, after the repetition of $N_b$ pulses.

### B. Stacking of multiple beams

Positron accumulation in the well is performed by repeating the same trapping operation as described above. As shown in Fig. 12, the trapped positron number $N^+$ observed after stacking $N_b$ pulses increases with $N_b$ until $N^+ \sim N_s^+ \tau_c/t_p$ where $\tau_c$ and $t_p$ are the confinement time and the pulse interval. The multiplication factor by the stacking is about 3 and the accumulated positron number is $1.2 \times 10^6$. The temperature of the trapped positrons at the saturation is about 1 eV, at which the Debye length is shorter than the axial length of this positron cloud but longer than its radius. The population trapped in the sloping region of the well is observed to increase with $N_b$, suggesting the presence of a cooling of positrons. If the confinement time becomes longer and an effective mechanism to cool down the trapped positrons is introduced, the accumulated number is expected to become much larger.

## V. POSSIBILITY TO CONFINE AMBIPLASMA (PAIR-PLASMA)

Laboratory experiments on neutral plasmas composed of positrons and electrons of equal densities, i.e., the simplest one of ambiplasmas or pair-plasma, have potentials to offer a lot of new understandings concerning plasma physics and also astrophysics. To do the experiments we have to confine such a plasma using appropriate systems. Configurations of systems suited to fundamental experiments should have a symmetry to certify invariants of motion which simplify analyses and also contribute to good confinement. Axisymmetric magnetic mirrors with open ends cannot confine plasmas stably, where flute- and loss cone-instabilities appear. Thus, a toroidal system: Spherator or Levitron with a current-carrying floating conductor ring (e.g., Freeman et al., 1969; Birdsall et al., 1966) as depicted in Fig. 13 seems to be the system most fitted to ambiplasma confinement (Yoshikawa, 1993). This configuration is formed without internal plasma currents, that is one of essential requirements for fundamental experiments. When the vacuum wall is cooled down to a low temperature, all surfaces facing the confined plasma becomes an absorber of radiations from the plasma since the floating ring is a cold super-conductor. Thus, radiation cooling is applicable there. Also, the system can be constructed so compact as to reduce the filling total particle number. How to inject charged particles across the field would be a problem to be solved. Plasma injection methods used in toroidal plasma experiments might be applied, e.g., dense plasma blob injection accompanied with hole currents (Ohkawa, 1961).

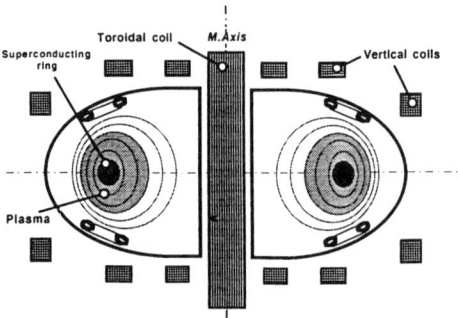

Fig. 13 Levitated Spherator useable to confine ambiplasmas.

## VI. SUMMARY

Methods to produce a large number of positrons have been explained based on the experiments which were performed with a LINAC of a medium energy output. Substantial efficiency enhancement in the positron production rate are demonstrated possible by (1) optimal setting to maximize the pair creation in the tungsten converter, (2) making the positron emission surface as wide as possible under the assurance of effective extraction of positrons, and (3) additive extraction of a low energy part of nonthermalized positrons. Though the LINAC beam energy in the experiments is as low as 30 MeV, the production rate of positrons with energies less than 3 keV reached $3 \times 10^{13}$ during LINAC beam pulse of 0.4 A with the $e^- \to e^+$ conversion efficiency of $5 \times 10^{-5}$.

Positron moderation using rare gas solid films exhibits useable especially when a wide area of moderation surface is necessary. Solid films of argon and nitrogen have higher vapor pressure at the temperature ($\sim 70$ K) of the radiation shroud surrounding the cold head plate. Therefore, gas atoms attached or frozen on the shroud during the solid film formation by gas puffing are quickly pumped out and the vacuum recovers soon. The experiment showed that the moderation efficiency changed from 10 % to 0.5 % as higher energy components of positrons ($< 3$ keV) were more included in the incident beam, and the energy spread was $5 \sim 7$ eV. The moderated positron yield per LINAC pulse in the case of argon film was $5 \times 10^5$ corresponding to the time-average of $6 \times 10^7$ [$e^+$/s].

Stacking of multi-slow positron beams in an electric well with 15 ms confinement time raised the stacked number by a factor 3 as $1.2 \times 10^6$. Stacked number will be increased much more when used a system with longer confinement time and a suitable cooling mechanism.

The authors hope that the efficiency enhancement methods shown in this article contribute to the future advanced experiments on positron plasma physics. They are much indebted to Professor S. Yoshikawa for his stimulated discussions on toroidal confinement of positron plasmas.

## REFERENCES

Birdsall D. H., R. J. Biggs, S. A. Colgate, H. P. Furth and C. W. Hartman, 1966, Proc. of Int. Conf. on Plasma Phys. and Controlled Nucl. Fusion (IAEA, Vienna), Vol. II, p.291.

Charlton, M. and G. Laricchia, 1992, Hyp. Int. **76**, 97.

Cowan, T. E., B. R. Beck, J. H. Hartley, R. H. Howell, R. R. Rohagi, J. Fajan and R. Gopalan, 1993, Hyp. Int., **76**, 135.

Dahn, J., R. Rey, K. D. Niebling, R. Schwarz and G. Werth, 1988, Hyp. Int. **44**, 141.

Freeman, R., M. Okabayashi, H. Pacher, S. Yoshikawa, 1969, Phys. Rev. Lett., **23**, 756.

Greaves, R. G., M. D. Tinkle and C. M. Surko, 1994, Phys. Plasmas, **1**, 1439.

Gullikson, E. M. and A. P. Mills, Jr., 1986, Phys. Rev. Lett., **57**, 376.

Ley, R., 1992, Mater. Sci., Forum 105-110, 1927.

Mills, Jr., A. P. and E. M. Gullikson, 1986, Appl. Phys. Lett., **49**, 1121.

Mohri, A., T. Michishita, T. Yuyama and H. Tanaka, 1991, Jpn. J. Appl. Phys., **30**, L936.
Mohri, A., H. Tanaka, T. Yuyama, T. Michishita, H. Higaki, Y. Yamazawa, Y. Kawase and K. Takami, 1994, 10th Int. Conf. on Positron Annihi., paper C1-1.
Murphy, T. J. and C. M. Surko, 1992, Phys. Rev., **46**, 5696.
Nelson, W. R., H. Hirayama and W. O. Rogers, 1985, EGS4 Code System, Stanford Linear Accel. Center Report SLAC-265.
Ohkawa, T. and W. D. Kerst, 1961, Il Nuovo Cimento, **XXII**, 784.
Shidara, T., A. Enomoto, T. Kamitani, H. Kobayashi, T. Kurihara, A. Shirakawa, H. Hirayama, I. Kanazawa, A. Asami and K. Nakahara, 1994, 10th Int. Conf. on Positron Annihi., paper C1-3.
Shlyapnikov, G. V., J. J. M. Walraven and E. L. Surkov, 1993, Hyp. Int., **76**, 31.
Surko, C. M., M. Leventhal and A. Passner, 1989, Phys. Rev. Lett., **62** 901.
Tanaka, H., T. Michishita, T. Yuyama, K. Takami, Y. Kawase and A. Mohri, 1992, Jpn. J. Appl. Phys., **31**, 4029.
Yoshikawa, S., 1993, private communication to A. M..

# Program of the Oji International Seminar on Elementary Processes in Dense Plasmas

# PROGRAM

— SUNDAY, June 26 —

**4:00 p.m. – Registration**
**7:00 p.m. – Reception**

— MONDAY, June 27 —

**8:00 a.m. – Opening**
**8:20 a.m. – 9:40 a.m.** (chairperson: Itoh)
  G. Baym (UIUC, Urbana) The quark-gluon plasma
  M. Fukugita (Inst. Fund. Phys., Kyoto U.) Neutrino cosmology
**9:40 a.m. – 10:20 a.m.   Break**
**10:20 a.m. – 12:20 p.m.** (chairperson: Ashcroft)
  H. Endo (Fukui Inst. Tech.) Liquids near the critical point
  F. Hensel (Philipps-U. Marburg) Interatomic interactions in dense metal vapors
  F. Yonezawa (Keio U., Yokohama) Supercooled fluids and anomalous relaxation
**12:20 p.m. – 2:00 p.m.   Lunch Break**
**2:00 p.m. – 3:20 p.m.** (chairperson: Kubono)
  T. Kajino (NAO, Mitaka) Cosmological phase transition and nucleosynthesis
  F.-K. Thielemann (U. Basel) Screening of thermonuclear reactions and abundances in nuclear statistical equilibrium
**3:20 p.m. – 3:50 p.m.   Break**
**3:50 p.m. – 5:10 p.m.** (chairperson: Surko)
  M. Yamada (Princeton U.) Review of recent laboratory experiments on magnetic reconnection in plasmas
  S. Tsuneta (U. Tokyo) Highlights of Yohkoh
**5:10 p.m. – 7:30 p.m.   Poster Session** (chairperson: Ito)
  T. Takatsuka (Iwate U., Morioka) Hot neutron stars at birth
  H. Suzuki (KEK, Tsukuba) Neutrino emission from protoneutron stars
  Y. Suzuki (Inst. Cosmic Rays, U. Tokyo) Solar neutrinos
  H. Kitamura (U. Tokyo) Electric and thermal resistivities in dense partially ionized high-$Z$ plasmas
  Y. Kato (ILE, Osaka U.) Atomic physics in x-ray lasers: recent experiments
  Y. Furutani (Okayama U.) Atomic states in a dense plasma
  T. Kato (NIFS, Nagoya) Nonequilibrium ionization in dense, hot plasmas
  I. Matsushima (Electrotech. Lab., Tsukuba) Temperature and density measurements of laser produced plasmas

H. Iyetomi (Energy Res. Lab., Hitachi) Antiferromagnetic spin correlations in layered electrons

S. Tanaka (Toshiba Adv. Res. Lab., Kawasaki) Variational quantum Monte Carlo approach to the cohesive properties of many-electron systems

K. Nishihara (ILE, Osaka U.) Instantaneous phase-space Lyapunov expansion rates in strongly coupled plasmas

H. Takabe (ILE, Osaka U.) Numerical study of compex hydrodynamics in laser driven implosion

Y. Ueshima (ILE, Osaka U.) 3-d particle simulation of Kelvin-Helmholtz instability in ion one component dense plasmas

— TUESDAY, June 28 —

**8:20 a.m. – 9:40 a.m.** (chairperson: Van Horn)
K. Sato (U. Tokyo) Mechanism of collapse-driven supernova explosion
K. Nomoto (U. Tokyo) White dwarfs and supernovae
**9:40 a.m. – 10:20 a.m.   Break**
**10:20 a.m. – 12:20 p.m.** (chairperson: Totsuka)
H. Shibahashi (U. Tokyo) Helioseismology and solar models
W. Däppen (USC, LA) Equation of state issues in helioseismology
W. Ebeling (Humboldt U., Berlin) Thermodynamics, kinetics, and phase transitions of dense plasmas
**12:20 p.m. – 2:00 p.m.   Lunch Break**
**2:00 p.m. – 3:20 p.m.** (chairperson: Furutani)
J.P. Wolfe (UIUC, Urbana) Electron-hole plasmas in elemental semiconductors
P. Vashishta (LSU, Baton Rouge) Computer simulation of materials on parallel architectures—glasses, solid $C_{60}$, and graphitic tubules
**3:20 p.m. – 3:50 p.m.   Break**
**3:50 p.m. – 5:10 p.m.** (chairperson: Makishima)
H.M. Van Horn (NSF, Wash. D.C.) Global oscillations of neutron stars as potential probes of microscopic physics at ultra-high densities
T. Murakami (ISAS, Sagamihara) Present status of $\gamma$-ray burst observations
**5:10 p.m. – 7:30 p.m.   Poster Session** (chairperson: Mima)
A. Förster (Humboldt U., Berlin) Ionization equilibria and plasma-phase transitions
H. Lehmann (Oxford U.) Statistical mechanics of atomic matter in intense electric fields
W. Kraeft (U. Greifswald) Density dependence of bound states in strongly correlated plasmas
M. Downer (UT, Austin) Quantitative optical properties of carbon, silicon, germanium, and iron under planetary interior conditions measured by femtosecond spectroscopy
T. Tajima (UT, Austin) Electron-neutrino phase separation instability
K. Tsuji (Keio U., Yokohama) Volume dependence of the structure of liquid metals
S. Hosokawa (Hiroshima U.) Singular diameter in the liquid-vapor coexistence curve of fluid selenium

S. Fujiwara (Keio U., Yokohama) Monte Carlo simulations of anomalous relaxation in disordered systems
S. Gomi (Keio U., Yokohama) Anomalous relaxation in fractal time random walk
K. Omata (Keio U., Yokohama) Determination of percoration thresholds with high accuracy by MC simulations
K. Minami (Niigata U.) Spectroscopy measurements of afterglow of pulsed discharges in liquid helium II
T. Michishita (Kyoto U.) Moderation and reemission of positrons incident into rare-gas solids

— WEDNESDAY, June 29 —

**8:20 a.m. – 9:40 a.m.** (chairperson: Endo)
R.J. Hemley (Carnegie Inst., Wash. D.C.) Materials at multimegabar pressures: symmetry breaking in dense hydrogen
N.W. Ashcroft (Cornell U., Ithaca) The dense hydrogen plasma: translational, orientational and electronic structure
**9:40 a.m. – 10:20 a.m.   Break**
**10:20 a.m. – 11:40 a.m.** (chairperson: Hubbard)
F. Rogers (LLNL, Livermore) Astrophysical opacity
C.F. Hooper (U. Florida, Gainsville) X-ray spectroscopy of dense, laser-produced plasmas
**12:00 noon – Excursion** (a bus tour in Shikotsu-Toya National Park area):
Lunch at Hotel Suimeikaku (by Lake Shikotsu) and Dinner at Kirin Garden House in Chitose

— THURSDAY, June 30 —

**8:20 a.m. – 9:40 a.m.** (chairperson: Nomoto)
S. Kubono (INS, U. Tokyo) Experimental approach to nucleosynthesis in the Universe
J. Meyer-ter-Vehn (MPG, Garching) The potential of high-power beams for studying multi-megabar matter, including low-entropy hydrogen compression
**9:40 a.m. – 10:20 a.m.   Break**
**10:20 a.m. – 12:20 p.m.** (chairperson: Fortov)
R. More (LLNL, Livermore) Electron transport phenomena in ultrashort pulse laser produced plasmas
W.B. Hubbard (UA, Tucson) Current uncertainties in the interior physics of brown dwarfs and giant planets
E. Ito (Res. Cent. Earth's Int., Okayama U., Misasa) Constitution and evolution of the Earth
**12:20 p.m. – 2:00 p.m.   Lunch Break**
**2:00 p.m. – 3:20 p.m.** (chairperson: Hooper)
C.M. Surko (UCSD, La Jolla) Physics with trapped positron gases and plasmas
A. Mohri (Kyoto U.) Production of intense positron beams and their moderation
**3:20 p.m. – 3:50 p.m.   Break**

**3:50 p.m. – 5:10 p.m.** (chairperson: Baym)
  J.C. Weisheit (Rice U., Houston) Intense magnetic field phenomena
  K. Makishima (U. Tokyo) Role of magnetized plasmas in production of cosmic x-ray
**5:10 p.m. – 7:30 p.m.  Poster Session** (chairperson: Mohri)
  T. Shigeyama (U. Tokyo) Light curve models for supernova 1993J in M81
  S. Ogata (U. Tokyo) Internal structure and energy transport in a white dwarf
  K. Iida (U. Tokyo) Magnetic phase diagrams of dense charged Fermi liquids
  T. Tsuji (U. Tokyo) Atmospheric structures of very low mass stars
  M. Kato (Keio U., Yokohama) Success of the new opacity in the light curve analysis of novae
  S. Kato (ILE, Osaka U.) Theories on ultra-short pulse laser interactions with solid density plasmas
  S. Miyamoto (ILE, Osaka U.) Analysis of MeV x-ray emission from ultra-intense laser irradiations with solid targets
  H. Furukawa (Laser Tech. Lab., Osaka) Spherical cell model including Fermi-degeneracy and discrete-ion effects and its application to hot dense plasmas
  X.-Z. Yan (Inst. Phys., Beijing) Ions in dense plasmas
  H. Totsuji (Okayama U.) Simulation of electron system in semiconductor microstructures
  M. Hasegawa (Iwate U., Morioka) Density functional theory of freezing of soft-core systems
  K. Tsuruta (U. Tokyo) Coulomb-cluster model for a dense classical one-component plasma

## — FRIDAY, July 1 —

**8:20 a.m. – 9:40 a.m.** (chairperson: Hensel)
  V. Fortov (IVTAN, Moscow) Shock compressed dense plasma
  K. Mima (ILE, Osaka U.) High density plasma physics in laser produced plasmas
**9:40 a.m. – 10:10 a.m.  Break**
**10:10 a.m. – 11:30 a.m.** (chairperson: Kubono)
  Y. Totsuka (Inst. Cosmic Rays, U. Tokyo) Neutrino astrophysics
  N. Itoh (Sophia U., Tokyo) Neutrino emission processes in stars
**11:30 a.m. – 11:50 noon  Closing**

# LIST OF PARTICIPANTS

**Neil W. ASHCROFT**
Cornell University, LASSP
Ithaca, New York 14853-2501, U. S. A.

**Gordon BAYM**
University of Illinois at Urbana-Champaign, Department of Physics
1110 W. Green St., Urbana, Illinois 61801, U. S. A.

**Werner DÄPPEN**
University of Southern California, Department of Physics and Astronomy
Los Angeles, California 90089-1342, U. S. A.

**Michael DOWNER**
University of Texas at Austin, Physics Department
Austin, Texas 78712, U. S. A.

**Werner EBELING**
Humboldt Universität zu Berlin, Department of Physics
Invalidenstr. 110, D-10099 Berlin, Germany

**Hirohisa ENDO**
Fukui Institute of Technology, Faculty of Engineering
3-6-1 Gakuen, Fukui city, Fukui 910, Japan

**Vladimir FORTOV**
Russian Academy of Sciences, High Energy Density Research Center
Izhorskaja Str. 13/19, Moscow 127412, Russia

**Andreas FÖRSTER**
Humboldt Universität zu Berlin, Department of Physics
Invalidenstr. 110, D-10099 Berlin, Germany

**Susumu FUJIWARA**
Keio University, Deparment of Physics
3-14-1 Hiyoshi, Kohoku, Yokohama, Kanagawa 223, Japan

**Masataka FUKUGITA**
Kyoto University, Yukawa Institute for Theoretical Physics
Oiwake, Kita-Shirakawa, Sakyo, Kyoto city, Kyoto 696-01, Japan

**Kurio FUKUSHIMA**
The Fujihara Foundation of Science
3-7-12 Ginza, Chuo, Tokyo 104, Japan

**Masao FUKUSHIMA**
The Fujihara Foundation of Science
3-7-12 Ginza, Chuo, Tokyo 104, Japan

**Hiroyuki FURUKAWA**
  Osaka University, Institute of Laser Engineering
  2-6 Yamada-oka, Suita, Osaka 565, Japan
**Yoichiro FURUTANI**
  Okayama University, Department of Electronics
  3-1-1 Tsushima-naka, Okayama city, Okayama 700, Japan
**Sohei GOMI**
  Keio University, Deparment of Physics
  3-14-1 Hiyoshi, Kohoku, Yokohama, Kanagawa 223, Japan
**Masayuki HASEGAWA**
  Iwate University, Department of Materials Science and Technology
  4-3-5 Ueda, Morioka, Iwate 020, Japan
**Russel J. HEMLEY**
  Carnegie Institution of Washington, Geophysical Laboratory
  5251 Broad Branch Rd. N.W., Washington, D.C. 20015, U. S. A.
**Friedrich HENSEL**
  Philipps-University of Marburg, Department of Physical Chemistry
  Hans-Meerwein-Strasse, 35032 Marburg, Germany
**Charles F., HOOPER, Jr.**
  University of Florida, Department of Physics
  215 Williamson Hall, Gainesville, FL 32611-8440, U. S. A.
**Shinya HOSOKAWA**
  Hiroshima University, Faculty of Science
  1-3-1 Kagamiyama, Higashi-Hiroshima, Hiroshima 724, Japan
**William B. HUBBARD**
  University of Arizona, Department of Planetary Sciences
  Tucson, Arizona 85721, U. S. A.
**Setsuo ICHIMARU**
  University of Tokyo, Department of Physics
  7-3-1 Hongo, Bunkyo, Tokyo 113, Japan
**Kei IIDA**
  University of Tokyo, Department of Physics
  7-3-1 Hongo, Bunkyo, Tokyo 113, Japan
**Eiji ITO**
  Okayama University, Institute for Study of the Earth's Interior
  827 Yamada, Misasa, Tohaku-gun, Tottori 682-01, Japan
**Naoki ITOH**
  Sophia University, Department of Physics
  7-1 Kioi, Chiyoda, Tokyo 102, Japan
**Hiroshi IYETOMI**
  Hitachi Ltd., Energy Research Laboratory
  7-2-1 Ohmika, Hitachi, Ibaraki 319-12, Japan
**Toshitaka KAJINO**
  National Astronomical Observatory
  2-12-1 Osawa, Mitaka, Tokyo 181, Japan

# LIST OF PARTICIPANTS

**Mariko KATO**
Keio University, Department of Astronomy
4-1-1 Hiyoshi, Kohoku, Yokohama, Kanagawa 223, Japan
**Susumu KATO**
Osaka University, Institute of Laser Engineering
2-6 Yamada-oka, Suita, Osaka 565, Japan
**Takako KATO**
National Institute for Fusion Science
Furocho, Chikusa, Nagoya, Aichi 464-01, Japan
**Yoshiaki KATO**
Osaka University, Institute of Laser Engineering
2-6 Yamada-oka, Suita, Osaka 565, Japan
**Ken KIKUCHI**
Japan Society for the Promotion of Science
6-26-3 Kioi, Chiyoda, Tokyo 102, Japan
**Hikaru KITAMURA**
University of Tokyo, Department of Physics
7-3-1 Hongo, Bunkyo, Tokyo 113, Japan
**Shozo KOH**
The Fujihara Foundation of Science
3-7-12 Ginza, Chuo, Tokyo 104, Japan
**Wolf KRAEFT**
Greifswald University, Department of Physics
Domstrasse 10a, 17489 Greifswald, Germany
**Shigeru KUBONO**
University of Tokyo, Institute for Nuclear Study
3-2-1 Midori, Tanashi, Tokyo 188, Japan
**Heiko LEHMANN**
Oxford University, Department of Theoretical Chemistry
5 South Parks Road, Oxford, OX1, 3UB, United Kingdom
**Kazuo MAKISHIMA**
University of Tokyo, Department of Physics
7-3-1 Hongo, Bunkyo, Tokyo 113, Japan
**Isao MATSUSHIMA**
Electrotechnical Laboratory
1-1-4 Umezono, Tsukuba, Ibaraki 305, Japan
**J. MEYER-TER-VEHN**
Max-Planck-Institute für Quantenoptik
Ludwig-Prandtl-Str. 10, Postfach 1513, D-85740, Garching, Germany
**Toshinori MICHISHITA**
Kyoto University, Faculty of Integrated Human Studies
Nihonmatsu-cyo, Yoshida, Sakyo, Kyoto city, Kyoto 606, Japan
**Kunioki MIMA**
Osaka University, Institute of Laser Engineering
2-6 Yamada-oka, Suita, Osaka 565, Japan

**Kazuo MINAMI**
 Niigata University, Faculty of Engineering
 8050 Nino-machi, Igarashi, Niigata city, Niigata 950-21, Japan
**Jun-ichi MISAWA**
 The Yomiuri Shinbun
 1-7-1 Otemachi, Chiyoda, Tokyo 100-55, Japan
**Seiji MIYAMOTO**
 Osaka University, Institute of Laser Engineering
 2-6 Yamada-oka, Suita, Osaka 565, Japan
**Akihiro MOHRI**
 Kyoto University, Faculty of Integrated Human Studies
 Nihonmatsu-cyo, Yoshida, Sakyo, Kyoto city, Kyoto 606, Japan
**Richard M. MORE**
 Lawrence Livermore National Laboratory, L-321
 East Avenue, Livermore, California 94550, U. S. A.
**Toshio MURAKAMI**
 Institute of Space and Astronomical Science
 3-1-1 Yoshinodai, Sagamihara, Kanagawa 229, Japan
**Yoshiji NAKAYAMA**
 Science and Technology Agency, Diffusion and Encouragement Department
 2-2-1 Kasumigaseki, Chiyoda, Tokyo 100, Japan
**Katsunobu NISHIHARA**
 Osaka University, Institute of Laser Engineering
 2-6 Yamada-oka, Suita, Osaka 565, Japan
**Seizou NOAKE**
 Japan Society for the Promotion of Science
 6-26-3 Kioi, Chiyoda, Tokyo 102, Japan
**Ken'ichi NOMOTO**
 University of Tokyo, Department of Astronomy
 7-3-1 Hongo, Bunkyo, Tokyo 113, Japan
**Shuji OGATA**
 University of Tokyo, Department of Physics
 7-3-1 Hongo, Bunkyo, Tokyo 113, Japan
**Kazumi OMATA**
 Keio University, Deparment of Physics
 3-14-1 Hiyoshi, Kohoku, Yokohama, Kanagawa 223, Japan
**Forrest ROGERS**
 Lawrence Livermore National Laboratory, L-296
 POBOX 808, Livermore, California 94550, U. S. A.
**Ken SAKAIZUMI**
 The Fujihara Foundation of Science
 3-7-12 Ginza, Chuo, Tokyo 104, Japan
**Katsuhiko SATO**
 University of Tokyo, Department of Physics
 7-3-1 Hongo, Bunkyo, Tokyo 113, Japan

## LIST OF PARTICIPANTS

**Hiromoto SHIBAHASHI**
University of Tokyo, Department of Astronomy
7-3-1 Hongo, Bunkyo, Tokyo 113, Japan

**Toshikazu SHIGEYAMA**
University of Tokyo, Department of Astronomy
7-3-1 Hongo, Bunkyo, Tokyo 113, Japan

**Clifford M. SURKO**
University of California at San Diego, Physics Department
9500 Glman Drive, La Jolla, California 92093-0319, U. S. A.

**Hideyuki SUZUKI**
National Laboratory for High Energy Physics
1-1 Ouho, Tsukuba, Ibaraki 305, Japan

**Yoichiro SUZUKI**
University of Tokyo, Institute for Cosmic Ray Research
3-2-1 Midori, Tanashi, Tokyo 188, Japan

**Toshiki TAJIMA**
University of Texas at Austin, Physics Department
Austin, Texas 78712, U. S. A.

**Hideaki TAKABE**
Osaka University, Institute of Laser Engineering
2-6 Yamada-oka, Suita, Osaka 565, Japan

**Tatsuyuki TAKATSUKA**
Iwate University, College of Humanities and Social Sciences
3-18-34 Ueda, Morioka, Iwate 020, Japan

**Shigenori TANAKA**
Toshiba Corporation, Research and Development Center
Saiwai, Kawasaki, Kanagawa 210, Japan

**Friedrich-K. THIELEMANN**
Universitaet Basel, Institute fuer Theoretische Physik
Klingelbergstrasse 82, CH-4056 Basel, Switzerland

**Hiroo TOTSUJI**
Okayama University, Department of Electronics
3-1-1 Tsushima-naka, Okayama city, Okayama 700, Japan

**Yoji TOTSUKA**
University of Tokyo, Institute for Cosmic Ray Research
3-2-1 Midori, Tanashi, Tokyo 188, Japan

**Kazuhiko TSUJI**
Deparment of Physics, Keio University
3-14-1 Hiyoshi, Kohoku, Yokohama, Kanagawa 223, Japan

**Takashi TSUJI**
University of Tokyo, Institute of Astronomy
2-12-1 Osawa, Mitaka, Tokyo 181, Japan

**Saku TSUNETA**
University of Tokyo, Institute of Astronomy
2-12-1 Osawa, Mitaka, Tokyo 181, Japan

**Kenji TSURUTA**
University of Tokyo, Department of Physics
7-3-1 Hongo, Bunkyo, Tokyo 113, Japan
**Yutaka UESHIMA**
Osaka University, Institute of Laser Engineering
2-6 Yamada-oka, Suita, Osaka 565, Japan
**Hugh M. VAN HORN**
National Science Foundation, Division of Astronomical Sciences
4201 Wilson Boulevard, Arlington, Virginia 22230, U. S. A.
**Priya VASHISHTA**
Louisiana State University, CCLMS, Department of Physics
Baton Rouge, Louisiana 70803-4001, U. S. A.
**Jon C. WEISHEIT**
Rice University, Space Physics and Astronomy Department
Houston, Texas 77251-1892, U. S. A.
**James P. WOLFE**
University of Illinois at Urbana-Champaign, Department of Physics
1110 W. Green St., Urbana, Illinois 61801, U. S. A.
**Masaaki YAMADA**
Princeton University, PPPL
POBOX 451, Princeton, New Jersey 08543, U. S. A.
**Xin-Zhong YAN**
Chinese Academy of Sciences, Institute of Physics
POBOX 603, Beijing 100080, China
**Fumiko YONEZAWA**
Keio University, Deparment of Physics
3-14-1 Hiyoshi, Kohoku, Yokohama, Kanagawa 223, Japan

# Index

accretion, 48
active galactic nuclei, 51
activity expansion, 184
adiabatic expansion, 393
adiabatic exponent, 209
Alfvén time, 436, 459
annihilation $\gamma$-ray, 387
anomalous relaxation, 339
anomalous skin effect, 422
atmosphere, 193, 234
atomic emission line, 51
atomic level, 150, 185
atomic partition function, 167, 395
atomic structure, 395
axial vector, 104

band gap, 416
baryogenesis, 77, 136
baryon inhomogeneity, 139
baryon number, 77
baryonic mass, 135
$\beta$ decay, 131
$\beta$ equilibrium, 18, 87
boiling curve, 399
bremsstrahlung, 386
brown dwarf, 193, 227

chain conformation, 310
chain length, 310
charge transfer, 280
charged hard spheres, 147, 174
chemical evolution, 239
chemical picture, 165, 221
chiral symmetry, 9, 74

coherent potential approximation, 266
collision induced absorption, 194
collisional excitation, 384
color charge, 6
    effective, 7
complex fluid, 339
complex Gaunt factor, 161
complex susceptibility, 339
compressed atom, 425
confined-atom model, 395
confinement time, 122
contact probability, 115
convection, 37, 195, 208, 219, 240, 380, 450
coordination number, 299, 321, 366
core collapse, 17, 85
corona, 47, 447
coronal hole, 456
correlation,
    electron-ion, 147
    interchain, 310
    many-particle, 116
cosmic background radiation, 76
cosmic x-ray, 47
Coulomb dissociation method, 128
Coulomb distortion effect, 106
crustal material, 26
cusp condition, 118, 255
cyclotron resonance, 47, 439

dark matter, 76, 135
density functional theory, 253, 351, 425
deuterium abundance, 135
deuterium fraction, 235
diamond-anvil cell, 272

diatomic molecule, 196, 428
direct correlation function, 352
dispersion (van der Waals) force, 151
    257, 399
dynamo action, 450

Earth's interior, 239
effective hard-sphere diameter, 301, 319
effective volume, 174
electric microfield, 404
electron
    incipient Rydberg state, 147
    inhomogeneous, 253
    relativistic, 101
electron capture, 35
electron collision time, 417
electron mean free path, 418
electron-hole recombination, 325
electron-positron pair production, 385
enhancement factor,
    density, 117
    thermal, 119
equation of state, 145, 184, 217, 260, 395
excluded volume, 149
exiton,
    bi-, 330
    poly-, 334
exiton binding energy, 326

femtosecond laser, 415
fluid mercury, 295
fluorescence, 319
fractal dimension, 343, 363
free-free opacity, 159
functional Taylor expansion, 351

Gamow rate, 115
giant planet, 227
gluon, 6
gradient correction, 425
graphitic tubule, 369
gravitational collapse, 36

gravitational lensing phenomena, 193,
    237

Hartree-Fock approximation, 18, 87
helioseismology, 203, 215
Homestake experiment, 94
Hubble constant, 135
hydrogen,
    metallic, 228, 254
    solid, 287
hydrogen burning, 126
hydrogen molecule, 257, 271

inertial confinement fusion, 283
inhomogeneous fluid, 351
instability,
    beam-plasma, 471
    MHD, 439, 448
    Rayleigh-Taylor, 38, 381
interaction,
    electroweak, 73
    inter-molecular, 367
    neutrino, 87
    positron-molecule, 471
    radiator-perturber, 405
    two-nucleon, 18
intermediate-range order, 361
internal rotation, 211
inverse power system, 353
ionization energy, 302
ionization equilibrium, 149, 176, 222
ionization process, 416
ionization zone, 221
iron group, 186
isentropic compression, 288
isotope effect, 274

j-j coupling, 188

Kamiokande experiment, 95
Knight shift, 297, 308

Landau orbital, 65

# INDEX 501

laser ablation, 376
laser-driven implosion, 407
laser implosion, 376
lattice gauge theory, 9
Lindemann ratio, 354
line spectrum, 408
linear accelerator (linac), 477
liquid,
    electron-hole, 327
    expanded, 320
liquid chalcogen, 310
local density approximation, 254, 425
local field correction, 160
LS coupling, 183
luminosity, 229
Lyman-$\alpha$, 384, 405

magma, 244
magnetic dipole moment, 74
magnetic field strength, 26, 47, 62
magnetic reconnection, 435, 448
magnetic Reynolds number, 451
magnetohydrodynamics (MHD), 51, 436
mantle, 85, 240
master equation, 173
maximum mass, 22
maximum rotation rate, 22
mercury cluster, 299
mercury vapor, 295
metal,
    alkali, 318
    expanded, 307
    liquid, 317, 416
    polyvalent, 322
    ultrahigh pressure, 120
metallicity, 186, 194, 231
metallization, 145, 280, 291
mineralogical constitution, 240
miscibility, 231
molecular binding, 428
molecular bonding, 271
molecular structure, 200
molten iron, 245

molten silicate, 245
Monte Carlo simulation, 118, 347
motional electric field, 64
Mott insulator, 255

neutrino burst, 88
neutrino degeneracy, 18, 36
neutrino diffusion, 17
neutrino emission, 35, 78, 85, 101
neutrino mass, 73
neutrino oscillation, 76, 98
neutron star,
    proto-, 35, 85
    hot, 17
non-gray model atmosphere, 194
nuclear cross section, 114, 125
nuclear level, 130
nuclear reaction,
    pycno-, 115
    thermo-, 114
nuclear reaction chain, 93
nuclear resonance, 127
nuclear symmetric energy, 88
nucleosynthesis, 41, 75, 125, 136
    cluster, 126

occupation probability formalism, 184, 221
opacity,
    astrophysical, 181
    molecular, 188, 195
    Rosseland mean, 161, 182
Opacity Project, 181, 221
opacity source, 194
OPAL project, 181, 221
optical (AC) conductivity, 311, 417
optical gap, 308
orientational ordering, 251, 271
oscillation mode, 28, 204
    acoustic, 28, 204
    interfacial, 28
    spheroidal, 28
    toroidal, 28

packing fraction, 319, 355
pair-distribution function, 366
parallel computer, 370
Penning trap, 463
percolation threshold, 347
phase diagram, 4, 149, 177, 276, 287
phase shift analysis, 156
phonon dispersion, 367
photoabsorption cross section, 182
photoexcitation, 325
photoionization, 400
photoluminescence, 325
photon-positron converter, 479
photosphere, 450
plane-wave approximation, 106
plasma
    ambi-, 484
    electron-ion two-component, 154
    electron-hole, 327
    electron-positron, 470
    interstellar, 51
    one-component, 166, 146, 355
    positron, 464, 477
    quark-gluon, 3, 136
plasma diagnostic, 413
plasma neutrino process, 101
plasmon,
    longitudinal, 101
    transverse, 101
plasmon cut-off, 162
plasmon dispersion relation, 101
polymorphic transformation, 240
porous silica, 362
positron accumulation, 482
positron affinity, 472
positron annihilation, 387, 471, 481
positron beam, 464
positron trap, 466
positronium, 472
positron-molecule resonance, 472
pseudogap, 297

quantum chromodynamics (QCD), 7, 136
quantum virial function, 169
quark core, 5
quasiperiodic fine structure, 28

radial distribution function, 298
radiation cooling, 484
radiative ablation, 284
random walk, 340
rapid (r) process, 38, 126
rapid proton (rp) process, 125
reaction time, 123
recombination neutrino process, 106
reconnection angle, 436
reconnection pattern, 442
reflectivity, 305, 311
relativistic positron, 475
relaxation function, 339
relaxation time, 339
resistivity (DC conductivity), 155, 295, 420
roton, 274
roughness exponent, 364

satellite,
    Asca, 54
    Ginga, 47
    Yohkoh, 447
scattering,
    electron-electron, 420
    electron-ion, 155
    inelastic neutron, 314
    neutrino, 35
    quark-gluon, 11
    s-wave, 114
screening,
    electron, 116
    gluon, 10
screening potential, 118
seismic tomography, 240
semiconductor, 325

shear modulus, 29
shock compression, 283, 389
shock wave, 39, 284, 393
short-range screening distance, 115
soft core system, 351
soft x-ray, 447
solar flare, 456
solar model, 203
solar neutrino problem, 80, 93, 128, 210
solar wind, 456
solid $C_{60}$, 366
sound speed, 208, 217
spectral line broadening, 183
spectral line profile, 64, 404
spheromak, 441
spinel, 241
spiral chain, 322
stagnation dynamics, 381
standard big-bang model, 135
standard solar model, 94, 203
Stark effect, 64, 404, 440
static structure factor, 160, 298, 313, 319, 365
stopping power, 378
structural relaxation, 340
structural transformation, 291, 321, 365
superfluidity, 30
supernova, 17, 35, 85
supernova remnant, 47
surface tension, 304, 349

thermonuclear outburst, 31
thermonuclear threshold, 234
thermopower, 308
tokamak, 438
transition
    band overlap, 259
    deconfining, 8
    freezing, 351
    liquid-vapor (gas), 296, 328
    metal-insulator (nonmetal), 251, 295, 308, 326
    QCD phase, 136
    wetting, 304, 348
transverse dielectric function, 159
triple point, 276
turbulent mixing, 380

unstable nucleus, 125

velocity fluctuation, 38
velocity mixing, 45
very low mass star, 193
vibron, 259, 273

Weinberg-Salam theory, 101

x-ray burster, 31
x-ray diffraction, 317
x-ray pulsar, 47